Large-Scale Magnetic Fields in the Universe

Space Sciences Series of ISSI

Volume 39

For other titles published in this series, go to
www.springer.com/series/6592

Rainer Beck • André Balogh • Andrei Bykov •
Rudolf A. Treumann • Lawrence Widrow
Editors

Large-Scale Magnetic Fields in the Universe

Previously published in *Space Science Reviews* Volume 166, Issues 1–4 and Volume 169, Issues 1–4, 2012

Springer

Editors
Rainer Beck
Max Planck Institute for Radio Astronomy
Bonn, Germany

André Balogh
Imperial College London
Berkhamsted, UK

Andrei Bykov
Russian Academy of Sciences
St. Petersburg, Russia

Rudolf A. Treumann
ISSI
Bern, Switzerland

Lawrence Widrow
Queen's University
Kingston, Canada

ISSN 1385-7525 Space Sciences Series of ISSI
ISBN 978-1-4614-5727-5
Springer New York Heidelberg Dordrecht London

Library of Congress Control Number: 2012949746

Cover illustration: Magnetic fields in the spiral galaxy M51. Total radio emission (contours) and magnetic field vectors at 4.8 GHz, combined from observations with the VLA and the Effelsberg telescope. The underlying image is from the Hubble Space Telecope. (Credit: A. Fletcher and R. Beck/Hubble Heritage Team, graphics by "Sterne und Weltraum", from R. Beck, Measurements of Cosmic Magnetism with LOFAR and SKA, Adv. Radio Sci., 5, 399–405, 2007.)

Printed on acid-free paper

Springer is part of Springer Science+Business Media (www.springer.com)

Contents

Space Sci Rev (2012) 166:1–35
DOI 10.1007/s11214-011-9839-z

Magnetic Fields in the Large-Scale Structure of the Universe

D. Ryu · D.R.G. Schleicher · R.A. Treumann · C.G. Tsagas · L.M. Widrow

Received: 12 October 2010 / Accepted: 14 September 2011 / Published online: 15 October 2011

Abstract Magnetic fields appear to be ubiquitous in astrophysical environments. Their existence in the intracluster medium is established through observations of synchrotron emission and Faraday rotation. On the other hand, the nature of magnetic fields outside of clusters, where observations are scarce and controversial, remains largely unknown. In this chapter, we review recent developments in our understanding of the nature and origin of intergalactic magnetic fields, and in particular, intercluster fields. A plausible scenario for the origin of galactic and intergalactic magnetic fields is for seed fields, created in the early universe, to be amplified by turbulent flows induced during the formation of the large scale structure. We present several mechanisms for the generation of seed fields both before and after recombination. We then discuss the evolution and role of magnetic fields during the formation of the first starts. We describe the turbulent amplification of seed fields during the formation of large scale structure and the nature of the magnetic fields that arise. Finally, we discuss implications of intergalactic magnetic fields.

Keywords Large-scale structure of the universe · Magnetic field · Turbulence

D. Ryu (✉)
Department of Astronomy and Space Science, Chungnam National University, Daejeon 305-764, Korea
e-mail: ryu@canopus.cnu.ac.kr

D.R.G. Schleicher
Institut für Astrophysik, Georg-August-Universität, Friedrich-Hund-Platz 1, 37077 Göttingen, Germany
e-mail: dschleic@astro.physik.uni-goettingen.de

R.A. Treumann
ISSI, Hallerstrasse 6, 3012 Bern, Switzerland
e-mail: treumann@issibern.ch

C.G. Tsagas
Department of Physics, Aristotle University of Thessaloniki, Thessaloniki 54124, Greece
e-mail: tsagas@astro.auth.gr

L.M. Widrow
Department of Physics, Queen's University, Kingston, Ontario K7L 3N6, Canada
e-mail: widrow@astro.queensu.ca

1 Introduction

In the highly successful ΛCDM cosmology, the large scale structure (LSS) forms through a process known as hierarchical clustering in which small-scale objects collapse first and merge to form systems of ever-increasing size. This scenario leads to a "cosmic web" of structure where galaxies reside mainly along filaments while galaxy clusters arise at the intersections of two or more filaments (see, e.g., Bond et al. 1996). Furthermore, a picture of a multi-phase intergalactic medium (IGM) has emerged. A hot phase, often referred to as the intracluster medium (ICM) because it is found inside and around clusters and groups of galaxies, has a temperature $T > 10^7$ K and is observable via X-ray emission. The warm-hot intergalactic medium (WHIM) ($10^5 < T < 10^7$ K) is found mainly in filaments of galaxies (Cen and Ostriker 1999; Kang et al. 2005). Gas in the IGM is heated by cosmological shocks which arise as objects in the hierarchy form. (Ryu et al. 2003; Pfrommer et al. 2006; Kang et al. 2007; Skillman et al. 2008; Vazza et al. 2009).

Magnetic fields are observed in galaxies of all types and in galaxy clusters. Moreover, there is some evidence that they permeate the filaments of the cosmic web. Various scenarios for the origin of galactic magnetic fields are discussed in Widrow (2002) and Widrow et al. (2010) of this issue. The basic idea is that weak fields, generated either by an exotic early universe mechanism or some astrophysical process, are amplified to μG strength during galaxy formation and by dynamo action during the subsequent quiescent phase of galaxy evolution. Similarly, weak seed magnetic fields can be amplified into the intergalactic magnetic field (IGMF) by turbulent flow motions during the formation of the LSS through a process known as small-scale turbulence dynamo. In a turbulence dynamo, kinetic energy of the fluid in converted to magnetic energy through stretching, twisting, and folding of the field (see, e.g., (Subramanian 1999; Cho and Vishniac 2000; Haugen et al. 2004a, 2004b, 2004c; Schekochihin et al. 2004; Brandenburg and Subramanian 2005; Cho et al. 2009; Brandenburg and Subramanian 2010), this issue). In addition, feedback from the black hole regions in active galactic nuclei (AGNs) can also contribute and amplify magnetic fields in the IGM.

This chapter will explore the structure formation-magnetic field connection. In particular, we will address two interrelated questions. First, can the structure formation itself generate new magnetic fields and amplify existing ones? Second, what role do magnetic fields play in structure formation and in other astrophysical processes? An outline of the chapter is as follows: In Sect. 2, we briefly review the observational evidence for magnetic fields in the clusters and filaments of the cosmic web. In Sect. 3, we describe various astrophysical mechanisms for the generation of seed magnetic fields, which would later be maintained and amplified within galaxies and clusters. In Sect. 4, we discuss the evolution and role of magnetic fields during the first star formation: their stabilization due to a non-zero ionization degree as well as their potential implication for star formation. We then describe how magnetic fields are amplified during structure formation in Sect. 5. We discuss the effects of magnetic fields on the propagation of cosmic rays in the IGM and also the Faraday rotation measure (RM) induced by magnetic fields in Sect. 6. Section 7 presents something of a departure from the more phenomenological and astrophysical discussion as it provides a formal treatment of the evolution of density perturbations in the presence of magnetic fields. Finally, a brief summary follows in Sect. 8.

2 Observational Evidence for Magnetic Fields in Clusters and Filaments

The existence and strength of the IGMF has been deduced from observations of synchrotron emission and confirmed directly from observations of RM (see, e.g., Carilli and Taylor 2002;

Govoni and Feretti 2004, for review). Synchrotron emission from a galaxy cluster was first discovered by Large et al. (1959) who surveyed the Coma cluster in the radio. Since then, it has been observed in numerous clusters, either as radio halos or as radio relics (see, e.g., Govoni and Feretti 2004; Cassano et al. 2008, and references therein). The diffuse radiation from radio halos is mostly unpolarized, and is thought to emerge from turbulent magnetic fields in the ICM. The radiation from radio relics is highly polarized and is believed to be emitted from shocked regions in the ICM. The strength of the magnetic field in radio halos is estimated to be ~ 1 μG, while stronger fields are found radio relics. These estimates either assume equipartition (also known as the minimum energy argument) or incorporate measurements of inverse-Compton emission.

Observations of the IGMF based on Faraday rotation have also been done, though mostly for magnetic fields in the ICM (see Carilli and Taylor 2002, and references therein). An RM map of the Coma cluster, for example, reveals a field with a strength $\sim$ μG and a coherence length of order ~ 10 kpc (Kim et al. 1990). For Abell clusters, the typical RM is ~ 100–$200\ \mathrm{rad\,m^{-2}}$ which indicates an average field strength of ~ 5–10 μG (Clarke et al. 2001; Clarke 2004). RM maps of clusters can be used to determine the power spectrum of turbulent magnetic fields in the ICM. For instance, a Kolmogorov-like spectrum with a bending at a few kpc scale is found in the cooled core region of the Hydra cluster (Vogt and Enßlin 2005), and spectra consistent with the Kolmogorov spectrum were reported in the wider ICM for the Abell 2382 cluster (Guidetti et al. 2008) and for the Coma cluster (Bonafede et al. 2010).

The nature of the IGMF in filaments, on the contrary, remains largely unknown, because the studies of synchrotron emission and RM outside clusters are still scarce and controversial. Although faint radio emission has been observed in the outskirts of clusters (see, e.g., Kim et al. 1989), there are no confirmed observations of synchrotron emission from filaments. Such measurements present a challenge for current facilities. As well, the removal of the galactic foreground is a non-trivial task (see, e.g., Brown et al. 2010). At present there is an upper limit of ~ 0.1 μG for the strength of the IGMF in filaments based on the observed limit of the RMs of background quasars (Ryu et al. 1998; Xu et al. 2006).

Recently, Neronov and Vovk (2010), Aleksić et al. (2010) reported a lower bound for the strength of the magnetic field in voids. Their claim, if true, is significant as it would represent the first evidence for magnetic fields in such low-density regions and on such large scales. The basic idea is that an IGMF will deflect charged particles that arise in an electromagnetic cascade whose source is the very high energy γ-rays produced in an AGN. A non-observation of γ-ray secondaries coincident with VHE γ-rays, is then taken as evidence of magnetic deflection along the path from the source to the observer. Neronov and Vovk (2010) quote a lower bound of $B \geq 3 \times 10^{-16}$ G based on observations by the Fermi and the High Energy Stereoscopic System (HESS) telescopes. A similar analysis was performed by Aleksić et al. (2010) using data from the MAGIC telescope though they are more cautious in presenting their conclusions and model dependencies. Note that Neronov and Vovk (2010) assume continuous emission of gamma-rays for 10^6 years, or longer which may not be realistic. If this assumption is relaxed, then one obtains a more conservative lower bound (Dermer et al. 2011).

The detection of the IGMF in filaments, if it exists, might be made with the next generation radio facilities. These facilities include the Square Kilometer Array (SKA), and upcoming SKA pathfinders, the Australian SKA Pathfinder (ASKAP) and the South African Karoo Array Telescope (MeerKAT), as well the Low Frequency Array (LOFAR) (see, e.g., papers in Carilli and Rawlings 2004).

For discussions on magnetic fields in the ICM and cluster outskirts, see Schekochihin et al. (2010) and Brüggen et al. (2010) of this volume.

3 Plasma Physics Mechanisms for Seed Fields

3.1 Biermann Battery

The Biermann battery is a promising mechanism for the creation of astrophysical magnetic fields. The mechanism arises in an ionized plasma whenever baroclinity exists, that is, when isodensity surfaces do not coincide with isobaric surfaces. This situation leads to an extra pressure gradient term in Ohm's law which drives currents. These currents, in turn, generate magnetic fields at a rate given by

$$\frac{d\mathbf{B}}{dt} \sim \frac{m_e c}{e} \frac{\nabla \rho_e \times \nabla p_e}{\rho_e^2}, \tag{1}$$

where ρ_e and p_e are the electron density and pressure, respectively (Biermann 1950). Though the mechanism was originally studied in the context of stars (Biermann 1950), it may arise during structure formation in cosmology whenever the electron pressure and density gradients are not aligned, as often occurs in shocks (Pudritz and Silk 1989; Kulsrud et al. 1997; Davies and Widrow 2000; Xu et al. 2008).

We note that vorticity, $\boldsymbol{\omega}$, is generated when the total pressure and density gradients are not aligned (baroclinity of flows – see (23) below). In an ionized plasma, a simple order of magnitude estimate yields

$$B \sim \frac{m_p c}{e} \omega \simeq 3 \times 10^{-21} \left(\frac{\omega}{\mathrm{km\,s^{-1}\,kpc^{-1}}} \right) \quad \text{Gauss.} \tag{2}$$

Since the present-day vorticity in the IGM is of order $\sim$ a few $\mathrm{km\,s^{-1}\,kpc^{-1}}$ (see Sect. 5.1 below), the seed fields from the Biermann battery will be rather small ($\sim 10^{-20}$ G). However, vorticity, and hence seed fields, were almost certainly larger at early times. The following argument, based on dimensional analysis, suggests that for the objects which dominate the structure hierarchy at a given epoch the ratio of the vorticity to the Hubble parameter, H, is roughly constant and of order a few hundred. Once an object collapses, $M \simeq v^2 R/G$ where M, v, and R are the characteristic mass, velocity and size of the object. The spherical collapse model, which serves as a useful toy-model for structure formation in an expanding universe, predicts that the mean density within an object is a factor $f_c \simeq 200$ times the critical density, ρ_c. Moreover, objects tend to follow the mass-size relation $M \propto R^2$. If we put all this together and use the fact that the critical density scales as the square of the Hubble parameter, we find that $\omega \simeq 300H \propto (1+z)^{3/2}$. Thus, we expect larger seed fields at early times. Similarly, the dynamical time for objects formed in the early universe is shorter than for objects today and hence the amplification of the magnetic fields by dynamo action will be more rapid. In short, the Biermann battery-dynamo amplification process may well operate at all stages of the structure-formation hierarchy.

Various alternatives have also been considered. For example, Lazarian (1992) considered a battery driven by electron diffusion. Subramanian et al. (1994) considered the Biermann effect during the epoch of reionization when ionization fronts sweep through the medium generating currents and magnetic fields. The fields are again fairly modest, but could be large enough to seed a dynamo which then amplifies them to an astrophysically interesting strength.

3.2 Thermal Fluctuations

At finite temperatures, a plasma exhibits a finite (though possibly low) level of thermal fluctuations at all scales $L = 2\pi/k$ and frequencies ω. Differences in the thermal motions between the different charges generate micro-currents and hence electromagnetic fields which, on average, provide a noisy electromagnetic background. In the early universe, the gas is both hot and dilute. For example, the gas temperature is $T \sim$ eV both at the recombination epoch and after reionization but before the formation of LSS. Therefore, the cosmic gas constitutes a classical plasma to which the classical fluctuation theory (Sitenko 1967) can be applied.

According to this theory, the power spectral density of magnetic field fluctuations $\mathbf{b}$ at wavenumber $\mathbf{k}$ and frequency ω is obtained from the spatial correlation function of the fluctuating magnetic fields and is given by

$$\frac{\langle b_i b_j \rangle_{\mathbf{k}\omega}}{\sqrt{2\pi}} = \frac{\mu_0 T}{\omega}\left(\delta_{ij} - \frac{k_i k_j}{k^2}\right)\frac{n^2 \mathrm{Im}\,\epsilon_\perp}{|n^2 - \epsilon_\perp|^2}. \tag{3}$$

Here, the fluctuating field is designated by lower case letters, $i, j = 1, 2, 3$, $n^2 = k^2c^2/\omega^2$ is the index of refraction and $\epsilon_\perp(\omega, \mathbf{k})$ is the complex transverse dielectric response function of the isotropic plasma. The latter, in (nonrelativistic) thermal equilibrium, is given by

$$\epsilon_\perp(\omega, \mathbf{k}) = 1 - \sum_{e,i} \frac{\omega_{e,i}^2}{2\omega k^2} \int \frac{\mathbf{v} \times (\mathbf{k} \times \mathbf{v}) F'_{0,e,i}(v)}{\mathbf{k}\cdot\mathbf{v} - \omega} dv^3, \tag{4}$$

where the sum is over negative (e) and positive (i) charges, $\omega_{e,i}$, the respective plasma frequencies, and $F'_0 \equiv \partial F_0/\partial \mathbf{v}$, the derivative of the equilibrium distribution F_0 which is assumed to be Maxwellian. The response function becomes

$$\epsilon_\perp = 1 - \frac{\omega_e^2}{\omega^2}\left[\Phi(\varpi) + \frac{\hat\theta}{M}\Phi(M\varpi) - i\varpi\sqrt{\pi}\left(\mathrm{e}^{-\varpi^2} + \frac{\hat\theta}{M}\mathrm{e}^{-M^2\varpi^2}\right)\right]. \tag{5}$$

Here, $M^2 \equiv (m_i/m_e)\hat\theta$, and $\Phi(\varpi) \approx 2\varpi^2 + \cdots$ is the Gordeyev integral of $\varpi = \sqrt{3/2}(\omega/kc_s)$ with $c_s^2 = 3T_e/m_e$, the square of the sound speed, and $\hat\theta = T_e/T_i$. Only at very large ion temperature $T_i \gg T_e$ do the ions contribute to the imaginary part.

Seed fields for a turbulence dynamo arise from the zero frequency contribution, $\varpi \to 0$ (index 0 in the following equation), which yields the magnetic power spectral density

$$\frac{\langle b_i b_j \rangle_{0k}}{\sqrt{2\pi}} \approx \frac{\mu_0 m_e c^2}{4\omega_e}\frac{\hat\Theta^{1/2} k\lambda_e}{(k^2\lambda_e^2 - m_e/m_i)^2} \simeq \frac{10^{-37}}{L_{\mathrm{kpc}} N_{[\mathrm{cm}^{-3}]}}\sqrt{T_{\mathrm{eV}}}\;\frac{\mathrm{V}^2\,\mathrm{s}^3}{\mathrm{m}}, \tag{6}$$

where $\hat\Theta \equiv T_e/m_ec^2$ is the normalized temperature T_e and $T_{e[\mathrm{eV}]}$ is temperature in eV, $\lambda_e = c/\omega_e \simeq 6/N^{1/2}$ is the electron inertial length, N is the number density, and $N_{[\mathrm{cm}^{-3}]}$ the density in cm^{-3}. To obtain the last equality we have ignored the term $k^2\lambda_e^2$ which is negligible on galactic and extragalactic scales. The magnetic spectral energy density thus grows linearly with wavenumber k, i.e. it decreases with the scale L.

For comparison, a $B = 1$ μG field has the spectral energy density $\langle B^2_{1\,\mu\mathrm{G}}\rangle_{0k} \approx 4.3 \times 10^{-14}\ \mathrm{V}^2\,\mathrm{s}^3/\mathrm{m}$. Were we to have integrated over all frequencies ϖ, we would have obtained the equipartition between magnetic and electron thermal energy densities $\langle b^2\rangle_{\mathbf{k}} = 2\mu_0 T_e$ which holds in thermal equilibrium. Evidently, the thermal fluctuation level in (6) is quite low. In order to obtain a $B = 1$ μG field, the spectral energy density must be amplified by the large factor of $\sim 5 \times 10^{23} L_{\mathrm{kpc}} N_{[\mathrm{cm}^{-3}]}/\sqrt{T_{\mathrm{eV}}}$.

3.3 Filamentation Instability

Dynamos on galactic or cluster scales need seed fields on comparably large scales to get them started. Moreover, large-scale dynamos require turbulence and/or helicity which may not exist at adequate levels. Recently, the so-called filamentation (or Weibel) instability (Weibel 1959; Fried 1959) has been discussed as a generator of the IGMF (Gruzinov and Waxman 1999; Gruzinov 2001; Medvedev and Loeb 1999) and a possible alternative to dynamos on galactic and extragalactic scales. The Weibel instability does require a strong departure of the plasma state from thermal equilibrium, which can be provided by fast beams on the plasma background (Achterberg and Wiersma 2007; Fried 1959; Sakai et al. 1999) or temperature anisotropies. The latter can be caused by pressure anisotropies (Weibel 1959; Treumann et al. 2010), shock waves (Jaroschek et al. 2005; Nishikawa et al. 2009), or thin current sheets. Turbulence, if it exists in the early universe, leads to thin current sheets which can accelerate particles by second-order Fermi acceleration (Jaroschek et al. 2008; Jaroschek and Hoshino 2009), the local electric field which is generated in the current sheets, or by reconnection. Here, we discuss the possibility that the instability is driven by kinetic pressure/temperature anisotropies in the absence of magnetic field, as a mechanism to create seed fields.

Pressure and temperature anisotropies are probably superior to the beam for driving the instability for the simple reason that beams are highly unstable with respect to high-frequency plasma instabilities (Langmuir and Buneman modes). If the beams have velocity V_b and are current-compensated (i.e., the velocities of particles with different charges are the same) then the two-stream instability is excited with growth rate $\gamma_{ts} \simeq \sqrt{3}(N_b/2N)^{1/3}\omega_e$. Current-compensated beams arise in the symmetric counter-streaming configurations found in shocks. For the reasonable beam density $N_b \sim 0.1N$ we find $\gamma_{ts} \approx 0.6\,\omega_e \sim$ few kHz with a wavenumber is $k_{ts} \sim \omega_e/V_b$. If the beams are not current compensated, which is the case in turbulence where a multitude of thin current sheets is generated, then the Buneman instability is excited with growth rate $\gamma_B \sim 0.03\,\omega_e \sim 10^2$ Hz, if the current velocity $V_c = V_e - V_i > v_e$ exceeds the electron thermal speed v_e. Otherwise, for $V_c < v_e$, the ion sound will be excited with growth rate roughly $\gamma_{ia} \sim \omega_p i \sim$ few 10^2 Hz. All these instabilities grow very fast and readily deplete the beams/streams, heating the electron plasma in the direction of flow and thus causing the pressure or temperature anisotropy of electrons with higher temperature along the flow direction, i.e. $\|$ to $\mathbf{V}_c$. If this is assumed to be the (readily achieved) final microscopic state, then the temperature anisotropy $A = T_{e\|}/T_e - 1 > 0$ becomes unstable with respect to the Weibel thermally anisotropic mode. The temperature anisotropy corresponds to a pressure anisotropy $\mathsf{P}_e = N[T_{e\|} + (T_{e\|} - T_e)\mathbf{V}_c\mathbf{V}_c/V_c^2]$. One should note that only the electrons, because of their large mobility, are important; the ions are much less active on these microscopic time scales and thus do not contribute.

In this thermally anisotropic case, the electrons obey a bi-Maxwellian equilibrium distribution function

$$f_e(v_\perp, v_\|) = \frac{(m_e/2\pi)^{3/2}}{T_e\sqrt{T_{e\|}}} \exp\left[-\frac{m_e v_\perp^2}{2T_e} - \frac{m_e v_\|^2}{2T_{e\|}}\right], \tag{7}$$

and the Weibel instability takes over with the linear electromagnetic dispersion relation $n^2 = \epsilon_\perp$ and the transverse dielectric response function

$$\epsilon_\perp \equiv 1 - \frac{\omega_e^2}{\omega^2}\{1 - (A+1)[1 + \zeta Z(\zeta)]\} - \frac{\omega_i^2}{\omega^2} = n^2. \tag{8}$$

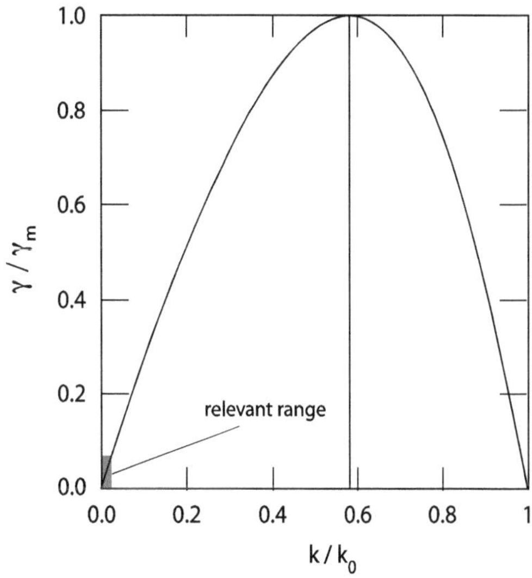

Fig. 1 The growth rate $\gamma = \gamma_W$ of the anisotropic-thermal Weibel instability normalized to the maximum growth, γ/γ_m, as a function of the normalized wavenumber k/k_0 (see the text for the definition of k_0). The *vertical line* indicates the position of the maximum growing wavenumber $k_m/k_0 = 1/\sqrt{3}$

Here, $Z(\zeta)$ is the plasma dispersion function, $\zeta = \omega/k_\perp v_e$, and $v_e = \sqrt{2T_e/m_e}$ is the electron thermal speed perpendicular to current. The Weibel instability grows in the plane perpendicular to the direction of higher thermal velocity, which in our case has been assumed as the parallel direction. Hence, $\mathbf{k} \equiv \mathbf{k}_\perp$. The contribution of the resting ions has been retained for completeness; because of the smallness of the ion plasma frequency $\omega_i \ll \omega_e$, being much less than the electron plasma frequency ω_e, it plays no role in the instability but is important in the discussion of the thermal level.

At zero real frequency, $\omega = i\gamma$, and the instability $\gamma(k_\perp) > 0$ sets on for phase velocities $\omega/k_\perp \ll v_e$ and wavenumbers $k_\perp < k_0$ with growth rate

$$\frac{\gamma_W}{\omega_e} \simeq \sqrt{\frac{2}{\pi}}\frac{v_{e\perp}}{c}\frac{k_\perp}{k_0}\left(1 - \frac{k_\perp^2}{k_0^2}\right)(A+1)(k_0\lambda_e)^3, \tag{9}$$

vanishing at infinite wavelength $k_\perp = 0$, where $k_0\lambda_e = \sqrt{A}$. The growth rate maximizes at wavenumber $k_m\lambda_e = k_0\lambda_e/\sqrt{3} = \sqrt{A/3}$ (see Fig. 1), where its value is

$$\gamma_m \approx 34\sqrt{N_{[\mathrm{cm}^{-3}]}T_{e[\mathrm{eV}]}}A^{3/2}(A+1)\ \mathrm{Hz}. \tag{10}$$

A substantial growth can be achieved and stationary (purely growing) non-fluctuating, short-wavelength magnetic fields are produced. The maximally growing wavelength is indeed very short. For a plasma density of $N \approx 0.01\ \mathrm{cm}^{-3}$, the wavelength is only of the order of $\lambda = (2\pi c/\omega_e)\sqrt{3/A} \sim 50\sqrt{3/A}$ km. Even very weak anisotropies of $A \sim 10^{-4}$ give $\lambda_m \sim 10^4$ km. Galactic or intergalactic scales L are far longer. For a realistic anisotropy of $A \lesssim 1$, at these large scales the growth rate scales with L as

$$\gamma_W \simeq 4A\sqrt{\pi}\hat{\Theta}(c/L) \simeq 10^{-13}A\sqrt{T_{\mathrm{eV}}}/L_{\mathrm{kpc}}\ \mathrm{Hz} \tag{11}$$

The above tells that mostly magnetic fields of small scales are excited and saturated by the instability, and those of galactic or intergalactic scales would not be substantial. Yet for $A \sim 0.1$ and $T \sim$ eV, the growth rate is $\gamma_W \sim 10^{-14}$ Hz on kpc-scales which is quite large.

Magnetic fields of $\sim 10^{-16}$ G on kpc-scales would arise in $\sim 10^7$ years or so provided the fields on smaller scales do not saturate the instability.

4 Magnetic Fields after Recombination

If magnetic fields were created before or during recombination they could have had a significant impact on the thermal and chemical evolution of gas during the dark ages, on the formation of first stars and galaxies, and on the epoch of reionization (e.g., Sethi and Subramanian 2005; Tashiro and Sugiyama 2006; Sethi et al. 2008; Schleicher et al. 2008, 2009) (see also Widrow et al. 2010, of this issue). On the other hand, the first stars and galaxies may have been a source of magnetic fields. Our discussion should be viewed with a critical eye since the existence, strength and scale of early universe fields is highly uncertain (see e.g. Grasso and Rubinstein 2001) as is the efficiency of dynamo action at early times.

4.1 Magnetic Fields During the Dark Ages

The dark ages designate the period between recombination and reionization, during which no astronomical objects exist. The only radiation produced during this epoch is the 21 cm line of neutral hydrogen. The primordial gas consists of about 75% hydrogen and 25% helium, as well as traceable amounts of lithium and beryllium (both of the order 10^{-9}) (see, e.g., Kolb and Turner 1990). The ionization degree freezes out at about 10^{-4} (Peebles 1968; Zeldovich et al. 1969; Seager et al. 1999). Though this ionization degree may seem rather low, it is significantly higher than the typical ionization degree in present-day molecular clouds (Myers and Khersonsky 1995). More to the point, it is high enough to prevent non-ideal magnetohydrodynamic (MHD) processes such as ambipolar or Ohmic diffusion from dissipating the magnetic fields.

The more relevant question is whether magnetic fields can be sustained during gravitational collapse when both the density and recombination rate increase and hence the ionization degree decreases. Gravitational collapse occurs on length scales where the thermal pressure can no longer balance the gravitational force. The critical length scale is given by the Jeans length

$$\lambda_J = \left(\frac{\pi c_s^2}{G\rho}\right)^{1/2}, \tag{12}$$

where c_s is the thermal sound speed and ρ the mass density. Previous studies by Maki and Susa (2004) and Glover and Savin (2009) followed the chemical evolution during gravitational collapse. Their one-zone model takes into account the density evolution in the central core, where the collapse takes place on the free-fall timescale $t_{ff} \sim 1/\sqrt{G\rho}$. As is well-known from analytic solutions and numerical simulations, the density in the central core is roughly homogeneous on the scale of the Jeans length λ_J (Larson 1969; Penston 1969; Abel et al. 2002; Bromm and Loeb 2003; Yoshida et al. 2008).

The calculations of Maki and Susa (2004) and Glover and Savin (2009) showed that the ionization degree evolves roughly as $\rho^{-1/2}$ until a density of $\sim 10^9$ cm^{-3} is reached. At this stage, the recombination rate becomes so high that the proton abundance becomes negligible. However, the presence of lithium and the low recombination rate of Li^+ stabilizes the ionization degree at a level of $\sim 10^{-10}$, and it stays roughly constant even at higher densities.

In order to assess the implications of this result, Schleicher et al. (2009) combine the chemical network of Glover and Savin (2009) with a state-of-the-art model for ambipolar and Ohmic diffusions, based on the work of Pinto et al. (2008) and Pinto and Galli (2008). In their multi-fluid approach, the ambipolar diffusion rate is given as

$$L_{\mathrm{AD}} = \frac{\eta_{\mathrm{AD}}}{4\pi} \left|(\nabla \times \mathbf{B}) \times \mathbf{B}/B\right|^2, \tag{13}$$

where the $\mathbf{B}$ denotes the magnetic field. The ambipolar diffusivity η_{AD} is given as

$$\eta_{\mathrm{AD}}^{-1} = \sum_i \eta_{\mathrm{AD},i}^{-1}, \tag{14}$$

with $\eta_{\mathrm{AD},i}$ denoting the ambipolar diffusivity due to an ionized species i. Similarly the Ohmic diffusion rate is given as

$$L_{\mathrm{Ohm}} = \frac{\eta_{\mathrm{O}}}{4\pi} \left|\nabla \times \mathbf{B}\right|^2, \tag{15}$$

with the Ohmic resistivity η_{O} given as

$$\eta_{\mathrm{O}}^{-1} = \sum_i \eta_{\mathrm{O},i}^{-1}. \tag{16}$$

The ambipolar resistivities $\eta_{\mathrm{AD},i}$ and the Ohmic resistivities $\eta_{\mathrm{O},i}$ due to ionized species i are calculated from the momentum-transfer coefficients as described in the appendix of Pinto et al. (2008).

To guide the physical intuition, we briefly summarize their results for the case of two charged fluids, e.g. electrons and positive ions, but note that the more detailed multi-fluid approach has been adopted in the calculations. The key quantities that regulate non-ideal MHD effects are the Hall parameters β_{sm}, defined as

$$\beta_{\mathrm{sn}} = \left(\frac{q_s B}{m_s c}\right) \frac{m_s + m_n}{\rho_n \langle \sigma v \rangle_{\mathrm{sn}}}. \tag{17}$$

Here, q_s denotes the charge of the species s, m_s its mass, m_n the mass of the neutrals, ρ_n the neutral mass density, c the speed of light and $\langle \sigma v \rangle_{sn}$ the momentum-transfer coefficients between the ions and the neutrals (for the expressions, see Pinto and Galli (2008) and the discussion below). As shown by Pinto et al. (2008), the ambipolar and Ohmic resistivities are then given as

$$\eta_{\mathrm{AD}} = \left(\frac{c^2}{4\pi\sigma}\right) \frac{\beta_{+n} |\beta_{-n}|}{\beta_{+n} + |\beta_{-n}|}, \tag{18}$$

$$\eta_{\mathrm{O}} = \left(\frac{c^2}{4\pi\sigma}\right) \frac{1}{\beta_{+n} + |\beta_{-n}|}, \tag{19}$$

where β_{+n}, β_{-n} denote the Hall parameters for the positively and negatively charged species, and σ is given as

$$\sigma = \frac{q_+ \rho_+ c}{m_+ B}. \tag{20}$$

In the latter expression, q_+ denotes the positive charge, and ρ_+ the mass density of the positively charged species.

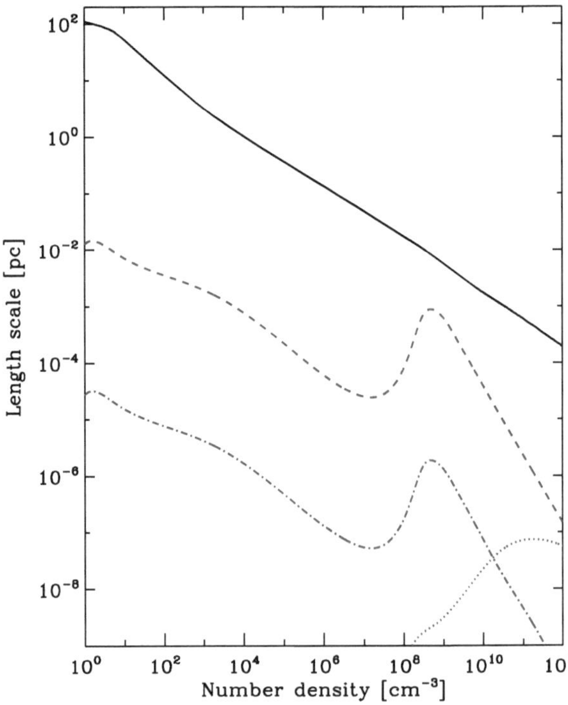

Fig. 2 Diffusion length scales during gravitational collapse. The *solid line* denotes the Jeans length as a function of central core density. The ambipolar diffusion scale for the equipartition magnetic field is given by the *dashed line*, while the *dashed-dotted line* corresponds to the scale for a weaker field (the magnetic energy equals $1/60$ of the kinetic energy). The *dotted line* gives the Ohmic diffusion scale, which is independent of the magnetic field strength. In all cases, the dissipation scales are smaller than the Jeans length

For collisions involving protons and electrons, the momentum-transfer coefficients of Pinto and Galli (2008) are adopted. For collisions involving Li^+, the momentum-transfer coefficients using the polarization approximation are calculated, as described by Schleicher et al. (2009). This is indeed justified for collisions with helium (Cassidy and Elford 1985) and H_2 (Dickinson et al. 1982; Røeggen et al. 2002). For collisions with atomic hydrogen, no detailed theoretical or experimental measurements are currently available. This process is of minor importance, however, as Li^+ becomes the dominant ionized species at densities when molecular hydrogen is the dominant neutral species.

Equations (13) and (15) show that the dissipation rates depend on the magnetic field strength itself, and we note that also the ambipolar resistivities increase with increasing field strength (Pinto et al. 2008). During gravitational collapse, an initially weak field, for instance due to the Biermann battery (Xu et al. 2008), may be amplified both due to gravitational compression as well as turbulence dynamo (Schleicher et al. 2010). If the amplification due to turbulence dynamo is efficient, saturation may occur at equipartition or a somewhat lower level (see, e.g., Subramanian 1999; Cho and Vishniac 2000; Haugen et al. 2004a, 2004b, 2004c; Schekochihin et al. 2004; Brandenburg and Subramanian 2005; Cho et al. 2009). As ambipolar and Ohmic diffusions are stronger for stronger fields, we will initially focus on equipartition magnetic fields and show that magnetic fields can be sustained. The same is then true also for weaker magnetic fields.

To have a definite model, we assume Kolmogorov turbulence, and thus that the equipartition field strength scales as $l^{1/3}$. This is a good approximation for subsonic turbulence where compressional effects are of minor importance, and it is this type of turbulence that is expected in the first star-forming systems (Abel et al. 2002). The Ohmic and ambipolar diffusion scales are calculated as a function of the central core density and compared them to the Jeans length (see Fig. 2). For clarity, the Ohmic/ambipolar diffusion scales are defined as the length scales on which the Ohmic or ambipolar diffusion time equals the eddy-turnover time. The calculation assumes that the turbulent velocity and the magnetic field follow a typical scaling law as expected for Kolmogorov turbulence. During collapse, the increased densities lead to a more efficient coupling between the ions and the neutrals, which largely compensates for the decreasing ionization degree. At a density of $10^9\ \text{cm}^{-3}$, the ambipolar diffusion scale increases due to a drop of the ionization degree at these high densities, but still stays an order of magnitude below the Jeans length. The Ohmic diffusion scale always stays orders of magnitude below the Jeans length. The figure also shows that when the magnetic field strength is below the equipartition value, the ambipolar diffusion scale is even small. The Ohmic dissipation scale, however, stays on the same low level, as the Ohmic dissipation rate scales roughly with the magnetic energy.

Numerical calculations suggest that the ionization degree may even be higher than expected from one-zone models, in particular due to the presence of shocks of Mach number ~ 1 (Clark et al. 2011; Turk et al. 2009). So we safely conclude that during the dark ages, the magnetic fields up to the equipartition strength can be sustained even at high densities in collapsed regions.

4.2 Magnetic Field Evolution in First Star-Forming Halos

We now turn to the amplification of magnetic fields during the formation of the first stars. The halos that harbor the first stars at redshifts of $z \sim 20-30$ typically have a total mass of $10^6\ M_\odot$, consist of primordial gas with a temperature of ~ 3000 K, and have a spatial extent of ~ 100 pc. Turbulence is expected to be present, but on a subsonic level (Abel et al. 2002; Bromm and Loeb 2003; Yoshida et al. 2008). Such turbulence may create strong magnetic fields in the first galaxies and around them (Ryu et al. 2008; Arshakian et al. 2009; de Souza and Opher 2010; Schleicher et al. 2010).

Consider a simple model in which magnetic fields are amplified and saturate at a level corresponding to an energy $1/60$ of the equipartition energy on an eddy turnover time t_{eddy}. We assume that turbulence is injected during the virialization phase on the length scale of the system (i.e., ~ 100 pc) with the velocity of the order of the sound speed. Both Kolmogorov and Burgers turbulences are considered; for Kolmogorov turbulence the velocity depends on the length scale as $l^{1/3}$, while for Burgers turbulence the scaling accords to $l^{1/2}$.

For magnetic fields to be relevant during the formation of the first stars, they must be amplified within the free-fall time scale $t_{ff} \sim 1/\sqrt{G\rho}$. The number of eddy turnovers is thus given by the ratio t_{ff}/t_{eddy}. Here, the eddy turnover time $t_{\text{eddy}} \equiv l/v$ has a characteristic dependence on length scale for Kolmogorov and Burgers turbulences. Then, the expected magnetic field strength is given in Fig. 3 as a function of the length scale of turbulence. The preliminary calculation shows that magnetic fields of the order 10^{-9} G can be reached in the first star-forming halos. For a comprehensive understanding of the strength and structure of the magnetic fields, the detailed implementation of turbulence dynamo will be required. Recent numerical MHD simulations suggest that the small-scale dynamo indeed operates during gravitational collapse, producing magnetic fields during the formation of the first structures (Sur et al. 2010; Federrath et al. 2011).

Additional amplification may occur after the formation of protostellar disks. In particular, a large-scale dynamo and the magneto-rotational instability may further enhance the magnetic fields in the first star-forming halos (Pudritz and Silk 1989; Tan and Blackman 2004; Silk and Langer 2006). For simplified field geometries, Machida et al. (2006, 2008) found that magnetic fields lead to the formation of jets and help to suppress fragmentation in protostellar disks. Fromang et al. (2004) studied self-gravitating magnetized disks and found

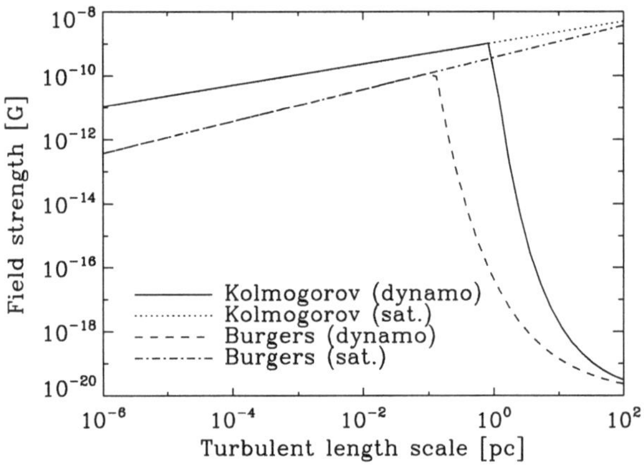

Fig. 3 The magnetic field strength after a free-fall time. The *horizontal axis* denotes the length scale of the turbulent eddies. We assume a power-law scaling relation for the rms turbulence on a given length scale, and explore the cases of Kolmogorov and Burgers-type turbulence (Schleicher et al. 2010). In this model, the turbulent injection scale is 100 pc. For every length scale, we show both the maximum field strength that could be obtained from the small-scale dynamo, as well as our estimate for the actual field strength that might be produced within a free-fall time. The model suggests the production of strong magnetic fields on small scales

that the interaction of turbulence created by the magneto-rotational instability may excite additional modes for the gravitational instability, and that the interaction of these modes reduces the accretion rate. This work suggests that magnetic fields may be amplified rapidly in the first star-forming halos and may become dynamically relevant though more realistic studies are necessary.

4.3 Seed Fields from Astrophysical Processes

The first stars probably possessed strong magnetic fields and therefore may have provided seed fields for dynamos in galaxies and in the IGM. If the stars subsequently explode as supernovae or lose a significant amount of mass through stellar winds, the fields ejected along with mass will find their way into the interstellar medium (ISM) and spread beyond galaxies into the IGM through galactic winds. Simple estimates by Syrovatskii (1970) illustrate the viability of the process. If there have been some 10^8 supernovae over the lifetime of galaxies, each of which spreads material through a $(10\ \mathrm{pc})^3$ volume. Using values for the field strength typical of the Crab nebula, one therefore expects galaxies to be filled by 10 pc regions with fields of strength $\sim 3\ \mu\mathrm{G}$. Assuming the same $L^{-3/2}$ scaling, one finds fields of strength of $\sim 10^{-11}$ G on 10 kpc scales. This value is significantly larger than those from the processes described in Sect. 3 and in Widrow et al. (2010) of this issue, although the filled volume is rather small. Recently, Donnert et al. (2009) suggested that galactic outflows during the starburst phase of galactic evolution can deposit a substantial amount of magnetic fields in the IGM.

Seed fields can be produced at cosmological shocks which were induced during the formation of the LSS of the universe (Ryu et al. 2003; Pfrommer et al. 2006; Kang et al. 2007; Skillman et al. 2008; Vazza et al. 2009) (see Sect. 5.1 for further description of cosmological shocks). Cosmological shocks are collisionless like shocks in other astrophysical environments, where CRs are accelerated at the same time as the gas is heated. During the process of acceleration, it was shown that the upstream magnetic field can be amplified nonlinearly by non-resonant growing modes (Bell 2004). Then, the magnetic field can have the energy up to

$$\varepsilon_B \sim \frac{1}{2}\frac{U_{n1}}{c}\varepsilon_{\rm CRs}. \tag{21}$$

With $U_{n1}/c \sim 10^{-3}$ and Mach number a few in cosmological shocks (Ryu et al. 2003; Kang et al. 2007), we get $\varepsilon_B \sim \varepsilon_{\rm CRs} \sim \varepsilon_{\rm therm}$. In addition, the Weibel instability described in Sect. 3.3 can operate and produce magnetic fields up to the level of $\varepsilon_B \sim 10^{-3}\varepsilon_{\rm sh}$ (Schlickeiser and Shukla 2003; Medvedev et al. 2006). Here, $\varepsilon_{\rm sh}$ is the energy density of upstream flow, and ε_B, $\varepsilon_{\rm CRs}$, and $\varepsilon_{\rm therm}$ are the energy densities of downstream magnetic fields, CRs, and gas random motion, respectively. These processes can potentially produce strong fields around cosmological shocks, although the volume filling is small and the coherence length of the resulting fields is expected to be microscopic.

Other mechanisms have been proposed. For instance, recently, Miniati and Bell (2011) suggested that the return current which is induced by cosmic-rays produced by early supernovae can deposit seed fields into the IGM.

5 Magnetic Fields in the IGM

With the processes described in Sects. 3 and 4 and also in Widrow et al. (2010) of this issue, there is no shortage of mechanisms to generate seed fields for the IGMF. Those fields are expected to be amplified by the turbulent flow motions which were induced during the formation of the LSS of the universe. The turbulence dynamo not only increases the strength of magnetic fields, but also produces the magnetic fields of large scales, up to the energy injection scale, through the inverse cascade. The beauty of the turbulence dynamo in LSS is that it erases the memory of weak seed fields and produces the IGMF, independent of the origin of seed fields.

5.1 Turbulence in the LSS of the Universe

Signatures of turbulence have been observed in the ICM. For example, Schuecker et al. (2004) analyzed gas pressure maps that were constructed from XMM-Newton X-ray data. They claimed that in the Coma cluster, which appears to be in a post-merger state, pressure fluctuations are consistent with Kolmogorov turbulence. The turbulence is likely subsonic but with an energy that is at least 10% of the thermal energy, i.e., $\varepsilon_{\rm turb} > 0.1\varepsilon_{\rm therm}$. The results agree with predictions from numerical simulations, namely that the flows in cluster scales have a power spectrum expected for Kolmogorov turbulence (Kulsrud et al. 1997; Kulsrud and Zweibel 2008), and even in relaxed clusters the flow motions have $\varepsilon_{\rm kin} \sim 0.1\varepsilon_{\rm therm}$ (Nagai et al. 2007). Turbulence in the ICM was also studied in RM maps of a few clusters (Vogt and Enßlin 2005; Guidetti et al. 2008; Bonafede et al. 2010).

Recently, Ryu et al. (2008) proposed a scenario in which vorticity is generated directly or indirectly at cosmological shocks and turbulence in the IGM is induced via the cascade of the vorticity. Here, we provide an estimate of the turbulence seen in simulations for the formation of the LSS. The results are based on a simulation described in Cen and Ostriker (2006) which includes the radiative processes of heating/cooling and feedback from galactic superwind. The work of Ryu et al. (2008) utilized a simulation, where only the gravitational and gas dynamical processes are included.

In the simulation of Cen and Ostriker (2006), the WMAP1-normalized ΛCDM cosmology was employed with the following parameters: $\Omega_b = 0.048$, $\Omega_m = 0.31$, $\Omega_\Lambda = 0.69$,

$h \equiv H_0/(100\ \text{km/s/Mpc}) = 0.69$, $\sigma_8 = 0.89$, and $n = 0.97$. A cubic box of comoving size $85\ h^{-1}$ Mpc was simulated using 1024^3 grid zones for gas and gravity and 512^3 particles for dark matter. It allows a uniform spatial resolution of $\Delta l = 83\ h^{-1}$ kpc. For detailed descriptions of input physics ingredients such as non-equilibrium ionization/cooling, photoionization/heating, star formation, and feedback processes, refer Cen et al. (2003) and Cen and Ostriker (2006). Feedback from star formation was treated in three forms: ionizing UV photons, galactic superwinds, and metal enrichment. Galactic superwinds were meant to represent cumulative supernova explosions, and modeled as outflows of several hundred km s^{-1}. The input of galactic superwind energy for a given amount of star formation was determined by matching the outflow velocities computed for star-burst galaxies in the simulation with those observed in the real world. The simulations were performed using a PM/Eulerian hydrodynamic cosmology code (Ryu et al. 1993).

In the IGM, vorticity, $\boldsymbol{\omega} \equiv \nabla \times \mathbf{v}$, can be generated directly at curved shocks and also by the baroclinity of flows. For uniform upstream flow, the vorticity produced behind curved shock surface is

$$\omega_{\text{cs}} = \frac{(\rho_2 - \rho_1)^2}{\rho_1 \rho_2} K \mathbf{U}_1 \times \hat{n}, \tag{22}$$

where ρ_1 and ρ_2 are the upstream and downstream gas densities, respectively, $\mathbf{U}_1$ is the upstream flow velocity in the shock rest frame, K is the curvature tensor of the shock surface, and $\hat{n}$ is the unit vector normal to the surface. If isopycnic surfaces do not coincide with isobaric surfaces, vorticity is generated with the rate given by

$$\dot{\omega}_{\text{bc}} = \frac{1}{\rho^2} \nabla \rho \times \nabla p. \tag{23}$$

Shock waves are ubiquitous in the IGM, as in other astrophysical environments. The spatial distribution and properties of cosmological shocks in the LSS of the universe have been studied quantitatively using simulations for the formation of LSS (Ryu et al. 2003; Pfrommer et al. 2006; Kang et al. 2007; Skillman et al. 2008; Vazza et al. 2009). In the cold dark matter universe with cosmological constant (ΛCDM), shocks with Mach number up to a few hundreds and speed up to a couple of thousand km s^{-1} are present at the present universe ($z = 0$). In the left panel of Fig. 4, the spatial distribution of cosmological shocks is shown. Numerous shocks are found. External shocks exist around sheets, filaments, and knots of mass distribution, which form when the gas in void regions accretes onto them. Within those nonlinear structures, internal shocks exist, which form by infall of previously shocked gas to filaments and knots, and during subclump mergers, as well as by chaotic flow motions. Due to the low temperature of the accreting gas, the Mach number of external shocks is high, extending up to $M \sim$ a few $\times 100$, while internal shocks have mostly low Mach number of $M \sim$ a few. The mean distance between shock surfaces is $\sim 3\ h^{-1}$ Mpc when averaged over all the universe, or $\sim 1\ h^{-1}$ Mpc inside nonlinear structures. Internal shocks of $M \sim 2$–4 formed with hot and high-density gas are responsible for most of shock dissipation into heat and CRs. It was shown that the shock dissipation can count most of the gas thermal energy in the IGM (Kang et al. 2005).

In the right panel of Fig. 4, the distribution of vorticity is shown. It closely matches that of shocks, suggesting that a substantial portion of the vorticity, if not all, has been generated at the shocks. As a matter of fact, as was noted in Ryu et al. (2008), the vorticity in the IGM can be accounted with that generated either directly at curved cosmological shocks or by the baroclinity of flows. The contributions from the two processes are comparable.

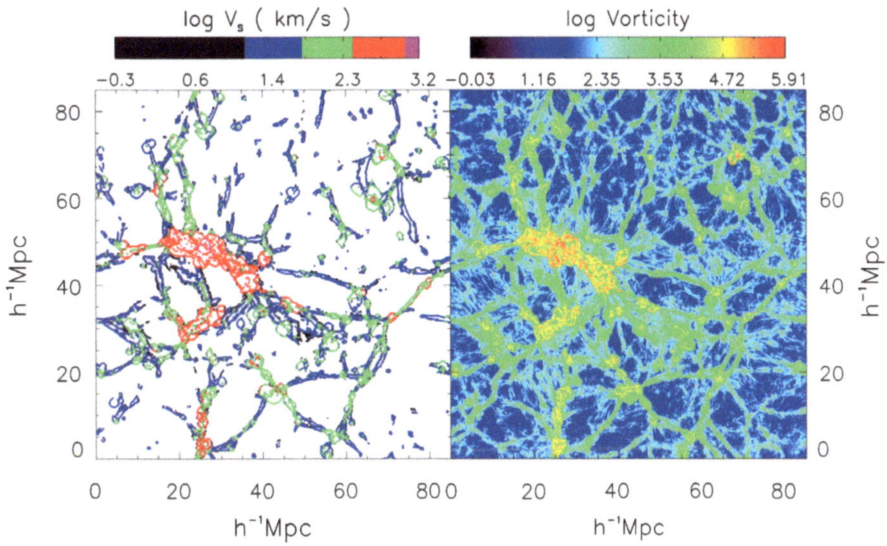

Fig. 4 Two-dimensional slice of $(85\ h^{-1}\ \mathrm{Mpc})^2$ showing shock locations with color-coded shock speed (*left panel*) and the magnitude of vorticity (*right panel*) at $z = 0$. The vorticity is given in units of $10^{-4}\ t_{\mathrm{age}}^{-1}$, where t_{age} is the age of the universe

The baroclinity resulted from the entropy variation induced at shocks. So all the vorticity generation also can be attributed to the presence of cosmological shocks.

For quantification of vorticity, the flow velocity is decomposed into

$$\mathbf{v} = \mathbf{v}_{\mathrm{div}} + \mathbf{v}_{\mathrm{curl}} + \mathbf{v}_{\mathrm{unif}}, \tag{24}$$

where the divergence and curl components are defined as $\nabla \cdot \mathbf{v}_{\mathrm{div}} \equiv \nabla \cdot \mathbf{v}$ and $\nabla \times \mathbf{v}_{\mathrm{curl}} \equiv \nabla \times \mathbf{v}$, respectively. That is, $\mathbf{v}_{\mathrm{div}}$ is associated to compressional motions, while $\mathbf{v}_{\mathrm{curl}}$ to incompressible shear motions. Here $\mathbf{v}_{\mathrm{unif}}$ is the component uniform across the computational box, whose magnitude is much smaller than the other two components. The decomposition is calculated exactly in Fourier space. We note with the above decomposition, locally

$$\mathbf{v}_{\mathrm{div}} \cdot \mathbf{v}_{\mathrm{curl}} \neq 0 \quad \text{so} \quad \frac{1}{2} v^2 \neq \frac{1}{2}\left(v_{\mathrm{div}}^2 + v_{\mathrm{curl}}^2\right). \tag{25}$$

However, globally

$$\int_{\mathrm{box}} \mathbf{v}_{\mathrm{div}} \cdot \mathbf{v}_{\mathrm{curl}} d^3\mathbf{x} = 0 \quad \text{so} \quad \int_{\mathrm{box}} \frac{1}{2} v^2 d^3\mathbf{x} = \int_{\mathrm{box}} \frac{1}{2}\left(v_{\mathrm{div}}^2 + v_{\mathrm{curl}}^2\right) d^3\mathbf{x}. \tag{26}$$

The power spectra for the gas velocity and its curl and divergence components at the present universe are shown in Fig. 5. At long wavelengths, the amplitude of perturbations are small, so that linear theory applies. That is, $P_{\mathrm{curl}}(k) \to 0$ as $k \to 0$, while $P_{\mathrm{div}}(k)$ follows the analytic theory expectation, $P_{\mathrm{div}}(k) \sim k^{-1}$. For wavelengths smaller than a few Mpc, nonlinearities dominate, and we see $P_{\mathrm{curl}}(k) \gtrsim P_{\mathrm{div}}(k)$. $P_{\mathrm{curl}}(k)$ peaks at $\sim 5\ h^{-1}$ Mpc, and for k somewhat larger than the peak wavenumber, the spectrum follows a power law of $k^{-5/3}$, the Kolmogorov spectrum. $P_{\mathrm{curl}}(k)$ has most power at ~ 2–$3h^{-1}$ Mpc, that indicates the typical scale of nonlinear structures in the simulation.

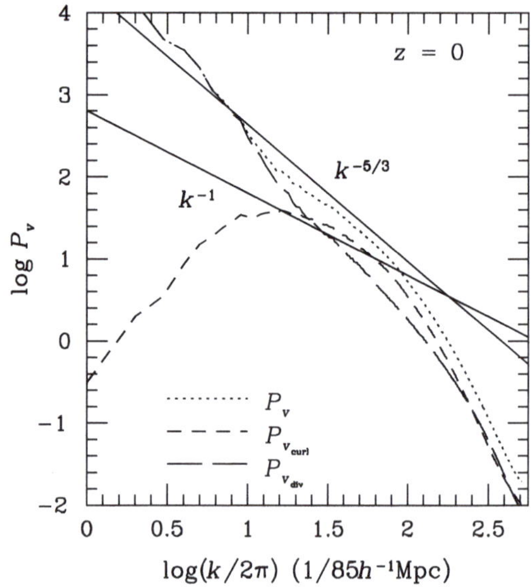

Fig. 5 Power spectra, $\int P_v dk = \langle (1/2)v^2 \rangle$, for the gas velocity and its curl and divergence components at $z = 0$. Two straight lines of slopes $-5/3$ and -1 are also drawn for comparison

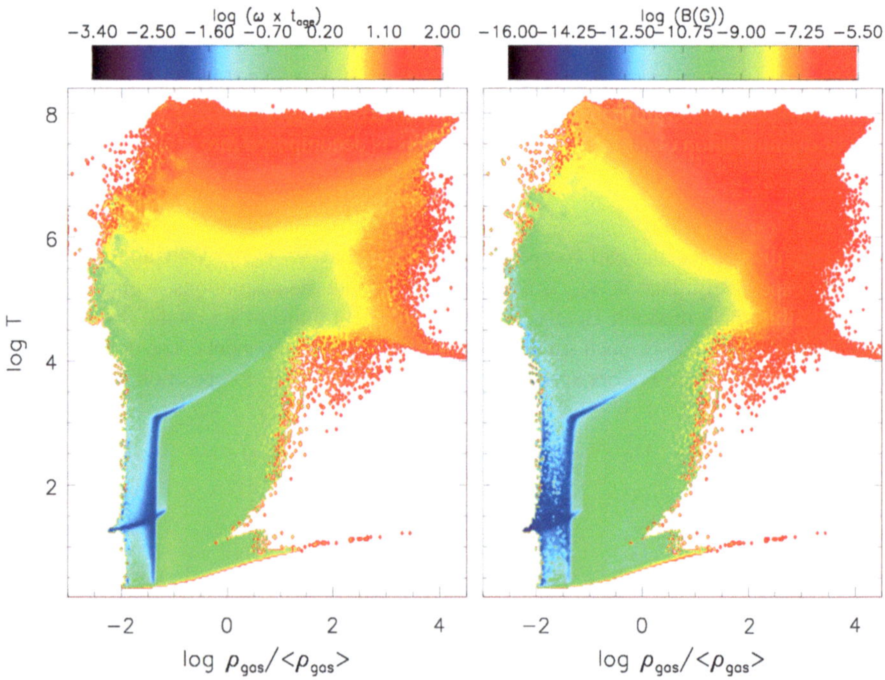

Fig. 6 Distribution of the magnitude of the vorticity in the IGM (*left panel*) and the strength of the IGMF (*right panel*) in the density-temperature plane at $z = 0$. The vorticity is given in units of $t_{\rm age}^{-1}$

The vorticity in the IGM as a function of gas density and temperature is shown in the left panel of Fig. 6. The figure exhibits a clear trend that the vorticity is larger in hotter

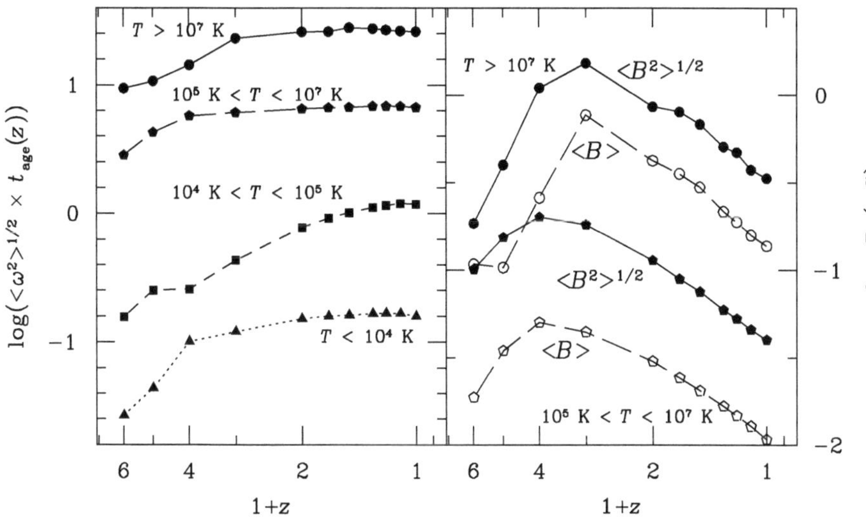

Fig. 7 The time evolution of the rms of the vorticity for four temperature phases of the IGM (*left panel*) and the time evolution of the averaged strengths of the IGMF for the hot IGM and WHIM (*right panel*) as a function of redshift z. The vorticity is given in units of $t_{\rm age}(z)^{-1}$, where $t_{\rm age}(z)$ is the age of the universe at z

regions. It is because hotter regions are occupied with the gas that has gone through shocks of larger speed. The vorticity generated at shocks of larger speed should be on average larger. Also, there are regions of high density ($\rho_{\rm gas}/\langle\rho_{\rm gas}\rangle \lesssim 100$) and warm temperature ($T \gtrsim 10^4$ K), where the vorticity is large. These regions contain the gas that was heated to high temperature and subsequently cooled down.

In the left panel of Fig. 7, the root-meas-square (rms) of the vorticity at a few redshifts for the gas in the four phases of the IGM is shown. Note that the vorticity shown was computed over the same comoving scales, and normalized with the age of the universe at given z. The vorticity increases as the universe evolves and the LSS of the universe develops. But over the period of the time presented in the figure, the vorticity, especially in the hot IGM and WHIM, has increased just by a factor of a few.

Figures 6 and 7 indicate that at the present epoch, $\omega_{\rm rms}t_{\rm age} \sim$ a few $\times 10$ inside and around clusters/groups with $T \gtrsim 10^7$ K and ~ 10 in filaments which is filled mostly with the WHIM. On the other hand, $\omega_{\rm rms}t_{\rm age}$ is on the order of unity in sheetlike structures and even smaller in voids. Here, $t_{\rm age}$ is the present age of the universe, so $\omega_{\rm rms}t_{\rm age}$ represents the number of eddy turnovers in the age of the universe. It takes a few turnover times for vorticity to decay and develop into turbulence. So it is likely that the flows in clusters/groups and filaments is in a turbulent state, while turbulence have not significantly developed in sheetlike structures and voids.

To estimate the energy associated with the turbulence induced in the IGM via the cascade of the vorticity, we assume that the energy of vortical motions, $(1/2)\rho_{\rm gas}v_{\rm curl}^2$, is transferred to that of turbulent motions. Then, we can regard it as the turbulence energy, $\varepsilon_{\rm turb}$. Figure 8 shows the ratio of the turbulence to thermal energies in clusters/groups and filaments as a function of temperature at the present universe. In clusters/groups with $T \gtrsim 10^7$ K, $\varepsilon_{\rm turb} < \varepsilon_{\rm therm}$. Particularly the mass averaged value shows that $\varepsilon_{\rm turb}/\varepsilon_{\rm therm}$ would be ~ 0.1–0.3 in cluster cores. The predicted $\varepsilon_{\rm turb}/\varepsilon_{\rm therm}$ is in a good agreement with observation (Schuecker et al. 2004). Note that $M_{\rm turb} \equiv v_{\rm curl}/c_s = \sqrt{1.8}\,(\varepsilon_{\rm turb}/\varepsilon_{\rm therm})^{1/2}$, where c_s is the sound speed.

Fig. 8 Turbulence to thermal energy ratio as a function of temperature at $z = 0$. The values shown are volume-averaged and mass-averaged over temperature bins

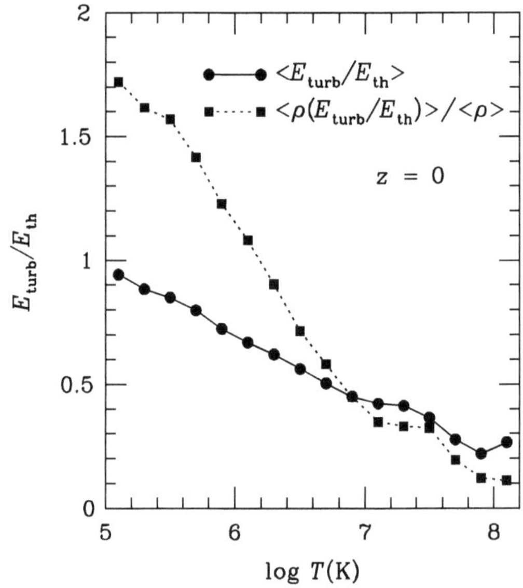

Therefore, overall turbulence is subsonic in clusters/groups, whereas it is transonic or mildly supersonic in filaments.

5.2 Amplification of the IGMF by the Turbulence in the IGM

In principle, if simulations for the formation of LSS in the universe includes magnetic fields, that is, if they are MHD, the amplification of the seed magnetic fields by the turbulent motions in the IGM should be able to be followed. But, in reality, simulations with the current capacity of computing power have too low a resolution to reproduce the full development of turbulence inside nonlinear structures. Also the numerical resistivity is larger than the physical resistivity by great many order of magnitude. As a result, the growth of magnetic fields is saturated before dynamo action become fully operative, and the amplification of magnetic fields can not be followed correctly, as was already pointed in Kulsrud et al. (1997).

In order to reproduce the growth of magnetic fields by dynamo action, a separate three-dimensional simulation of MHD turbulence in a controlled box was performed; incompressible, driven turbulence with initially very weak or zero magnetic fields was simulated using a pseudospectral code (Cho and Vishniac 2000). Hyperviscosity and hyperresistivity with the Prandtl number of unity were used. The advantage of performing incompressible simulations using a pseudospectral code is that the intrinsic numerical viscosity and resistivity are virtually zero. And by using hyperviscosity and hyperresistivity, the inertial range can be maximized. We point that it would take much higher resolution to achieve the same growth rate in simulations of compressible MHD turbulence.

The top-left panel of Fig. 9 shows the time evolution of kinetic and magnetic energies for four different simulations: 256H8-$B_0 10^{-3}$, 256H3-$B_0 10^{-3}$, 256H8-$B_0 0$, and 256H8-$B_0 10^{-6}$. Simulations are denoted with 256Y-B_0Z, where 256 refers to the number of grid points in each spatial direction, Y refers to hyperdissipation (H) and its order, and Z refers to the strength of the external magnetic fields. The turbulence was driven at the scale of $L_0 \sim (1/2)L_{\rm box}$ where $L_{\rm box}$ is the computational box size, and the driving strength was set so that the total energy is $E_{\rm tot} \equiv E_{\rm kin} + E_B \sim 1$ at saturation. The time is given in

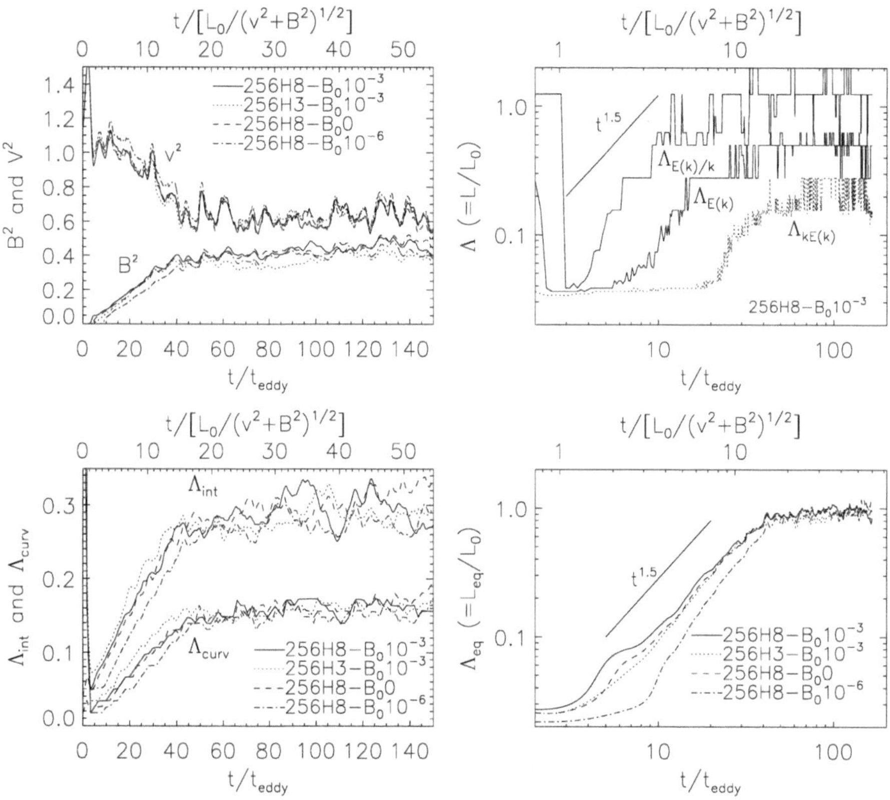

Fig. 9 *Top-left panel*: The time evolution of V^2 and B^2 for four different runs. Here, the kinetic energy and magnetic energy densities are $V^2/2$ and $B^2/2$, respectively. *Top-right panel*: Time evolution of peak scales of magnetic field spectrum, $L_{E(k)}$, $L_{kE(k)}$, and $L_{E(k)/k}$. *Bottom-right panel*: Time evolution of the energy equipartition scale, $L_{\rm eq}$. *Bottom-left panel*: Time evolution of the integral scale, $L_{\rm int}$, and the curvature scale, $L_{\rm curv}$. Λ's are the scales normalized with the energy injection scale L_0. See the text for details

units of the eddy turnover time that is defined as the inverse of vorticity at driving scale, $t_{\rm eddy} \equiv 1/\omega_{\rm driving}$, at saturation. See Cho et al. (2009) for details of simulations.

The amplification of magnetic fields by turbulence dynamo is shown in the figure. It does not sensitively depend on the initial magnetic field strength once it is sufficiently weak as well as details of simulations including the dissipation prescription. The evolution of magnetic fields goes through three stages: the initially exponential growth when the back reaction of magnetic fields is negligible, then the linear growth when the back reaction starts to operate, and the final saturation. By fitting the evolution, we model the growth and saturation of magnetic field energy as

$$E_B = \begin{cases} 0.04 \times \exp[(t/t_{\rm eddy} - 4)/0.36] & \text{for } t/t_{\rm eddy} < 4, \\ (0.36/41) \times (t/t_{\rm eddy} - 4) + 0.04 & \text{for } 4 < t/t_{\rm eddy} < 45, \\ 0.4 & \text{for } t/t_{\rm eddy} > 45. \end{cases} \tag{27}$$

Along with the amplification of magnetic field strength, the magnetic fields become coherent through the inverse cascade. For the quantification, different characteristic lengths of magnetic fields can be defined in MHD turbulence: the peak scale of the spectrum of magnetic fields, $L_{E(k)}$, the scale containing the largest energy of magnetic fields, $L_{kE(k)}$, the peak scale of the spectrum of projected magnetic fields, $L_{E(k)/k}$, the energy equipartition scale, $L_{\rm eq}(=2\pi/k_{\rm eq})$, defined as

$$\int_{k_{\rm eq}}^{k_{\rm max}} E_v(k)\,dk = \int_0^{k_{\rm max}} E_b(k)\,dk, \tag{28}$$

the integral scale, $L_{\rm int}$, defined as

$$L_{\rm int} = 2\pi \frac{\int E_b(k)/k\,dk}{\int E_b(k)\,dk}, \tag{29}$$

and the curvature scale, $L_{\rm curv}$, defined as a typical radius of curvature of field lines.

The rest of Fig. 9 show the time evolution of characteristic lengths of magnetic fields. At saturation, the peak of magnetic field spectrum, $L_{E(k)}$, occurs at $\sim L_0/2$, where L_0 is the energy injection scale, while the most energy containing scale, $L_{kE(k)}$ is $\sim L_0/5$. During the stage of magnetic field amplification, the energy equipartition scale, $L_{\rm eq}$, shows a power-law increase of $\sim t^{1.5}$, while the integral scale, $L_{\rm int}$, and the curvature scale, $L_{\rm curv}$, show a linear increase. The equipartition, integral, and curvature scales saturate at $\sim L_0$, $\sim 0.3L_0$, and $\sim 0.15L_0$, respectively. See Cho and Ryu (2009) for further details of characteristic lengths in MHD turbulence with very weak or zero mean magnetic fields.

The results of the incompressible MHD turbulence simulation were convoluted to the data of the stimulation for the formation of the LSS of the universe to get the IGMF. For the estimation of the magnetic field strength, it was assumed that a fraction of the turbulence energy is converted into the magnetic energy. The fraction was expressed as a function of the number of local eddy turns over the age of the universe, so

$$\varepsilon_B = \varepsilon_{\rm curl} \cdot \phi(\omega \times t_{\rm age}). \tag{30}$$

For the fraction $\phi(\omega \times t_{\rm age})$, the fitting formula in (27) was used. Then, the magnetic field strength was calculated as $(8\pi\varepsilon_B)^{1/2}$.

The resulting magnetic field strength as a function of gas density and temperature is shown in the right panel of Fig. 6. On average, the IGMF is predicted to be stronger in hotter and denser regions. It is because with the turbulence energy is larger in hotter and denser regions. Also the conversion factor, ϕ, is larger in hotter and denser regions.

In the right panel of Fig. 7, the volume-averaged and rms values of the magnetic field strength at a few redshifts for the gas in the hot IGM and WHIM are shown. Our scenario predicts that the magnetic field strength would be $\langle B\rangle \gtrsim 1$ μG inside clusters/groups, ~ 0.1 μG around clusters/groups, and ~ 10 nG in filaments at the present universe. The magnetic fields should be much weaker in sheetlike structures and voids. But as noted in Sect. 5.1, turbulence is not fully developed there. So our estimation of the IGMF in those regions should not be applicable.

In each temperature range, the magnetic fields were stronger in the past at $z \sim$ a few. Our calculation indicates that the vorticity has not changed much since $z \sim$ a few (see the left panel of Fig. 7), nor the vortical component of flow velocity (Ryu and Kang 2008). So the stronger magnetic fields in the past should be due to the higher density. We point that clusters were virialized around $z \sim 1$ or so, so the density within the virial radius of ~ 1 Mpc has not changed much since then. But our estimation of the magnetic field strength for the hot IGM with $T > 10^7$ K also includes the gas in cluster outskirt, which extends far beyond the virial radius, up to several Mpc or even larger (see, e.g. Ryu et al. 2003); the density of the gas there was higher in the past. We also note that the magnetic fields, when averaged all over the computational box, were weaker in the past, because the fraction of the strong field regions was smaller.

Based on a kinetic theory, assuming the Kolmogorov spectrum for turbulence flow motions, Kulsrud et al. (1997) and Kulsrud and Zweibel (2008) estimated that the magnetic field strength in clusters would be a few μG. Our result agrees with the previous work. It also matches well with the observed strength of magnetic fields in the ICM, which is discussed in Sect. 1. On the other hand, our prediction of $B \sim 10$ nG in filaments is within but lower than the upper limit of ~ 0.1 μG which is imposed from RMs outside of clusters (Ryu et al. 1998; Xu et al. 2006).

The characteristic lengths of the IGMF in our scenario can be conjectured. In clusters, turbulence is near the saturation stage with $t/t_{eddy} \sim 30$ (see Figs. 7 and 9). Then, for instance, we may estimate that the peak scale of magnetic field spectrum and the integral scale would $L_{E(k)}/L_0 \sim 0.4$ and $L_{\rm int}/L_0 \sim 0.2$. If we take the energy injection scale $L_0 \sim 100$ kpc, which is approximately the scale height of cluster core, $L_{E(k)} \sim 40$ kpc and $L_{\rm int} \sim 20$ kpc, respectively. In filaments, on the other hand, with $t/t_{eddy} \sim 10$, turbulence is expected to be still in the linear growth stage, and $L_{E(k)}/L_0 \sim 1/10$ and $L_{\rm int}/L_0 \sim 1/15$. We may take the energy injection scale $L_0 \sim 5$ Mpc, which is the typical thickness of filaments. Then, $L_{E(k)} \sim 0.5$ Mpc and $L_{\rm int} \sim 0.3$ Mpc, respectively.

5.3 Contribution from AGNs

There is a possibility that the IGMF is further strengthened by the magnetic fields ejected through jets from back holes in AGNs (see, e.g., Kronberg et al. 2001). Strong magnetic fields almost certainly arise in accretion disks surrounding black holes. These fields may find their way into the IGM via magnetically dominated jets. The potential field strength due to this process can be estimated as follows (see, e.g., Hoyle 1969). The rotational energy associated with the central compact objects of mass M which power AGNs can be parametrized as fMc^2 where $f < 1$. If we assume equipartition between rotational and magnetic energies within a central volume V_c, we find

$$B_c \sim \left(\frac{8\pi f M c^2}{V_c}\right)^{1/2}. \tag{31}$$

If this field then expands adiabatically to fill a volume V in the IGM, one finds $B \sim B_c(V_c/V)^{2/3}$. Considering the values of $M = 10^9\, M_\odot$, $f = 0.1$, and $V \simeq (1\ \text{Mpc})^3$, we get $B \sim 10$ nG. Note this field strength is comparable to that estimated in filaments with turbulence dynamo. So this process may enhance the strength of the IGMF by a factor of two or so.

Although there are some RM observations that indicate strong magnetic fields in some of jets, however, it is not yet clear whether all jets are magnetically dominated. Also the details of the population of AGN jets and the volume filling fraction of magnetic fields in the universe by this process need to be further worked out.

6 Astrophysical Implications of the IGMF

The existence of the IGMF, especially in filaments, can have noticeable implications on a variety of astrophysical phenomena. In this section, we discuss two: the effect on the propagation of ultra-high energy cosmic rays (UHECRs) and the inducement of Faraday rotation.

6.1 Propagation of Ultra-high Energy Cosmic Rays

UHECRs are known to originate from extragalactic sources. Hence, on their path through the intergalactic space, the trajectories of UHECRs are deflected by the magnetic fields between sources and us, the IGMF as well as the galactic magnetic field (GMF). Das et al. (2008) studied the effect of the IGMF described in Sect. 5 on the propagation of UHECRs. Under the premise that the sources of UHECRs are strongly associated with the LSS of the universe, super-GZK protons of $E \geq 10^{19}$ eV were injected by AGN-like sources located inside clusters of galaxies. Then, the trajectories of the protons were followed, while taking account of the energy loss due to interactions with the cosmic microwave background (CMB) radiation. Das et al. (2008) found that the deflection of UHECR trajectories is caused mostly by the IGMF in filaments, rather than the intracluster magnetic field in clusters; it is because filaments fill a larger fraction of volume, although their magnetic fields are weaker, than clusters. With the gyroradius of protons

$$r_g = 1\ \mathrm{Mpc}\left(\frac{E_{\mathrm{UHECR}}}{10^{19}\ \mathrm{eV}}\right)\left(\frac{B}{10\ \mathrm{nG}}\right)^{-1}, \tag{32}$$

that is, with the gyroradius corresponding typical filament size for the magnetic fields typical in filaments, the deflection due to the IGMF in filaments is expected to be significant. Indeed, the deflection angle between the arrival direction of super-GZK protons and the sky position of their actual sources was found to be quite large with the mean value of $\langle\theta\rangle \sim 15°$ and the median value of $\tilde{\theta} \sim 7$–$10°$.

The above deflection is much larger than the deflection by the GMF; the deflection angle due to the GMF was predicted to be a few degree (see, e.g., Takami and Sato 2008). As a matter of fact, the deflection angle of $\langle\theta\rangle \sim 15°$ is also much than the angular window of 3.1° used by the Auger collaboration in the study of the correlation between their highest energy UHECR events and nearby AGN (1). Although the deflection angle is large, Ryu et al. (2010) noticed that in the work of Das et al. (2008), the separation angle between the arrival direction of super-GZK protons and the sky position of nearest AGNs is substantially smaller with $\langle S\rangle \sim 3.5$–$4°$, which is similar to the mean angular distance in the sky to nearest neighbors among AGNs. This mean separation angle is comparable to the angle used in the correlation study by the Auger collaboration. This is a direct consequence of the fact that the sources and us, as well as the IGMF, all trace the matter distribution of the universe. That is, although the IGMF described in Sect. 5 predicts larger deflection of UHECRs, it is not necessarily inconsistent with the intervening magnetic fields implied in the Auger experiment.

6.2 Faraday Rotation Induced by the IGMF

The IGMF described in Sect. 5 induces RM. With characteristic lengths smaller than the dimension of clusters or filaments, the inducement of RM is a random walk process; the standard deviation of RM is

$$\sigma_{\mathrm{RM}} = 0.81\ \bar{n}_e\ B_{\|\mathrm{rms}}\sqrt{\left(\frac{3L_{\mathrm{int}}}{4}\right)L}\ \mathrm{rad}\,m^{-2}, \tag{33}$$

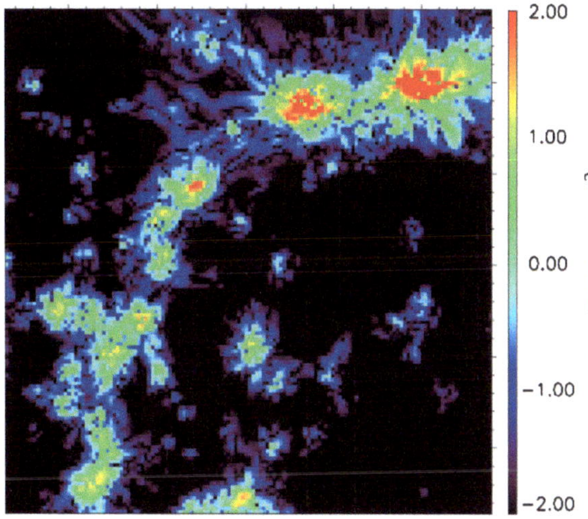

Fig. 10 RM map of path length of $L = 100\,h^{-1}$ in the local universe o $(28\,h^{-1}\,\text{Mpc})^2$ area at $z = 0$

where n_e, $B_{\|\text{rms}}$, and L are in units of cm^{-3}, μG, and pc, respectively (Cho and Ryu 2009). $B_{\|\text{rms}}$ is the rms strength of line-of-sight magnetic field and L is the path length. Here, the coherence length for RM is given as $l = (3/4)L_{\text{int}}$. For clusters, with $\bar{n}_e \sim 10^{-3}\ \text{cm}^{-3}$, $B_{\text{rms}} \sim$ a few μG, $L \sim 1$ Mpc, and $L_{\text{int}} \sim 20$ kpc (see Sect. 5), we get $\sigma_{\text{RM}} \sim 100\ \text{rad m}^{-2}$, which agrees with the observed RM in clusters (Clarke et al. 2001). The magnetic field strength in filaments quoted in Sect. 5 is $\langle B\rangle \sim 10$ nG. But the value depends on how it is averaged; that is, $\langle B^2\rangle^{1/2} \sim$ a few $\times$ 10 nG, $\langle\rho B\rangle/\langle\rho\rangle \sim 0.1\ \mu$G, and $\langle(\rho B)^2\rangle^{1/2}/\langle\rho^2\rangle^{1/2} \sim$ a few $\times$ 0.1 μG, in the warm-hot intergalactic medium (WHIM) with $T = 10^5$–10^7 K which mostly composes filaments. The average value of $\langle(\rho B)^2\rangle^{1/2}/\langle\rho^2\rangle^{1/2}$ should be relevant to RM. With $\bar{n}_e \sim 10^{-5}\ \text{cm}^{-3}$, $B_{\text{rms}} \sim 0.3$ μG, $L \sim 5$ Mpc, and $L_{\text{int}} \sim 300$ kpc, we get $\sigma_{\text{RM}} \sim 1$ rad m^{-2} through filaments.

Akahori and Ryu (2010) studied in details the RM due to the magnetic fields in filaments using the IGMF described in Sect. 5. Figure 10 shows a typical RM map in $(28\,h^{-1}\,\text{Mpc})^2$ area; the spatial distribution of RM traces the large-scale distribution of matter, showing two clusters and a filamentary structure containing several groups. The resultant RM is dominantly contributed by the density peak along line of sight. The rms of RM through filaments at the present universe was predicted to be $\sim 1\ \text{rad m}^{-2}$, which agrees with the estimation above (Cho and Ryu 2009). Figure 11 shows the two-dimensional power spectrum of the RM in the local universe within $100\,h^{-1}$ Mpc; $P_{\text{RM}}(k) \sim |\text{RM}(\mathbf{k})|^2 k$, where $\text{RM}(\mathbf{k})$ is the Fourier transform of $\text{RM}(\mathbf{x})$. The power spectrum of RM peaks at $k \sim 100$, which corresponds to $\sim 1\,h^{-1}$ Mpc. In addition, Akahori and Ryu (2010) predicted that the probability distribution function (PDF) of |RM| through filaments follows the log-normal distribution. We note that RM $\sim 1\ \text{rad m}^{-2}$ is an order of magnitude smaller than the values of |RM| toward the Hercules and Perseus-Pisces superclusters reported in Xu et al. (2006). The difference is mostly due to the mass-weighted path length; the value quoted by Xu et al. (2006) is about two orders of magnitude larger than ours.

RM $\sim 1\ \text{rad m}^{-2}$ due to the IGMF in filaments is too small to be confidently observed with currently available facilities. In addition, the galactic foreground of $\sim 10\ \text{rad m}^{-2}$ (toward halo) poses an additional challenge for its observation. The next generation radio interferometers, however, are expected to be able to observe the RM. Particularly, the SKA could measure RM for $\sim 10^8$ polarized extragalactic sources across the sky with an aver-

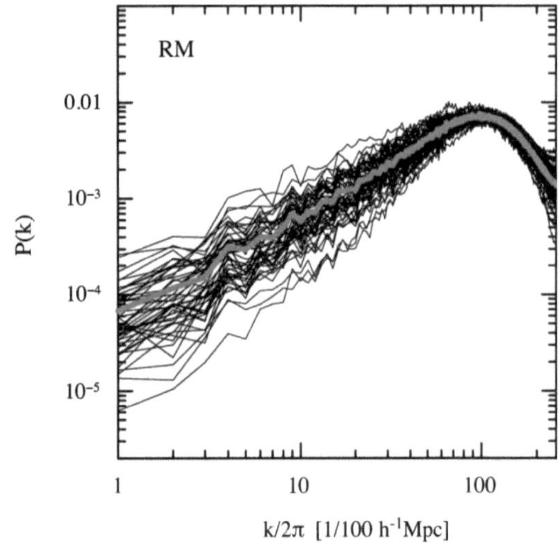

Fig. 11 Two-dimensional power spectra of RM for 3 × 16 two-dimensional projected maps from 16 simulations with different realizations of initial condition to compensate cosmic variance. The *thick gray line* shows the average

age spacing of ~ 60 arcsec between lines of sight (papers in Carilli and Rawlings 2004; Krause et al. 2009), enabling us to investigate the IGMF in the LSS of the universe.

7 Structure Formation and Magnetic Fields

The existence of magnetic fields in the early universe, although probably weak, can have consequences on the structure formation itself. In this section, we briefly review recent developments in our understanding of magnetized structure formation.

7.1 The Linear Regime

Studies of magnetized structure formation go back to the late 1960s with the early efforts based on Newtonian gravity and the relativistic approaches being a relatively recent addition. All treatments typically work within the ideal MHD approximation and look at the effects of the magnetic Lorentz force on density inhomogeneities (Ruzmaikina and Ruzmaikin 1971; Wasserman 1978; Fennelly and Evans 1980; Kim et al. 1996; Battaner et al. 1997; Tsagas and Barrow 1997; Tsagas and Maartens 2000a). These effects generally come in the form of scalar, vector and (trace-free) tensor distortions. The former are those commonly referred to as density perturbations and represent over-densities or under-densities in the matter distribution. Vector inhomogeneities describe rotational (i.e. vortex-like) density perturbations. Finally, tensor-type density inhomogeneities correspond to shape distortions.[1] Following (Tsagas and Barrow 1997; Barrow et al. 2007), we define the scalar

$$\Delta \equiv \frac{a^2}{\rho} \mathrm{D}^2 \rho, \tag{34}$$

which describes linear density perturbations and corresponds to the more familiar density contrast $\delta\rho/\rho$. Note that positive values for Δ indicate over-densities and negative ones

[1] It should be made clear that trace-free tensor inhomogeneities (i.e. shape deformations) and pure-tensor distortions (i.e. gravitational waves) are two different types of perturbations.

under-densities. Also, a is the cosmological scale factor and $\mathrm{D}^2=\mathrm{D}^a\mathrm{D}_a$ is the 3-D Laplacian operator that corresponds to an observer moving with 4-velocity u_a.[2] In a perturbed, weakly magnetized and spatially flat Friedmann-Robertson-Walker (FRW) universe, the above defined scalar evolves according to

$$\dot{\Delta}=3wH\Delta-(1+w)\mathcal{Z}+\frac{3}{2}c_{\mathrm{a}}^2(1+w)H\mathcal{B}, \tag{35}$$

where over-dots denote proper-time derivatives (relative to the u_a-frame). Also, $w=p/\rho$ is the (constant) barotropic index of the matter, $H=\dot{a}/a$ is the background Hubble parameter (with $\Theta=3H$ there) and $c_{\mathrm{a}}^2=B^2/\rho(1+w)$ is the square of the Alfvén speed. The variables $\mathcal{Z}=a^2\mathrm{D}^2\Theta$ and $\mathcal{B}=(a^2/B^2)\mathrm{D}^2B^2$ describe linear inhomogeneities in the smooth Hubble expansion and in the magnetic energy density respectively. Then, to first order,

$$\dot{\mathcal{Z}}=-2H\mathcal{Z}-\frac{1}{2}\rho\Delta+\frac{1}{4}c_{\mathrm{a}}^2(1+w)\rho\mathcal{B}-\frac{c_s^2}{1+w}\mathrm{D}^2\Delta-\frac{1}{2}c_{\mathrm{a}}^2\mathrm{D}^2\mathcal{B} \tag{36}$$

and

$$\dot{\mathcal{B}}=\frac{4}{3(1+w)}\dot{\Delta}, \tag{37}$$

respectively. Note that $c_s^2=\dot{p}/\dot{\rho}$ is the square of the adiabatic sound speed and we have assumed that $B^2\ll\rho$, given the relative weakness of the magnetic fields.

Equation (35) shows that magnetic fields are generic sources of linear density perturbations. Indeed, even if Δ and $\mathcal{Z}$ are zero initially, $\dot{\Delta}$ will take nonzero values solely due to the magnetic presence. Also, (37) ensures that perturbations in the magnetic field energy density evolve in tune with their matter counterparts (i.e. $\mathcal{B}\propto\Delta$). Finally, we should emphasize that only the pressure part of the Lorentz force contributes to the linear relations (35) and (36).[3]

The system of (35)–(37) has analytical solutions in the radiation and dust eras (Tsagas and Barrow 1997). Before equipartition, when $w=1/3=c_s^2$, $H=1/2t$, $\rho=3/4t^2$ and $c_{\mathrm{a}}^2=3B^2/4\rho=$ constant, large-scale magnetized density perturbations obey a power-law solution. In particular, on super-horizon scales and keeping only the dominant growing and decaying modes, one arrives at (Tsagas and Maartens 2000a; Barrow et al. 2007)

$$\Delta=\mathcal{C}_1t^{-1/2+10c_{\mathrm{a}}^2/9}+\mathcal{C}_2t^{1-4c_{\mathrm{a}}^2/9}. \tag{38}$$

In the absence of magnetic fields, we recover the standard growing and decaying modes of $\Delta\propto t$ and $\Delta\propto t^{-1/2}$ respectively. So, the magnetic presence has reduced the growth rate of the density contrast by $4c_{\mathrm{a}}^2/9$.

---

[2] For an observer with 4-velocity u_a (so that $u_au^a=-1$), the tensor $h_{ab}=g_{ab}+u_au_b$ projects orthogonal to u_a and $\mathrm{D}_a=h_a{}^b\nabla_b$ defines the covariant derivative operator of the spatial hypersurfaces (g_{ab} is the space-time metric and ∇_a the associated covariant derivative).

[3] The Lorentz force splits as $\varepsilon_{abc}B^b\mathrm{curl}B^c=\mathrm{D}_a/2B^2-B^b\mathrm{D}_bB_a$, with the former term corresponding to the magnetic pressure and the latter to the field's tension. The effects of the B-field on (scalar) density perturbations propagate via the divergence of the Lorentz force. To leading order, the latter is given by $\mathrm{D}^a(\varepsilon_{abc}B^b\mathrm{curl}B^c)=\mathrm{D}^2B^2/2-B^b\mathrm{D}_b\mathrm{D}^aB_a-KB^2/a^2$, where $K=0,\pm1$ represents the background 3-curvature index. Given that $\mathrm{D}^aB_a=0$ at the ideal-MHD limit, the magnetic tension effects are not included in this perturbative level, unless the spatial curvature of the FRW model is accounted for (Tsagas and Maartens 2000a, 2000b).

Well inside the horizon we can no longer ignore the role of the pressure gradients. There, the k-mode oscillates like a magneto-sonic wave with

$$\Delta_{(k)} \propto \sin\left[c_s\left(1+\frac{2}{3}c_{\rm a}^2\right)\left(\frac{\lambda_H}{\lambda_k}\right)_0\sqrt{\frac{t}{t_0}}\right], \tag{39}$$

where $\lambda_k = a/k$ is the perturbed scale and $\lambda_H = 1/H$ the Hubble horizon (Tsagas and Maartens 2000a; Barrow et al. 2007). Here, the magnetic pressure increases the effective sound speed and therefore the oscillation frequency. The former makes the Jeans length larger than in non-magnetized models. The latter brings the peaks of short-wavelength oscillations in the radiation density closer, leaving a potentially observable signature in the CMB spectrum (Adams et al. 1996).

When dust dominates, $w = 0 = c_s^2$, $H = 2/3t$, $\rho = 4/3t^2$ and $c_{\rm a}^2 = B^2/\rho \propto t^{-2/3}$. Then, on superhorizon scales, the main growing and decaying modes of the density contrast are (Tsagas and Barrow 1997; Barrow et al. 2007)

$$\Delta = \mathcal{C}_1 t^{\alpha_1} + \mathcal{C}_2 t^{\alpha_2}, \tag{40}$$

with $\alpha_{1,2} = -[1 \pm 5\sqrt{1-(32/75)(c_{\rm a}\lambda_H/\lambda_k)_0^2}]/6$. In the absence of magnetic fields we recover again the standard solution with $\alpha_1 = 2/3$ and $\alpha_2 = -1$. Thus, as with the radiation era before, the magnetic presence slows down the growth rate of density perturbations. Also, since $\mathcal{B} \propto \Delta$ [see (37)], the above describes the linear evolution of the magnetic energy-density perturbations as well. This means that cosmological magnetic fields trapped inside an overdense region of the post-recombination universe could grow by approximately two to three orders of magnitude. Note that the aforementioned increase is different from the one occurring during the subsequent, nonlinear contraction of a protogalactic cloud (see Sect. 7.2 below).

The field pressure also leads to a magnetically induced Jeans length, below which density perturbations cannot grow (Subramanian and Barrow 1998; Sethi and Subramanian 2005). As a fraction of the Hubble radius, this purely magnetic Jeans scale is (Tsagas and Maartens 2000a; Barrow et al. 2007)

$$\lambda_J \sim c_{\rm a}\lambda_H. \tag{41}$$

Setting $B \sim 10^{-9}$ G, which is the maximum homogeneous field strength allowed by the CMB (Zeldovich 1970a; Barrow et al. 1997), we find that $\lambda_J \sim 10$ kpc. Alternative, magnetic fields close to 10^{-7} G, like those found in galaxies and galaxy clusters, give $\lambda_J \sim 1$ Mpc. The latter lies intriguingly close to the size of a cluster of galaxies.

Overall, the magnetic effect on density perturbations is rather negative. Although magnetic fields generate this type of distortions, they do not help them to grow. Instead, the magnetic presence either suppresses the growth rate of density perturbations, or increases the effective Jeans length and therefore the domain where these inhomogeneities cannot grow. This negative role of magnetic fields, which was also observed in the Newtonian treatment of Ruzmaikina and Ruzmaikin (1971), reflects the fact that only the pressure part of the Lorentz force has been incorporated into the equations. When the tension component (i.e. the elasticity of the field lines) is also accounted for, the overall magnetic effect can change and in some cases it could even reverse (Tsagas and Maartens 2000b).

Magnetic fields also induce and affect rotational, vortex-like, density inhomogeneities (Wasserman 1978; Tsagas and Maartens 2000a). To linear order, these are described by the vector $\mathcal{W}_a = -(a^2/2\rho)\varepsilon_{abc}\mathrm{D}^b\mathrm{D}^c\rho$, with ε_{abc} representing the 3-D Levi-Civita tensor. Then, on an spatially flat FRW background,

$$\ddot{\mathcal{W}}_a = -4H\dot{\mathcal{W}}_a - \frac{1}{2}\rho\mathcal{W}_a + \frac{1}{3}c_{\mathrm{a}}^2\mathrm{D}^2\mathcal{W}_a, \tag{42}$$

after matter-radiation equality (Tsagas and Maartens 2000a; Barrow et al. 2007). Defining $\lambda_{\mathrm{a}} = c_{\mathrm{a}}\lambda_H$ as the Alfvén horizon, we may write the associated solution in the form

$$\mathcal{W}_{(k)} = \mathcal{C}_1 t^{\alpha_1} + \mathcal{C}_2 t^{\alpha_2}, \tag{43}$$

with $\alpha_{1,2} = -[5 \pm \sqrt{1-(48/9)(\lambda_{\mathrm{a}}/\lambda_k)_0^2}]/6$. On scales far exceeding the Alfvén horizon, $\lambda_{\mathrm{a}} \ll \lambda_k$ and the perturbed mode decays as $\mathcal{W} \propto t^{-2/3}$. This rate is considerably slower than $\mathcal{W} \propto t^{-1}$, the decay rate associated with magnetic-free dust cosmologies. Well inside λ_{a}, on the other hand, magnetized vortices oscillate like Alfvén waves, with (Tsagas and Maartens 2000a)

$$\mathcal{W}_{(k)} \propto t^{-5/6}\cos\left[\frac{2\sqrt{3}}{9}\left(\frac{\lambda_{\mathrm{a}}}{\lambda_k}\right)_0 \ln t\right]. \tag{44}$$

Thus, the effect of magnetic fields on a given vortex mode is to reduce its standard depletion rate. Analogous is the magnetic effect on ω_a, namely on the vorticity proper. Hence, magnetized cosmologies appear to rotate faster than their magnetic-free counterparts. In contrast to density perturbations, magnetic fields seem to favor the presence of vorticity. This qualitative difference should probably be attributed to the fact that the tension part of the Lorentz force also contributes to (42).

In addition to scalar and vector perturbations, magnetic fields also generate and affect tensor-type inhomogeneities that describe shape-distortions in the density distribution (Tsagas and Maartens 2000a). An initially spherically symmetric inhomogeneity, for example, will change shape due to the magnetically induced anisotropy. All these effects result from the Lorentz force. Even when the latter is removed from the system, however, magnetic fields remain active. Due to its energy density and anisotropic nature, for example, magnetism affects both the local and long-range gravitational fields. The anisotropic magnetic pressure, in particular, leads to shear distortions and subsequently to gravitational-wave production (Caprini and Durrer 2002; Tsagas 2002; Wang 2010). Overall, magnetic fields are a very versatile source. They are also rather unique in nature, since they are the only known vector source of energy. An additional unique magnetic feature, which remains relatively unexplored, is the field's tension. When we add to all these the widespread presence of magnetic fields in the universe, it is not unreasonable to say that no realistic structure formation scenario should a priori exclude them.

7.2 Aspects of the Nonlinear Regime

The evolution of large-scale magnetic fields during the nonlinear stages of structure formation is addressed primarily by means of numerical simulations. The reason is the high complexity of the nonlinear MHD equations, which considerably hampers analytical studies, unless certain simplifying assumptions are imposed.

The simplest approximation is to assume spherically symmetric compression. Realistic collapse, however, is not isotropic. In fact, when magnetic fields are present, their generically anisotropic nature makes the need to go beyond spherical symmetry greater. Certain aspects of anisotropic contraction can be analytically studied within the Zeldovich approximation (Zeldovich 1970b). The latter is based on a simple ansatz, which extrapolates to the nonlinear regime a well known linear result. The assumption is that the irrotational and acceleration-free linear motion of dust, also holds during the early nonlinear stages of galaxy formation. This allows the analytical treatment of the nonlinear equations, leading to solutions that describe anisotropic (one dimensional) collapse and to the formation of the well-known Zeldovich pancakes.

Suppose that magnetic fields are frozen into a highly conductive protogalactic cloud that is falling into the (Newtonian) potential wells formed by the Cold Dark Matter (CDM) sector.[4] Relative to the physical coordinate system $\{r^\alpha\}$, the motion of the fluid velocity is $u_\alpha = 3Hr_\alpha + v_\alpha$, where $H = \dot{a}/a$ is the Hubble parameter of the unperturbed FRW background and v_α is the peculiar velocity of the fluid (with $\alpha = 1, 2, 3$). Then, the magnetic field induction equation reads (Bruni et al. 2003)

$$\dot{B}_\alpha = -2HB_\alpha - \frac{2}{3}\vartheta B_\alpha + \sigma_{\alpha\beta}B^\beta, \tag{45}$$

where $\vartheta = \partial^\alpha v_\alpha$ and $\sigma_{\alpha\beta} = \partial_{\langle\beta}v_{\alpha\rangle}$ are the peculiar volume scalar and the peculiar shear tensor respectively.[5] The former takes negative values (i.e. $\vartheta < 0$), since we are dealing with a protogalactic cloud that has started to turn around and collapse. Also note that the first term in the right-hand side of (45) reflects the background expansion, the second is due to the peculiar contraction and the last carries the anisotropic effects. Introducing the rescaled magnetic field $\mathcal{B}_\alpha = a^2B_\alpha$, the above expression recasts into

$$\mathcal{B}'_\alpha = -\frac{2}{3}\tilde{\vartheta}\mathcal{B}_\alpha + \tilde{\sigma}_{\alpha\beta}\mathcal{B}^\beta, \tag{46}$$

with primes indicating differentiation with respect to the scale factor. Also $\vartheta = aH\tilde{\vartheta}$ and $\sigma_{\alpha\beta} = aH\tilde{\sigma}_{\alpha\beta}$, where $\tilde{\vartheta} = \partial^\alpha\tilde{v}_\alpha$ and $\tilde{\sigma}_{\alpha\beta} = \partial_{\langle\beta}\tilde{v}_{\alpha\rangle}$ (with $\tilde{v}_\alpha = ax'_\alpha$ and $v_\alpha = aH\tilde{v}_\alpha$). In the shear eigen-frame, $\tilde{\sigma}_{\alpha\beta} = (\tilde{\sigma}_{11}, \tilde{\sigma}_{22}, \tilde{\sigma}_{33})$ and (46) leads to

$$\mathcal{B}'_1 = -\frac{2}{3}\tilde{\vartheta}\mathcal{B}_1 + \tilde{\sigma}_{11}\mathcal{B}_1, \tag{47}$$

with exactly analogous relations for the rest of the magnetic components. The resulting system describes the second-order evolution of magnetic fields, which is frozen-in with the highly conductive matter of a collapsing protogalaxy, within the limits of the Zeldovich approximation. To obtain analytical solutions, we recall that in the absence of rotation and acceleration, the peculiar volume scalar is given by

$$\tilde{\vartheta} = \frac{\lambda_1}{1 + a\lambda_1} + \frac{\lambda_2}{1 + a\lambda_2} + \frac{\lambda_3}{1 + a\lambda_3}. \tag{48}$$

At the same time,

$$\tilde{\sigma}_{11} = \frac{\lambda_1}{1 + a\lambda_1} - \frac{1}{3}\vartheta, \tag{49}$$

[4]The Newtonian theory is a very good approximation, since we are dealing with non-relativistic matter and the scales of interest are well inside the curvature radius of the universe.

[5]When dealing with purely baryonic collapse, the Zeldovich ansatz only holds during the early stages of the nonlinear regime, when the effects of the fluid pressure are negligible. Assuming that the contraction is driven by non-baryonic CDM, means that we can (in principle) extend the domain of the Zeldovich approximation further.

while analogous expressions hold for the other two shear eigenvalues. Note that λ_1, λ_2 and λ_3 are the eigenvalues of the initial tidal field that determine the nature of the collapse (Matarrese 1996; Bruni 1996). One-dimensional collapse along, say, the third eigen-direction is characterized by $\lambda_1 = 0 = \lambda_2$ and by $\lambda_3 < 0$. In that case, the pancake singularity is reached as $a \to -1/\lambda_3$. Spherically symmetric collapse, on the other hand, has $\lambda_1 = \lambda_2 = \lambda_3 = \lambda < 0$. Then, we have a point-like singularity when $a \to -1/\lambda$.

Substituting, (48) and (49) into the right-hand side of (47) we arrive at the solution

$$B_1 = B_1^0 \left[\frac{(1 + a_0\lambda_2)(1 + a_0\lambda_3)}{(1 + a\lambda_2)(1 + a\lambda_3)} \right] \left(\frac{a_0}{a} \right)^2, \tag{50}$$

for the first of the magnetic components. A similar calculation leads to exactly analogous equations for B_2 and B_3. The zero suffix in the above indicates a given time during the protogalactic collapse. The ratio a_0/a in parentheses reflects the magnetic dilution due to the background expansion and the brackets monitor the increase in the field's strength caused by the collapse of the protogalactic cloud. According to (50), when dealing with pancake collapse along the third eigen-direction, the B_3-component decays as a^{-2}, while the other two increase arbitrarily as $a \to -1/\lambda_3$. Alternatively, during a spherically symmetric contraction, the magnetic field evolve as

$$B = B_0 \left(\frac{1 + a_0\lambda}{1 + a\lambda} \right)^2 \left(\frac{a_0}{a} \right)^2. \tag{51}$$

Here, all the magnetic components diverge as we approach the point singularity (i.e. for $a \to -1/\lambda$). Comparing the two results, we deduce that the anisotropic (pancake) collapse leads to a stronger increase as long as $\lambda_3 < \lambda$. The later is always satisfied, provided that the initial conditions are the same for both types of collapse, given that $\lambda_3 = \tilde{\vartheta}_0/(1 - a_0\tilde{\vartheta}_0)$ and $\lambda = \tilde{\vartheta}_0/(3 - a_0\tilde{\vartheta}_0)$ [see expression (48) above].

Our qualitative analysis indicates that magnetic fields trapped in an anisotropically contracting protogalactic cloud will increase beyond the limits of the idealized spherically symmetric collapse. Also, the amplified magnetic fields will end up essentially confined to the galactic plane. Quantitatively, the achieved final strength depends on the time the magnetic back-reaction has grown strong enough to halt the collapse (Zeldovich et al. 1983). Thus, the longer the anisotropic collapse persists, the stronger the residual magnetic fields. The analytical study of Bruni et al. (2003), in particular, showed that (realistically speaking) the anisotropy could add one or two orders of magnitude to the magnetic strength achieved through conventional isotropic compression. These results appear in very good agreement with numerical studies simulating shear and tidal effects on the magnetic field evolution in galaxies and galaxy clusters (Roettiger et al. 1999; Dolag et al. 2002).

8 Summary

The cosmic web, a network of filaments and nodes wherein most galaxies reside, is a prediction of the highly successful ΛCDM cosmology and appears to be borne out by simulations and observations. The web is filled with an ionized plasma, the IGM, which is expected to be permeated by magnetic fields. In this chapter, we reviewed recent developments in our theoretical understanding of the nature and origin of IGMF with a special focus on fields outside of clusters. We addressed two basic questions: First, "Can the process of structure formation generate seed magnetic fields and amplify them?" and second, "What role do magnetic fields play in structure formation and in other astrophysical processes?" We presented several plasma physics mechanisms for the generation of seed magnetic fields, and showed that these fields could be amplified during the first star formation and later during the formation of the cosmic web. We saw that magnetic fields with a strength of $\langle B\rangle \sim 10$ nG are expected in filaments at the present universe, while magnetic fields should be stronger in and around clusters/groups. We then discussed the effects and implications of magnetic fields on the formation of the first stars. We also presented a formal treatment of the evolution of density perturbations in the presence of magnetic fields and their effects on the formation of structures.

Magnetic fields in the IGM are difficult to be observed, mainly due to the weak nature of them, as well as due to the diffuseness of the media. However, the next generation radio interferometers including the Square Kilometer Array (SKA), and upcoming SKA pathfinders, the Australian SKA Pathfinder (ASKAP) and the South African Karoo Array Telescope (MeerKAT), as well the Low Frequency Array (LOFAR) will enable us to investigate magnetic fields outside clusters with high-sensitivity observations of synchrotron radiation and RM (see, e.g., papers in Carilli and Rawlings 2004).

Acknowledgements The work of DR was supported by the National Research Foundation of Korea through grant 2007-0093860. The work of DS was supported by a funding from the European Community's Seventh Framework Programme (/FP7/2007-2013/) under grant agreement No 229517. The work of LMW was supported by a Discovery Grant from the Natural Sciences and Engineering Research Council of Canada.

References

T. Abel, G.L. Bryan, M.L. Norman, The formation of the first star in the universe. Science **295**, 93–98 (2002). doi:10.1126/science.1063991

A. Achterberg, J. Wiersma, The Weibel instability in relativistic plasmas. I. Linear theory. Astron. Astrophys. **475**, 1–36 (2007). doi:10.1051/0004-6361:20065365

J. Adams, U.H. Danielson, D. Grasso, H.R. Rubinstein, Distortion of the acoustic peaks in the CMBR due to a primordial magnetic field. Phys. Lett. B **388**, 253–258 (1996). doi:10.1016/S0370-2693(96)01171-9

T. Akahori, D. Ryu, Faraday rotation measure due to the intergalactic magnetic field. Astrophys. J. **723**, 467–481 (2010). doi:10.1088/0004-637X/723/1/476

J. Aleksić et al.and MAGIC Collaboration, Search for an extended VHE γ-ray emission from Mrk 421 and Mrk 501 with the MAGIC Telescope. Astron. Astrophys. **524**, A77 (2010). doi:10.1051/0004-6361/201014747

T.G. Arshakian, R. Beck, M. Krause, D. Sokoloff, Evolution of magnetic fields in galaxies and future observational tests with the Square Kilometre Array. Astron. Astrophys. **494**, 21–32 (2009). doi:10.1051/0004-6361:200810964

J.D. Barrow, P.G. Ferreira, J. Silk, Constraints on a primordial magnetic field. Phys. Rev. Lett. **78**, 3610–3613 (1997). doi:10.1103/PhysRevLett.78.3610

J.D. Barrow, R. Maartens, C.G. Tsagas, Cosmology with inhomogeneous magnetic fields. Phys. Rep. **449**, 131–171 (2007). doi:10.1016/j.physrep.2007.04.006

E. Battaner, E. Florido, J. Jimenez-Vicente, Magnetic fields and large scale structure in a hot universe. I. General equations. Astron. Astrophys. **326**, 13–22 (1997)

A.R. Bell, Turbulent amplification of magnetic field and diffusive shock acceleration of cosmic rays. Mon. Not. R. Astron. Soc. **353**, 550–558 (2004). doi:10.1111/j.1365-2966.2004.08097.x

L. Biermann, Über den Ursprung der Magnetfelder auf Sternen und im interstellaren Raum. Z. Naturforsch. A **5**, 65–71 (1950)

A. Bonafede, L. Feretti, M. Murgia, F. Govoni, G. Giovannini, D. Dallacasa, K. Dolag, G.B. Taylor, The Coma cluster magnetic field from Faraday rotation measures. Astron. Astrophys. **513**, A30 (2010). doi:10.1051/0004-6361/200913696

J.R. Bond, L. Kofman, D. Pogosyan, How filaments of galaxies are woven into the cosmic web. Nature **380**, 603–606 (1996). doi:10.1038/380603a0

A. Brandenburg, K. Subramanian, Astrophysical magnetic fields and nonlinear dynamo theory. Phys. Rep. **417**, 1–209 (2005). doi:10.1016/j.physrep.2005.06.005

A. Brandenburg, K. Subramanian, D.D. Sokoloff, Turbulent dynamos. Space Sci. Rev. (2010, this issue)

V. Bromm, A. Loeb, Formation of the first supermassive black holes. Astrophys. J. **596**, 34–46 (2003). doi:10.1086/377529

S. Brown, D. Farnsworth, L. Rudnick, Cross-correlation of diffuse synchrotron and large-scale structures. Mon. Not. R. Astron. Soc. **402**, 2–6 (2010). doi:10.1111/j.1365-2966.2009.15867.x

M. Brüggen, A. Bykov, D. Ryu, H. Röttgering, Magnetic fields, relativistic particles, and shock waves in cluster outskirts. Space Sci. Rev. (2010, this issue)

M. Bruni, Cosmological collapses of irrotational dust, in *Mapping, measuring and modelling the universe*, ed. by P. Coles, V. Martinez, M.-J. Pons-Borderia. ASP Conference Series, vol. 94 (1996), pp. 31–36

M. Bruni, R. Maartens, C.G. Tsagas, Magnetic field amplification in cold dark matter anisotropic collapse. Mon. Not. R. Astron. Soc. **338**, 785–789 (2003). doi:10.1046/j.1365-8711.2003.06095.x

C. Caprini, R. Durrer, Gravitational wave production: a strong constraint on primordial magnetic fields. Phys. Rev. D **65**, 023517 (2002). doi:10.1103/PhysRevD.65.023517

C.L. Carilli, S. Rawlings, *Science with the Square Kilometre Array*. New Astronomy Reviews, vol. 48. (Elsevier, Amsterdam, 2004)

C.L. Carilli, G.B. Taylor, Cluster magnetic fields. Annu. Rev. Astron. Astrophys. **40**, 319–348 (2002). doi:10.1146/annurev.astro.40.060401.093852

R. Cassano, G. Brunetti, T. Venturi, G. Setti, D. Dallacasa, S. Giacintucci, S. Bardelli, Revised statistics of radio halos and the reacceleration model. Astron. Astrophys. **480**, 687–697 (2008). doi:10.1051/0004-6361:20078986

R.A. Cassidy, M.T. Elford, The mobility of Li^+ ions in helium and argon. Aust. J. Phys. **38**, 587–601 (1985)

R. Cen, J.P. Ostriker, Where are the baryons? Astrophys. J. **514**, 1–6 (1999). doi:10.1086/306949

R. Cen, J.P. Ostriker, Where are the baryons? II. Feedback effects. Astrophys. J. **650**, 560–572 (2006). doi:10.1086/506505

R. Cen, J.P. Ostriker, J.X. Prochaska, A.M. Wolfe, Metallicity evolution of damped Ly systems in ΛCDM cosmology. Astrophys. J. **598**, 741–755 (2003). doi:10.1086/378881

J. Cho, D. Ryu, Characteristic lengths of magnetic field in magnetohydrodynamic turbulence. Astrophys. J. **705**, L90–L94 (2009). doi:10.1088/0004-637X/705/1/L90

J. Cho, E.T. Vishniac, The generation of magnetic fields through driven turbulence. Astrophys. J. **538**, 217–225 (2000). doi:10.1086/309127

J. Cho, E.T. Vishniac, A. Beresnyak, A. Lazarian, D. Ryu, Growth of magnetic fields induced by turbulent motions. Astrophys. J. **693**, 1449–1461 (2009). doi:10.1088/0004-637X/693/2/1449

P.C. Clark, S.C.O. Glover, R.S. Klessen, V. Bromm, Gravitational fragmentation in turbulent primordial gas and the initial mass function of population III stars. Astrophys. J. **727**, 110 (2011). doi:10.1088/0004-637X/727/2/110

T.E. Clarke, Faraday rotation observations of magnetic fields in galaxy clusters. J. Korean Astron. Soc. **37**, 337–342 (2004)

T.E. Clarke, P.P. Kronberg, H. Böhringer, A new radio-X-ray probe of galaxy cluster magnetic fields. Astrophys. J. **547**, L111–L114 (2001). doi:10.1086/318896

S. Das, H. Kang, D. Ryu, J. Cho, Propagation of ultra-high-energy protons through the magnetized cosmic web. Astrophys. J. **682**, 29–38 (2008). doi:10.1086/588278

G. Davies, L.M. Widrow, A possible mechanism for generating galactic magnetic fields. Astrophys. J. **540**, 755–764 (2000). doi:10.1086/309358

C.D. Dermer, M. Cavadini, S. Razzaque, J.D. Finke, J. Chiang, B. Lott, Time delay of cascade radiation for TeV blazars and the measurement of the intergalactic magnetic field. Astrophys. J. Lett. **733**, L21–L24 (2011). doi:10.1088/2041-8205/733/2/L21

A.S. Dickinson, M.S. Lee, W.A. Lester, Jr., Close-coupling calculation of Li^+-H_2 diffusion cross sections. J. Phys. B **15**, 1371–1376 (1982). doi:10.1088/0022-3700/15/9/013

K. Dolag, M. Bartelmann, H. Lesch, Evolution and structure of magnetic fields in simulated galaxy clusters. Astron. Astrophys. **387**, 383–395 (2002). doi:10.1051/0004-6361:20020241

J. Donnert, K. Dolag, H. Lesch, E. Müller, Cluster magnetic fields from galactic outflows. Mon. Not. R. Astron. Soc. **392**, 1008–1021 (2009). doi:10.1111/j.1365-2966.2008.14132.x

A.J. Fennelly, C.R. Evans, Magnetohydrodynamic perturbations of Robertson-Walker universes and of anisotropic Bianchi type-I universes. Nuovo Cimento A **60**, 1–45 (1980). doi:10.1007/BF02723065

C. Federrath, S. Sur, D.R.G. Schleicher, R. Banerjee, R.S. Klessen, A new jeans resolution criterion for (M)HD simulations of self-gravitating gas: application to magnetic field amplification by gravity-driven turbulence. Astrophys. J. **71**, 62 (2011). doi:10.1088/0004-637X/731/1/62

B.D. Fried, Mechanism for instability of transverse plasma waves. Phys. Fluids **2**, 337 (1959). doi:10.1063/1.1705933
S. Fromang, S.A. Balbus, C. Terquem, J. De Villiers, Evolution of self-gravitating magnetized disks. II. Interaction between magnetohydrodynamic turbulence and gravitational instabilities. Astrophys. J. **616**, 364–375 (2004). doi:10.1086/424829
S.C.O. Glover, D.W. Savin, Is H_+^3 cooling ever important in primordial gas? Mon. Not. R. Astron. Soc. **393**, 911–948 (2009). doi:10.1111/j.1365-2966.2008.14156.x
F. Govoni, L. Feretti, Magnetic fields in clusters of galaxies. Int. J. Mod. Phys. D **13**, 1549–1594 (2004). doi:10.1142/S0218271804005080
D. Grasso, H.R. Rubinstein, Magnetic fields in the early universe. Phys. Rep. **348**, 163–266 (2001). doi:10.1016/S0370-1573(00)00110-1
A. Gruzinov, Gamma-ray burst phenomenology, shock dynamics, and the first magnetic fields. Astrophys. J. Lett. **563**, L15–L18 (2001). doi:10.1086/324223
A. Gruzinov, E. Waxman, Gamma-ray burst afterglow: polarization and analytic light curves. Astrophys. J. **511**, 852–861 (1999). doi:10.1086/306720
D. Guidetti, M. Murgia, F. Govoni, P. Parma, L. Gregorini, H.R. deRuiter, R.A. Cameron, R. Fanti, The intracluster magnetic field power spectrum in Abell 2382. Astron. Astrophys. **483**, 699–713 (2008). doi:10.1051/0004-6361:20078576
N.E.L. Haugen, A. Brandenburg, W. Dobler, Simulations of nonhelical hydromagnetic turbulence. Phys. Rev. E **70**, 016308 (2004a). doi:10.1103/PhysRevE.70.016308
N.E.L. Haugen, A. Brandenburg, W. Dobler, High-resolution simulations of nonhelical MHD turbulence. Astrophys. Space Sci. **292**, 53–60 (2004b). doi:10.1023/B:ASTR.0000045000.08395.a3
N.E.L. Haugen, A. Brandenburg, A.J. Mee, Mach number dependence of the onset of dynamo action. Mon. Not. R. Astron. Soc. **353**, 947–952 (2004c). doi:10.1111/j.1365-2966.2004.08127.x
F. Hoyle, Magnetic fields and highly condensed objects. Nature **223**, 936 (1969). doi:10.1038/223936a0
C.H. Jaroschek, M. Hoshino, Radiation dominated relativistic current sheets. Phys. Rev. Lett. **103**, 075002 (2009). doi:10.1103/PhysRevLett.103.075002
C.H. Jaroschek, H. Lesch, R.A. Treumann, Ultra-relativistic plasma shell collision s in gamma-ray burst sources: dimensional effects and the final steady state magnetic field. Astrophys. J. **618**, 822–831 (2005). doi:10.1086/426066
C.H. Jaroschek, M. Hoshino, H. Lesch, R.A. Treumann, Stochastic particle acceleration by the forced interaction of relativistic current sheets. Adv. Space Res. **41**, 481–490 (2008). doi:10.1016/j.asr.2007.07.001
H. Kang, D. Ryu, R. Cen, D. Song, Shock-heated gas in the large-scale structure of the universe. Astrophys. J. **620**, 21–30 (2005). doi:10.1086/426931
H. Kang, D. Ryu, R. Cen, J.P. Ostriker, Cosmological shock waves in the large-scale structure of the universe: nongravitational effects. Astrophys. J. **669**, 729–740 (2007). doi:10.1086/521717
K.T. Kim, P.P. Kronberg, G. Giovannini, T. Venturi, Discovery of intergalactic radio emission in the Coma-A1367 supercluster. Nature **341**, 720–723 (1989). doi:10.1038/341720a0
K.T. Kim, P.P. Kronberg, P.D. Dewdney, T.L. Landecker, The halo and magnetic field of the Coma cluster of galaxies. Astrophys. J. **355**, 29–37 (1990). doi:10.1086/168737
E.-J. Kim, A.V. Olinto, R. Rosner, Generation of density perturbations by primordial magnetic fields. Astrophys. J. **468**, 28–50 (1996). doi:10.1086/177667
E.W. Kolb, M.S. Turner, *The Early Universe* (Addison-Wesley, Redwood City, 1990)
M. Krause, P. Alexander, R. Bolton, J. Geisbüsch, D.A. Green, J. Riley, Measurements of the cosmological evolution of magnetic fields with the Square Kilometre Array. Mon. Not. R. Astron. Soc. **400**, 646–656 (2009). doi:10.1111/j.1365-2966.2009.15489.x
P.P. Kronberg, Q.W. Dufton, H. Li, S.A. Colgate, Magnetic energy of the intergalactic medium from galactic black holes. Astrophys. J. **560**, 178–186 (2001). doi:10.1086/322767
R.M. Kulsrud, E.G. Zweibel, On the origin of cosmic magnetic fields. Rep. Prog. Phys. **71**, 046901 (2008). doi:10.1088/0034-4885/71/4/046901
R.M. Kulsrud, R. Cen, J.P. Ostriker, D. Ryu, The protogalactic origin for cosmic magnetic fields. Astrophys. J. **480**, 481–491 (1997). doi:10.1086/303987
M.I. Large, D.S. Mathewson, C.G.T. Haslam, A high-resolution survey of the Coma cluster of galaxies at 408 Mc/s. Nature **183**, 1663–1664 (1959). doi:10.1038/1831663a0
R.B. Larson, Numerical calculations of the dynamics of collapsing proto-star. Mon. Not. R. Astron. Soc. **145**, 271–295 (1969)
A. Lazarian, Diffusion-generated electromotive force and seed magnetic field problem. Astron. Astrophys. **264**, 326–330 (1992)
M.N. Machida, K. Omukai, T. Matsumoto, S. Inutsuka, The first jets in the universe: protostellar jets from the first stars. Astrophys. J. Lett. **647**, L1–L4 (2006). doi:10.1086/507326

M.N. Machida, S. Inutsuka, T. Matsumoto, T. High- and low-velocity magnetized outflows in the star formation process in a gravitationally collapsing cloud. Astrophys. J. **676**, 1088–1108 (2008). doi:10.1086/528364

H. Maki, H. Susa, Dissipation of magnetic flux in primordial gas clouds. Astrophys. J. **609**, 467–473 (2004). doi:10.1086/421103

S. Matarrese, Relativistic cosmology: from superhorizon to small scales, in *Dark Matter in the Universe*, ed. by S. Bonometto, J.R. Primack, A. Provenzale (IOS Press, Oxford, 1996), pp. 601–628

M.V. Medvedev, A. Loeb, Generation of magnetic fields in the relativistic shock of gamma-ray burst sources. Astrophys. J. **526**, 697–706 (1999). doi:10.1086/308038

M.V. Medvedev, L.O. Silva, M. Kamionkowski, Cluster magnetic fields from large-scale structure and galaxy cluster shocks. Astrophys. J. Lett. **642**, L1–L4 (2006). doi:10.1086/504470

F. Miniati, A.R. Bell, Resistive magnetic field generation at cosmic dawn. Astrophys. J. **729**, 73 (2011). doi:10.1088/0004-637X/729/1/73

P.C. Myers, V.K. Khersonsky, On magnetic turbulence in interstellar clouds. Astrophys. J. **442**, 186–196 (1995). doi:10.1086/175434

D. Nagai, A. Vikhlinin, A.V. Kravtsov, Testing X-ray measurements of galaxy clusters with cosmological simulations. Astrophys. J. **655**, 98–108 (2007). doi:10.1086/509868

A. Neronov, I. Vovk, Evidence for strong extragalactic magnetic fields from Fermi observations of TeV Blazars. Science **328**, 73–75 (2010). doi:10.1126/science.1184192

K.I. Nishikawa, J. Niemiec, P.E. Hardee, M. Medvedev, H. Sol, Y. Mizuno, B. Zhang, M. Pohl, M. Oka, D.H. Hartmann, Weibel instability and associated strong fields in a fully three-dimensional simulation of a relativistic shock. Astrophys. J. Lett. **698**, L10–L13 (2009). doi:10.1088/0004-637X/698/1/L10

P.J.E. Peebles, Recombination of the primeval plasma. Astrophys. J. **153**, 1–11 (1968). doi:10.1086/149628

M.V. Penston, Dynamics of self-gravitating gaseous spheres. III. Analytical results in the free-fall of isothermal cases. Mon. Not. R. Astron. Soc. **144**, 425–448 (1969)

C. Pfrommer, V. Springel, T.A. Enßlin, M. Jubelgas, Detecting shock waves in cosmological smoothed particle hydrodynamics simulations. Mon. Not. R. Astron. Soc. **367**, 113–131 (2006). doi:10.1111/j.1365-2966.2005.09953.x

C. Pinto, D. Galli, Three-fluid plasmas in star formation. II. Momentum transfer rate coefficients. Astron. Astrophys. **484**, 17–28 (2008). doi:10.1051/0004-6361:20078819

C. Pinto, D. Galli, F. Bacciotti, Three-fluid plasmas in star formation. I. Magneto-hydrodynamic equations. Astron. Astrophys. **484**, 1–15 (2008). doi:10.1051/0004-6361:20078818

R.E. Pudritz, J. Silk, The origin of magnetic fields and primordial stars in protogalaxies. Astrophys. J. **342**, 650–659 (1989). doi:10.1086/167625

I. Røeggen, H.R. Skullerud, T.H. Løvaas, D.K. Dysthe, The Li^+-H_2 system in a rigid-rotor approximation: potential energy surface and transport coefficients. J. Phys. B **35**, 1707–1725 (2002). doi:10.1088/0953-4075/35/7/309

K. Roettiger, J.M. Stone, J.O. Burns, Magnetic field evolution in merging clusters of galaxies. Astrophys. J. **518**, 594–602 (1999). doi:10.1086/307298

D. Ryu, H. Kang, Vorticity and turbulence in the large-scale structure of the universe, in *Numerical Modeling of Space Plasma Flows (Astronum 2007)*, ed. by N.V. Pogorelov, E. Audit, G.P. Zank. ASP Conference Series, vol. 385 (2008), pp. 44–49

D. Ryu, J.P. Ostriker, H. Kang, R. Cen, A cosmological hydrodynamic code based on the total variation diminishing scheme. Astrophys. J. **414**, 1–19 (1993). doi:10.1086/173051

D. Ryu, H. Kang, P.L. Biermann, Cosmic magnetic fields in large scale filaments and sheets. Astron. Astrophys. **335**, 19–25 (1998)

D. Ryu, H. Kang, E. Hallman, T.W. Jones, Cosmological shock waves and their role in the large-scale structure of the universe. Astrophys. J. **593**, 599–610 (2003). doi:10.1086/376723

D. Ryu, H. Kang, J. Cho, S. Das, Turbulence and magnetic fields in the large-scale structure of the universe. Science **320**, 909–912 (2008). doi:10.1126/science.1154923

D. Ryu, S. Das, H. Kang, Intergalactic magnetic field and arrival direction of ultra-high-energy Protons. Astrophys. J. **710**, 1422–1431 (2010). doi:10.1088/0004-637X/710/2/1422

T.V. Ruzmaikina, A.A. Ruzmaikin, Gravitational stability of an expanding universe in the presence of a magnetic field. Sov. Astron. **14**, 963–966 (1971)

J.I. Sakai, R. Schlickeiser, P.K. Shukla, Simulation studies of magnetic field generation in cosmological plasmas. Phys. Lett. A **330**, 384–389 (1999). doi:10.1016/j.physleta.2004.08.007

A.A. Schekochihin, S.C. Cowley, S.F. Taylor, J.L. Maron, J.C. McWilliams, Simulations of the small-scale turbulent dynamo. Astrophys. J. **612**, 276–307 (2004). doi:10.1086/422547

A.A. Schekochihin, M. Brüggen, L. Feretti, M.W. Kunz, L. Rudnick, Magnetic fields in galaxy clusters: why bother. Space Sci. Rev. (2010, this issue)

D.R.G. Schleicher, R. Banerjee, R.S. Klessen, Reionization: a probe for the stellar population and the physics of the early universe. Phys. Rev. D **78**, 083005 (2008). doi:10.1103/PhysRevD.78.083005
D.R.G. Schleicher, D. Galli, S.C.O. Glover, R. Banerjee, F. Palla, R. Schneider, R.S. Klessen, The influence of magnetic fields on the thermodynamics of primordial star formation. Astrophys. J. **703**, 1096–1106 (2009). doi:10.1088/0004-637X/703/1/1096
D.R.G. Schleicher, R. Banerjee, S. Sur, T.G. Arshakian, R.S. Klessen, R. Beck, M. Spaans, Small-scale dynamo action during the formation of the first stars and galaxies. I. The ideal MHD limit. Astron. Astrophys. **522**, A115 (2010). doi:10.1051/0004-6361/201015184
R. Schlickeiser, P.K. Shukla, Cosmological magnetic field generation by the Weibel instability. Astrophys. J. Lett. **599**, L57–L60 (2003). doi:10.1086/381246
P. Schuecker, A. Finoguenov, F. Miniati, H. Böhringer, U.G. Briel, Probing turbulence in the Coma galaxy cluster. Astron. Astrophys. **426**, 387–397 (2004). doi:10.1051/0004-6361:20041039
S. Seager, D.D. Sasselov, D. Scott, A new calculation of the recombination epoch. Astrophys. J. Lett. **523**, L1–L5 (1999). doi:10.1086/312250
S.K. Sethi, K. Subramanian, Primordial magnetic fields in the post-recombination era and early reionization. Mon. Not. R. Astron. Soc. **356**, 778–788 (2005). doi:10.1111/j.1365-2966.2004.08520.x
S.K. Sethi, B.B. Nath, K. Subramanian, Primordial magnetic fields and formation of molecular hydrogen. Mon. Not. R. Astron. Soc. **387**, 1589–1596 (2008). doi:10.1111/j.1365-2966.2008.13302.x
J. Silk, M. Langer, On the first generation of stars. Mon. Not. R. Astron. Soc. **371**, 444–450 (2006). doi:10.1111/j.1365-2966.2006.10689.x
A.G. Sitenko, *Electromagnetic Fluctuations in Plasma* (Academic Press, New York, 1967). Chap. 4
S.W. Skillman, B.W. O'Shea, E. Hallman, J.O. Burns, M.L. Norman, Cosmological shocks in adaptive mesh refinement simulations and the acceleration of cosmic rays. Astrophys. J. **689**, 1063–1077 (2008). doi:10.1086/592496
R.S. de Souza, R. Opher, Origin of magnetic fields in galaxies. Phys. Rev. D **81**, 067301 (2010). doi:10.1103/PhysRevD.81.067301
K. Subramanian, Unified treatment of small- and large-scale dynamos in helical turbulence. Phys. Rev. Lett. **83**, 2957–2960 (1999). doi:10.1103/PhysRevLett.83.2957
K. Subramanian, J.D. Barrow, Magnetohydrodynamics in the early universe and the damping of nonlinear Alfvén waves. Phys. Rev. D **58**, 083502 (1998). doi:10.1103/PhysRevD.58.083502
K. Subramanian, D. Narashimha, S.M. Chitre, Thermal generation of cosmological seed magnetic fields in ionization fronts. Mon. Not. R. Astron. Soc. **271**, L15–L18 (1994)
S. Sur, D.R.G. Schleicher, R. Banerjee, C. Federrath, R.S. Klessen, The generation of strong magnetic fields during the formation of the first stars. Astrophys. J. Lett. **721**, L134–L138 (2010). doi:10.1088/2041-8205/721/2/L134
S.I. Syrovatskii, in *Interstellar Gas Dynamics*, ed by H.J. Habing (Springer, New York, 1970)
H. Takami, K. Sato, Distortion of ultra-high-energy sky by galactic magnetic field. Astrophys. J. **681**, 1279–1286 (2008). doi:10.1086/588513
J.C. Tan, E.G. Blackman, Protostellar disk dynamos and hydromagnetic outflows in primordial star formation. Astrophys. J. **603**, 401–413 (2004). doi:10.1086/381668
H. Tashiro, N. Sugiyama, Early reionization with primordial magnetic fields. Mon. Not. R. Astron. Soc. **368**, 965–970 (2006). doi:10.1111/j.1365-2966.2006.10178.x
(The Pierre Auger Collaboration) Correlation of the highest-energy cosmic rays with nearby extragalactic objects. Science **318**, 938–943 (2007). doi:10.1126/science.1151124
R.A. Treumann, R. Nakamura, W. Baumjohann, Collisionless reconnection: mechanism of self-ignition in thin plane current homogeneous sheets. Ann. Geophys. **28**, 1935–1943 (2010). doi:10.5194/angeo-28-1935-2010
C.G. Tsagas, Gravitational waves and cosmic magnetism: a cosmological approach. Class. Quantum Gravity **19**, 3709–3722 (2002). doi:10.1088/0264-9381/19/14/311
C.G. Tsagas, J.D. Barrow, A gauge-invariant analysis of magnetic fields in general-relativistic cosmology. Class. Quantum Gravity **14**, 2539–2562 (1997). doi:10.1088/0264-9381/14/9/011
C.G. Tsagas, R. Maartens, Magnetized cosmological perturbations. Phys. Rev. D **61**, 083519 (2000a). doi:10.1103/PhysRevD.61.083519
C.G. Tsagas, R. Maartens, Cosmological perturbations on a magnetized Bianchi I background. Class. Quantum Gravity **17**, 2215–2241 (2000b). doi:10.1088/0264-9381/17/11/305
M.J. Turk, T. Abel, B. O'Shea, The formation of population III binaries from cosmological initial conditions. Science **325**, 601–605 (2009). doi:10.1126/science.1173540
F. Vazza, G. Brunetti, C. Gheller, Shock waves in Eulerian cosmological simulations: main properties and acceleration of cosmic rays. Mon. Not. R. Astron. Soc. **395**, 1333–1354 (2009). doi:10.1111/j.1365-2966.2009.14691.x

C. Vogt, T. Enßlin, A Bayesian view on Faraday rotation maps seeing the magnetic power spectra in galaxy clusters. Astron. Astrophys. **434**, 67–76 (2005). doi:10.1051/0004-6361:20041839

S. Wang, New primordial-magnetic-field limit from the latest LIGO S5 data. Phys. Rev. D **81**, 023002 (2010). doi:10.1103/PhysRevD.81.023002

I. Wasserman, On the origins of galaxies, galactic angular momenta and galactic magnetic fields. Astrophys. J. **224**, 337–343 (1978). doi:10.1086/156381

E. Weibel, Spontaneously growing transverse waves in a plasma due to anisotropic velocity distribution. Phys. Rev. Lett. **2**, 83–84 (1959). doi:10.1103/PhysRevLett.2.83

L.M. Widrow, Origin of galactic and extragalactic magnetic fields. Rev. Mod. Phys. **74**, 775–823 (2002). doi:10.1103/RevModPhys.74.775

L.M. Widrow, D. Ryu, D.R.G. Schleicher, K. Subramanian, R.A. Treumann, C. Tsagas, The first magnetic fields. Space Sci. Rev. (2010, this issue)

H. Xu, B.W. O'Shea, D.C. Collins, M.L. Norman, H. Li, S. Li, The Biermann battery in cosmological MHD simulations of population III star formation. Astrophys. J. Lett. **688**, L57–L60 (2008). doi:10.1086/595617

Y. Xu, P.P. Kronberg, S. Habib, Q.W. Dufton, A Faraday rotation search for magnetic fields in large-scale structure. Astrophys. J. **637**, 19–26 (2006). doi:10.1086/498336

N. Yoshida, K. Omukai, L. Hernquist, Protostar formation in the early universe. Science **321**, 669–671 (2008). doi:10.1126/science.1160259

Y.B. Zeldovich, The hypothesis of cosmological magnetic inhomogeneity. Sov. Astron. **13**, 608–611 (1970a)

Y.B. Zeldovich, Separation of uniform matter into parts under the action of gravitation. Astrofizika **6**, 319–335 (1970b)

Y.B. Zeldovich, V.G. Kurt, R.A. Sunyaev, Recombination of hydrogen in the hot model of the universe. Sov. Phys. JETP **28**, 146–150 (1969)

Y.B. Zeldovich, A.A. Ruzmaikina, D.D. Sokolov, *Magnetic Fields in Astrophysics* (Gordon & Breach, New York, 1983)

Space Sci Rev (2012) 166:37–70
DOI 10.1007/s11214-011-9833-5

The First Magnetic Fields

**Lawrence M. Widrow · Dongsu Ryu ·
Dominik R.G. Schleicher · Kandaswamy Subramanian ·
Christos G. Tsagas · Rudolf A. Treumann**

Received: 11 October 2010 / Accepted: 14 September 2011 / Published online: 7 October 2011

Abstract We review current ideas on the origin of galactic and extragalactic magnetic fields. We begin by summarizing observations of magnetic fields at cosmological redshifts and on cosmological scales. These observations translate into constraints on the strength and scale magnetic fields must have during the early stages of galaxy formation in order to seed the galactic dynamo. We examine mechanisms for the generation of magnetic fields that operate prior during inflation and during subsequent phase transitions such as electroweak symmetry breaking and the quark–hadron phase transition. The implications of strong primordial magnetic fields for the reionization epoch as well as the first generation of stars are discussed in detail. The exotic, early-Universe mechanisms are contrasted with astrophysical processes that generate fields after recombination. For example, a Biermann-type battery

L.M. Widrow (✉)
Department of Physics, Engineering Physics and Astronomy, Queen's University, Kingston, Ontario, Canada K7L 3N6
e-mail: widrow@astro.queensu.ca

D. Ryu
Department of Astronomy and Space Science, Chungnam National University, Daejeon 305-764, Korea
e-mail: ryu@canopus.cnu.ac.kr

D.R.G. Schleicher
Institut für Astrophysik, Georg-August-Universität, Friedrich-Hund-Platz 1, 37077 Göttingen, Germany
e-mail: dschleic@astro.physik.uni-goettingen.de

K. Subramanian
IUCAA, Pune University Campus, Post Bag 4, Ganeshkhind, Pune 411 007, India
e-mail: kandu@iucaa.ernet.in

C.G. Tsagas
Department of Physics, Aristotle University Thessaloniki, Thessaloniki 54124, Greece
e-mail: tsagas@astro.auth.gr

R.A. Treumann
ISSI, Hallerstrasse 6, 3012 Bern, Switzerland
e-mail: treumann@issibern.ch

can operate in a proto-galaxy during the early stages of structure formation. Moreover, magnetic fields in either an early generation of stars or active galactic nuclei can be dispersed into the intergalactic medium.

Keywords Magnetic fields · Inflation · Early Universe · Quark–hadron transition

> *There is much to be learned about cosmic magnetic fields. We have a rather sketchy information about the field distribution on the largest scales, and the origin of the magnetic fields remains a mystery.*
>
> Alexander Vilenkin 2009

1 Introduction

Magnetic fields are observed in virtually all astrophysical systems, from planets to galaxy clusters. This fact is not surprising since gravitational collapse and gas dynamics, the key processes for structure formation, also amplify and maintain magnetic fields. Moreover, the conditions necessary for a magnetic dynamo, namely differential rotation and turbulence, exist in galaxies, which are the building blocks for large scale structure. The one notable example where magnetic fields are searched for but not yet found is in the surface of last scattering. All this raises an intriguing question: When did the first magnetic fields arise?

Numerous authors have suggested that magnetic fields first appeared in the very early Universe. (For recent reviews, see Grasso and Rubinstein 2001, Widrow 2002). There is strong circumstantial evidence that large scale structure formed from the amplification of linear density perturbations that originated as quantum fluctuations during inflation. It is therefore natural to consider whether quantum fluctuations in the electromagnetic field might similarly give rise to large-scale magnetic fields. Indeed, magnetic fields were almost certainly generated during inflation, the electroweak phase transition, and the quark–hadron phase transition but with what strength and on what scale? More to the point, what happened to these fields as the Universe expanded? Were these early Universe fields the seeds for the magnetic fields observed in present-day galaxies or clusters? And even if not, did they leave an observable imprint on the cosmic microwave background?

Exotic early universe mechanisms for field generation stand in contrast with mechanisms that operate in the post-recombination Universe. There are several ways to generate magnetic fields during the epoch of structure formation. At some level, all mechanisms begin with a battery, a process that treats positive and negative charges differently and thereby drives currents. A Biermann battery, for example, can (in fact *must*) operate during the formation of a proto-galaxy. While angular momentum in proto-galaxies is generated by the tidal torques due to nearby systems, vorticity arises from gasdynamical processes. The same processes almost certainly drive currents and hence generate fields, albeit of small amplitude.

The Biermann battery also operates in compact objects such as accretion disks and stars. Since the dynamical timescales for these systems is relatively short, tiny seed fields are rapidly amplified. The magnetic fields in AGN and/or Pop III stars can be expelled into the proto-galactic medium to provide another source of seed fields for subsequent dynamo action.

In this review, we survey ideas on the generation of magnetic fields. Our main focus is on mechanisms that operate in the early Universe, either during inflation, or during the phase transitions that follow. We contrast these mechanisms with ones that operate during the early stages of structure formation though we leave the details of those ideas for the subsequent chapter on magnetic fields and the formation of large scale structure. The outline of the chapter is as follows: In Sect. 2, we summarize observational evidence for magnetic fields at cosmological redshifts and on supercluster scales and beyond. In Sect. 3, we discuss the generation of magnetic fields during inflation. We make the case that inflation is an attractive arena for magnetic-field generation but for the fundamental result that electromagnetic fields in the standard Maxwell theory and in an expanding, spatially flat, inflating spacetime are massively diluted by the expansion of the Universe. However, one can obtain astrophysically interesting fields in spatially curved metrics or with non-standard couplings between gravity and electromagnetism. Section 4 addresses the question of whether fields can be generated during a post-inflation phase transition. We will argue that strong fields almost certainly arise but that their scales are limited by the Hubble radius at these early times. Only through some dynamical process such as an inverse cascade (which requires appreciable magnetic helicity) can one obtain astrophysically interesting fields. In Sect. 5 we take, as given, the existence of strong fields from an early Universe phase transitions and explore their impact on the post-recombination Universe. In particular, we discuss the implications of strong primordial fields on the first generation of stars and on the reionization epoch. Finally, in Sect. 6 we briefly review field-generation mechanisms that operate after recombination. A summary and some conclusions are presented in Sect. 7.

2 Cosmological Magnetic Fields Observed

The existence of microgauss fields in present-day galaxies and galaxy clusters is well established. These fields can be explained by the amplification of small seed fields over the 13.7 Gyr history of the Universe. However, there is mounting evidence that microgauss fields existed in galaxies when the Universe was a fraction of its present age. Moreover, there are hints that magnetic fields exist on supercluster scales. Both of these observations present challenges for the seed field hypothesis. In this section, we summarize observational evidence for magnetic fields at early times and on cosmological scales and briefly discuss the implications for the seed field hypothesis.

2.1 Galactic Magnetic Fields at Intermediate Redshifts

Microgauss fields are found in present-day galaxies of all types as well as galaxy clusters (see, for example, Kronberg 1994; Widrow 2002; Carilli and Menten 2002; Kulsrud and Zweibel 2008). Perhaps more significant, at least for our purposes, are observations of magnetic fields in intermediate and high redshift galaxies. For example, Kronberg et al. (1992) found evidence for a magnetized galaxy at a redshift of $z = 0.395$. To be specific, they mapped the rotation measure (RM) across the absorption-line quasar, PKS 1229-021 ($z = 1.038$) and determined the residual rotation measure (RRM—defined to be the observed rotation measure minus the Galactic rotation measure). The RRM was then identified with an intervening galaxy whose magnetic properties were inferred by detailed modelling.

Along similar lines, Bernet et al. (2008) provide a compelling argument for magnetic fields at $z \sim 1.3$. Earlier work by Kronberg (2008) found a correlation between the *spread* in the quasar RM distribution and redshift. The naive expectation is that the spread in the distribution should decrease with redshift. Recall that the polarization angle is proportional to the square of the wavelength; the proportionality constant is what we define as the RM. As

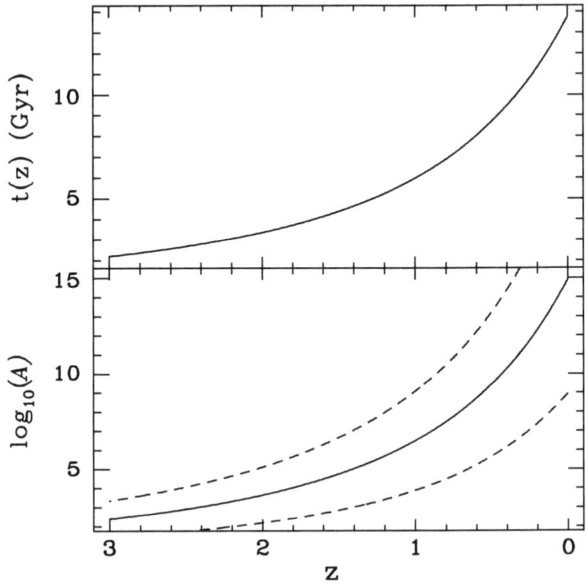

Fig. 1 *Upper panel*: Age of the Universe as a function of redshift for the cosmology described in the text and in Komatsu et al. (2010). *Lower panel*: Amplification factor for a seed magnetic field assuming exponential growth with one of three choices for the growth rate: $\Gamma = 1.5\ \mathrm{Gyr}^{-1}$, $\Gamma = 2.5\ \mathrm{Gyr}^{-1}$ or $\Gamma = 1.5\ \mathrm{Gyr}^{-1}$

electromagnetic radiation propagates from source to observer, the rotation angle is preserved by the wavelength increases as $(1+z)^{-1}$. Hence, the RM is diluted by a factor $(1+z)^{-2}$. In principle, the change in the RM distribution with redshift could be indicative of a redshift dependence in quasar magnetic fields. However, Bernet et al. (2008) sorted the sample according to the presence of MgII absorption lines and showed that the RM spread for set of objects with one or more lines was significantly greater than for the objects with no absorption lines. MgII absorption lines arise as the quasar light passes through the halos of normal galaxies. The implication is that intervening galaxies produce both large RMs and MgII absorption lines and hence, the intervening galaxies must have strong magnetic fields. Simple estimates suggest that the fields are comparable to those in present-day galaxies and that these galaxies are at a redshift of $z \sim 1.3$.

Athreya et al. (1998) studied 15 radio galaxies with redshifts between $z \simeq 2$ and $z \simeq 3.13$ and found significant RM's in almost all of them. Moreover, the RM's were found to differ significantly between the two radio lobes, which suggests that they are due to fields intrinsic to the object rather than due to the Faraday screen of the Galaxy. The RM's, corrected for cosmological expansion and with the Galactic contribution removed, typically range from 100–6000 rad m^{-2}, which implies microgauss fields.

Observations of magnetic fields at intermediate redshift imply a shorter time over which the dynamo can operate. Consider the standard ΛCDM cosmological model with $H_0 = 70.5\ \mathrm{km\,s^{-1}\,Mpc^{-1}}$, $\Omega_m = 0.272$ and $\Omega_\Lambda = 0.728$ where H_0 is the Hubble constant, and Ω_m and Ω_Λ are the density, in units of the critical density, for matter (both baryons and dark matter) and dark energy (Komatsu et al. 2010). In Fig. 1, we show the age of the Universe as a function of redshift for this cosmology. We also show the amplification factor for a seed magnetic field where we assume exponential growth and one of three choices for the growth rate, $\Gamma = 1.5\ \mathrm{Gyr}^{-1}$, $2.5\ \mathrm{Gyr}^{-1}$, or $3.5\ \mathrm{Gyr}^{-1}$. We see that a seed field of only 10^{-21} G is required to reach microgauss strength assuming $\Gamma \simeq 2.5\ \mathrm{Gyr}^{-1}$. However, for the same growth rate, a 10^{-11} G seed field is required to reach microgauss strengths by a redshift $z = 1.3$.

2.2 Magnetic Fields on Supercluster Scales and Beyond

Kim et al. (1989) used the Westerbork Synthesis Radio Telescope to map the Coma cluster and its environs at 326 MHz. Their results provide what remains the best direct evidence for magnetic fields on supercluster scales.

Recently, Neronov and Vovk (2010) argued that the deficit of GeV gamma-rays in the direction of TeV gamma-ray sources yields a *lower bound* of 3×10^{-16} G on the strength of intergalactic magnetic fields. The reasoning goes as follows: TeV gamma-rays and photons from the diffuse extragalactic background light produce $e^{\pm}$ pairs which, in turn, inverse Compton scatter off photons from the cosmic microwave background (CMB). The scattered CMB photons typically have energies in the GeV range. In the absence of appreciable magnetic fields, these secondary photons contribute to the overall emission toward the original TeV source. However, magnetic fields will deflect the intermediate $e^{\pm}$ pairs. Comparison of model predictions with the observed spectrum from HESS Cherenkov Telescopes and upper limits from the NASA Fermi Gamma-Ray Telescope hint at just such a deficit and lead Neronov and Vovk (2010) to derive their lower limit on the magnetic fields. It is important to note that the results of Neronov and Vovk (2010) rely on the assumption that the TeV sources emit gamma-rays continuously for 10^5 years, or longer. This point is stressed in Dermer et al. (2011) who derive a more conservative lower bound of 10^{-18} G based on the assumption that the TeV flux remains constant over the 3–4 year period during which the source has been observed.

Along similar lines, Ando and Kusenko (2010) pointed out that the deflection of $e^{\pm}$ pairs by an intergalactic magnetic field leads to gamma-ray halos around AGN. Indeed, they claim to have found evidence for just such halos in the stacked images of AGN from the Fermi Gamma-ray Telescope and suggested that these halos could be explained by a 10^{-15} Gauss intergalactic magnetic field. Finally, Takahashi et al. (2011) proposed a means to infer the existence of cosmic intergalactic magnetic fields via pair echos from gamma-ray bursts (GRBs). A fraction of the primary gamma-rays from a GRB interact with low-energy photons of the diffuse intergalactic radiation field and produce secondary electron–positron pairs. These pairs produce secondary gamma-rays by inverse Compton upscattering cosmic microwave background (CMB) photons. However, if an intergalactic magnetic field exists, the electrons and positrons will be deflected before interacting with the CMB photons. Thus, the secondary photons will be delayed by an amount that depends on the strength of the magnetic field. Pair echos may be detectable in experiments such as the Cherenkov Telescope Array or the Advanced Gamma-ray Imaging System. Takahashi et al. (2011) suggest that magnetic fields at $z \sim 5$–10 with strength $B \sim 10^{-16}$–10^{-15} G would then produce a measurable effect.

3 Magnetic Fields from Inflation

3.1 General Considerations

The hierarchical clustering scenario provides a compelling picture for the formation of large-scale structure. Linear density perturbations from the early Universe grow via gravitational instability. Small-scale objects form first and merge to create systems of ever-increasing size. The spectrum of the primordial density perturbations, as inferred from the CMB anisotropy spectrum and various statistical measures of large scale structure (e.g., the galaxy two-point correlation function) is generally thought to be scale-invariant and very close to the form

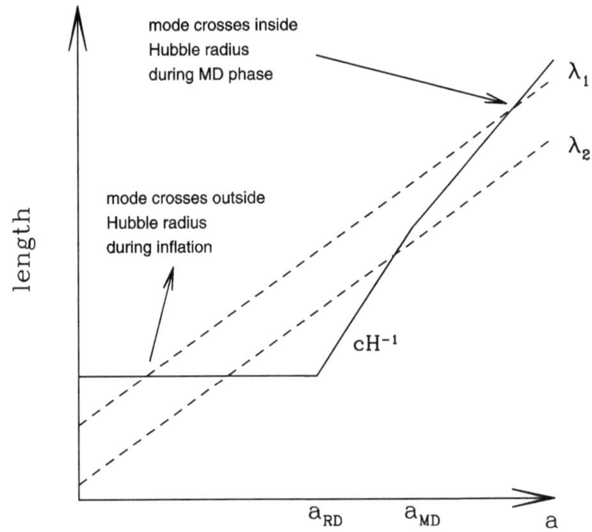

Fig. 2 Evolution of the physical size for the Hubble radius (*solid curve*) and two scales, λ_1 and λ_2 (*dashed curves*) as a function of scale factor a. Shown is the point at which the scale λ_1 crosses outside the Hubble radius during inflation and back inside the Hubble radius during the matter-dominated phase. The scale factors a_{RD} and a_{MD} correspond to the start of the radiation and matter dominated epochs, respectively. Here, λ_1 enters the Hubble volume during the matter-dominated epoch while λ_2 enters the Hubble volume during the radiation-dominated epoch

proposed by Zel'dovich (1970). One of the great successes of inflation is that it leads to just such a spectrum (Guth and Pi 1982; Hawking 1982; Starobinskii 1982). It is therefore natural to ask whether a similar mechanism might generate large-scale magnetic fields.

In order to understand the meaning and significance of the results for density perturbations, we must say a few words about horizons in cosmology. The Hubble radius, essentially, the speed of light divided by the Hubble parameter, sets the maximum scale over which microphysical processes can operate. In a radiation or matter-dominated Universe, the Hubble scale is proportional to the causal scale, that is, the distance over which a photon could have propagated since the Big Bang. The Hubble scale grows linearly with time t in a radiation or matter-dominated Universe. On the other hand, the physical size of an object associated with a fixed comoving (or present-day) scale grows as the scale factor a, which is proportional to $t^{1/2}$ during the radiation-dominated phase and $t^{2/3}$ during the matter-dominated phase. Thus, a physical scale crosses inside the Hubble radius (that is, becomes causally connected) after the Big Bang, with smaller objects crossing earlier than larger ones. This point is illustrated in Fig. 2 and explains why it is so difficult to generate *large-scale* magnetic fields in the early Universe but after inflation.

During inflation, the Hubble parameter is approximately constant (the spacetime is approximately de Sitter) and a physical scale that is initially inside the Hubble radius will cross outside the Hubble radius or Hubble scale, at least until some time later in the history of the Universe (Fig. 2). We therefore have the potential for microphysical processes to operate on large scales during inflation with the consequences of these processes becoming evident much later when the scale re-enters the Hubble radius.

Inflation provides the dynamical means for generating density perturbations: quantum mechanical fluctuations in the de Sitter phase excite modes on the Hubble scale with an energy density set by the Hubble parameter. Therefore, to the extent that the Hubble parameter is constant during inflation, the energy density in modes as they crosses outside the Hubble radius will be scale-independent.

There is one further and crucial part to the story. During inflation, the energy density of the Universe is approximately constant. It is, indeed, the constant energy density that drives the exponential expansion of the de Sitter phase. Naively, we expect the energy density in (relativistic) fluctuations to scale as a^{-4} and to therefore be diluted by an enormous factor. However, once produced, the fluctuations are stretched to length scales much larger than the Hubble radius. The energy density, $\delta\rho$, in these super-horizon-sized modes is not a gauge invariant quantity in that it depends on ones choice of the surface of simultaneity. However, the quantity $\xi \equiv \delta\rho/(\rho + p)$ is gauge invariant. Moreover, for super-horizon-sized energy density fluctuations, $\xi =$ constant. Thus, the ratio, r, of the energy density in the fluctuation relative to the background density, is the same (up to a constant of order unity) at the time when the mode re-enters the Hubble radius as when it crossed outside the Hubble radius during inflation. This behaviour is often referred to as super-adiabatic growth since it implies that the energy density in the fluctuation grows relative to the energy density in the radiation field.

The equation of motion for a scalar field, ϕ, in a curved spacetime is

$$\nabla^\alpha \nabla_\alpha \phi - \xi R\phi - \frac{dV}{d\phi} = 0, \tag{1}$$

where R is the Ricci scalar, V is the scalar field potential and ξ is a dimensionless constant. Super-adiabatic growth occurs for a minimally-coupled scalar field ($\xi = 0$). On the other hand, the field evolves adiabatically (energy density scales in the same way as radiation) if it is conformally coupled ($\xi = 1/6$).

The equation of motion for energy density perturbations is identical to that for a minimally coupled scalar field if one ignores the scalar potential term. On the other hand, electromagnetism is conformally-coupled to gravity (at least in the simplest version of the theory). The equation of motion for the gauge field therefore resembles (1) with $\xi = 1/6$. Let us assume that there are de Sitter-induced quantum fluctuations in the electromagnetic field. As with other quantum fields, the energy density of these fluctuations at the time when they are produced, is set by the Hubble parameter. Roughly speaking, $\rho_B \sim H^4$ for modes with physical wavelength $\lambda_{\rm phys} \sim H^{-1}$. Once produced the energy density in these modes scales as a^{-4}, that is, scales in the same way as radiation. Meanwhile, the total energy density of the Universe is constant during inflation and scales as a^{-3} during both the reheating and matter-dominated phases of the Universe. We therefore find that the relative strength a magnetic mode at the end of inflation is

$$\frac{\rho_B}{\rho_t} \simeq 10^{-78} \left(\frac{M}{m_{Pl}}\right)^4 \left(\frac{M}{10^{14}\ \text{GeV}}\right)^{-8/3} \left(\frac{T_{RH}}{10^{10}\ \text{GeV}}\right)^{-4/3} \left(\frac{\lambda}{\text{Mpc}}\right)^{-4}, \tag{2}$$

where λ is the comoving scale of the mode and $m_{Pl} \simeq 10^{19}$ GeV is the Planck mass. The above ratio also depends on the energy scale of our inflation model (M) and on the associated reheating temperature (T_{RH}). During reheating $\rho_B \propto a^{-4}$ and $\rho_t \propto a^{-3}$, which means that $\rho_B/\rho_t \propto a^{-1} \propto T$ between the end of inflation and the radiation era. Therefore, recalling that $\rho_t \simeq M^4$ throughout the de Sitter phase and $\rho_t \simeq \rho_{RH} \simeq T_{RH}^4$ by the end of reheating, expression (2) yields

$$\frac{\rho_B}{\rho_t} \simeq 10^{-104} \left(\frac{\lambda}{\text{Mpc}}\right)^{-4}, \tag{3}$$

at the beginning of the radiation epoch (when $\rho_t \simeq \rho_\gamma$). From then on, the high conductivity of the matter is restored and the magnetic flux is conserved (i.e. $B \propto a^{-2}$ and $\rho_B \propto a^{-4}$). As a result, the dimensionless ratio

$$r \equiv \frac{\rho_B}{\rho_\gamma} \simeq 10^{-104} \left(\frac{\lambda}{\text{Mpc}}\right)^{-4}, \tag{4}$$

remains constant until today (recall that $\rho_\gamma \propto a^{-4}$ at all times). This result implies a present-day field strength no greater than 10^{-50} G on comoving scales of order 10 kpc, which are the scales relevant for the galactic dynamo.

We are lead to the conclusion that inflation-produced magnetic fields are astrophysically uninteresting. However, this 'negative' result holds for the standard formulation of Maxwell's equations and under the assumption of a spatially-flat FLRW cosmology. In the next sections, we show that super-adiabatic growth can occur in various "open" cosmologies and in models where certain additional couplings between electromagnetism and gravity are included.

3.2 Maxwell's Equations

The Maxwell field may be invariantly described by the antisymmetric Faraday tensor, F_{ab}. Relative to an observer moving with 4-velocity u_a, we can write $F_{ab} = 2u_{[a}E_{b]} + \varepsilon_{abc}B^c$, where $E_a = F_{ab}u^b$ and $B_a = \varepsilon_{abc}F^{bc}/2$ respectively represent the electric and magnetic components of the EM field and ε_{abc} is the 3-dimensional Levi–Civita tensor. Maxwell's equations split into two pairs of propagation and constraint equations (Tsagas 2005). The former consists of

$$\dot{E}_{\langle a\rangle} = -\frac{2}{3}\Theta E_a + (\sigma_{ab} + \omega_{ab})E^b + \varepsilon_{abc}A^b B^c + \text{curl}\, B_a - \mathcal{J}_a, \tag{5}$$

and

$$\dot{B}_{\langle a\rangle} = -\frac{2}{3}\Theta B_a + (\sigma_{ab} + \omega_{ab})B^b - \varepsilon_{abc}A^b E^c - \text{curl}\, E_a, \tag{6}$$

which may be seen as the $1+3$ covariant analogues of the Ampère and the Faraday laws respectively. The constraints, on the other hand, read

$$\mathrm{D}^a E_a = \mu - 2\omega^a B_a \quad \text{and} \quad \mathrm{D}^a B_a = 2\omega^a E_a, \tag{7}$$

providing the $1+3$ forms of Coulomb's and Gauss' laws respectively. In the above Θ, σ_{ab}, ω_{ab} and A_a respectively represent the volume expansion, the shear, the vorticity and the 4-acceleration associated with the observer's motion.[1] In addition, $\mathcal{J}_a$ and μ are the 3-current and the charge densities respectively. We also note that overdots indicate proper-time derivatives and D_a is the 3-dimensional covariant derivative operator. Finally, $\dot{E}_{\langle a\rangle} = h_a{}^b\dot{E}_b$ and $\text{curl}\, B_a = \varepsilon_{abc}\mathrm{D}^b B^c$ by definition, where the tensor h_{ab} projects orthogonal to u_a.

These equations, together with the Einstein equations and the Ricci identities, lead to wave equations for the electric and magnetic fields. For example, linearised on a Friedmann–Lemaître–Robertson–Walker (FLRW) background, the wave equations of the electric and the magnetic components of the Maxwell field read (Tsagas 2005)

$$\ddot{E}_a - \mathrm{D}^2 E_a = -5H\dot{E}_a + \frac{1}{3}\kappa(\rho + 3p)E_a - 4H^2 E_a - \frac{1}{3}\mathcal{R}E_a - \dot{\mathcal{J}}_a - 3H\mathcal{J}_a \tag{8}$$

[1] By construction $\Theta = \mathrm{D}^a u_a$ is the volume scalar, which determines whether we have expansion ($\Theta > 0$), or contraction ($\Theta < 0$). The shear tensor ($\sigma_{ab} = \mathrm{D}_{(b}u_{a)} - (\Theta/3)h_{ab}$) monitors changes in the shape of the fluid element under constant volume, while the vorticity tensor ($\omega_{ab} = \mathrm{D}_{[b}u_{a]}$) governs the rotational behaviour of the medium. Finally, the 4-acceleration, $A_a = \dot{u}_a$, reflects the presence of non-gravitational forces and vanishes when the motion is driven by gravity alone (see Tsagas et al. 2008 for details). Note that the antisymmetry of the vorticity tensor allows us to define the vorticity vector, by means of $\omega_a = \varepsilon_{abc}\omega^{bc}/2$, which also determines the rotational axis.

and

$$\ddot{B}_a - \mathrm{D}^2 B_a = -5H\dot{B}_a + \frac{1}{3}\kappa(\rho+3p)B_a - 4H^2 B_a - \frac{1}{3}\mathcal{R}B_a + \operatorname{curl}\mathcal{J}_a, \tag{9}$$

respectively. Here, $H = \dot{a}/a = \Theta/3$ is the background Hubble parameter, $\mathcal{R} = 6K/a^2$ is the Ricci scalar of the spatial sections (with $K = 0, \pm 1$), $\kappa = 8\pi G$ is the gravitational constant and $\mathrm{D}^2 = \mathrm{D}^a\mathrm{D}_a$ is the 3-D covariant Laplacian. Note the 3-Ricci term, which reflects a purely relativistic/geometrical coupling between the electromagnetic field and the spacetime curvature. We will return to this particular interaction to examine the way it can affect the evolution of cosmological magnetic fields.

3.3 Ohm's Law in the Expanding Universe

The literature contains various expressions of Ohm's law, which provides the propagation equation of the electric 3-current. For a single fluid at the limit of resistive magnetohydrodynamics (MHD), Ohm's law takes the simple form (Greenberg 1971; Jackson 1975)

$$\mathcal{J}_a = \sigma E_a, \tag{10}$$

where σ represents the electric conductivity of the matter (not to be confused with the shear tensor σ_{ab}). In highly conducting environments, $\sigma \to \infty$ and the electric field vanishes. This is the familiar ideal-MHD approximation where the electric currents keep the magnetic field frozen-in with the charged fluid. Conversely, when the conductivity is very low, $\sigma \to 0$. Then, the 3-currents vanish despite the presence of nonzero electric fields. Here, we will consider these two limiting cases. For any intermediate case, one needs a model for the electrical conductivity of the cosmic medium.

3.4 Adiabatic Decay of Magnetic Fields in a Spatially Flat FLRW Cosmology

Consider the case of a poorly conductive environments where there are no 3-currents. The wave equation (9), then reduces to

$$\ddot{B}_a - \mathrm{D}^2 B_a = -5H\dot{B}_a + \frac{1}{3}\kappa(\rho+3p)B_a - 4H^2 B_a - \frac{1}{3}\mathcal{R}B_a. \tag{11}$$

To simplify the above, we introduce the rescaled the magnetic field $\mathcal{B}_a = a^2 B_a$ and employ conformal time, η, where $\dot{\eta} = 1/a$. Then, on using the harmonic splitting $\mathcal{B}_a = \sum_n \mathcal{B}_{(n)}\mathcal{Q}_a^{(n)}$—so that $\mathrm{D}_a\mathcal{B}_{(n)} = 0 = \dot{\mathcal{Q}}_a^{(n)} = \mathrm{D}^a\mathcal{Q}_a^{(n)}$ and $\mathrm{D}^2\mathcal{Q}_a^{(n)} = -(n/a)^2\mathcal{Q}_a^{(n)}$, expression (11) takes the compact form

$$\mathcal{B}''_{(n)} + n^2\mathcal{B}_{(n)} = -2K\mathcal{B}_{(n)}, \tag{12}$$

with the primes denoting conformal-time derivatives, n representing the comoving wavenumber of the mode and $K = 0, \pm 1$ (Tsagas 2005). Note the magneto-curvature term in the right-hand side of (12), which shows that the magnetic evolution also depends on the spatial geometry of the FLRW spacetime.

When the background has Euclidean spatial hypersurfaces, the 3-curvature index is zero (i.e. $K = 0$) and expression (12) assumes the Minkowski-like form

$$\mathcal{B}''_{(n)} + n^2\mathcal{B}_{(n)} = 0. \tag{13}$$

This equation accepts the oscillatory solution $\mathcal{B}_{(n)} = \mathcal{C}_1 \sin(n\eta) + \mathcal{C}_2 \cos(n\eta)$, which recasts into

$$B_{(n)} = \left[\mathcal{C}_1 \sin(n\eta) + \mathcal{C}_2 \cos(n\eta)\right]\left(\frac{a_0}{a}\right)^2, \tag{14}$$

for the actual B-field. In other words, the adiabatic ($B_{(n)} \propto a^{-2}$) depletion of the magnetic component is guaranteed, provided the background spacetime is a spatially flat and the electrical conductivity remains very poor. Applied during inflation, this result leads to (2).

The adiabatic decay-law also holds in highly conductive environments. There, $\sigma \to \infty$ and, according to Ohm's law (see (10)) the electric field vanishes in the frame of the fluid. As a result, Faraday's law (see (6)) linearises to

$$\dot{B}_a = -2HB_a, \tag{15}$$

around an FLRW background. The above ensures that $B_a \propto a^{-2}$ on all scales, regardless of the equation of state of the matter and of the background 3-curvature. Applied after inflation and combined with (3), this result leads directly to (4).

3.5 Superadiabatic Magnetic Amplification in Spatially Open FLRW Models

The "negative" results discussed at the beginning of this section have been largely attributed to the conformal invariance of Maxwell's equations and to the conformal flatness of the Friedmannian spacetimes. The two are thought to guarantee an adiabatic decay-rate for all large-scale magnetic fields at all times. Equation (14), however, only holds for the spatially flat FLRW cosmology. Although all three FLRW universes are conformally flat, only the spatially flat model is globally conformal to Minkowski space. For the rest, the conformal mappings are local. Put another way, in spatially curved Friedmann universes, the conformal factor is no longer the cosmological scale factor but has an additional spatial dependence (e.g. see Stefani 1990; Keane and Barrett 2000). This means that the wave equation of the rescaled magnetic field ($\mathcal{B}_a = a^2 B_a$) takes the simple Minkowski-like form (13) only on FLRW backgrounds with zero 3-curvature. In any other case, there is an additional curvature-related term (see expressions (11) and (12)), which reflects the non-Euclidean spatial geometry of the host spacetime. As a result, when linearised around an FLRW background with nonzero spatial curvature, the magnetic wave equation (see (12)) reads

$$\mathcal{B}''_{(n)} + \left(n^2 \pm 2\right)\mathcal{B}_{(n)} = 0, \tag{16}$$

where the plus and the minus signs refer to spatially closed and open models, respectively. Recall that in the former case the eigenvalues are discrete (with $n^2 \geq 3$) and in the latter continuous (with $n^2 \geq 0$). As expected, the curvature-related effects fade away as we move to progressively smaller scales (i.e. for $n^2 \gg 2$).

Equation (16) shows that on FLRW backgrounds with spherical spatial hypersurfaces, the B-field still decays adiabatically. The picture changes when the background FLRW model is open. There, the hyperbolic geometry of the 3-D hypersurfaces alters the nature of the magnetic wave equation on large enough scales (i.e. when $0 < n^2 < 2$). These wavelengths include what we may regard as the largest subcurvature modes (i.e. those with $1 \leq n^2 < 2$) and the supercurvature lengths (having $0 < n^2 < 1$). Note that eigenvalues with $n^2 = 1$ correspond to the curvature scale with physical wavelength $\lambda = \lambda_K = a$ (e.g. see Lyth and Woszczyna 1995). Here, we will focus on the largest subcurvature modes.

We proceed by introducing the scale-parameter $k^2 = 2 - n^2$, with $0 < k^2 < 2$. Then, $k^2 = 1$ indicates the curvature scale, the range $0 < k^2 < 1$ corresponds to the largest subcurvature modes and their supercurvature counterparts are contained within the $1 < k^2 < 2$ interval. In the new notation and with $K = -1$, (16) recasts into

$$\mathcal{B}''_{(n)} - k^2\mathcal{B}_{(n)} = 0, \tag{17}$$

with the solution given by $\mathcal{B}_{(k)} = \mathcal{C}_1 \sinh(|k|\eta) + \mathcal{C}_2 \cosh(|k|\eta)$. Written in terms of the actual magnetic field, the latter takes the form (Tsagas and Kandus 2005; Barrow and Tsagas 2008)

$$B_{(k)} = \left[\mathcal{C}_1 \mathrm{e}^{|k|(\eta-\eta_0)} + \mathcal{C}_2 \mathrm{e}^{-|k|(\eta-\eta_0)}\right]\left(\frac{a_0}{a}\right)^2. \tag{18}$$

Magnetic fields that obey the above evolution law can experience superadiabatic amplification without modifying conventional electromagnetism and despite the conformal flatness of the FLRW host. For instance, during the radiation era, the scale factor of an open FLRW universe evolves as $a \propto \sinh(\eta)$. Focusing on the curvature length, for simplicity, we may set $|k| = 1$ in (18). On that scale, the dominant magnetic mode never decays faster than $B_{(1)} \propto a^{-1}$. The B-field has been superadiabatically amplified.

Analogous amplification also occurs during the dust and the reheating eras (when $p = 0$ and $a = \sinh^2(\eta/2)$), as well as in open FLRW universes with an inflationary (i.e. $p = -\rho$) equation of state. In fact, the superadiabatic amplification of large-scale magnetic fields is essentially independent of the equation of state of the matter and appears to be a generic feature of the open Friedmann models (see Barrow and Tsagas 2011 for further discussion and details). During slow-roll inflation, in particular, the scale factor evolves as (Tsagas et al. 2008)

$$a = a_0\left(\frac{1-\mathrm{e}^{2\eta_0}}{1-\mathrm{e}^{2\eta}}\right)\mathrm{e}^{\eta-\eta_0}, \tag{19}$$

where $\eta, \eta_0 < 0$. Substituting the above into (18), we find that near the curvature scale (i.e. for $|k| \to 1$) the magnetic evolution is given by

$$B_{(1)} = \mathcal{C}_3\left(1-\mathrm{e}^{2\eta}\right)\left(\frac{a_0}{a}\right) + \mathcal{C}_4 \mathrm{e}^{-\eta}\left(\frac{a_0}{a}\right)^2, \tag{20}$$

with $\mathcal{C}_3$, $\mathcal{C}_4$ depending on the initial conditions. This also implies superadiabatic amplification for the B-field, since the dominant mode never decays faster than $B_{(1)} \propto a^{-1}$. The adiabatic decay rate is only recovered at the end of inflation, as $\eta \to 0$.[2]

The strength of the residual magnetic strength is calculated in a way analogous to that given in the previous section. This time, however, the B-field has been superadiabatically amplified throughout the lifetime of the universe. Then, close to the curvature scale, where $B \propto a^{-1}$, we find that

[2]The magnetogeometrical interaction and the resulting effects are possible because, when applied to spatially curved FLRW models, inflation does not lead to a globally flat de Sitter space. Although the inflationary expansion dramatically increases the curvature radius of the universe, it does not change its spatial geometry. Unless the universe was perfectly flat from the beginning, there is always a scale where the 3-curvature effects are important. It is on these lengths that primordial B-fields can be superadiabatically amplified.

$$r = \frac{\rho_B}{\rho_\gamma} \simeq 10^{-8}\left(\frac{M}{10^{14}\ \text{Gev}}\right)^4 \left(\frac{\lambda}{\text{Mpc}}\right)^{-2}$$
$$\simeq 10^{-8}\left(\frac{M}{10^{14}\ \text{GeV}}\right)^4 \left(\frac{\lambda_H}{\text{Mpc}}\right)^{-2} (1-\Omega), \qquad (21)$$

at present (Barrow and Tsagas 2011). Note that $\lambda \to \lambda_K = \lambda_H/\sqrt{1-\Omega}$, with λ_K, λ_H and Ω representing the curvature scale the Hubble length and the density parameter of the universe respectively. According to (21), the higher the scale of inflation, the stronger the amplification. On the other hand, the larger the curvature scale, namely the higher the number of e-folds during inflation, the weaker the residual magnetic field (see Barrow and Tsagas 2011 for further discussion). Setting $M \sim 10^{14}$ GeV and $\lambda_H \sim 10^3$ Mpc in the right-hand side of (21), we find a current (comoving) magnetic strength of approximately 10^{-14} Gauss on scales close to the present curvature length. The latter is approximately 10^4 Mpc if we assume that $1-\Omega \sim 10^{-2}$ today (Komatsu et al. 2010). These lengths are far larger than 10 kpc; the minimum magnetic size required for the dynamo to work. Nevertheless, once the galaxy formation starts, the field lines should break up and reconnect on scales similar to that of the collapsing protogalactic cloud.

Galactic-scale magnetic fields of strength 10^{-14} G are stronger than those generated by many of the other scenarios considered in the literature (see below) and may be strong enough to seed the galactic dynamo. Recall, however, that inflation was introduced to avoid various shortcomings of the standard cosmological model including the so-called flatness problem. Essentially, inflation is meant to inflate away (push to extremely large scales) any curvature that might exist in the pre-inflationary Universe. Evidently, for superadiabatic growth of magnetic field without explicit conformal symmetry breaking (as described in the next section) one requires enough inflation to push the curvature scale beyond our present Hubble volume, but not too far beyond. To be quantitative, the most recent analysis by the WMAP group (Komatsu et al. 2010) finds that the effective energy density parameter for curvature, $\Omega_K \equiv -K/(aH)^2 = 1 - \Omega_\Lambda - \Omega_m$, is constrained between $-0.0133 < \Omega_K < 0.0084$. Should future measurements find Ω_K to be more consistent with negative curvature, rather than zero or positive curvature, they would lend credence to the mechanism for magnetic amplification described above.

3.6 Inflation-Produced Magnetic Fields via Non-conformal Couplings

As pointed out above, the electromagnetic field is conformally coupled to gravity and therefore, in a spatially flat, FLRW cosmology, the magnetic fields generated during inflation decay adiabatically and are therefore of negligible astrophysical importance. Turner and Widrow (1988) pointed out that by adding additional terms to the Lagrangian such as RA^2 and $R_{\mu\nu}A^\mu A^\nu$ one explicitly breaks conformal invariance and can essentially force the magnetic field to behave like a minimally-coupled scalar field. These terms also break gauge invariance and hence induce an effective mass for the photon whose size depends on the spacetime curvature. In fact, the effective mass-squared for the photon is negative for the case where the magnetic field behaves as a minimally-coupled scalar field. A negative mass-squared signals an instability in the semi-classical equations for the field and can be viewed as the origin of the superadiabaticity. Such terms also lead to potential theoretical difficulties in the particle theory and numerous authors have attempted to find more effective and more natural ways to break conformal invariance. Indeed, Turner and Widrow (1988), recognizing the potential for such difficulties, considered terms such as RF^2 which break conformal invariance but not gauge invariance. They also briefly considered $\varphi F\tilde{F} \sim \varphi \mathbf{E}\cdot\mathbf{B}$ terms where φ was a pseudo-scalar (i.e., axion-like terms). Ratra (1992) demonstrated that appreciable magnetic fields couple be produced during inflation if the electromagnetic field couples to the inflaton field, Φ, through a term of the form $e^{\alpha\Phi}F_{\mu\nu}F^{\mu\nu}$ where α is a constant. In his model, the inflaton potential is also an exponential: $V(\Phi) \propto \exp(-q\Phi)$. An attractive feature of this model is that is preserves gauge invariance since the additional term is constructed from the Maxwell tensor, F, rather than the gauge field, A. Along similar lines Gasperini et al. (1995) demonstrated that magnetic fields of sufficient strength to seed the galactic dynamo could be produced in a string-inspired cosmology. In their model, the electromagnetic field is coupled to the dilaton which, in turn, is coupled to gravity. The dilaton is a scalar field that naturally arises in theories with extra dimensions and whose vacuum expectation value effectively controls Newton's constant.

The RA^2 terms give rise to "ghosts" in the theory which signal an instability of the vacuum (Himmetoglu et al. 2009a, 2009b). A theory with ghosts is internally consistent only as an effective low-energy theory and hence is only valid below some energy scale, Λ. Himmetoglu et al. (2009b) argue that $\Lambda \lesssim \text{MeV}$ but this scale is well below the scales assumed in models for inflation-produced magnetic fields. Thus, their results call into question the viability of the mechanism. A detailed and critical analysis of magnetic-field production in string-inspired models of inflation can be found in Martin and Yokoyama (2008) (see, also, the review by Subramanian 2010). There, particular attention was paid to a potential backreaction problem which arises if the *electric* fields produced along with the magnetic fields have an energy density comparable to the background energy density. Similar conclusions were reached in Demozzi et al. (2009). In models the RA^2 terms, the longitudinal degree of freedom of the **A**-field becomes dynamical. Demozzi et al. (2009) argued that the energy density associated with this component grows during inflation and could, in principle, shut off inflation. Indeed, they found that the magnetic field should not exceed 10^{-32} G on Mpc-scales today. Demozzi et al. (2009) also considered backreaction from F^2 terms and found that interesting magnetic fields required large couplings where perturbation theory could not be trusted. Durrer et al. (2011) studied field generation from $\varphi F\tilde{F}$ terms and found that while backreaction does not appear to be a problem in this scenario, the fields that arose were generally too weak to seed a galactic dynamo. In summary, while inflation remains an attractive arena for the production of seed magnetic fields, there appear to be significant, if not insurmountable, technical difficulties with the scenario.

4 Magnetic Fields from Early Universe Phase Transitions

The early universe was characterized by a series of phase transitions in which the nature of particles and fields changed in fundamental ways (see Fig. 3). For example, electroweak symmetry breaking marked the transition from a high-energy regime in which the W and Z bosons and the photon were effectively massless and interchangeable to one in which the W and Z bosons were heavy while the photon remained massless. The transition also marked the emergence of two distinct forces: electromagnetism and the weak nuclear force. Likewise, the Quark–Hadron phase transition marked the transition from the free quark–gluon phase to one in which quarks were locked into baryons. Both of these transitions had the potential to generate strong magnetic fields since they involved the release of an enormous amount of free energy and since they involve charged particles which could, in turn, drive currents. Indeed, strong magnetic fields are almost certainly generated. The question is one of physical scale since the Hubble radius was so small at these early times.

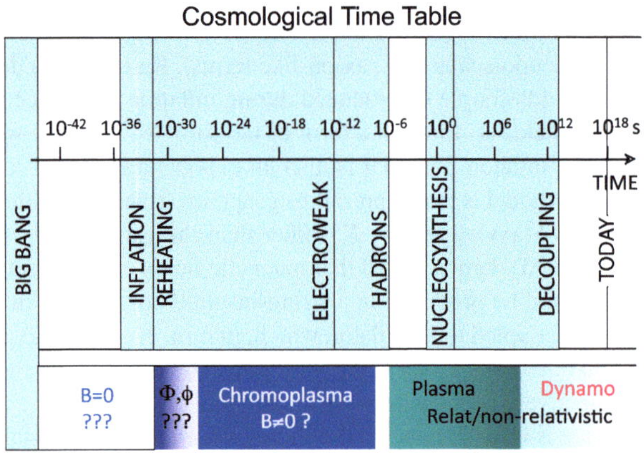

Fig. 3 The cosmological periods. Starting with reheating the quantum chromodynamic quark–gluon–plasma period lasts until hadron formation. Non-Abelian Weibel instabilities may be responsible for generating seed magnetic fields during this time, which in the presence of free energy grow from non-Abelian thermal fluctuations. At later times the period of relativistic and classical plasma sets on. Here seed magnetic fields can be generated from thermal background electrodynamic fluctuations if thermal anisotropies or beams can exist in this regime

4.1 General Considerations

At any phase in the history of the Universe, the strength of the fields generated by some microphysical process is limited by equipartition with the background energy density. Moreover, the maximum scale for magnetic fields generated by a microphysical process is set by the Hubble radius. For a radiation-dominated Universe, the energy density of the Universe is given by $\rho \propto g_* T^4$ where T is the temperature of the thermal bath and g_* counts the effective number of relativistic degrees of freedom (see, for example, Kolb and Turner 1990; Peacock 1999). The Hubble radius is given by $L_H = c/H$ where $H = a^{-1}da/dt \propto T^2$ is the Hubble parameter. Numerically, we have

$$B_{\max}\left(l = cH^{-1},\, T\right) = B_{\text{equipartition}}(T) \simeq 10^{18}\ \text{Gauss}\left(\frac{T}{150\ \text{MeV}}\right)^2 \tag{22}$$

for

$$l \simeq 100\ \text{cm}\left(\frac{150\ \text{MeV}}{T}\right)^2. \tag{23}$$

In practice, the strength of fields generated during some early Universe phase transition will be well below the value set by equipartition and have a length scale significantly smaller than the Hubble radius. We therefore write $B(l,T) = f B_{\max}$ for $l = gc/H$ and f and g are constants.

Typically, we are interested in fields on scales much larger than the characteristic scales of the original field. On purely geometric grounds, Hogan (1983) argued that the field strength on some large scale L due to small-scale cells of size l with field strength $B(l)$ will be $B(L) = B(l)(l/L)^{3/2}$. However, Durrer and Caprini (2003) showed that the divergence-free nature of the magnetic field places a constraint on the statistical properties of the field and that the most natural choice of scaling is $B(L) = B(l)(l/L)^{5/2}$. Their derivation assumes a scale-free energy spectrum for the magnetic field which may not apply for fields produced during an early Universe phase transition. Nevertheless, the argument does call into question Hogan's result.

In the absence of some dynamical effect, the field strength will be diluted by the expansion as a^{-2} between the time when it is generated and some later time. Thus, a field $B(l, T)$ generated prior to recombination will lead to a field at recombination with strength

$$B(L, T_{\rm rec}) = B(l, T)\left(\frac{T_{\rm rec}}{T}\right)^2 \left(\frac{l}{L}\right)^\beta \tag{24}$$

where $\beta = 5/2$, according to Durrer and Caprini (2003). More generally we have the following scaling:

$$B(L, T) \sim f g^\beta T^{-\beta} L_{\rm cm}^{-\beta}, \tag{25}$$

where we emphasize that L is the comoving length scale.

4.2 First-Order Phase Transitions

Detailed calculations of magnetic field generation during the electroweak and QCD phase transitions have been carried out by numerous authors. By and large, these groups assume that the transitions are first-order, that is characterized by a mixed-phase regime in which bubbles of the new phase nucleate and expand, eventually filling the volume. The energy associated with the bubble walls is released as a form of latent heat. Quashnock et al. (1989) demonstrated that a Biermann battery can operate during the QCD phase transition. The up, down, and strange quarks (the three lightest quarks) have charges $2/3$, $-1/3$, and $-1/3$ respectively. If these quarks were equal in mass, the quark–gluon plasma would be electrically neutral. However, the strange quark is heavier and therefore less abundant. The implication is that there is a net positive charge in the quark–gluon plasma and a net negative charge in the lepton sector. Electric currents are therefore generated at the bubble walls that separate the quark phase from the baryon phase. Quashnock et al. (1989) found that 5 G fields could be generated on scales of 100 cm at the time of the QCD phase transition. Following the arguments outlined above, the field strength on galactic scales at the time of recombination would be (a disappointingly small) $\sim 10^{-31}$ G.

Somewhat larger estimates were obtained by Cheng and Olinto (1994) and Sigl et al. (1997) who realized that as the hadronic regions grow, baryons would concentrate on the bubble walls due to a "snowplow" effect. (Sigl et al. 1997 also showed that fluid instabilities could give rise to strong magnetic fields during the QCD phase transition.) For reasonable parameters, they obtained fields about seven orders of magnitude larger than those found by Quashnock et al. (1989).

Magnetic fields can arise during cosmological phase transitions even if they are second order—that is, phase transitions signaled by the smooth and continuous transition of an order parameter. Vachaspati (1991), for example, showed that gradients in the Higgs field vacuum expectation value (the order parameter for the electroweak phase transition) induce magnetic fields on a scale $\sim T_{EW}^{-1}$ with strength of order $q_{EW}^{-1} T_{EW}^{-2}$ where T_{EW} is the temperature of the electroweak phase transition and q_{EW} is the Higgs field coupling constant. To estimate the field on larger scales, Vachaspati (1991) assumed that the Higgs field expectation value executed a random walk with step size equal to the original coherence length. The field strengths were small (10^{-23} G on 100 kpc scales) but not negligible.

4.3 Inverse Cascade

The discussion above suggests that strong magnetic fields are likely to have been generated in the early Universe but that their coherence length is so small, the effective large-scale fields are inconsequential for astrophysics. However, dynamical mechanisms may lead to an increase in the coherence length of magnetic fields produced at early times. Chief among these is an inverse cascade of magnetic energy from small to large scales which occurs when there is substantial magnetic helicity (Frisch et al. 1975). The effect was investigated in the context of primordial magnetic fields (see, for example, Cornwall 1997; Son 1999; Field and Carroll 2000; Brandenburg 1996; Banerjee and Jedamzik 2004). As the Universe expands, magnetic energy shifts to large scales as the field attempts to achieve equilibrium while conserving magnetic helicity. Under suitable conditions, Field and Carroll (2000) showed that astrophysically interesting fields with strength 10^{-10} G 10 kpc scales could be generated during the electroweak phase transition.

4.4 Plasma Processes Capable of Generating Magnetic Fields

4.4.1 Chromodynamic Magnetic Fields?

The QCD regime lasts from the end of reheating until hadron formation, roughly $t_{ns} \sim 10^{-6}$ s after the Big Bang (see Fig. 3). In this regime, matter comprises massive bosons, gluons, quarks and leptons and forms a hot dense chromoplasma or quark–gluon plasma (QGP). At the higher temperatures, that is, not too long after reheating, the QGP is asymptotically free and can be considered collisionless. Since many of the particles in the QGP carry electric charge, under certain conditions, they can generate induced Yang–Mills currents $j_a^\mu(x) = D_\mu F^{\mu\nu}(x)$, with Yang–Mills field $F_{\mu\nu} = A_{\nu,\mu} - A_{\mu,\nu} - ig[A_\mu, A_\nu]$ expressed through the non-Abelian gauge field $A_{a;\nu}(x)$. The colour index a corresponds to the $N^2 - 1$ colour channels. These currents couple to the electromagnetic gauge field and consequently are accompanied by magnetic fields. Several mechanisms associated with the QCD phase transition were discussed in the previous section. Here we explore whether plasma instabilities in the thermal QGP could possibly lead to appreciable fields. In the context of magnetic field generation in the early universe and cosmology this possibility has not yet been considered; it has, however, been investigated in high-energy physics not working in a comoving system.

The simplest plasma mechanism capable of magnetic field generation is the Weibel (current filamentation)[3] instability first discovered in classical plasma (Weibel 1959). Its free energy is provided by a local pressure anisotropy $A = P_\| / P_\perp - 1 \neq 0$ in the non-magnetic plasma. The subscripts $\|, \perp$ refer to the two orthogonal directions $\hat{\|}, \hat{\perp}$ of the pressure tensor $\mathsf{P} = P_\perp \mathsf{I} + (P_\| - P_\perp)\hat{\|}\hat{\|}$ whose non-diagonal (dissipative) elements are small (Blaizot and Iancu 2002). Pressure anisotropy creates microscopic currents and hence microscopic magnetic fields. Free energy can also be provided by partonic beams passing the QGP. Such beams, when produced by some independent process naturally introduce a preferred direction and may cause additional pressure anisotropy by dissipating their momentum in some (collisionless) way.

Assuming that a cold partonic beam passes the QGP, both analytical theory and numerical simulations (Arnold and Moore 2006a, 2006b; Arnold 2007; Arnold and Moore 2007;

[3]For a discussion of its physics in classical plasmas see Fried (1959).

Arnold and Leang 2007; Rebhan et al. 2008; Romatschke and Rebhan 2006; Schenke et al. 2008; Strickland 2007a, 2007b) prove that the QCD beam-driven Weibel instability excites magnetic fields which subsequently scatter and thermalize the beams by magnetizing the partons. The linear waves, that is, oscillations of the effective quark phase-space momentum distribution $\Phi_{\text{eff}}(p)$ as function of the 4-momentum, $p \equiv (p^0, \mathbf{p})$, are solutions of the semi-classical QGP dispersion relation (Pokrovsky and Selikhov 1988; Mrówczyński 1988, 1993) in Fourier 4-space $k = (\omega/c, \mathbf{k})$

$$\det\left[\mathbf{k}^2\delta^{ij} - k^i k^j - \omega^2\epsilon^{ij}(|\mathbf{k}|)\right] = 0 \tag{26}$$

with ϵ^{ij} the permeability ($v^i = p^i/\sqrt{p^l p_l}$ is the 4-velocity) given as functional of the effective phase space density $\Phi_{\text{eff}}(\mathbf{p})$, which does not depend any more on the colour index

$$\epsilon^{ij}(\omega, \mathbf{k}) = \delta^{ij} + \frac{g^2}{2\omega^2}\int \frac{d^3p}{8\pi^3}\frac{v^i[\partial\Phi_{\text{eff}}(\mathbf{p})/\partial p^l]}{\omega - \mathbf{k}\cdot\mathbf{v} + i0}\left[(\omega - \mathbf{k}\cdot\mathbf{v})\delta^{lj} + k^l v^j\right]. \tag{27}$$

Instability is found for low frequency ($\omega \approx 0$) non-oscillatory (filamentation) modes with wave vectors, $k_\perp$, perpendicular to the parton-beam 4-velocity, U. The implication is that stationary magnetic fields are generated in this process.

Analytical growth rates of transverse modes, where $\operatorname{Im}\omega > 0$, have been obtained for simple Gaussian and other mock equilibrium particle distributions and nuclear physics parameters. All calculations performed refer to non-comoving physical systems. Arnold (2007) has shown that the breakdown of perturbation theory at momenta $p \sim g^2 T$ and the fact that theory becomes non-perturbative at high T implies that it can be treated as if one had a cold plasma, $T = 0$, plus weak coupling. Thus, the semi-classical approach describes long-range properties.[4] In the numerical simulations one takes advantage of this fact, linearises around a stationary homogeneous locally colourless state (Blaizot and Iancu 2002), and considers the evolution of the fluctuation $W^\mu(v, x)$ of the distribution function according to the non-Abelian Vlasov equation and Yang–Mills current density:

$$v_\mu D^\mu W^\nu(\mathbf{v}, x) = -v_\sigma F^{\sigma\nu}(x), \qquad j^\nu(x) = -g^2\int \frac{d^3p}{8\pi^3}\frac{p^\nu}{|\mathbf{p}|}\frac{\partial\Phi_{\text{eff}}(\mathbf{p})}{\partial p^\sigma}W^\sigma(\mathbf{v}, x), \tag{28}$$

where the 4-velocity v^ν enters the field equations which close the system and describe the evolution of fields and particles,

$$\left.\begin{aligned} v_\mu D^\mu W^\nu &= -g(\mathbf{E} + \mathbf{v}\times\mathbf{B})\cdot\nabla_{\mathbf{p}}\Phi_{\text{eff}} \\ D_\mu F^{\mu\nu} &= -g^2(2\pi)^{-3}\textstyle\int d^3p\, v^\nu (D^\nu)^{-1}(\mathbf{E} + \mathbf{v}\times\mathbf{B})\cdot\nabla_{\mathbf{p}}\Phi_{\text{eff}} \end{aligned}\right\}. \tag{29}$$

Instability requires anisotropy to lowest order O(1) in $\mathbf{p}$ of the effective distribution $\Phi_{\text{eff}}(\mathbf{p})$ (Arnold and Moore 2006a). The maximum unstable wave vector k_m and growth rate $\gamma_m \equiv +(\operatorname{Im}\omega)_m$ of the QCD Weibel mode scale as

$$k_m^2 \sim \gamma_m^2 \sim m_\infty^2 \sim g^2\int_{\mathbf{p}_>} \Phi_{\text{eff}}(\mathbf{p})/|\mathbf{p}| \tag{30}$$

[4] The scalar potential A^0 picks up a Debye screening mass m_D and decouples at distances $\gg m_D^{-1} \sim (gT)^{-1}$ leaving the vector potential (transverse chromo-electromagnetic) fields, which is equivalent to classical plasmas where at frequencies $\omega > \omega_p$ above the plasma frequency ω_p any propagating perturbation is purely electromagnetic.

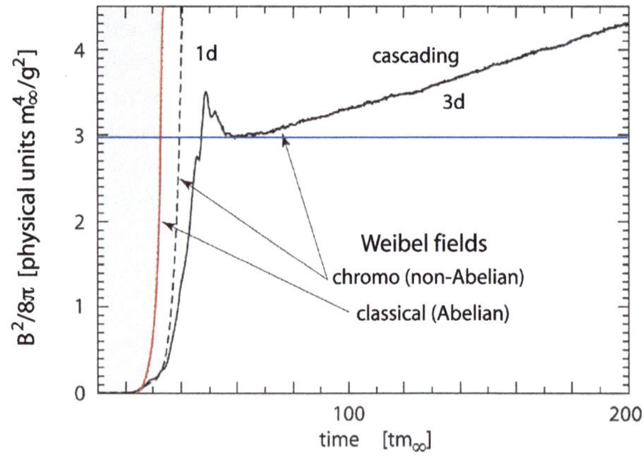

Fig. 4 Numerical simulation results of Weibel fields under Abelian and non-Abelian (QGP) conditions (after Arnold and Moore 2006a). The Abelian and spatially 1d-cases have a long linear phase of exponential growth. The corresponding linear phase in the spatially 3d-case is very short (*shaded region*), followed by a nonlinearly growing phase which reaches maximum field strength and afterward increases weakly and linearly. Closer investigation has shown that this further linear growth is caused by cascading to shorter wavelength (larger k) related to increasing frequencies thereby dispersing the free energy that feeds the nonlinear phase into a broad spectrum of magnetic oscillations. The *blue horizontal line* is the nonlinear saturation level of the low-frequency long-wavelength Weibel magnetic field that persists during the cascade and presumably survives it

where $m_\infty \sim g\sqrt{N_>/p_>}$ is the "effective mass" scale defined by the spatial beam number density $N_>$ of particles with momentum $p_>$ which contribute to the anisotropy, i.e. $N_> \equiv \int_{p_>} \Phi_{\rm eff}(\mathbf{p}) d^3p/8\pi^3$.

Figure 4 plots three simulation results: an Abelian run, a spatially 1d, and a spatially 3d-non-Abelian run. Shown is the magnetic fluctuation energy density $B^2/8\pi$ as function of time (all in proper simulation units). The Abelian and the 1d-non-Abelian cases cover just their linear (exponentially growing) phases. The 3d-non-Abelian case differs from these in several important respects. Its linear phase is very short, shown as the shaded region. It is followed by a longer nonlinear growth phase when the magnetic energy density increases at a slower and time dependent rate until reaching maximum, when it starts decaying. Afterwards it recovers to end up in a further slow but now purely algebraic linear growth.

4.4.1.1 Weibel Saturation Level The nonlinear phase wave-particle interaction terms come progressively into play when the field energy increases. The subsequent slow linear growth phase results from the sudden onset of a (turbulent) cascade from 'long' to short wavelengths and higher frequencies. This cascade has been demonstrated in simulations (Arnold and Moore 2006b; Arnold and Leang 2007). The settings of the simulations are physical not taking into account any general relativistic effects (which should be weak compared to QCD interactions) nor cosmological expansion. The cascade distributes the energy that is fed into the longest unstable scales over a broad range of short scales and higher frequencies $\omega \neq 0$, thereby identifying the nonlinear state of the QGP Weibel instability as short-scale turbulence hosting a mixture of propagating waves. So far it lacks any process (e.g. an inverse cascade) in which the Weibel fields produce large-scale stationary cosmological magnetic fields. In the average the energy density in the Weibel 'long' wavelength domains stays at a constant level

$$\mathcal{E}_{\rm Weibel}^{\rm sat} \lesssim 3m_\infty^4/g^2. \tag{31}$$

For the coupling constant one may use (Bethke 2009) the "world-average value" $g^2 = 4\pi\alpha_s(M_Z^2) \approx 4\pi \times 0.12 = 0.48\pi$, where M_Z is the mass of the Z boson, while for m_∞^2 additional knowledge is required of the effective parton distribution Φ_{eff}, i.e. the state of the undisturbed distribution including the anisotropy.

Unfortunately, the above formula cannot be used directly to estimate the long-wavelength Weibel field saturation value in the early universe. If we accept that the saturation level is robust within few orders of magnitude, then, because $m_\infty^2 \sim \omega_p^2/\sqrt{\theta}$, one has as for an estimate (in physical units)

$$B^{\text{sat}} \approx \omega_p^2 \sqrt{2\mu_0 \hbar / c^3 \theta} \tag{32}$$

where ω_p is the effective chromo-plasma frequency $\omega_p^2 = gT/m_D\lambda_D^2$ which is related to the Debye mass m_D, screening length λ_D, and $\theta \sim \tan^{-1}(A^{-1})$ is a given effective anisotropy angle (Arnold and Leang 2007). The definition of θ holds for both, thermal anisotropies $P_\perp/P_\| \equiv T_\perp/T_\|$ in the thermal Weibel case, and parton beam anisotropies where $A \propto T^{-1}\int_{\mathbf{p}_>} \Gamma(\mathbf{p})\Phi_{\text{eff}}(\mathbf{p}) d^3p/8\pi^3$ is the 'equivalent' thermal anisotropy caused by the parton beam of Lorentz factor Γ in QGP of isotropic temperature T.

With these numbers one estimates quite a strong (and thus cosmologically unrealistic) saturation magnetic field

$$B^{\text{sat}} \approx 10^{-19}(m_e/M_Z\sqrt{\theta})N_{\text{eff}} \sim 2 \times 10^{-25} N_{\text{eff}}/\sqrt{\theta}\ \text{G} \tag{33}$$

a value that depends on the effective parton number density to anisotropy ratio holding for small finite anisotropies. Taking, say, $N_{\text{eff}} \sim 10^{10}\ \text{cm}^{-3}$, it yields a substantial saturated low-frequency QGP seed magnetic field of the order of $B^{\text{sat}} \sim 10^{-5}/\sqrt{\theta}\ \mu\text{G}$.

Thus, the Weibel mechanism is doubtlessly capable of generating substantially strong magnetic fields during the QGP phase. However, these fields have very short correlation lengths, of the order of the parton skin depth or mean free paths and do probably not contribute to the large-scale field in the cosmological evolution.

4.4.1.2 Thermal Fluctuation Level When the QGP phase lacks any thermal anisotropy thermal fluctuations still provide seed magnetic fields at the thermal chromo-dynamic magnetic fluctuation level far below the magnetic saturation level of the Weibel instability. It results from the thermal fluctuations in the microscopic chromo-plasma currents. For its estimation the unknown effective *isotropic equilibrium* distribution $\Phi_{\text{eff}}(\mathbf{p})$ and the complex response function, i.e. the QGP polarisation tensor $\Pi^{\mu\nu}$, are needed whose determination requires solution of the equilibrium chromo-Vlasov equation.

For our purposes, a rough estimate of the thermal level can be obtained from the above simulations taking advantage of the evolution equation of the average magnetic energy density $\langle B^2(t)\rangle$ in the 'zero-frequency' domain. The thermal level is not subject to the Weibel mode instability; however the magnetic energy density measured in the Weibel mode at a certain time t_1 in the linear phase has evolved from the background magnetic thermal fluctuation level $\langle b^2(t=0)\rangle$ at time $t=0$ according to

$$\langle B^2(t_1)\rangle \simeq \langle b^2(t=0)\rangle \exp(2\gamma_m t) \tag{34}$$

From the linear phase in Fig. 4 we find for the numerically determined linear chromo-Weibel growth rate that $\gamma_m \approx 0.28/m_\infty$. This value used in the last equation yields an approximate (initial) thermal magnetic energy density level of $\langle b^2\rangle \approx 2.88 \times 10^{-8} m_\infty^4 g^{-2}$, which corresponds to a thermal magnetic fluctuation level independent of the Weibel mode of average

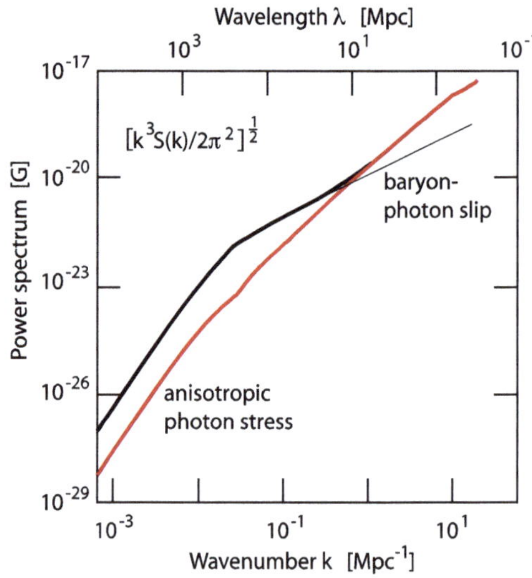

Fig. 5 The spectrum $S(k)$ of magnetic fields in the pre-recombination era generated from cosmological density perturbations (after Ichiki et al. 2006) plotted in units of magnetic field (G) and shown to be composed of contributions from baryon–phonon slip and anisotropic photon stresses. Below roughly $k < 1/\mathrm{Mpc}$ the baryon slip dominates the spectrum while for larger k (shorter wavelength than roughly 10 Mpc) the anisotropic stresses contribute most

amplitude

$$\langle b^{\mathrm{th}} \rangle \sim 3.4 \times 10^{-29} (N_{\mathrm{eff}}/N) \quad \mathrm{G} \tag{35}$$

during the corresponding QCD phase, depending on the ratio of the effective number density of energetic particles to chromo-plasma density N. This is a low though not unreasonable upper limit on the thermal fluctuation level of stationary ($\omega \approx 0$) QGP collective magnetic seed fields which, however, is probably too low to be of direct astrophysical importance.

4.4.2 Thermal Fluctuations Shortly Before Photon-Matter Decoupling

The previous section dealt with the possible generation of low frequency seed magnetic fields during the QCD phase of the early universe. A more conventional proposal has been elaborated recently (Ichiki et al. 2006, 2006) and is proposed to work during the classical plasma phase before recombination. These authors make the reasonable assumption that during this epoch the coupling between photons and electrons by Compton scattering is much stronger than the coupling between photons and ions. Cautious examination of the particle–photon interaction in a cosmological density fluctuation field indeed shows that pressure anisotropy and currents are induced. The generalised Ohm's law that includes the photon interaction allows for a finite electron current flow because of the differences in the bulk electron and proton velocities caused. This effect produces the desired magnetic fields. However, only the second order density perturbations in the Compton scattering terms lead to fields that survive on a range of spatial scales. The power spectrum $S(k)$ of the fields in the long wavelength range (see Fig. 5) scales as $\sqrt{k^3 S(k)} \propto k$ Here the photon-caused anisotropic stress dominates.

Seed magnetic fields reach values of $B \sim 10^{-18}$ G on scales of $\sim$1 Mpc and $B \sim 10^{-14}$ G on $\sim$10 kpc scales. After decoupling these fields decay adiabatically with expansion of the universe. In a standard cosmology they should today be of strength $B(t_0) \sim 10^{-24}$ G at 1 Mpc and $\sim 10^{-20}$ G at 10 kpc and may have played a role in structure formation after recombination.

These seed fields are surprisingly strong, however. Recently, Fenu et al. (2011) performed numerical simulations taking into account all relevant general relativistic effects. Their estimates based on the WMAP7 parameters suggest that the generation of seed magnetic fields on cosmological scales is mainly related to Compton drag by photons on baryons whereby vorticity is exchanged and magnetic fields are generated still after recombination. The power spectra obtained are $\propto k^4$ for $k \ll k_{eq}$ and $\propto \sqrt{k}$ for $k \gg k_{eq}$ yielding substantially lower comoving fields $B \sim 10^{-29}$ G at 1 Mpc.

5 Magnetic Fields and the Cosmic Microwave Background

The field of cosmology has witnessed a revolution lead, in large part, by detailed measurements of the CMB anisotropy and polarization spectra. Fundamental cosmological parameters such density parameters of baryons, dark matter, and dark energy and the Hubble constant are now known to a precision unimaginable just two decades ago.

If magnetic fields were present at the time of matter-radiation decoupling or soon after, then they would have an effect on the anisotropy and polarization of the CMB (see Subramanian 2006; Durrer 2007 for reviews). First, a very large scale (effectively homogeneous) field would lead to anisotropic expansion and hence to a quadrupole anisotropy in the CMB (see, for example, Thorne 1967). The degree of isotropy of the CMB was used to place a limit of several nG on the strength of such a field redshifted to the current epoch (Barrow et al. 1997). Note, however, that the results of Barrow et al. (1997) assume that magnetic fields are the only source of large-scale anisotropy. Recently, Adamek et al. (2011) showed that neutrinos, which free-stream so long as they are relativistic, generate an anisotropic pressure which can counteract the anisotropic pressure generated by a homogeneous magnetic field. This result depends on the neutrino masses but may damp the effect of very large scale magnetic fields on the CMB.

Primordial magnetogenesis scenarios on the other hand generally lead to tangled fields, plausibly Gaussian random, characterized by say a spectrum $M(k)$. This spectrum is normalized by giving the field strength B_0, at some fiducial scale, and as measured at the present epoch, assuming it decreases with expansion as $B = B_0/a^2(t)$. Since magnetic and radiation energy densities both scale with expansion as $1/a^4$, we can characterize the magnetic field effect by the ratio $B_0^2/(8\pi\rho_{\gamma 0}) \sim 10^{-7} B_{-9}^2$ where $\rho_{\gamma 0}$ is the present day energy density in radiation, and $B_{-9} = B_0/(10^{-9}$ G). Magnetic stresses are therefore small compared to the radiation pressure for nano Gauss fields.

Nevertheless, the scalar, vector and tensor parts of the perturbed stress tensor associated with primordial magnetic fields lead to corresponding metric perturbations, including gravitational waves. Further the compressible part of the Lorentz force leads to compressible (scalar) fluid velocity and associated density perturbations, while its vortical part leads to vortical (vector) fluid velocity perturbation. These magnetically induced metric and velocity perturbations lead to both large and small angular scale anisotropies in the CMB temperature and polarization.

The scalar contribution has been the most subtle to calculate, and has only begun to be understood by several groups (Giovannini and Kunze 2008; Yamazaki et al. 2008; Finelli et al. 2008; Shaw and Lewis 2010). The anisotropic stress associated with the magnetic field leads in particular to the possibility of two types of scalar modes, a potentially dominant mode which arises before neutrino decoupling sourced by the magnetic anisotropic stress. And a compensated mode which remains after the growing neutrino anisotropic stress has compensated the magnetic anisotropic stress (cf. Shaw and Lewis 2010; Bonvin and Caprini 2010 for detailed discussion). The magnetically induced compressible fluid perturbations, also changes to the acoustic peak structure of the angular anisotropy power spectrum (see, for example, Adams et al. 1996). However, for nano Gauss fields, the CMB anisotropies due to the magnetized scalar mode are grossly subdominant to the anisotropies generated by scalar perturbations of the inflaton.

Potentially more important is the contribution of the Alfvén mode driven by the rotational component of the Lorentz force (Subramanian and Barrow 1998; Mack et al. 2002; Subramanian et al. 2003; Lewis 2004). Unlike the compressional mode, which gets strongly damped below the Silk scale, L_S due to radiative viscosity (Silk 1968), the Alfvén mode behaves like an over damped oscillator. This is basically because the phase velocity of oscillations, in this case the Alfvén velocity, is $V_A \sim 3.8 \times 10^{-4} c B_{-9}$ much smaller than the relativistic sound speed $c/\sqrt{3}$. Note that for an over damped oscillator there is one normal mode which is strongly damped and another where the velocity starts from zero and freezes at the terminal velocity till the damping becomes weak at a latter epoch. The net result is that the Alfvén mode survives Silk damping down to much smaller scales; $L_A \sim (V_A/c)L_S \ll L_S$, the canonical Silk damping scale (Jedamzik et al. 1998; Subramanian and Barrow 1998). The resulting baryon velocity leads to a CMB temperature anisotropy, $\Delta T \sim 5\mu K(B_{-9}/3)^2$ for a scale invariant spectrum, peaked below the Silk damping scale (angular wavenumbers $l > 10^3$). This rough estimate has also been borne out by the detailed calculations of Lewis (2004).

The magnetic anisotropic stress also induces tensor perturbations, resulting in a comparable CMB temperature anisotropy, but now peaked on large angular scales of a degree or more (Durrer et al. 2000). Both the vector and tensor perturbations lead to ten times smaller B-type polarization anisotropy, at respectively small and large angular scales (Seshadri and Subramanian 2001; Subramanian et al. 2003; Mack et al. 2002; Lewis 2004). Note that inflationary generated scalar perturbations only produce the E-type mode. The small angular scale vector contribution in particular can potentially help to isolate the magnetically induced signals (Subramanian et al. 2003).

A crucial difference between the magnetically induced CMB anisotropy signals compared to those induced by inflationary scalar and tensor perturbations, concerns the statistics associated with the signals. Primordial magnetic fields lead to non-Gaussian statistics of the CMB anisotropies even at the lowest order, as magnetic stresses and the temperature anisotropy they induce depend quadratically on the magnetic field. In contrast, CMB non-Gaussianity due to inflationary scalar perturbations arises only as a higher order effect. A computation of the non-Gaussianity of the magnetically induced signal has begun (Seshadri and Subramanian 2009; Caprini et al. 2009; Cai et al. 2010), based on earlier calculations of non-Gaussianity in the magnetic stress energy (Brown and Crittenden 2005). This new direction of research promises to lead to tighter constraints or a detection of strong enough primordial magnetic fields.

A primordial magnetic field leads to a number of other effects on the CMB which can probe its existence: Such a field in the inter galactic medium can cause Faraday rotation of the polarized component of the CMB, leading to the generation of new B-type signals from the inflationary E-mode signal (Kosowsky and Loeb 1996). Any large-scale helical component of the field leads to a parity violation effect, inducing non-zero T-B and E-B cross-correlations (Kahniashvili and Ratra 2005); such cross-correlations between signals of even and odd parity are necessarily zero in standard inflationary models. The damping of primordial fields in the pre-recombination era can lead to spectral distortions of the CMB (Jedamzik et al. 2000), while their damping in the post-recombination era can change the ionization and thermal history of the universe and hence the electron scattering optical depth as a function of redshift (see Sethi and Subramanian 2005; Tashiro and Sugiyama 2006a; Schleicher et al. 2008, and Sect. 5.3). Future CMB probes like PLANCK can potentially detect the modified CMB anisotropy signal from such partial re-ionization. In summary primordial magnetic fields of a few nG lead to a rich variety of effects on the CMB and thus are potentially detectable via observation of CMB anisotropies.

6 Implications of Strong Primordial Fields in the Post-recombination Universe

If strong magnetic fields have been produced during phase-transitions in the early universe, and if these fields had some non-zero helicity, they may have remained strong until recombination and beyond (Christensson et al. 2001; Banerjee and Jedamzik 2004). They could then affect the thermal and chemical evolution during the dark ages of the Universe, the formation of the first stars, and the epoch of reionization.

6.1 Implications During the Dark Ages

At high redshift $z > 40$, the universe is close to homogeneous, and the evolution of the temperature, T, is governed by the competition of adiabatic cooling, Compton scattering with the CMB and, in the presence of strong magnetic fields, ambipolar diffusion. It is thus given as

$$\frac{dT}{dz} = \frac{8\sigma_T a_R T_{\rm rad}^4}{3H(z)(1+z)m_e c} \frac{x_e(T - T_{\rm rad})}{1 + f_{\rm He} + x_e} + \frac{2T}{1+z} - \frac{2(L_{\rm AD} - L_{\rm cool})}{3nk_B H(z)(1+z)}, \tag{36}$$

where $L_{\rm AD}$ is the heating function due to ambipolar diffusion (AD), $L_{\rm cool}$ the cooling function (Anninos et al. 1997), σ_T the Thomson scattering cross section, a_R the Stefan–Boltzmann radiation constant, m_e the electron mass, c the speed of light, k_B Boltzmann's constant, n the total number density, $x_e = n_{\rm e}/n_H$ the electron fraction per hydrogen atom, $T_{\rm rad}$ the CMB temperature, $H(z)$ is the Hubble factor and $f_{\rm He}$ is the number ratio of He and H nuclei.

AD occurs due to the friction between ionized and neutral species, as only the former are directly coupled to the magnetic field. Primordial gas consists of several neutral and ionized species, and for an appropriate description of this process, we thus adopt the multi-fluid approach of Pinto et al. (2008), defining the AD heating rate as

$$L_{\rm AD} = \frac{\eta_{\rm AD}}{4\pi} \left|(\nabla \times \mathbf{B}) \times \mathbf{B}/B\right|^2, \tag{37}$$

where $\eta_{\rm AD}$ is given as

$$\eta_{\rm AD}^{-1} = \sum_n \eta_{{\rm AD},n}^{-1}. \tag{38}$$

In this expression, the sum includes all neutral species n, and $\eta_{{\rm AD},n}$ denotes the AD resistivity of the neutral species n. We note that the AD resistivities themselves are a function of magnetic field strength, temperature and chemical composition.

In the primordial IGM, the dominant contributions to the total resistivity are the resistivities of atomic hydrogen and helium due to collisions with protons. These are calculated based on the momentum transfer coefficients of Pinto and Galli (2008). As the power spectrum of the magnetic field is unknown, we estimate the expression in (37) based on the coherence length L_B, given as the characteristic scale for Alfvén damping (Jedamzik et al. 1998;

Subramanian 1998; Seshadri and Subramanian 2001). Contributions from decaying MHD turbulence may also be considered, but are negligible compared to the AD heating (Sethi and Subramanian 2005).

The additional heat input provided by AD affects the evolution of the ionized fraction of hydrogen, x_p, which is given as

$$\begin{aligned}\frac{dx_\mathrm{p}}{dz} = &\frac{[x_e x_p n_H \alpha_H - \beta_H (1-x_p) e^{-h_p \nu_{H,2s}/kT}]}{H(z)(1+z)[1+K_H(\Lambda_H+\beta_H)n_H(1-x_p)]} \\ &\times \left[1+K_H \Lambda_H n_H (1-x_p)\right] - \frac{k_\mathrm{ion} n_H x_p}{H(z)(1+z)}.\end{aligned} \tag{39}$$

Here, n_H is the number density of hydrogen atoms and ions, h_p Planck's constant, k_ion is the collisional ionization rate coefficient (Abel et al. 1997). Further details of notation, as well as the parametrized case B recombination coefficient for atomic hydrogen α_H, are given by Seager et al. (1999). The chemical evolution of the primordial gas is solved with a system of rate equations for the chemical species H^-, H_2^+, H_2, HeH^+, D, D^+, D^-, $HD^+$and HD based on the primordial rate coefficients tabulated by Schleicher et al. (2008). For the mutual neutralization rate of H^- and H^+, we use the more recent result of Stenrup et al. (2009). The evolution of the magnetic energy density $E_B = B^2/8\pi$ is given as

$$\frac{dE_B}{dt} = \frac{4}{3}\frac{\partial \rho}{\partial t}\frac{E_B}{\rho} - L_\mathrm{AD}. \tag{40}$$

The first term describes the evolution of the magnetic field in a homogeneous universe in the absence of specific magnetic energy generation or dissipation mechanisms. The second term accounts for corrections due to energy dissipation via AD.

The dynamical implications of magnetic fields can be assessed from the magnetic Jeans mass, the critical mass scale for gravitational forces to overcome magnetic pressure. In the cosmological context, the magnetic Jeans mass is given as (Subramanian 1998; Sethi and Subramanian 2005)

$$M_J^B \sim 10^{10} M_\odot \left(\frac{B_0}{3\ \mathrm{nG}}\right)^3. \tag{41}$$

We note that this equation is derived in the cosmological context and is only valid on cosmological scales. A more general expression for the magnetic Jeans length is given below.

Due to ambipolar diffusion, strong magnetic fields also affect the gas temperature and thus the thermal Jeans mass, i.e. the critical mass scale required to overcome gas pressure. It is defined as

$$M_J = \left(\frac{4\pi\rho}{3}\right)^{-1/2}\left(\frac{5k_B T}{2\mu G m_P}\right)^{3/2} \tag{42}$$

with Boltzmann's constant k_B, the mean molecular weight μ and the proton mass m_p.

To describe virialization in the first minihalos, we employ the spherical collapse model of Peebles (1993) for pressureless dark matter until an overdensity of $\sim$200 is reached. Equating cosmic time with the timescale from the spherical collapse model allows one to calculate the overdensity ρ/ρ_b as a function of time or redshift. In this model, we further assume that the formation of the protocloud will reduce the coherence length of the magnetic field to the size of the cloud.

The evolution of the magnetic field strength, the IGM temperature and the chemical abundances of different species have been calculated by Schleicher et al. (2009b) using an

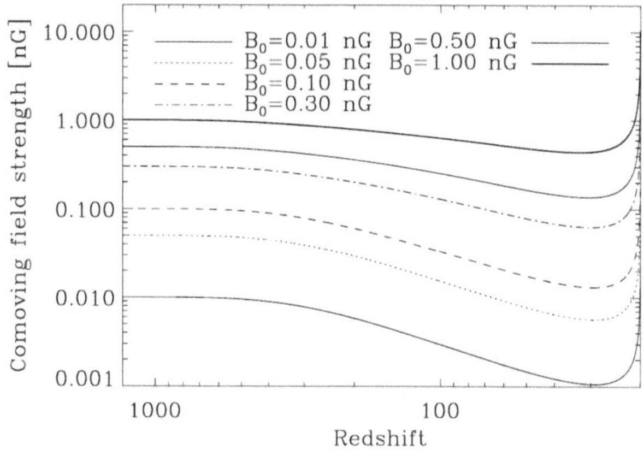

Fig. 6 The evolution of the comoving magnetic field strength due to AD as a function of redshift for different initial comoving field strengths, from the homogeneous medium at $z = 1300$ to virialization at $z = 20$

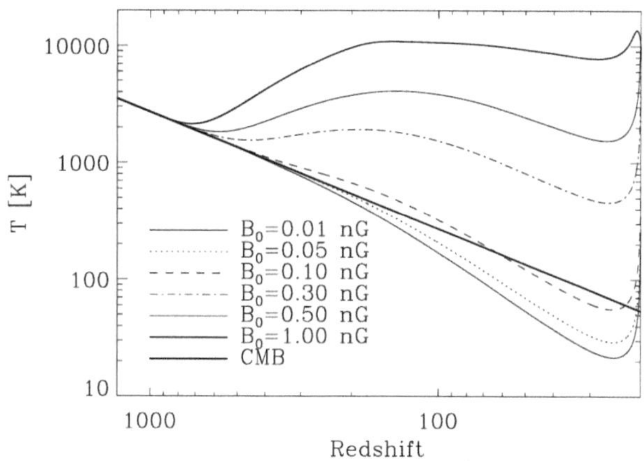

Fig. 7 The gas temperature evolution in the IGM as a function of redshift for different comoving field strengths, from the homogeneous medium at $z = 1300$ to virialization at $z = 20$. For the case with $B_0 = 0.01$ nG, we find no difference in the thermal evolution compared to the zero-field case

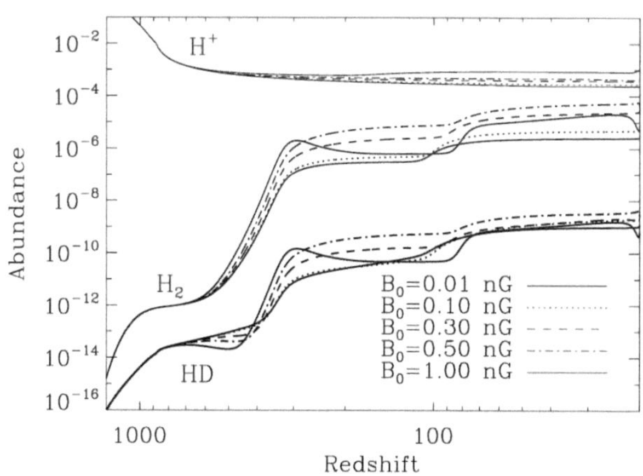

Fig. 8 The evolution of ionization degree, H_2 and HD abundances as a function of redshift for different comoving field strengths, from the homogeneous medium at $z = 1300$ to virialization at $z = 20$. For the case with $B_0 = 0.01$ nG, we find no difference in the chemical evolution compared to the zero-field case

extension of the recombination code RECFAST (Seager et al. 1999). The results are shown in Figs. 6, 7 and 8. As shown in Fig. 6, ambipolar diffusion primarily affects magnetic fields with initial comoving field strengths of 0.2 nG or less. For stronger fields, the dissipation of only a small fraction of their energy increases the temperature and the ionization fraction of the IGM to such an extent that AD becomes less effective. For comoving field strengths up to ~0.1 nG, the additional heat from ambipolar diffusion is rather modest and the gas in the IGM cools below the CMB temperature due to adiabatic expansion. However, it can increase significantly for stronger fields and reaches $\sim 10^4$ K for a comoving field strength of 1 nG, where Lyman α cooling and collisional ionization become efficient and prevent a further increase in temperature. The increased temperature enhances the ionization fraction and leads to larger molecule abundances at the onset of star formation.

6.2 Implications for the Formation of the First Stars

We now explore in more detail the consequences of magnetic fields for the formation of the first stars, during the protostellar collapse phase. For this purpose, a model describing the chemical and thermal evolution during free-fall collapse was developed by Glover and Savin (2009) and extended by Schleicher et al. (2009b) for the effects of magnetic fields. Particularly important with respect to this application is the fact that it correctly models the evolution of the ionization degree and the transition at densities of $\sim 10^8$ cm^{-3} where Li^+ becomes the main charge carrier. Based on the Larson–Penston type self-similar solution (Larson 1969; Penston 1969; Yahil 1983), we evaluate how the collapse timescale is affected by the thermodynamics of the gas.

During protostellar collapse, magnetic fields are typically found to scale as a power-law with density ρ. Assuming ideal MHD with flux freezing and spherical symmetry, one expects a scaling with $\rho^{2/3}$ in the case of weak fields. Deviations from spherical symmetry such as expected for dynamically important fields give rise to shallower scalings, e.g. $B \propto \rho^{0.6}$ (Banerjee et al. 2004; Banerjee and Pudritz 2006), $B \propto \rho^{1/2}$ (Hennebelle and Fromang 2008; Hennebelle and Teyssier 2008). Based on numerical simulations of Machida et al. (2006), we find an empirical power-law relation $B \propto \rho^{\alpha}$ where

$$\alpha = 0.57 \left(\frac{M_J}{M_J^B} \right)^{0.0116}. \tag{43}$$

In a collapsing cloud, the more general expression for the magnetic Jeans mass,

$$M_J^B = \frac{\Phi}{2\pi \sqrt{G}}, \tag{44}$$

is adopted. In this prescription, $\Phi = \pi r^2 B$ denotes the magnetic flux, G the gravitational constant, r an appropriate length scale. The calculation of the magnetic Jeans mass thus requires an assumption regarding the size of the dense region. Numerical hydrodynamics simulations show that they are usually comparable to the thermal Jeans length (Abel et al. 2002; Bromm and Larson 2004). This is also suggested by analytical models for gravitational collapse (Larson 1969; Penston 1969; Yahil 1983).To account for magnetic energy dissipation via AD, we calculate the AD heating rate from (37) and correct the magnetic field strength accordingly. We note that due to the large range of densities during protostellar collapse, additional processes need to be taken into account to calculate the AD resistivity correctly. In particular, at a density of $\sim 10^9$ cm^{-3}, the three-body H_2 formation rates start to increase the H_2 abundance significantly, such that the gas is fully molecular at densities of $\sim 10^{11}$ cm^{-3}. As a further complication, the proton abundance drops considerably at densities of $\sim 10^8$ cm^{-3}, such that Li^+ becomes the main charge carrier (Maki and Hajime 2004; Glover and Savin 2009). These effects are incorporated in our multi-fluid approach.

Table 1 The physical field strength B at beginning of collapse as a function of the comoving field strength B_0 used to initialize the IGM calculation at $z = 1300$

B_0 [nG]	B [nG]
1	4.8×10^3
0.3	6.3×10^2
0.1	1.3×10^2
0.01	1.0×10^1

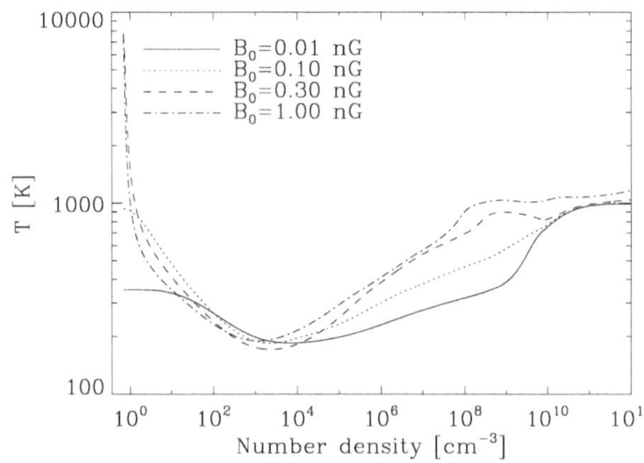

Fig. 9 The gas temperature as a function of density for different comoving field strengths. For $B_0 = 0.01$ nG, the thermal evolution corresponds to the zero-field case

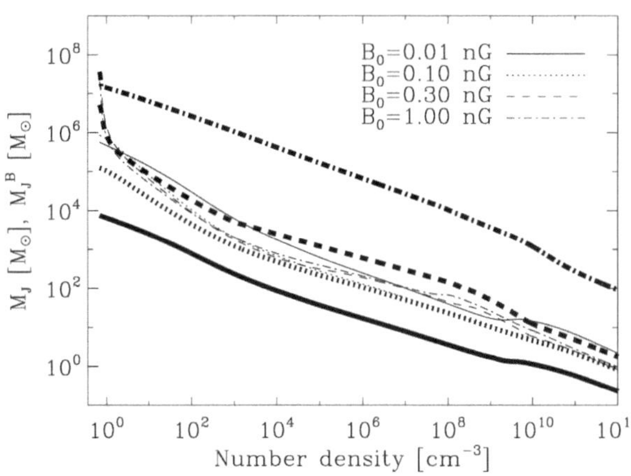

Fig. 10 Thermal (*thin lines*) and magnetic (*thick lines*) Jeans mass as a function of density for different comoving field strengths. For $B_0 = 0.01$ nG, the thermal evolution and thus the thermal Jeans mass corresponds to the zero-field case

As an initial condition for these model calculations, we use the physical field strength and the chemical abundances obtained from the spherical collapse model in Sect. 6.1. The relation between co-moving and physical field strength at the beginning of this calculation is thus given in Table 1.

Figure 9 shows the calculated temperature evolution as a function of density for different comoving field strengths. For comoving fields of 0.01 nG or less, there is virtually no difference in the temperature evolution from the zero-field case. For comoving fields of $\sim$0.1 nG, cooling wins over the additional heat input in the early phase of collapse, and the temperature decreases slightly below the zero-field value at densities of 10^3 cm^{-3}. At higher densities, the additional heat input dominates over cooling and the temperature steadily increases. At densities of $\sim 10^9$ cm^{-3}, the abundance of protons drops considerably and increases the AD resistivity defined in (38) and the heating rate until Li^+ becomes the main charge carrier. In particular for comoving fields larger than $\sim$0.1 nG, this transition is reflected by a small bump in the temperature evolution due to the increased heating rate in this density range.

Apart from the transition where Li^+ becomes the dominant charge carrier, the magnetic field strength usually increases more rapidly than $\rho^{0.5}$, and weak fields increase more rapidly than strong fields. This is what one naively expects from (43), and it is not significantly affected by magnetic energy dissipation. Another important point is that comoving fields of only 10^{-5} nG are amplified to values of $\sim$1 nG at a density of 10^3 cm^{-3}. Such fields are required to drive protostellar outflows that can magnetize the IGM (Machida et al. 2006).

Figure 10 shows the evolution of the thermal and magnetic Jeans mass during collapse. The thermal Jeans masses are quite different initially, but as the temperatures reach the same order of magnitude during collapse, the same holds for the thermal Jeans mass. The thermal Jeans mass in this late phase has only a weak dependence on the field strength. As expected, the magnetic Jeans masses are much more sensitive to the magnetic field strength, and initially differ by about two orders of magnitude for one order of magnitude difference in the field strength. For comoving fields of $\sim$1 nG, the magnetic Jeans mass dominates over the thermal one and thus determines the mass scale of the protocloud. For $\sim$0.3 nG, both masses are roughly comparable, while for weaker fields the thermal Jeans mass dominates. The magnetic Jeans mass shows features both due to magnetic energy dissipation, but also due to a change in the thermal Jeans mass, which sets the typical length scale and thus the magnetic flux in the case that $M_J > M_J^B$.

The uncertainties in these models have been explored further by Schleicher et al. (2009b), and an independent calculation including stronger magnetic fields has been provided by Sethi et al. (2010). We also note that the consequences of initially weak magnetic fields for primordial star formation are explored in more detail in the next chapter.

6.3 Implications for Reionization

Strong magnetic fields may influence the epoch of reionization in various ways. As discussed above, they may affect the formation of the first stars and change their mass scale, and thus their feedback effects concerning cosmic reionization and metal enrichment. They may further give rise to fluctuations in the large-scale density field and potentially enhance high-redshift structure formation (Kim et al. 1996; Sethi and Subramanian 2005; Tashiro and Sugiyama 2006a). On the other hand, the increased thermal and magnetic pressure may indeed suppress star formation in small halos and thus delay reionization (Schleicher et al. 2008; Rodrigues et al. 2010). In both cases, unique signatures of the magnetic field may become imprinted in the 21 cm signature of reionization, which may help to constrain or detect such magnetic fields with LOFAR,[5] EDGES[6] or SKA[7] (Tashiro and Sugiyama 2006b; Schleicher et al. 2009a). Upcoming observations of these facilities may thus provide a unique opportunity to probe high-redshift magnetic fields in more detail.

[5]LOFAR homepage: http://www.lofar.org/.

[6]EDGES homepage: http://www.haystack.mit.edu/ast/arrays/Edges/.

[7]SKA homepage: http://www.skatelescope.org/.

7 Seed Fields in the Post-recombination Universe

We have seen that magnetic fields arise naturally during inflation and during phase transitions after inflation but before recombination. However, the effective field strength on galactic scales may be exceedingly small. Indeed, the seed fields for the galactic dynamo may well arise from astrophysical processes rather than exotic early-Universe ones. In this section, we review three processes that can generate magnetic field in the post-recombination Universe.

7.1 Biermann Battery

In the hierarchical clustering scenario, proto-galaxies acquire angular momentum from tidal torques produced by neighboring systems (Hoyle 1949; Peebles 1969; White 1984). However, these purely gravitational forces cannot generate vorticity (the gravitational force can be written as the gradient of a potential whose curl is identically zero) and therefore the existence of vorticity must arise from gasdynamical processes such as those that occur in shocks. More specifically, vorticity is generated whenever one has crossed pressure and density gradients. In an ionized plasma, this situation drives currents which, in turn, generate magnetic field. This mechanism, known as the Biermann battery and originally studied in the context of stars (Biermann 1950) was considered in the cosmological context by Pudritz and Silk (1989), Kulsrud et al. (1997), Davies and Widrow (2000), and Xu et al. (2008).

In magnetohydrodynamics, the time dependence of the magnetic field can be written as:

$$\frac{\partial \mathbf{B}}{\partial t} = \nabla \times (\mathbf{v} \times \mathbf{B}) + \frac{c^2}{4\pi\sigma}\nabla^2\mathbf{B}, \tag{45}$$

where σ is the conductivity, which is assumed to be constant in space. The Biermann effect leads to an additional source term for $\partial\mathbf{B}/\partial t$ of the form

$$\frac{m_e c}{e}\frac{\nabla p_e \times \nabla \rho_e}{\rho_e^2}, \tag{46}$$

where p_e and ρ_e are the electron pressure and density.

The equation for the vorticity, ω, has a similar source term:

$$\frac{\partial \omega}{\partial t} = \nabla \times (\mathbf{v} \times \omega) + \frac{\nabla p \times \nabla \rho}{\rho^2}, \tag{47}$$

where, here, p and ρ refer to the whole fluid. The implication is that when vorticity is generated, so it magnetic field. A simple order of magnitude estimate yields

$$B_{\text{Biermann}} \simeq \frac{m_p c}{e}\omega \simeq 3 \times 10^{-21}\left(\frac{\omega}{\text{km}\,\text{s}^{-1}\,\text{kpc}^{-1}}\right)\text{Gauss}. \tag{48}$$

We note that the vorticity in the solar neighborhood is of order 30 $\text{km}\,\text{s}^{-1}\,\text{kpc}^{-1}$ so for galactic-scale fields, we may expect seed fields of order 10^{-19} G from the Biermann mechanism.

The Biermann effect is expected to arise in a wide range of astrophysical settings. Lazarian (1992) showed that a Biermann-type effect operates in the interstellar medium, which is, in general, non-uniform, multiphase, and clumpy. Of particular interest is the epoch of reionization, which occurs at a redshift $z \sim 10$. During this epoch, photons from descrete sources such as QSOs and first-generation stars, reionize the intergalactic medium. Reionization begins when ionization fronts propagate away from these sources. Since the pressure and density (or temperature) gradients in these fronts are, in general, misaligned, one can expect the Biermann mechanism to operate and generate seed magnetic fields (Subramanian et al. 1994; Gnedin et al. 2000).

Any force that acts differently on electrons and protons will drive currents and hence generate magnetic fields. The Biermann battery is but one example. Radiation pressure, which will also be important during reionization, acts more strongly on electrons than protons and will therefore drive electric currents (Langer et al. 2003; Ando et al. 2010).

7.2 First-Generation Stars

The first generation of stars provides another potential source of seed fields for the galactic dynamo. Even if stars are born without magnetic fields, a Biermann battery will generate weak fields which can then be rapidly amplified by a stellar dynamo. If the star subsequently explodes or loses a significant amount of mass through stellar winds, the fields will find their way into the interstellar medium and spread throughout the (proto) galaxy. Simple estimates by Syrovatskii (1970) illustrate the viability of the mechanism. There have been some 10^8 supernovae over the lifetime of the galaxy, each of which spreads material through a $(10\ \mathrm{pc})^3$ volume. Using values for the field strength typical of the Crab nebula, one therefore expects the galaxy to be filled by 10 pc regions with field strengths $\sim 3\ \mu\mathrm{G}$. Assuming the same $L^{-\beta}$ scaling discussed above, one finds a field strength of 10^{-13} G on 10 kpc scales (assuming $\beta = 5/2$ as in Durrer and Caprini 2003), a value significantly larger than the ones obtained by more exotic early Universe mechanisms.

The strong amplification of seed magnetic fields during primordial star formation has been suggested in a number of works. Analytical estimates by Pudritz and Silk (1989), Tan and Blackman (2004), and Silk and Langer (2006) suggest that large-scale dynamos as well as the magneto-rotational instability could significantly amplify weak magnetic seed fields until saturation. Even before, during the protostellar collapse phase, the small-scale dynamo leads to an exponential growth of the magnetic fields, as found in both semi-analytic and numerical studies (Schleicher 2010; Sur et al. 2010).

7.3 Active Galactic Nuclei

Strong magnetic fields almost certainly arise in active galactic nucleii (AGN). These fields will find their way into the intergalactic medium via jets thereby providing a potential source of magnetic field for normal galaxies (Hoyle 1969; Rees 1987, 1994; Daly and Loeb 1990). The potential field strengths due to this mechanism may be estimated as follows: The rotational energy associated with the central compact (mass M) object which powers the AGN can be parametrized as fMc^2 where $f < 1$. If we assume equipartition between rotational and magnetic energy within a central volume V_c, we find

$$B_c = \left(\frac{8\pi f M c^2}{V_c}\right)^{1/2}. \tag{49}$$

If this field then expands adiabatically to fill a "galactic" volume V_g one finds $B_g = B_c(V_c/V_g)^{2/3}$. Hoyle (1969) considered values $M = 10^9\,M_\odot$, $f = 0.1$, $V_g \simeq (100\ \mathrm{kpc})^3$ and found $B_c \simeq 10^9$ G and $B_g \simeq 10^{-5}$ G.

8 Conclusions

The origin of the seed magnetic fields necessary to prime the galactic dynamo remains a mystery and one that has become more, rather than less, perplexing over the years as observations have pushed the epoch of microgauss galactic fields back to a time when the Universe was a third its present age. Numerous authors have explored the possible that the galactic fields observed today and at intermediate redshift have their origin in the very early Universe.

The impetus for the study of early Universe magnetic fields came from the successful marriage of ideas from particle physics and cosmology that occurred during the latter half of the last century. It is a remarkable prediction of modern cosmology that the particles and fields of the present-day Universe emerged during phase transitions a fraction of a second after the Big Bang. The electromagnetic and weak interactions became distinct during the electroweak phase transitions at $t \simeq 10^{-12}$ s while baryons replaced the quark–gluon plasma at $t \simeq 10^{-6}$ s. Perhaps more fantastical is the notion that galaxies, clusters, and superclusters arose from quantum-produced density perturbations that originated during inflation at even earlier times. This idea is supported by strong circumstantial from the CMB anisotropy spectrum, so much so, that it is now part of the standard lore of modern cosmology.

Both inflation and early Universe phase transitions have many of the ingredients necessary for the creation of magnetic fields. If our understanding of inflation-produced density perturbations is correct, the similar quantum fluctuations in the electromagnetic field will lead to fields on the scales of galaxies and beyond. As well, electromagnetic currents, and hence fields, will almost certainly be driven during both the electroweak and quark–hadron phase transitions.

Early Universe schemes for magnetic field generation, however, face serious challenges. In the standard electromagnetic theory, and in a flat or closed FLRW cosmology, inflation-produced fields at diluted by the expansion to utterly negligible levels. One may address this issue by considering fields in an open Universe and "just-so" inflation scenario, that is, a scenario with just enough e-folds of inflation to solve the flatness problem. Alternatively, one may incorporate additional couplings of the field to gravity into the theory though many terms lead to unwanted consequences which render the theory unphysical. Observations and advances in theoretical physics may settle the issue. For example, if future determinations find that the Universe has a slight negative curvature (density parameter for matter and dark energy slightly less than one), it would give some credence to the idea of superadiabatic field amplification in an open Universe. On the other hand, string theory may point us to couplings between gravity and electromagnetism that naturally generate fields during inflation.

The main difficulty with fields generated from phase transitions after inflation arises from the small Hubble scale in the very early Universe. Strong fields are almost certainly produced by one of a number of mechanism. But the coherence length is so small, the effective field strength on galactic scales is likely to be well-below the level of interest for astrophysics. An inverse cascade may help; if the field has a net helicity, the magnetic field energy will be efficiently transferred from small to large scales. But even if fields are uninteresting for the galactic dynamo, they may have an effect on processes in the post-recombination Universe such as reionization and the formation of the first generation of stars.

Where, if not the early Universe, did the seed fields for the galactic dynamo arise? Astrophysics provides a number of promising alternatives. Galactic disks have angular momentum *and* vorticity. While the former is generated by the gravitational interaction between neighboring protogalaxies, the latter comes form gasdynamical effects which necessarily generate magnetic fields via the Biermann battery. As well, magnetic fields, rapidly created and amplified inside some early generation of stars or in active galactic nucleii, can be dispersed throughout the intergalactic medium.

Acknowledgements DR is supported by National Research Foundation of Korea through grant 2007-0093860. LMW is supported by a Discovery Grant from the Natural Sciences and Engineering Research Council of Canada.

References

T. Abel, P. Anninos, Y. Zhang, M.L. Norman, New Astron. **2**, 181 (1997)
T. Abel, G.L. Bryan, M.L. Norman, Science **295**, 93 (2002)
J. Adamek, R. Durrer, F. Elisa, M. Vonlanthen, J. Cosmol. Astropart. Phys. **6**, 17 (2011)
J. Adams, U.H. Danielsson, D. Grasso, H. Rubinstein, Phys. Lett. B **388**, 253 (1996)
S. Ando, A. Kusenko, arXiv:1005.1924 (2010)
M. Ando, K. Doi, H. Susa, Astrophys. J. **716**, 1566 (2010)
P. Anninos, Y. Zhang, T. Abel, M.L. Norman, New Astron. **2**, 209 (1997)
P. Arnold, G.D. Moore, Phys. Rev. D **73**, 025006 (2006a)
P. Arnold, G.D. Moore, Phys. Rev. D **73**, 025013 (2006b)
P. Arnold, Int. J. Mod. Phys. E **16**, 2555–2594 (2007)
P. Arnold, G.D. Moore, Phys. Rev. D **76**, 045009 (2007)
P. Arnold, P.S. Leang, Phys. Rev. D **76**, 065012 (2007)
R.M. Athreya, V.K. Kapahi, P.J. McCarthy, W. van Breugel, Astron. Astrophys. **329**, 809 (1998)
R. Banerjee, K. Jedamzik, Phys. Rev. D **70**, 123003 (2004)
R. Banerjee, R.E. Pudritz, Astrophys. J. **641**, 949 (2006)
R. Banerjee, R.E. Pudritz, L. Holmes, Mon. Not. R. Astron. Soc. **355**, 248 (2004)
J.D. Barrow, C.G. Tsagas, Phys. Rev. D **77**, 107302 (2008)
J.D. Barrow, C.G. Tsagas, Mon. Not. R. Astron. Soc. **414**, 312 (2011)
J.D. Barrow, P.G. Ferreira, J. Silk, Phys. Rev. Lett. **78**, 3610–3613 (1997)
A. Brandenburg, Phys. Rev. D **54**, 1291 (1996)
I. Brown, R. Crittenden, Phys. Rev. D **72**, 063002 (2005)
M.L. Bernet, F. Miniati, S. J Lilly, P.P. Kronber, M. Dessauges-Zavadsky, Nature **454**, 302 (2008)
S. Bethke, Eur. Phys. J. C **64**, 689–703 (2009)
L. Biermann, Z. Naturforsch. A **5**, 65 (1950)
J.-P. Blaizot, E. Iancu, Phys. Rep. **359**, 355 (2002)
C. Bonvin, C. Caprini, J. Cosmol. Astropart. Phys. **5**, 022 (2010)
V. Bromm, R.B. Larson, Annu. Rev. Astron. Astrophys. **42**, 79 (2004)
R.-G. Cai, B. Hu, H.-B. Zhang, J. Cosmol. Astropart. Phys. **8**, 025 (2010)
C. Caprini, F. Finelli, D. Paoletti, A. Riotto, J. Cosmol. Astropart. Phys. **6**, 021 (2009)
C.L. Carilli, K.M. Menten, Annu. Rev. Astron. Astrophys. **40**, 319 (2002)
B. Cheng, A.V. Olinto, Phys. Rev. D **50**, 2421 (1994)
M. Christensson, M. Hindmarsh, A. Brandenburg, Phys. Rev. E **64**, 056405 (2001)
J.M. Cornwall, Phys. Rev. D **56**, 6146 (1997)
R.A. Daly, A. Loeb, Astrophys. J. **355**, 416 (1990)
G. Davies, L.M. Widrow, Astrophys. J. **540**, 755 (2000)
V. Demozzi, V. Mukhanov, H. Rubinstein, J. Cosmol. Astropart. Phys. **8**, 25 (2009)
C.D. Dermer et al., Astrophys. J. **733L**, 21 (2011)
R. Durrer, L. Hollenstein, R.K. Jain, J. Cosmol. Astropart. Phys. **3**, 37 (2011)
R. Durrer, New Astron. Rev. **51**, 275 (2007)
R. Durrer, C. Caprini, J. Cosmol. Astropart. Phys. **11**, 10 (2003)
R. Durrer, P.G. Ferreira, T. Kahniashvili, Phys. Rev. D **61**, 043001 (2000)
E. Fenu, C. Pitrou, R. Maartens, Mon. Not. R. Astron. Soc. **414**, 580 (2011). doi:10.1111/j.1365-2966.2011.18554.x
G.B. Field, S.M. Carroll, Phys. Rev. D **62**, 103008 (2000)
F. Finelli, F. Paci, D. Paoletti, Phys. Rev. D **78**, 023510 (2008)
B.D. Fried, Phys. Fluids **2**, 337 (1959)
U. Frisch et al., J. Fluid Mech. **68**, 769 (1975)
M. Gasperini, M. Giovannini, G. Veneziano, Phys. Rev. Lett. **75**, 3796 (1995)

M. Giovannini, K.E. Kunze, Phys. Rev. D **77**, 063003 (2008)
S.C.O. Glover, D.W. Savin, Mon. Not. R. Astron. Soc. **393**, 911 (2009)
N.Y. Gnedin, A. Ferrara, E.G. Zweibel, Astrophys. J. **539**, 505 (2000)
D. Grasso, H.R. Rubinstein, Phys. Rep. **248**, 163 (2001)
P.J. Greenberg, Astrophys. J. **164**, 589 (1971)
A.H. Guth, S.-Y. Pi, Phys. Rev. Lett. **49**, 1110 (1982)
S.W. Hawking, Phys. Lett. B **115**, 295 (1982)
P. Hennebelle, S. Fromang, Astron. Astrophys. **477**, 9 (2008)
P. Hennebelle, R. Teyssier, Astron. Astrophys. **477**, 25 (2008)
B. Himmetoglu, C.R. Contaldi, M. Peloso, Phys. Rev. Lett. **102**, 111301 (2009a)
B. Himmetoglu, C.R. Contaldi, M. Peloso, Phys. Rev. D **81**, 063528 (2009b)
C.J. Hogan, Phys. Rev. Lett. **51**, 1488 (1983)
F. Hoyle, in *Proceedings of the Symposium on Motion of Gaseous Masses of Cosmical Dimensions*, ed. by J.M. Burgers, H.C. van de Hulst (Central Air Documents Office, Dayton, 1949), p. 195
F. Hoyle, Nature **223**, 936 (1969)
K. Ichiki, K. Takahashi, H. Ohno, H. Hanayama, N. Sugiyama, Science **311**, 827 (2006)
K. Ichiki, K. Takahashi, N. Sugiyama, H. Hanayama, H. Ohno, astro-ph/0701329v1 (2006)
J.D. Jackson, *Classical Electrodynamics* (Wiley, New York, 1975)
K. Jedamzik, V. Katalinić, A.V. Olinto, Phys. Rev. D **57**, 3264 (1998)
K. Jedamzik, V. Katalinić, A.V. Olinto, Phys. Rev. Lett. **85**, 700 (2000)
T. Kahniashvili, B. Ratra, Phys. Rev. **71**, 103006 (2005)
A.J. Keane, R.K. Barrett, Class. Quantum Gravity **17**, 201 (2000)
K.-T. Kim, P.P. Kronber, G. Giovannini, T. Vernturi, Nature **341**, 720 (1989)
E.-J. Kim, A.V. Olinto, R. Rosner, Astrophys. J. **468**, 28 (1996)
E.W. Kolb, M.S. Turner, *The Early Universe, Frontiers in Physics* (Addison-Wesley, Reading, 1990)
E. Komatsu, K.M. Smith, J. Dunkley, C.L. Bennett, B. Gold, G. Hinshaw, N. Jarosik, D. Larson, M.R. Nolta, L. Page, D.N. Spergel, M. Halpern, R.S. Hill, A. Kogut, M. Limon, S.S. Meyer, N. Odegard, G.S. Tucker, J.L. Weiland, E. Wollack, E.L. Wright, arXiv:1001.4538 (2010)
A. Kosowsky, A. Loeb, Astrophys. J. **469**, 1 (1996)
P.P. Kronberg, Rep. Prog. Phys. **57**, 325 (1994)
P.P. Kronberg, Astrophys. J. **676**, 70 (2008)
P.P. Kronberg, J.P. Perry, E.L.H. Zukowski, Astrophys. J. **387**, 528 (1992)
R.M. Kulsrud, E.G. Zweibel, Rep. Prog. Phys. **71**, 046901 (2008)
R.M. Kulsrud, R. Cen, J.P. Ostriker, D. Ryu, Astrophys. J. **480**, 481 (1997)
M. Langer, J.-L. Puget, N. Aghanim, Phys. Rev. D **67**, 3505 (2003)
R.B. Larson, Mon. Not. R. Astron. Soc. **145**, 271 (1969)
A. Lazarian, Astron. Astrophys. **264**, 326 (1992)
A. Lewis, Phys. Rev. D **70**, 043011 (2004)
D.H. Lyth, A. Woszczyna, Phys. Rev. D **52**, 3338 (1995)
M.N. Machida, K. Omukai, T. Matsumoto, S. Inutsuka, Astrophys. J. Lett. **647**, L1 (2006)
A. Mack, T. Kahniashvili, A. Kosowsky, Phys. Rev. D **65**, 123004 (2002)
H. Maki, S. Hajime, Astrophys. J. **609**, 467 (2004)
J. Martin, J. Yokoyama, J. Cosmol. Astropart. Phys. **1**, 25 (2008)
St. Mrówczyński, Phys. Lett. B **214**, 587 (1988)
St. Mrówczyński, Phys. Lett. B **314**, 118 (1993)
A. Neronov, I. Vovk, Science **328**, 73 (2010)
J.A. Peacock, *Cosmological Physics* (Cambridge University Press, Cambridge, 1999)
P.J.E. Peebles, Astrophys. J. **155**, 393 (1969)
P.J.E. Peebles, *Principles of Physical Cosmology*. Princeton Series in Physics (Princeton University Press, Princeton, 1993)
M.V. Penston, Mon. Not. R. Astron. Soc. **144**, 425 (1969)
C. Pinto, D. Galli, Astron. Astrophys. **484**, 17 (2008)
C. Pinto, D. Galli, F. Bacciotti, Astron. Astrophys. **484**, 1 (2008)
Yu.E. Pokrovsky, A.V. Selikhov, Pis'ma Zh. Eksp. Teor. Fis. **47**, 11 (1988) [English Translation JETPL **47**, 12 (1988)]
R.E. Pudritz, J. Silk, Astrophys. J. **342**, 650 (1989)
J.A. Quashnock, A. Loeb, D. Spergel, Astrophys. J. Lett. **344**, L49 (1989)
B. Ratra, Astrophys. J. **391**, L1 (1992)
A. Rebhan, M. Strickland, M. Attems, Phys. Rev. D **78**, 045023 (2008)
M.J. Rees, Q. J. R. Astron. Soc. **28**, 197 (1987)
M.J. Rees, in *Cosmical Magnetism*, ed. by D. Lynden-Bell (Kluwer Academic, Dordrecht, 1994), p. 155

L.F.S. Rodrigues, R.S. de Souza, R. Opher, Mon. Not. R. Astron. Soc. **406**, 482 (2010)
P. Romatschke, A. Rebhan, Phys. Rev. Lett. **97**, 252301 (2006)
B. Schenke, A. Dumitru, Y. Nara, M. Strickland, J. Phys. G **35**, 104101 (2008)
D.R.G. Schleicher, D. Galli, F. Palla, M. Camenzind, R.S. Klessen, M. Bartelmann, S.C.O. Glover, Astron. Astrophys. **490**, 521 (2008)
D.R.G. Schleicher, R. Banerjee, R.S. Klessen, Astrophys. J. **692**, 236 (2009a)
D.R.G. Schleicher, D. Galli, S.C.O. Glover, R. Banerjee, F. Palla, R. Schneider, R.S. Klessen, Astrophys. J. **703**, 1096 (2009b)
D.R.G. Schleicher, Astron. Astrophys. arXiv:1008.3481 (2010)
S. Seager, D.D. Sasselov, D. Scott, Astrophys. J. Lett. **523**, L1 (1999)
T.R. Seshadri, K. Subramanian, Phys. Rev. Lett. **87**, 101301 (2001)
T.R. Seshadri, K. Subramanian, Phys. Rev. Lett. **103**, 081303 (2009)
S.K. Sethi, K. Subramanian, Mon. Not. R. Astron. Soc. **356**, 778 (2005)
S.K. Sethi, Z. Haiman, K. Pandey, arXiv:1005.2942 (2010)
J.R. Shaw, A. Lewis, Phys. Rev. D **81**, 043517 (2010)
G. Sigl, A.V. Olinto, K. Jedamski, Phys. Rev. D **55**, 4582 (1997)
J. Silk, Astrophys. J. **151**, 431 (1968)
J. Silk, M. Langer, Mon. Not. R. Astron. Soc. **371**, 444 (2006)
D.T. Son, Phys. Rev. D **59**, 063008 (1999)
A.A. Starobinskii, Phys. Lett. B **117**, 175 (1982)
H. Stefani, *Introduction to General Relativity* (Cambridge University Press, Cambridge, 1990)
M.A. Stenrup, A. Larson, N. Elander, Phys. Rev. A **79**, 012713 (2009)
M. Strickland, Braz. J. Phys. **37**, 762 (2007a)
M. Strickland, J. Phys. G, Nucl. Part. Phys. **34**, S429 (2007b)
K. Subramanian, Astron. Nachr. **327**, 333 (2006)
K. Subramanian, Mon. Not. R. Astron. Soc. **294**, 718 (1998)
K. Subramanian, Astron. Nachr. **331**, 110 (2010)
K. Subramanian, J.D. Barrow, Phys. Rev. D **58**, 083502 (1998)
K. Subramanian, J.D. Barrow, Phys. Rev. Lett. **81**, 3575 (1998)
K. Subramanian, D. Narasimha, S.M. Chitre, Mon. Not. R. Astron. Soc. **271**, 15 (1994)
K. Subramanian, T.R. Seshadri, J.D. Barrow, Mon. Not. R. Astron. Soc. **344**, L31 (2003)
S. Sur, et al., Astrophys. J. Lett. **721**, L134 (2010)
S.I. Syrovatskii, in *Interstellar Gas Dynamics*, ed. by H.J. Habbing. IAU Symposium No. 39 (Springer, New York, 1970)
K. Takahashi et al., Mon. Not. R. Astron. Soc. **410**, 247 (2011)
J.C. Tan, E.G. Blackman, Astrophys. J. **603**, 401 (2004)
H. Tashiro, N. Sugiyama, Mon. Not. R. Astron. Soc. **368**, 965 (2006a)
H. Tashiro, N. Sugiyama, Mon. Not. R. Astron. Soc. **372**, 1060 (2006b)
K. Thorne, Astrophys. J. **148**, 51 (1967)
C.G. Tsagas, Class. Quantum Gravity **22**, 393 (2005)
C.G. Tsagas, A. Kandus, Phys. Rev. D **71**, 123506 (2005)
C.G. Tsagas, A. Challinor, R. Maartens, Phys. Rep. **465**, 61 (2008)
M.S. Turner, L.M. Widrow, Phys. Rev. D **37**, 2743 (1988)
T. Vachaspati, Phys. Lett. B **265**, 258 (1991)
E. Weibel, Phys. Rev. Lett. **2**, 83 (1959)
S.D.M. White, Astrophys. J. **286**, 38 (1984)
L.M. Widrow, Rev. Mod. Phys. **74**, 775 (2002)
H. Xu et al., Astrophys. J. Lett. **688**, L57 (2008)
A. Yahil, Astrophys. J. **265**, 1047 (1983)
D.G. Yamazaki, K. Ichiki, T. Kajino, G.J. Mathews, Phys. Rev. D **77**, 043005 (2008)
Ya.B. Zel'dovich, Astron. Astrophys. **5**, 84 (1970)

Space Sci Rev (2012) 166:71–95
DOI 10.1007/s11214-011-9761-4

Magnetic Fields in Cosmic Particle Acceleration Sources

Andrei M. Bykov · Donald C. Ellison · Matthieu Renaud

Received: 30 September 2010 / Accepted: 11 February 2011 / Published online: 18 March 2011

Abstract We review here some magnetic phenomena in astrophysical particle accelerators associated with collisionless shocks in supernova remnants, radio galaxies and clusters of galaxies. A specific feature is that the accelerated particles can play an important role in magnetic field evolution in the objects. In particular, we discuss a number of cosmic-ray (CR) driven, magnetic field amplification processes that are likely to operate when diffusive shock acceleration (DSA) becomes efficient and nonlinear. The turbulent magnetic fields produced by these processes determine the maximum energies of accelerated particles and result in specific features in the observed photon radiation of the sources. Equally important, magnetic field amplification by the CR currents and pressure anisotropies may affect the shocked gas temperatures and compression, both in the shock precursor and in the downstream flow, if the shock is an efficient CR accelerator. Strong fluctuations of the magnetic field on scales above the radiation formation length in the shock vicinity result in intermittent structures observable in synchrotron emission images. The finite size twinkling, intermittent structures—dots, clumps, and filaments—are most apparent in the cut-off region of the synchrotron spectrum. Even though these X-ray synchrotron structures result from turbulent magnetic fields, they could still be highly polarized providing an important diagnostic of the spectrum of the turbulence. We discuss both the thermal and non-thermal observational consequences of magnetic field amplification in supernova remnants and radio-galaxies. Resonant and non-resonant CR streaming instabilities in the shock precursor can generate mesoscale magnetic fields with scale-sizes comparable to supernova remnants and

A.M. Bykov (✉)
Ioffe Institute for Physics and Technology, 194021 St. Petersburg, Russia
e-mail: byk@astro.ioffe.ru

D.C. Ellison
Physics Department, North Carolina State University, Box 8202, Raleigh, NC 27695, USA
e-mail: don_ellison@ncsu.edu

M. Renaud
Laboratoire de Physique Theorique et Astroparticules (LPTA), Universite Montpellier II, Montpellier, France
e-mail: mrenaud@lpta.in2p3.fr

even superbubbles. This opens the possibility that magnetic fields in the earliest galaxies were produced by the first generation Population III supernova remnants and by clustered supernovae in star forming regions.

Keywords Radiation mechanisms: non-thermal · X-rays: ISM · (ISM:) supernova remnants · Clusters of galaxies · Shock waves · Magnetic fields

1 Introduction

Particle acceleration takes place in many active astrophysical objects of very different nature and scales. Magnetic fields play the central role in charged particle acceleration either as a intermediary between the plasma flows and energetic particles, as is the case for Fermi acceleration (e.g., in collisionless shocks), or as a source of free energy to be converted into energetic particles (i.e., magnetic field reconnection processes).

The existence of the highly amplified magnetic fields in the range of 0.1–1 mG in the shells of young supernova remnants (SNRs) was established assuming equipartition between relativistic particles and magnetic fields in the synchrotron radio emitting shells (see e.g., Ginzburg and Syrovatskii 1964, and the references therein). The topology of the magnetic field in SNRs inferred from the observations of synchrotron emission differ strongly between young and old SNRs. Radio polarization studies reveal super-adiabatic magnetic field amplification and a net radial orientation of the magnetic fields in young SNRs, while it is often just shocked interstellar field, mainly tangential, in old SNRs (e.g., Milne 1990). Milne has also pointed out that as the resolution increases, the polarization structure becomes more complex.

A pixel-by-pixel map of Faraday rotation has been produced applying rotation measure (RM) synthesis to the data observed with the *Australia Telescope Compact Array*, for the entire supernova remnant G296.5 + 10.0 by Harvey-Smith et al. (2010). A highly ordered rotation measure structure, with an anti-symmetric rotation measure pattern, was observed. The authors propose that the observed rotation measures are the imprint of an azimuthal magnetic field in the stellar wind of the progenitor star. A swept-up magnetized wind from a red supergiant can produce an azimuthal pattern of the magnetic field at large distances from the star and can naturally produce the observed anti-symmetric RM pattern. Supernova expansion into such a wind could account for the apparent bilateral structure of the SNR's radio and X-ray morphologies. In the case of SN1006, a comparison between observed and synthesized radio maps, making different assumptions about the dependence of electron injection efficiency on the shock obliquity, allowed Petruk et al. (2009) to constrain possible nonthermal electron injection models.

A number of possible amplification mechanisms were considered. Rayleigh-Taylor instabilities in the shell have been considered for many years as a potential source of the turbulent magnetic fields. The Rayleigh-Taylor instability at the interface of the ejecta and the shocked ambient medium was proposed to explain these observations. Jun and Norman (1996) performed multi-dimensional MHD simulations of the instability in the shell and its effect on the local magnetic field. They found that the evolution of the instability is very sensitive to the deceleration of the ejecta and the evolutionary stage of the remnant. The Rayleigh-Taylor and Kelvin-Helmholtz instabilities amplify ambient magnetic fields in the simulations locally by as much as a factor of 60 around dense fingers due to stretching, winding, and compression. Globally, the amount of magnetic-field amplification was nevertheless low and the magnetic energy density reaches only about 0.3% of the turbulent energy density at the end of simulation. Strong magnetic field lines draped around the fingers produce the radial B-vector polarization. Magnetic Rayleigh-Taylor instabilities in three dimensions were simulated by Stone and Gardiner (2007), with a focus on the nonlinear structure and evolution that results from different initial field configurations. They found that strong magnetic fields do not suppress the instability but reduce mixing between the heavy and light fluids and cause the rate of growth of bubbles and fingers to increase in comparison to hydrodynamics. Schure et al. (2009) suggested that higher compression ratios and additional turbulence due to the dominant presence of CRs can be important in attempts to reproduce the observed magnetic field morphology with the Rayleigh-Taylor instability simulations. Both numerical simulations and laboratory laser experiments are ongoing to study the Rayleigh-Taylor instability problem. The possible effects of magnetic fields on laser experiments of Rayleigh-Taylor instabilities in supernova-like setups was discussed recently by Fryxell et al. (2010). Comparing numerical simulations, as well as experimentally using the Omega Laser at the University of Rochester, the authors discussed a possibility that the morphology of the instability is significantly altered by the generation of very strong magnetic fields during the laser experiments.

On the other hand, Chevalier (1977) shown that radio observations of Tycho's supernova remnant indicated the presence of a collisionless shock wave undergoing turbulent magnetic field amplification by a factor of about 20. He noted that the amplification resembled some phenomena in heliospheric collisionless shocks. In fact, direct support for the self-generation of MHD turbulence has been seen in heliospheric shocks, e.g., Bamert et al. (2004, 2008). In the upstream region of a shock associated with the Bastille Day coronal mass ejection, the excitation of hydromagnetic waves with power spectral density levels well above the levels in the ambient solar wind were observed by Bamert et al. (2004) with the magnetometer on board the *Advanced Composition Explorer*. Achterberg et al. (1994) set upper limits on the scattering mean free paths of radio emitting electrons in front of supernova remnant shock fronts using high-resolution radio observations of four Galactic supernova remnants. They found that, for the sharpest synchrotron radio rims, the mean free path is typically less than one percent of that derived for cosmic rays of similar rigidity in the interstellar medium, implying that enhanced hydromagnetic wave intensity is generated by diffusive shock acceleration.

Recent observations of X-ray synchrotron radiation from several young SNRs revealed thin X-ray filaments in the vicinity of blast waves. This provides indirect, but rather strong evidence for strong super-adiabatic magnetic field amplification (MFA) in the blast wave associated with CR production (see e.g., Vink and Laming 2003; Bamba et al. 2005, 2006; Uchiyama et al. 2007; Patnaude and Fesen 2009; Reynolds 2008; Vink 2008). The synchrotron filamentary structures could be due to a narrow spatial extension of the highest energy electron population in the shock acceleration region limited by rapid synchrotron energy losses. In this case, strong magnetic field amplification in the shock vicinity is required to match the observed thickness of the non-thermal filaments (e.g., Vink and Laming 2003; Bamba et al. 2005), thus supporting the case for highly efficient DSA. An alternative interpretation, that the observed narrow filaments are limited by magnetic field damping and not by the energy losses of the radiating electrons, has also been proposed (see Pohl et al. 2005; Butt et al. 2008, for example). Cassam-Chenaï et al. (2007, 2008) have studied this effect in detail for Tycho's SNR and SN1006.

Diffusive shock acceleration is a very promising mechanism for producing superthermal and relativistic particles in a wide range of astrophysical objects ranging from the Earth bow shock (e.g., Ellison et al. 1990) to shocks in galaxy clusters (Blandford and Eichler 1987; Jones and Ellison 1991; Malkov and Drury 2001; Bykov et al. 2008a). This mechanism is believed to be efficient (e.g., Helder et al. 2009) and capable of producing CRs to energies well above 10^{15} eV in young SNRs (e.g., Ptuskin et al. 2010), and even higher in active radio-galaxies such as Cen A (see, for example, Croston et al. 2009).

Fast and efficient DSA demands that particles are multiply scattered by magnetic fluctuations in the shock vicinity. The amplitude of the required MHD turbulence is substantially higher than the ambient MHD turbulence suggesting that the turbulence is generated by the shock. In order for DSA to accelerate particles to high energies, the energetic particles must be able to interact with magnetic turbulence over a broad wavelength range. The weakly anisotropic distribution of accelerated particles, is believed capable of producing this turbulence in a symbiotic relationship where the magnetic turbulence required to accelerate the CRs is created by the accelerated CRs themselves. In efficient DSA, this wave-particle interaction can be strongly nonlinear where CRs modify the plasma flow and the specific mechanisms of magnetic field amplification.

2 Magnetic Field Amplification by CR-driven Instabilities

The non-equilibrium distributions of accelerated particles in efficient DSA are believed capable of producing magnetic turbulence and magnetic field amplification. For the forward shock in a young, isolated SNR, the accelerated CR pressure can be a sizeable fraction of the shock ram pressure and far exceed the background plasma pressure. The anisotropic distribution of CRs in the shock precursor results in ponderomotive forces on the background plasma that can cause the amplification of magnetic fluctuations in a certain wavevector range and polarization. We briefly discuss in Sect. 2.1 the kinetic background for the description of this CR anisotropy and then outline a number of instabilities related to the anisotropic CR distributions.

2.1 Anisotropic CR Distribution in DSA

First-order Fermi acceleration (also called diffusive shock acceleration) of CRs by non-relativistic magnetized flows, is characterized by a nearly isotropic CR momentum distribution in a broad energy range. Two notable exceptions to this are injected, barely superthermal particles, and the highest energies CRs. The CR distribution is formed by the multiple scattering of charged particles across the shock by magnetic fluctuations over a wide dynamic range. Cosmic-ray particle dynamics in DSA is governed by electromagnetic fields both regular and stochastic. The distribution function of energetic particles, $f(\mathbf{r},\mathbf{p},t) = \langle F(\mathbf{r},\mathbf{p},t)\rangle_c$, averaged over electro-magnetic fluctuations with scales below some value l_c, appropriate for a particular problem, satisfies the kinetic equation

$$\frac{\partial f}{\partial t} + \mathbf{v}\cdot\frac{\partial f}{\partial \mathbf{r}} + e\mathbf{E}\cdot\frac{\partial f}{\partial \mathbf{p}} - \frac{ec}{\mathcal{E}}\mathbf{B}\cdot\widehat{\mathcal{O}}f = \widehat{I}_c[f], \tag{1}$$

where $\widehat{\mathcal{O}}$ is the momentum rotation operator, defined by

$$\widehat{\mathcal{O}} = \mathbf{p}\times\frac{\partial}{\partial \mathbf{p}}, \tag{2}$$

$\mathcal{E}$ is the particle energy, and $\mathbf{E} = \langle \mathbf{E}_0\rangle_c$ and $\mathbf{B} = \langle \mathbf{B}_0\rangle_c$ are the electric and magnetic fields averaged over electro-magnetic fluctuations with scales below l_c. The fluctuating parts of the electromagnetic fields are $\mathbf{b} = \mathbf{B}_0 - \langle \mathbf{B}_0\rangle_c$ and $\mathbf{e} = \mathbf{E}_0 - \langle \mathbf{E}_0\rangle_c$. The fields are self-consistent since the sources in Maxwell's equations include the CR currents determined by $F(\mathbf{r},\mathbf{p},t)$.

We should stress here that it is not trivial to get a closed system of equations for the averaged distribution function $f(\mathbf{r},\mathbf{p},t) = \langle F(\mathbf{r},\mathbf{p},t)\rangle_c$ since the higher order moments of F may contribute through $\langle \mathbf{b}(\mathbf{r},t)F(\mathbf{r},\mathbf{p},t)\rangle_c$ and $\langle \mathbf{e}(\mathbf{r},t)F(\mathbf{r},\mathbf{p},t)\rangle_c$. In fact, even the existence of a well defined averaging procedure, and the averages themselves, is assumed rather than proved.

The collision operator

$$\widehat{I}_c[f] = -Ze\left\langle \mathbf{e}\cdot\frac{\partial F}{\partial \mathbf{p}} - \frac{c}{\mathcal{E}}\mathbf{b}\cdot\widehat{\mathcal{O}}F\right\rangle_c, \tag{3}$$

is, in general, a nonlinear operator if the effect of energetic particles on the electromagnetic fields is non-negligible.

Different approaches are used to get a closed analytic expression for $\widehat{I}_c[f]$ including Fokker-Planck type equations and quasi-linear theory (see e.g., Toptygin 1985; Blandford and Eichler 1987; Berezinskii et al. 1990; Schlickeiser 2002). The simplest case for the collision operator can be deduced from (3) for particles scattered by magnetic fluctuations with scales, l_c, smaller than the particle gyroradius, $r_g = cp/eB$, in the total magnetic field B (random plus coherent). In the high-energy limit, $r_g \gg l_c$, the particle momentum change over the field correlation length l_c is small ($\propto l_c/r_g$) and the scattering rate in (3) reduces to $v/2\Lambda(p)\widehat{\mathcal{O}}^2$ with $\Lambda(p) = g\,R^2_{\rm st}(p)/l_c$. Here, $R_{\rm st}(p)$ is the particle gyroradius in the stochastic magnetic field, and g is a numerical factor depending on the short-scale fluctuation spectral index. The strong momentum dependence, $\Lambda \propto p^2$, makes this scenario unfavorable for confining and accelerating high-energy particles in DSA.

It is important to note that $\widehat{\mathcal{O}}^2$ can be reduced to the angular part of the Laplace operator with well-known spherical harmonics as the eigenfunctions. The high-energy asymptotic range is valid even in the case of strong (but quasi-static) magnetic fluctuations of amplitudes larger than the mean magnetic field. However, for particles of intermediate and low energies, with $r_g \lesssim l_c$ and scattered by strong magnetic fluctuations, the collision operator is no longer simple and only some phenomenological scheme like the *relaxation time approximation* can be used. These schemes do not have a microscopic justification.

The momentum distribution of accelerated particles, $f(\mathbf{r},\mathbf{p},t)$, of a weakly anisotropic distribution can be presented as

$$f(\mathbf{p}) = \frac{n_{\rm cr}}{4\pi}N(p)[1+\delta_{\rm a}(\mathbf{p})], \tag{4}$$

where $n_{\rm cr}$ is the CR number density, $N(p)$ is the isotropic and $\delta_{\rm a}(\mathbf{p})$ is the anisotropic part of the distribution function defined to satisfy

$$\int \delta_{\rm a}(\mathbf{p})\,d\Omega_{\mathbf{p}} = 0, \qquad \int f(\mathbf{p})p^2\,dp\,d\Omega_{\mathbf{p}} = n_{\rm cr}.$$

Due to CR particle scattering by magnetic fluctuations of different scales, the distribution function is nearly isotropic on scales larger than the CR mean free path, $\Lambda(p)$, and, therefore, its anisotropic part is small, i.e., $\delta_{\rm a}(\mathbf{p}) \ll 1$ on scales $k\Lambda(p) < 1$. The actual angular dependence of $\delta_{\rm a}(\mathbf{p})$, and the momentum dependence of $\Lambda(p)$, are determined by the exact form of the collision operator, the structure of regular fields, and the boundary conditions. It can be presented in the spherical harmonic expansion

$$\delta_{\rm a}(\mathbf{p}) = \sum_{l,m}\Delta_{l,m}(p)\,Y_{l,m}(\vartheta,\varphi), \quad |m| \le l, \tag{5}$$

where ϑ and φ are the pitch and azimuthal angles of the particle momentum, correspondingly. Since the isotropic part of the distribution function is separated from the anisotropic part in (4), $\Delta_{0,0} = 0$. The current-type anisotropy that is determined by $\Delta_{1,m} < 1$, is usually considered to be the dominant term. That results in the standard diffusion approximation and the transport equation for the isotropic part $N(\mathbf{r}, p, t)$. In the case of a high-energy particle scattering off of small-scale fluctuations, $r_g \gg l_c$, the spherical functions are the eigenfunctions of the collision operator since $\widehat{\mathcal{O}}^2 Y_{l,m} = -l(l+1)Y_{l,m}$. In that case, indeed, $\Delta_{2,m} \sim \Delta_{1,m}^2$. However, the high multipole components of the distribution can be substantial in the case of anisotropic scattering, or in plasma flows with a complex structure of quasi-regular magnetic fields (e.g., in the presence of regular or stochastic magnetic traps). In the latter case one can expect $\Delta_{2,m} \sim \Delta_{1,m}$ and the transport equation for the isotropic part of the distribution function $N(\mathbf{r}, p, t)$ can differ from the standard advection-diffusion equation that is used in the analytic DSA models.

Standard analytic DSA models use a simplified description of particle diffusion. Generally, the diffusion approximation is made where it is assumed that a small anisotropy of the momentum distribution can be completely described by the CR current and higher moments of the momentum angular distribution can be neglected (see e.g., Blandford and Eichler 1987; Jones and Ellison 1991; Malkov and Drury 2001). Then, further assuming that magnetic fields are frozen-in the plasma, the kinetic (1) can be reduced to a transport equation for the isotropic part of the distribution function $N(\mathbf{r}, p, t)$. This approach can account for particle advection, diffusion and acceleration by the first-order Fermi mechanism in non-relativistic shocks. The diffusion coefficient in these analytical models is parameterized as $D(p) \propto r_g^{\mathrm{a}}$, where the numerical index is generally in the range $0 \le \mathrm{a} \le 2$. The widely used phenomenological *Bohm diffusion model* assumes $\mathrm{a} = 1$, i.e., the mean free path $\Lambda(p) = \eta\, r_g(p)$, where the factor $\eta \ge 1$.

The CR current and anisotropy in the rest frame of a non-relativistic shock is $|\Delta_{1,m}| \sim v_s/c \ll 1$, where v_s is the shock velocity. The CR anisotropy can also be parameterized by the CR drift velocity v_{drift} relative to the background plasma in the rest frame of the upstream plasma. Various parameterizations of the CR current in DSA were used in recent studies of magnetic field amplification mechanisms by (e.g., Bell 2004; Pelletier et al. 2006; Amato and Blasi 2009; Luo and Melrose 2009) based on the CR transport equation. Monte Carlo simulations of DSA provide an alternative approach to the kinetic theory. In Monte Carlo techniques, particle scattering rates are parameterized as a function of energy and then the CR current (and other CR anisotropy moments) are derived in nonlinear shock models accounting for the CR backreaction on the plasma flow (e.g., Ellison et al. 1996; Vladimirov et al. 2006, 2008).

2.2 Resonant CR-streaming Instability

Resonant interactions have long been known to amplify magnetic fluctuations on the scale of the CR gyroradius (e.g., Wentzel 1974; Zweibel 2003; Kulsrud 2005). The simplest case of interest for DSA is Alfvén wave amplification by streaming CRs. If the streaming speed of the cosmic rays through the ambient medium exceeds the Alfvén speed, then the amplitude of resonant Alfvén waves grows exponentially with time. The waves are emitted by the CR particles gyrating with gyrofrequency Ω as coherent cyclotron radiation. The resonance condition is $k\, v \cos(\vartheta) = \Omega$. This resonance condition is typical for DSA. The growth rate is positive if, in the wave frame, the particles stream in the opposite direction to the motion of the background fluid. Equivalently, the waves grow if, in the fluid frame, the mean streaming velocity of the particles exceeds the wave velocity, i.e., $v_{\rm drift} > v_{\rm ph}$. The growth rate estimated by Blandford and Eichler (1987) is

$$\gamma^{\rm res} \approx \Omega\left[\frac{v_{\rm drift}}{v_{\rm ph}} - 1\right]\frac{n_{\rm cr}}{n_{\rm a}}, \tag{6}$$

where $n_{\rm cr}$ is the total number density of resonant cosmic rays and $n_{\rm a}$ is the background ion density.

2.3 Non-resonant Streaming Instabilities: Short-scale

In addition to resonant amplification, the CR current can also efficiently amplify non-resonant magnetic fluctuations with scales smaller than the CR gyroradius (Bell 2004). The following two points are important to produce strong, short-scale turbulence.

(i) The cosmic-ray current has only a weak response to the fluctuations with scales shorter than the gyroradii, r_{g0}, of CR particles dominating the CR current (short-scale fluctuations), i.e., $kr_{g0} \gg 1$ (see discussion in Sect. 2.1). In this case, the electric current of accelerated particles $\mathbf{J}^{\rm cr}$ is an external current in the momentum equation of the background plasma. The current is governed by sources of energetic particles and by local electromagnetic fields. The current $\mathbf{J}^{\rm cr}$ initiates a compensating return current in the background plasma.

(ii) The ponderomotive force of the CR current, $\mathbf{J}^{\rm cr} \times \mathbf{B}/c$, is large enough to dominate the magnetic field tension in the momentum equation of the background plasma. Then the $\mathbf{J}^{\rm cr} \times \mathbf{B}/c$ force from the CR current drives the Bell short-scale instability at scales below the CR gyroradius. This condition is satisfied for scales $k < k_1$ where

$$k_1 = \frac{4\pi}{c}\frac{J_0^{\rm cr}}{B_0}. \tag{7}$$

Note that the condition for mode growth, $r_{g0}k > 1$, together with (7), implies that the CR current that determines k_1 must be large enough to drive the non-resonant instability. The anisotropy of the relativistic particle distribution, $|\Delta_{1,0}|$, should exceed the ratio of the mean magnetic field energy density to the energetic particle energy density $E_{\rm cr}$. That is

$$|\Delta_{1,0}| > \frac{B^2}{4\pi E_{\rm cr}}, \tag{8}$$

where the CR current is given by $J_0^{\rm cr} \approx |\Delta_{1,0}|en_{cr}c$.

Bell (2004, 2005) discovered that the system is unstable against linear perturbations that are $\propto \exp(\gamma t + i\mathbf{kr})$ and the return current drives nearly purely growing modes. Here, γ is the linear growth rate.

In a cold plasma with sound speed a_0, well below the Alfvén velocity v_a, the linear growth rate obtained by Bell (2005) depends only on the wavevector projection k_z on the local mean magnetic field, i.e.,

$$\gamma \approx \gamma_{\rm max}k_z/k, \tag{9}$$

where

$$\gamma_{\rm max} = v_a\sqrt{k_1|k| - k^2}\,. \tag{10}$$

According to the linear analysis of Bell (2005), the wavenumber of a growing mode must satisfy the condition $r_{g0}^{-1} < k < k_1$. Therefore, the instability growth rate is higher than the

Alfvén frequency $v_a k$. The amplitudes of the growing mode in the linear approximation can be expressed as

$$|\mathbf{v}(\mathbf{k})|^2 \approx \frac{1}{4\pi\rho}\frac{k_1}{|k_z|}|\mathbf{b}(\mathbf{k})|^2, \tag{11}$$

provided that the kinetic energy density in the growing mode dominates over the magnetic energy density because $k_1 > k_z$. This is in contrast to Alfvén modes where the energy densities are equal.

The cold plasma approximation used in the analysis of Bell turbulence is valid if

$$\left(\frac{v_a}{v_{Ti}}\right)^2 > k_1 r_{g0}\frac{v_a}{c}, \tag{12}$$

where v_{Ti} is the thermal ion velocity. The approximation is typically good for galactic SNRs in a warm interstellar medium where the plasma parameter $\beta = a_0^2/v_a^2 \sim 1.0$. In the case of DSA by large-scale shocks in superbubbles (see e.g., Bykov and Toptygin 1993, 2001), or in hot intercluster plasma, the thermal corrections to the wave dispersion relations can be essential. A thorough discussion of the effects of a hot plasma on the short-scale modes was given by Zweibel and Everett (2010).

Particle-in-cell simulations by Riquelme and Spitkovsky (2009, 2010) showed that the back-reaction of the amplified field on CRs would limit the amplification factor of the Bell instability to less than a factor of about 10 in the upstream flows of galactic SNRs. The authors studied the possibility of further amplification driven near shocks by magnetized CRs, whose Larmor radii are smaller than the length scale of the field that was previously amplified by the Bell instability in the upstream flow. They found that additional amplification can occur due to the CR current perpendicular to the field. The maximum amplification of the instability is determined by the disruption of the CR current, which happens when CR Larmor radii in the amplified field become comparable to the length scale of the instability. Amplification factors up to $\sim$45 were estimated in that case.

An important feature of the instability driven by the perpendicular current established by Riquelme and Spitkovsky (2010) is the characteristic dependence of the amplified field on the shock velocity, $B^2 \propto v_s^2$, which contrasts with the Bell instability acting alone where $B^2 \propto v_s^3$. Different scalings can, in principle, be constrained in statistical studies of radio SNRs, as done by Bandiera and Petruk (2010), but the results are sensitive to other parameters such as ambient density. It is also possible to constrain these scalings by comparing radio and X-ray synchrotron lightcurves of individual young supernova shells. In an amplified magnetic field, the synchrotron lightcurve will differ from the radio if strong magnetic field damping behind the shock occurs.

2.4 Non-resonant Streaming Instabilities: Long-wavelength

The nonresonant, short-scale instability introduced by Bell (2004), and discussed in Sect. 2.3, is fast and can strongly amplify short-scale magnetic field fluctuations. These strong, short-scale fluctuations are important for the dynamics of CR modified shocks and their emission properties, however, to study the maximum energies of CR particles in DSA one needs to study the amplification mechanisms for long-wavelength fluctuations of scales $r_{g0}k < 1$. Specifically, what are the consequences of long-wavelength fluctuations propagating in highly turbulent plasma with much shorter scale fluctuations? Also, contrary to the short-scale case, the response of the magnetized CR current on magnetic field fluctuations is not small in the limit $r_{g0}k < 1$.

Using a multi-scale, quasi-linear analysis, Bykov et al. (2011) showed that the presence of turbulence with scales shorter than the CR gyroradius enhances the growth of modes with scales longer than the gyroradius, at least for particular polarizations. They used a mean-field approach to average the equation of motion and the induction equation over the ensemble of magnetic field oscillations accounting for the anisotropy of relativistic particles on the background plasma. Bykov et al. (2011) derived the response of the magnetized CR current on magnetic field fluctuations and showed that, in the presence of short-scale Bell-type turbulence, long wavelength modes are amplified. In general, the growth rates depend on the mode propagation angle. For modes propagating parallel to the initial magnetic field, the growth rate is

$$\gamma^{lw}(k) \approx \sqrt{\frac{\pi N_B}{2}} \sqrt{\frac{kk_0}{\eta}} v_a, \tag{13}$$

and these modes have the fastest growth rates for the Bohm diffusion regime with $\eta \sim 1$.

The propagation angle $\theta_{\max}$ of the mode of maximum growth for $\eta > 1$ is

$$\cos\theta_{\max} = 1/\eta, \tag{14}$$

and the maximum growth rate at $\eta \gg 1$ is determined by

$$\gamma(k) \approx \sqrt{\frac{\pi N_B}{4}} \sqrt{kk_0} v_a. \tag{15}$$

The above results were obtained assuming $v_{\mathrm{ph}}/v_s \ll 1$. In the case under consideration $\gamma(k) \sim \omega(k)$, and therefore from (15) one may get the validity condition in the form $\sqrt{\pi N_B/4}\sqrt{k_0/k}\, M_{\mathrm{a}}^{-1} \ll 1$, where M_{a} is the Alfvén Mach number of the shock. The angular dependence of the growth rate in the hydrodynamical regime depends on the dimensionless collision strength η, as is shown in Fig. 1 of Bykov et al. (2011). The growth rates of resonant, non-resonant Bell's, and non-resonant long wavelength instabilities are illustrated in Fig. 2 taken from Bykov et al. (2011) where the polarization, helicity, and angular dependence of the growth rates are calculated for obliquely propagating modes for wavelengths both below and above the CR mean free path. The long-wavelength growth rates estimated for typical supernova remnant parameters are sufficiently fast to suggest a fundamental increase in the maximum CR energy a given shock can produce. It should be noted that the phase velocity of the long-wavelength modes is a growing function of the wavelength and it is larger than the Alfvén speed. That may steepen the particle spectra at the highest energy end by reducing the effective shock compression at high energies. Also, the high phase velocity of the modes could increase the role of Fermi-II acceleration.

An important point is that the short-scale turbulence can influence the large-scale dynamics through the ponderomotive forces imposed on the plasma by the turbulence and the CR current. To derive the mean ponderomotive force one must average the momentum equation over the ensemble of short-scale fluctuations. When this is done, it is seen that the ponderomotive force due to the CR current response may result in a long-wavelength instability in a way somewhat similar to Bell's instability.

2.5 Non-resonant Firehose Instability

The two CR streaming instabilities, resonant and non-resonant, despite being rather different physical mechanisms, both rely on the non-zero, first moment of the CR distribution, $\Delta_{1,m}$, i.e., the CR current. The anisotropic CR distributions, however, have non-vanishing higher order harmonics, $\Delta_{l,m}$. The well-known example of the instability due to the high-order anisotropy is the weakly magnetized thermal background plasma where the gas pressure anisotropy can be large. The systems we consider are known to be unstable if the anisotropy is large enough providing $P_{\parallel} > P_{\perp} + B^2/4\pi$ (firehose instability) or $P_{\perp} > P_{\parallel} + B^2/4\pi$ (mirror instability) (see e.g., Treumann and Baumjohann 1997, and references therein), and can be important in astrophysical shocks (see, for example, Blandford and Eichler (1987, 2007)). Since the growth rate is $\propto k[(P_{\parallel} - P_{\perp} - B^2/4\pi)/\rho]^{1/2}$, the instability operates in hot, high β plasmas.

In a strong supernova shock, the far upstream plasma is cold unless the SNR is located in a hot superbubble. Then, the free energy of the background plasma pressure anisotropy is not an efficient driver of magnetic field amplification. However, the CR pressure in the shock precursor close to the shock is expected to reach a sizeable fraction of the shock ram pressure, i.e., $P^{\rm CR} = \zeta\rho_a v_s^2/2$. In this case, the anisotropy of the CR pressure defined by $\Delta_{2,m}$ should play a role. In the hydrodynamical regime, where the wavelength $2\pi/k$ is longer than the CR mean free path, Λ, the growth rate of the firehose instability can reach

$$\gamma^{fh}(k) \approx \sqrt{\zeta\,|\Delta_{2,0}|/2}\; v_s\, k \sim \sqrt{\zeta/2}\frac{v_s^2 k}{c}. \tag{16}$$

In this estimate, we assumed that $\Delta_{2,0} \propto \Delta_{1,0}^2$. However, as we have pointed out earlier, in some plasma flows with a complex structure of quasi-regular magnetic fields, or in the flows with regular or stochastic magnetic traps, one can expect much larger pressure anisotropy with $\Delta_{2,0} \propto \Delta_{1,0}$ and, therefore, higher growth rates. The mirror instability would occur with a growth rate similar to (16) if the perpendicular component of CR pressure dominates over the parallel pressure.

2.6 CR-pressure Gradient Driven Acoustic Instability

Drury and collaborators (Dorfi and Drury 1985; Drury and Falle 1986) showed that acoustic perturbations in a two-fluid plasma with a CR pressure gradient can grow if the scale of the CR gradient is shorter than D/a_0, where a_0 is the thermal gas sound speed and D is the diffusion coefficient (see also Chalov 1988; Zank et al. 1990; Malkov and Drury 2001; Malkov et al. 2010). It is argued by Malkov and Diamond (2009) that this acoustic instability can be important for DSA. The growth rate of the acoustic instability can be estimated as

$$\gamma^{\rm ac} \approx \frac{|\nabla P^{CR}|}{2\rho_a\, a_0} \tag{17}$$

for $k > \gamma^{\rm ac}/a_0$ (see, e.g., Malkov et al. 2010, for a recent discussion). Since $|\nabla P^{CR}| \sim \zeta v_s^3/(\Lambda c)$, then $k_{\rm min} \sim \zeta(M_s^2\, v_s/c)/\Lambda$. Thus, the long wavelength fluctuations of $k\Lambda < 1$ that are the most important to increase the maximum energies of CRs accelerated by DSA, can grow only if $\zeta(M_s^2\, v_s/c) < 1$. This condition is rather restrictive for young SNRs in the warm interstellar medium. On the other hand, the short-scale fluctuations amplified by the instability are important for heating the plasma upstream of the shock and can contribute to the ponderomotive force discussed in Sect. 2.4.

Beresnyak et al. (2009) proposed a DSA model in which stochastic magnetic fields in the shock precursor are amplified through small-scale dynamo effects. The solenoidal velocity perturbations that are required for the dynamo to work are produced in the model through the interactions of the pressure gradient of the CR precursor and the density perturbations in the inflowing fluid.

2.7 Instabilities in Partially-ionized Plasmas

The H_α emission from SN1006 and some other SNRs implies the presence of a sizable fraction of partially-ionized plasma in the circumstellar medium. Using a kinetic description of CRs, combined with a fluid description of the background plasma, Bykov and Toptygin (2005) and Reville et al. (2007) investigated the growth of hydro-magnetic waves driven by CR streaming in the partially-ionized precursor environment of a supernova remnant shock. In this case, modifications of Ohm's law in the magnetized plasma can be important for the instability, as can ion-neutral damping. If the cosmic ray acceleration is efficient, large neutral fractions are required to compensate the growth of the non-resonant mode. For modest acceleration rates, the ion-neutral damping can dominate over the instability of weakly driven modes even at modest ionization fractions. In the case of a supernova shock interacting with a molecular cloud, such as in IC 443, W28, W44, or W51C, the low ionization in the ambient medium could limit the maximum energy of accelerated particles since this is largely determined by the upstream magnetic turbulence (e.g., Bykov et al. 2000). Observationally, SNRs interacting with molecular clouds are indeed found to be sources of high-energy gamma-ray emission (see e.g., Albert et al. 2007; Aharonian et al. 2008; Abdo et al. 2010; Castro and Slane 2010). This emission is likely due to some combination of pion-decay and relativistic bremsstrahlung (e.g., Bykov et al. 2000; Uchiyama et al. 2010).

Charge exchange reactions in the downstream region, from neutrals crossing the shock and interacting, could result in a fast, cold ion beam. Ohira et al. (2009) presented a linear analysis of collisionless plasma instabilities between the cold beam and the hot downstream plasma. They found that, under SNR conditions, either the resonant instability or the Weibel-type instability are growing. The authors concluded that the mechanism may amplify the downstream magnetic field to more than 100 μG, changing the shock structure and the synchrotron spectra and profiles. However, as we have pointed out *the turbulence amplification in the shock precursor* is the main factor determining the maximum CR energy and the efficiency of energy conversion to accelerated particles.

3 Magnetic Field Amplification in Diffusive Shock Acceleration

The plasma instabilities in particle acceleration sources discussed above may result in magnetic turbulence and strong magnetic field amplification. The high efficiency of the acceleration means that nonlinear models are needed to predict the statistical characteristics of the amplified magnetic fields and their observational appearances. Diffusive shock acceleration models in supernova remnants imply wide dynamic ranges for the fluctuations and the particle momentum spectra extending for more than four decades. These wide dynamic ranges present a severe problem for the direct use of particle-in cell plasma simulations and some simplifications are required.

The nonlinear modeling of the short-scale Bell instability was done using MHD simulations (e.g., Bell 2004; Zirakashvili and Ptuskin 2008; Zirakashvili et al. 2008) assuming a fixed CR current as an external parameter. The MHD approach is justified since the CR current has only a weak response to the short-scale fluctuations. These models studied the spectral evolution in the short-scale range as well as the transformation of the turbulence through the subshock. The evolution of the Bell modes downstream from the shock was addressed by Pelletier et al. (2006), Marcowith et al. (2006), and Marcowith and Casse (2010).

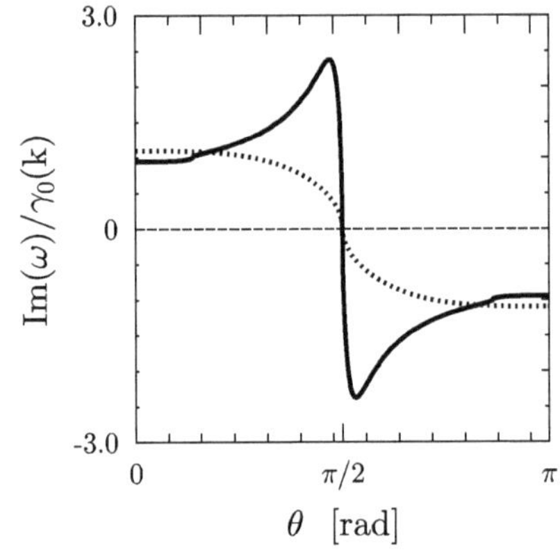

Fig. 1 The angular dependence of the growth rates of the long-wavelength unstable modes in the hydrodyamical regime where $kr_{g0} < \eta^{-1}$, for $\eta = 10$. The angle θ is between the wavevector and the CR current in the upstream rest frame of a parallel shock. The direction of the magnetic field determines the polarization (helicity) of the modes. The model parameters are $k_1 r_{g0} = 100$, the particle distribution index $\alpha = 4.0$, and the plasma parameter $\beta = a_0^2/v_a^2 = 1.0$. The *solid* and *dot-dashed curves* show the two unstable modes. The normalizing parameter γ_0 is determined by (13). See Bykov et al. (2011) for a comparison with the Bohm diffusion case $\eta = 1$

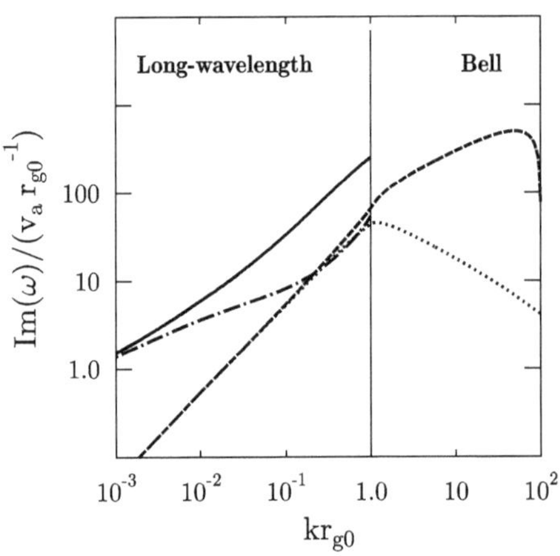

Fig. 2 The figure, from Bykov et al. (2011), shows growth rates of the parallel propagating modes as a function of the wavenumber to illustrate the effect of short-scale turbulence on the long-wavelength instability. The model parameters are $k_0 r_{g0} = 100$ and $\alpha = 4.0$. The *solid* and *dot-dashed curves* are simulated for two modes in the model with short-scale turbulence of $\xi = 5$ and $\eta = 10$ to demonstrate the behavior of the modes in the intermediate regime. For comparison, the dashed and dotted curves are calculated for the model without the short-scale turbulence, i.e., with $N_B = 0$ and $\eta \to \infty$ (cf., Bell 2004)

A nonlinear Monte Carlo model of strong forward shocks in young supernova remnants with efficient particle acceleration, where a nonresonant instability driven by the CR current amplifies magnetic turbulence in the shock precursor, was developed by Vladimirov et al. (2009). Particle injection, magnetic field amplification, and the nonlinear feedback of particles and fields on the bulk flow were derived consistently in the Monte Carlo model as presented by Vladimirov et al. (2006, 2008). It was found that the shock structure de-

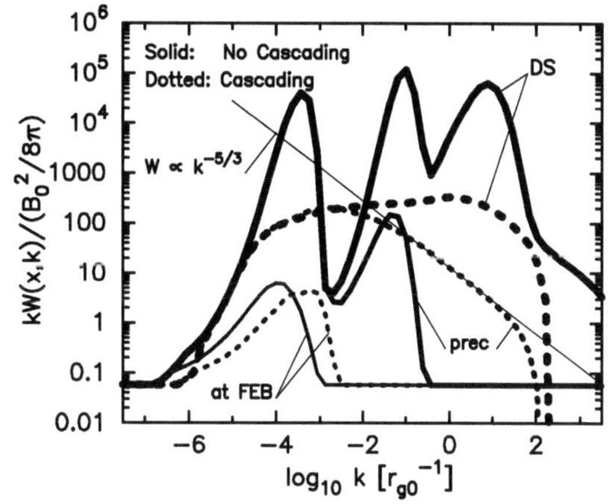

Fig. 3 Turbulence spectra, kW, at different locations relative to the subshock position as simulated by Vladimirov et al. (2009). The far upstream seed turbulence (at all k) is at the level indicated by the horizontal lines

pends critically on the efficiency of turbulence cascading. If cascading is suppressed, MFA is strong, the shock precursor is stratified, and the turbulence spectrum contains several discrete peaks. In Fig. 3, simulated turbulence spectra, kW, are shown at different locations relative to the subshock position.

In the solid curve, cascading is fully suppressed. In the dotted curve, the cascading from large to small scales is efficient and has the differential form corresponding to Kolmogorov's model. In the case with no cascading, the dissipation was assumed to be zero. With cascading, viscous dissipation was assumed. The peaks, apparent in the model with suppressed cascading, should influence synchrotron X-ray images of SNR shells, allowing observational tests of cascading and other assumptions intrinsic to the nonlinear model of nonresonant wave growth. In the next section, we discuss some models of X-ray synchrotron imaging of SNR shells.

Two opposite limits of cascading were considered in the model by Vladimirov et al. (2009) because our knowledge of turbulent cascading in collisionless plasmas is still very limited. It is limited, in part, because of the rather narrow dynamical range of the available nonlinear simulations. Information from direct measurements in the interplanetary medium can help constrain models of turbulent cascading, but this is also limited. The cascading of plasma wave spectra upstream of interplanetary shocks driven by coronal mass ejection events were analyzed by Bamert et al. (2008). They studied the competition between the upstream wave generation by superthermal protons accelerated at the shock, and the cascading of wave energy in the inertial range of solar wind turbulence. Bamert et al. (2008) concluded that amplified solar wind turbulence upstream of interplanetary traveling shocks is better described by a Iroshnikov-Kraichnan-type cascade rather than the Kolmogorov-type.

The turbulent cascade is expected to be anisotropic, as is the case in the interplanetary medium (see e.g., Matthaeus et al. 1996; Horbury et al. 2008; Podesta 2009, and the references therein). Turbulent cascade dissipation will heat ions and electrons both in the upstream and downstream regions of the shock. Since the magnetic field amplification region in the shock precursor is of a finite size $\sim (c/v_s)\Lambda$, the upstream plasma temperature depends strongly on the heating rate and the plasma advection time. The cascade rate τ_c^{-1} depends on the field amplitude, growing mode polarization, and the plasma parameters. If $\tau_c^{-1}(c/v_s^2)\Lambda < 1$, the upstream plasma heating is inefficient. However, in the opposite limit with strong upstream turbulence, the plasma ion heating via an appropriate Landau resonance can be strong if the plasma parameter $\beta = a_0^2/v_a^2 > 1$. Electron heating is dominant in the highly magnetized upstream plasma if $\beta < 1$ (e.g., Quataert and Gruzinov 1999;

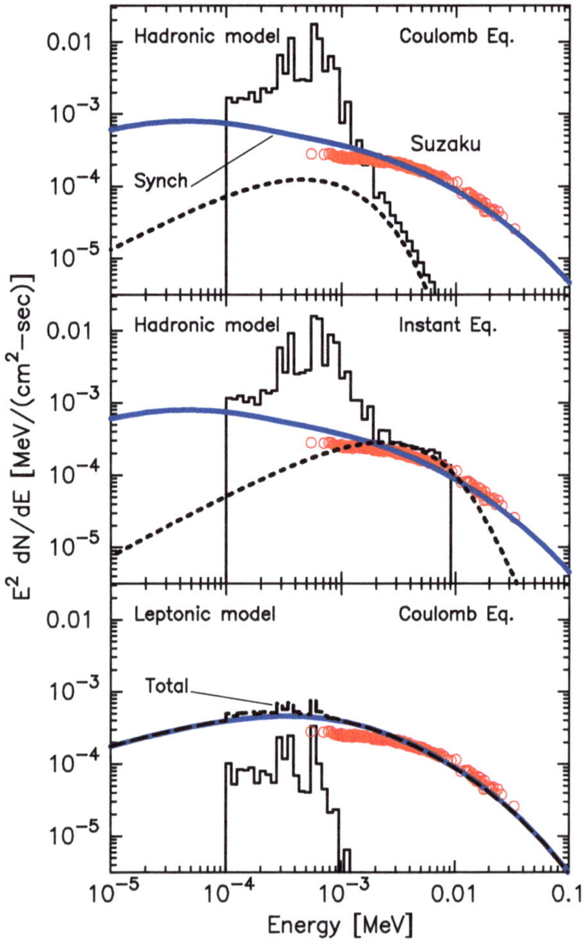

Fig. 4 The *top two panels* show fits to the *Suzaku* SNR J1713 observations with a hadronic model for both Coulomb and instant temperature equilibration but ignoring the X-line emission. The *blue* (*heavy wt.*) *solid curve* is the synchrotron continuum, the *black solid curve* is the thermal emission (only lines above 10^{-4} MeV are included), and the *dotted curve* is the underlying bremsstrahlung continuum. The observed emission would be the sum (not shown in the top two panels) of the *solid black* and *blue curves*. The *bottom panel* shows the leptonic model (with Coulomb equilibration) where parameters have been chosen to be consistent with the *Suzaku* observations. For the hadronic model, the radiation intensity is multiplied by 0.95 to match the observations. For the leptonic model, a normalization factor of 0.2 is required to match the observations. The *Suzaku* data have been adjusted for interstellar extinction so no extinction is applied to the model in this plot. See Ellison et al. (2010) for full details

Howes 2008). The Monte-Carlo simulations of the nonlinear structure of CR modified shocks by Vladimirov et al. (2008) demonstrated that the CR acceleration efficiency and the shock modification are sensitive to ion heating by the upstream turbulence. X-ray line emission from shocked shells of SNRs can be used to constrain the DSA models (e.g., Ellison et al. 2010).

There is an emerging class of SNRs, such as SN1006, RXJ1713.72-3946, Vela Jr, and others, that are *dominated* by non-thermal emission in X-rays, probably of synchrotron origin. Such emission results from electrons and/or positrons that are accelerated to well above TeV energies. Models of supernova shells show that these leptons radiate in strong magnetic fields and the radiation losses produce a marked cut off in the lepton spectrum. Figure 4 (from, Ellison et al. 2010) shows the X-ray emission from RXJ1713.72-3946 as observed by the Suzaku spacecraft. This figure, and the models given in Ellison et al. (2010), indicate that the lack of thermal X-ray emission lines can be important for determining the origin of the non-thermal emission and the strength of the magnetic fields in the SNR. Since the amplified magnetic fields in DSA must be fluctuating, we next discuss the effect of fluctuations on the synchrotron images of SNRs.

4 Synchrotron Images, Spectra and Polarization in Supernova Shells

Non-thermal emission in many sources of interest such as SNRs, gamma-ray bursts, and AGNs originates from synchrotron radiation of ultra-relativistic particles in turbulent magnetic fields. The effect of a random magnetic field on synchrotron images, emission spectra, and polarization can be very substantial.

4.1 Synchrotron Mapping of Twinkling Supernova Shells

It is instructive to illustrate first the effect of intermittency caused by the random magnetic field in terms of a power-law electron spectrum with spectral index Γ. The synchrotron emissivity $\tilde{I}(\mathbf{r}, t, \nu) \propto B^{(\Gamma+1)/2}$ shows that the local emissivity is relatively very high for large B and large Γ (a large Γ implies a steep spectrum). Therefore, the high-order statistical moments of the field dominate the synchrotron emissivity in the spectral cut-off regime and it is possible for a single strong local field maximum to produce a feature (dot or clump) that stands out on the map even after integrating the local emissivity over the line of sight. In lower energy maps, the contribution of a single maximum can be smoothed or washed out by contributions from a number of weaker field maxima integrated over the line of sight. A high-energy map, however, can be highly intermittent because the synchrotron emissivity depends on high-order moments of the random magnetic field at the cut-off frequency regime. The electron/positron spectra in the cut-off regime are usually exponential, and the exact shapes of the spectra dominated by synchrotron/Compton losses depend on the diffusion coefficient. A power-law approximation can only apply for a narrow electron energy range in the cut-off regime, where the effective spectral index Γ is increasing with the electron energy.

In order to calculate synchrotron emission maps and spectra, Bykov et al. (2008b) modeled a system of finite size, filled with a random magnetic field and electron spectra simulated with a DSA model accounting for synchrotron/Compton losses (see Bykov et al. 2000). A forward shock of spherical geometry was assumed in the image construction. The random field was composed of a superposition of magnetic fluctuations (plane waves) with random phases and a given spectrum of amplitudes. *Non-steady*, localized structures (dots, clumps and filaments), in which the magnetic field reaches exceptionally high values, typically arise in the statistically stationary random field sample. These non-steady magnetic field concentrations dominate the synchrotron emission (integrated along the line of sight) from the highest energy electrons in the cut-off regime of the electron distribution, resulting in an intermittent, twinkling, clumpy appearance even for a steady distribution of ultra-relativistic electrons. The spectral and temporal evolution of these high intensity synchrotron events, and the synchrotron images they produce, vary drastically at different photon energies.

In Fig. 5 the synchrotron emission of a small region around a dot labeled D1 is shown for emission at 0.5, 5, 20 and 50 keV. There are observable differences in the synchrotron maps indicating that some features are bright at high energies and much less prominent at lower energies and vice versa. The physical reason for this difference comes from the fact that in the high-energy cut-off region of the electron spectrum, the local synchrotron emissivity depends strongly on the local field value as $B^{(\Gamma+1)/2}$, where the effective index Γ is increasing with the photon energy in the cut-off spectral regime.

The simulated light curves of X-ray clumps may explain those observed in X-ray images of some supernova remnants by Uchiyama et al. (2007) and Patnaude and Fesen (2009). It is important to note that if the clumps are due to magnetic field intermittency, the time scale of their variation is not reflecting the ultra-relativistic electron loss time. The distinct

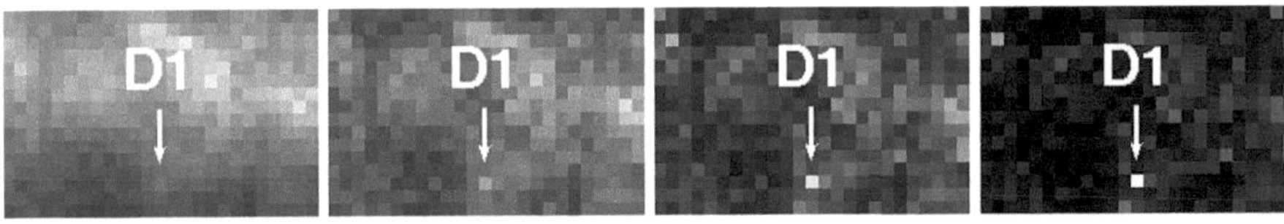

Fig. 5 Emission in the vicinity of a bright synchrotron dot D1 at different photon energies (see Bykov et al. 2008a, 2008b for details). The *four panels* show, from left to right, $\nu^2 \cdot I(\mathbf{R}_\perp, t, \nu)$ maps of synchrotron emission at 0.5, 5, 20 and 50 keV, respectively

characteristic of the modeled synchrotron emission is that its strong intermittency results directly from the exceptionally high magnetic field amplifications that randomly occur. The peaks in synchrotron emission maps occur due to high-order moments of the magnetic field probability distribution function (PDF). Even if the PDF of projections of the local magnetic field are nearly Gaussian, the corresponding PDF of synchrotron peaks, simulated for a spatially homogeneous relativistic particle distribution, has strong departures from the Gaussian at large intensity amplitudes. This is because of the nonlinear dependence of the synchrotron emissivity on the local magnetic field in the high-energy cut-off regime of the electron spectrum. Intermittency of this nature, with very different appearances, is rather a common phenomenon in stochastic media (see e.g., Zel'dovich et al. 1987).

The models by Bykov et al. (2008b, 2009) make a few basic predictions. One is that the overall efficiency of synchrotron radiation from the cut-off regime in the electron spectrum can be strongly enhanced in a turbulent field with some value of $\sqrt{\langle B^2\rangle}$, compared to emission from a uniform field, B_0, where $B_0 = \sqrt{\langle B^2\rangle}$. The second is that strong variations in the brightness of small structures can occur on time scales much shorter than variations in the underlying particle distribution. The variability time scale is shorter for higher energy synchrotron images. An estimate of the time scales of these intensity variations is consistent with the rapid time variability seen in some young SNRs by *Chandra*. The strong energy dependence predicted may be important for the future missions *NuSTAR*, *Astro-H* and others that will image SNR shells up to 50 keV.

4.2 Synchrotron X-ray Polarization in DSA

Synchrotron radiation in a regular, quasi-homogeneous magnetic field is polarized (e.g., Ginzburg and Syrovatskii 1965). It has long been known that random directions of magnetic fields, in addition to Faraday rotation, may strongly reduce the *average* polarization of synchrotron emission sources (e.g., Westfold 1959; Crusius and Schlickeiser 1986; Stroman and Pohl 2009). This explains the relatively low polarization frequently observed for radio synchrotron sources. However, as it was shown by Bykov et al. (2009), the turbulent magnetic fields that reduce the *average* polarization can result in highly polarized, intermittent, patchy structures potentially observable in high resolution X-ray images. In terms of a power-law electron spectrum with spectral index Γ, the degree of polarization is given by $\tilde{\Pi} \approx (\Gamma + 1)/(\Gamma + 7/3)$ and, therefore, $\tilde{\Pi}$ increases with Γ. Thus the synchrotron images will be highly polarized at high energies.

In order to construct maps of polarized synchrotron emission from SNR shells, it is convenient to use the local densities of the Stokes parameters (see e.g., Ginzburg and Syrovatskii 1965). Because of the additive property of the Stokes parameters $\tilde{I}, \tilde{Q}, \tilde{U}, \tilde{V}$ for incoherent photons, we can integrate these weighted with the distribution function of radiating particles over the line of sight across the source. The degree of polarization is then determined in

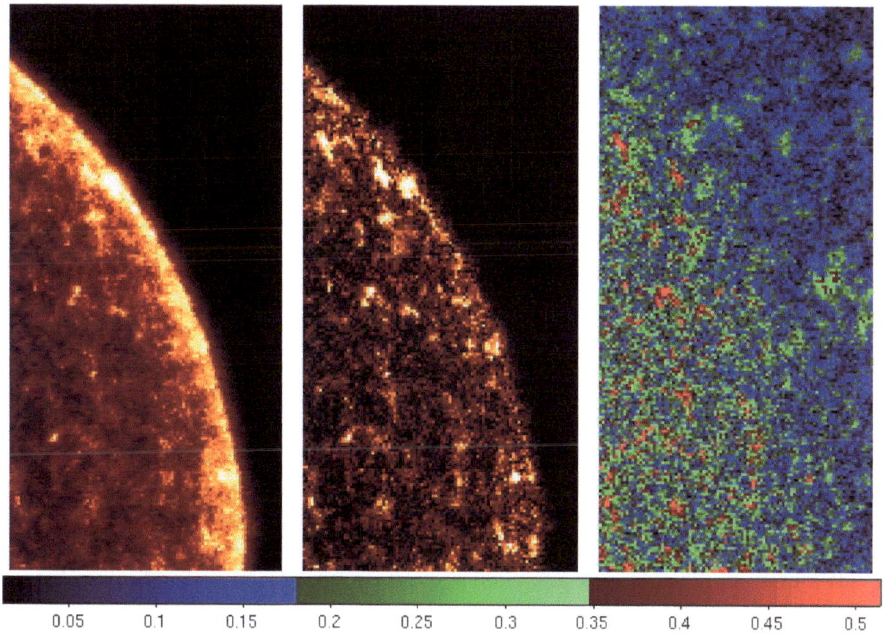

Fig. 6 Simulated maps of polarized synchrotron emission in a random magnetic field at 5 keV. Intensity, $\nu^2 \cdot I(\mathbf{R}_\perp, t, \nu)$, is shown with a *linear color scale* in the *left panel*. The *central panel* shows the product of intensity and polarization degree. The *right panel* shows the degree of polarization indicated by the *color bar*. The stochastic magnetic field sample has $\sqrt{\langle B^2 \rangle} = 3 \times 10^{-5}$ G and spectral index of magnetic fluctuations is $\delta = 1.0$ (from Bykov et al. (2009))

a standard way as $\Pi = \sqrt{Q^2 + U^2 + V^2}/I$. Starting from the simulated random magnetic-field, Bykov et al. (2009) have constructed maps of *polarized* X-ray emission of SNR shells. These are highly clumpy with polarizations up to 50% for energetic >TeV electrons in the cut-off regime.

The detection of polarization in X-ray sources as faint as 1 milliCrab is an aim of the *Gravity and Extreme Magnetism Small Explorer* (*GEMS*) mission, recently selected for flight in 2014 by NASA. It will make a sensitive search for X-ray polarization using thin foil mirrors and Time Projection Chamber polarimeters with high efficiency in the 2–10 keV band (Kallman et al. 2010). The polarimeters under consideration, like *XPOL* aboard the planned *IXO* mission[1] and *POLARIX* proposed to measure X-ray polarization with an angular resolution of about 20 arcsec in a field of view of 15 × 15 arcmin with the minimum detectable polarization of 12% for a source of 1 mCrab within 100 ks of observing time (Costa et al. 2010), have good perspectives to detect polarized synchrotron emission from SNR shells.

The intermittent appearance of the polarized X-ray emission maps of young SNR shells can be studied in detail with imagers of a few arcsecond resolution, though even arcmin resolution images can provide important information as is illustrated in Fig. 6 (see Bykov et al. 2009). The polarized emission clumps of arcsecond scales are time variable on a year or longer (depending on the observed photon energy, magnetic field amplification factor,

[1] http://ixo.gsfc.nasa.gov/science/performanceRequirements.html.

and the plasma density in the shell) allowing for rather long exposures even in the hard X-ray energy band. Hard X-ray observations in the spectral cut-off regime are the most informative to study the magnetic fluctuation spectra and the acceleration mechanisms of ultra-relativistic particles.

5 Magnetic Field Amplification in Star Forming Galaxies

The origin and evolution of magnetic field structures in galaxies is an important unsolved problem (e.g., Ruzmaikin et al. 1988; Rees 2006; Kulsrud and Zweibel 2008). Radio observations of the diffuse polarized radio emission from the disks of some spiral galaxies, as reviewed by Beck (2008), revealed large-scale coherent magnetic patterns that could be generated by dynamos, and dynamo models have been developed to explain the main observational features of the global magnetic fields of spiral galaxies (e.g., Ruzmaikin et al. 1988; Beck et al. 1996; Brandenburg and Subramanian 2005). In most galaxies, however, the field structure is more complicated. For example, radio synchrotron observations of nearby galaxies reveal dynamically important magnetic fields of 10–30 μG total strength in the spiral arms. These spiral arm fields, however, tend to have random orientations, while ordered fields (observed in radio polarization) are strongest in interarm regions and follow the orientation of the adjacent gas spiral arms.

A variety of observations (see Widrow 2002, for a review) suggest that magnetic fields are present in all galaxies including ellipticals and in galaxy clusters. Furthermore, it remains to be understood, in the context of the dynamo models, why well organized fields of surprisingly high strengths are observed in normal galaxies when the Universe was much younger than its present age (see e.g., Bernet et al. 2008; Wolfe et al. 2008).

5.1 Magnetic Fields in Irregular Galaxies

Another issue to be addressed concerns the magnetic fields of dwarf irregular galaxies. Studies of magnetic fields in nearby dwarfs, as discussed recently by Chyzy (2010), revealed typically weak fields with the mean value of the total field strength about three times smaller than in normal spirals. The slow rotation rate in many irregular galaxies makes standard dynamo-type amplification of the magnetic field unlikely. On the other hand, dwarfs with stronger fields are associated with violent star-forming activity and these tend to be more massive and evolved systems. Their magnetic fields are thought to be regulated mainly by the surface density of the galactic star-formation rate (e.g., Chyzy 2010).

Star forming activity with clustered supernova explosions in superbubbles observed in galaxies (see e.g., Heiles 1990) may affect the mean magnetic field as suggested by Ferriere (1992). Moreover, numerical magnetohydrodynamical models by Siejkowski et al. (2010) of cosmic-ray driven dynamos in the interstellar medium of an irregular galaxy indicated the possibility of magnetic field amplification. The process of cosmic-ray acceleration may play a role in generating galactic magnetic fields at different stages of cosmic evolution. Non-resonant mechanisms of magnetic field amplification allow field enhancements on scales much larger than the gyroradii of the ultra-relativistic particles accelerated in the sources. Such fields amplified by the CR current and CR pressure anisotropy will have scales well above a parsec in the star forming regions, and well above a kiloparsec in clusters of galaxies. Observations of supernova remnants and galactic CRs suggest that there is a high efficiency of conversion of the kinetic power of supernova shocks into CRs. Therefore, the specific mechanisms discussed above may convert a sizable fraction of the power released by SNRs into intermediate scale magnetic fields that are much larger than stellar scale fields. Some large-scale dynamo mechanism may also be necessary to produce the largest scale coherent fields seen in spiral galaxies.

5.2 Magnetic Fields in the First Generation Pop III Supernova Remnants

Simulations of nonlinear cosmic structure formation by Abel et al. (2002) and Xu et al. (2008) showed that the high-redshift analog of a molecular cloud is able to produce individual massive primordial stars, offering a natural explanation for the absence of purely metal free low mass stars in the Milky Way. The production of small-scale dynamos by turbulence created by accretion shocks during gravitational collapse may amplify initially weak magnetic fields in the protostellar cloud in the formation process of the first stars and galaxies (i.e., Schleicher et al. 2010). Massive stars exploding as supernovae provide energy, momentum, entropy, and metals at the very earliest stages of galactic evolution. Observations of high-redshift gamma-ray bursts likely indicate that relativistic particles and magnetic fields are produced during the collapse of massive primordial stars.

An important problem is whether there is a way to transfer a significant fraction of the high kinetic energy of the supernova ejecta into the magnetic fields of the remnant. The questions to be addressed are the following.

(i) Can *collisionless* shocks be created by the supersonic ejecta in the very weakly magnetized plasma of $\beta \gg 1$?
(ii) Can relativistic particles be accelerated in Pop III supernova remnants?
(iii) Are there effective mechanisms of magnetic field amplification in Pop III supernova remnants?

Two-dimensional particle-in-cell simulations performed by Kato and Takabe (2010) to study weakly magnetized perpendicular shocks demonstrated the development of a *collisionless* shock with $\beta \approx 26$ (and an Alfvén Mach number of about 100). They showed that current filaments form in the foot region of the shock due to the ion-beam-Weibel instability (or the ion filamentation instability) and these filaments generated a strong magnetic field there. Strong fluctuating field generation is necessary in the standard paradigm of a collisionless shock since the only way to decelerate and thermalize the cold upstream flow is through the electromagnetic fields. In the downstream region, these current filaments also generated a tangled magnetic field that was about 15 times stronger than the upstream magnetic field. The shock simulations by Kato and Takabe (2010) reproduced well the Rankine-Hugoniot relations and indicated a substantial damping of the strong Weibel-type field fluctuations in the shock ramp (in the case of $\beta \approx 26$, the magnetic field contribution in the Rankine-Hugoniot relations is small). The authors noticed that a fraction of the ions are accelerated slightly on reflection at the shock.

If the temporal and spatial scales of the simulation were increased, it is expected that this fraction of reflected superthermal particles would continue to be accelerated by the Fermi-type DSA process. In this case, the magnetic field would continue to be amplified in a symbiotic relationship where the magnetic turbulence required to scatter and accelerate the CRs is created by the accelerated CRs themselves (e.g., Bell 2004; Vladimirov et al. 2008; Bykov et al. 2011). If this amplification operates as expected, a sizable fraction of the energy released by the supernova explosion can be transferred to energetic particles even with the initially weak magnetic field fluctuations of $B_0 \sim 0.1$–1 nG expected in the protostellar cloud.

The largest scale magnetic fluctuations amplified by the long wavelength non-resonant instabilities (that are discussed above in Sect. 2) can approach the scale of the shock. Therefore, the supernova remnant and its vicinity, with a scale size of about 30 pc, can be filled with a magnetic field of strength $\sim$0.1 μG generated by instabilities driven by CR acceleration. The longest scale fields generated by the non-resonant CR instabilities in the shock *precursor* are in the collisional plasma regime since their wavelengths are longer than the Coulomb mean free path of the ambient thermal plasma. The long-wavelength fields should have, by many orders of magnitude, longer damping times than the short-lived Weibel-type fluctuations that are located just in a very narrow vicinity of the shock jump. The magnetic fields amplified in the SNR shock precursor can be considered as mesoscale fields since their scales are much larger than the stellar field sizes and larger than the CR gyroradii, but are below the observed largest scale coherent galactic magnetic fields in spiral galaxies presumably generated by galactic dynamos or by a global galactic CR current. The mesoscale fields can be the dominant fields in irregular galaxies and can be spread by the galactic winds through galactic groups and clusters. Furthermore, CR-driven winds in high-redshift galaxies may have important implications for the metal and magnetic field enrichment of the intergalactic medium (see e.g., Samui et al. 2010; Dubois and Teyssier 2010). In contrast, magnetic fields much lower than expected from CR instabilities were predicted in MHD simulations of the Biermann battery mechanism in the first generation of supernova remnants (i.e., Hanayama et al. 2009).

Magnetic field amplification by CRs produced by the first supernova explosions during the re-ionization epoch of the universe was discussed by Miniati and Bell (2010). The CRs escape into the intergalactic medium carrying an electric current that results in magnetic field amplification. The authors found that magnetic fields are robustly generated throughout intergalactic space at a rate of 10^{-17}–10^{-16} Gauss/Gyr, until the temperature of the intergalactic medium is raised by cosmic re-ionization. In another shock-related process, Ryu et al. (2008) discussed a scenario in which turbulent-flow motions are induced via the cascade of vorticity generated at cosmological shocks during the formation of large-scale structure. It was argued that this turbulent dynamo model might provide average magnetic field strengths of a few microgauss inside clusters and groups, approximately 0.1 μG around clusters and groups, and approximately 10 nG in filaments.

6 Discussion

Magnetic fields, both regular and stochastic, play a key role in charged particle accelerators providing both particle confinement and energy gain. The enormous energy release of gravitational (AGN, GRB, SNRs, clusters of galaxies, etc.) or thermonuclear origin (Type Ia SNRs, etc.), produces collisionless shocks and accelerates particles in many sources. It has long been known that strong shocks can only form in collisionless plasmas if magnetic turbulence is generated and some thermal particles are injected and accelerated to superthermal energies. The strong interrelation between shock formation and structure, magnetic turbulence, and energetic particles makes understanding the self-consistent production of magnetic turbulence in diffusive shock acceleration both essential and difficult.

While a great deal of progress has been made in this field, important challenges remain. In this brief review, we have discussed some recent work including resonant streaming and non-resonant CR-current driven instabilities. An important example of this is the Bell mechanism that efficiently amplifies short-scale turbulence by the CR current. Magnetic turbulence amplification may govern the nonlinear structure of CR-dominated flows, heat the upstream plasma, enhance synchrotron radiation, and limit high-energy electron energies. Furthermore, this mechanism might operate hand-in-hand with an instability for producing long-scale turbulence. The long-scale turbulence can be responsible for the highest energy particles accelerated by DSA. The basic DSA prediction that the highest energy accelerated ions can have a hard spectrum implies that the highest energy CRs may contain a sizeable fraction of the shock ram pressure. This implies that a consistent model of magnetic fields in the highly supersonic flow with accelerated particles is nonlinear and multi-scale.

The field of nonlinear shock acceleration is evolving rapidly with several prominent areas of active study. The long-wavelength instabilities discussed in Sect. 2 are yet to be combined with the resonant and short-wavelength instabilities in a nonlinear shock simulation. Such a combination will determine the relative importance of each and how the overall shock structure is influenced by them. The instabilities produced by escaping cosmic rays are only just now starting to be explored. Only with a consistent model combining the turbulence produced by escaping cosmic rays on the escaping cosmic rays themselves, can a reliable estimate of the maximum particle energy a given shock can produce be obtained.

In principle, the plasma physics of shock structure, magnetic turbulence, particle injection and acceleration, indeed, all aspects collisionless plasmas, can be determined with particle-in-cell (PIC) simulations. A great deal has been done in this field and we have mentioned a small fraction of this work in Sect. 2.3. The computational difficulties in producing PIC simulations with parameters close enough to systems such as SNRs are severe and we see future progress coming from synthesis of PIC models with semi-analytic and Monte Carlo techniques.

We have also discussed some recent work on the observational consequences of magnetic turbulence and synchrotron emission. The important result is that fluctuating magnetic fields strongly modify the spectra of synchrotron emission in the case of steep electron spectra that are typical for X-ray emitting, TeV particles in young SNRs. X-ray images of synchrotron dominated SNR shells are predicted to be highly intermittent with polarized twinkling structures of different scales if the strong amplification of magnetic turbulence is indeed operating in DSA.

The current generation of X-ray telescopes—*Chandra* with its superb angular resolution and *XMM-Newton* with its high sensitivity—has uncovered a new world of synchrotron structures in young SNR shells. Further progress is certain to come from improved polarization and timing observations of SNRs at X-ray energies where the Faraday rotation is not essential and characteristic time scales are short enough. Observations at GeV–TeV energies are becoming more sensitive and theories of the highest energy cosmic rays must be improved accordingly. In all aspects of theory and observations, the turbulent magnetic field is the complicated glue tying things together.

Acknowledgements We thank the referee for a constructive comment. A.M.B. was supported in part by RBRF grants 09-02-12080, 11-02-00429 and by the RAS Presidium Programm, and the Russian government grant 11.G34.31.0001 to Sankt-Petersburg State Politechnical University. He performed some of the simulations at the Joint Supercomputing Centre (JSCC RAS) and the Supercomputing Centre at Ioffe Institute, St. Petersburg. D.C.E acknowledges support from NASA grants ATP02-0042-0006, NNH04Zss001N-LTSA, and 06-ATP06-21.

References

A.A. Abdo, M. Ackermann, M. Ajello, L. Baldini et al., Observation of supernova remnant IC 443 with the Fermi large area telescope. Astrophys. J. **712**, 459–468 (2010). arXiv:1002.2198

T. Abel, G.L. Bryan, M.L. Norman, The formation of the first star in the universe. Science **295**, 93–98 (2002). arXiv:astro-ph/0112088

A. Achterberg, R.D. Blandford, S.P. Reynolds, Evidence for enhanced MHD turbulence outside sharp-rimmed supernova remnants. Astron. Astrophys. **281**, 220–230 (1994)

F. Aharonian, A.G. Akhperjanian, A.R. Bazer-Bachi, B. Behera et al., Discovery of very high energy gamma-ray emission coincident with molecular clouds in the W 28 (G6.4–0.1) field. Astron. Astrophys. **481**, 401–410 (2008). arXiv:0801.3555

J. Albert, E. Aliu, H. Anderhub, P. Antoranz et al., Discovery of very high energy gamma radiation from IC 443 with the MAGIC telescope. Astrophys. J. **664**, L87–L90 (2007). arXiv:0705.3119

E. Amato, P. Blasi, A kinetic approach to cosmic-ray-induced streaming instability at supernova shocks. Mon. Not. R. Astron. Soc. Lett. **392**, 1591–1600 (2009). arXiv:0806.1223

A. Bamba, R. Yamazaki, T. Yoshida, T. Terasawa et al., A spatial and spectral study of nonthermal filaments in historical supernova remnants: observational results with Chandra. Astrophys. J. **621**, 793–802 (2005). arXiv:astro-ph/0411326

A. Bamba, R. Yamazaki, T. Yoshida, T. Terasawa et al., Small-scale structure of non-thermal X-rays in historical SNRs. Adv. Space Res. **37**, 1439–1442 (2006)

K. Bamert, R. Kallenbach, J.A. le Roux, M. Hilchenbach et al., Evidence for Iroshnikov-Kraichnan-type turbulence in the solar wind upstream of interplanetary traveling shocks. Astrophys. J. **675**, L45–L48 (2008)

K. Bamert, R. Kallenbach, N.F. Ness, C.W. Smith et al., Hydromagnetic wave excitation upstream of an interplanetary traveling shock. Astrophys. J. **601**, L99–L102 (2004)

R. Bandiera, O. Petruk, A statistical approach to radio emission from shell-type SNRs. I. Basic ideas, techniques, and first results. Astron. Astrophys. **509**, A34 (2010). arXiv:0911.0829

R. Beck, Galactic and extragalactic magnetic fields, in *American Institute of Physics Conference Series*, ed. by F.A. Aharonian, W. Hofmann, F. Rieger. American Institute of Physics Conference Series, vol. 1085 (2008), pp. 83–96. arXiv:0810.2923

R. Beck, A. Brandenburg, D. Moss, A. Shukurov et al., Galactic magnetism: recent developments and perspectives. Annu. Rev. Astron. Astrophys. **34**, 155–206 (1996)

A.R. Bell, Turbulent amplification of magnetic field and diffusive shock acceleration of cosmic rays. Mon. Not. R. Astron. Soc. **353**, 550–558 (2004)

A.R. Bell, The interaction of cosmic rays and magnetized plasma. Mon. Not. R. Astron. Soc. Lett. **358**, 181–187 (2005)

A. Beresnyak, T.W. Jones, A. Lazarian, Turbulence-induced magnetic fields and structure of cosmic ray modified shocks. Astrophys. J. **707**, 1541–1549 (2009). arXiv:0908.2806

V.S. Berezinskii, S.V. Bulanov, V.A. Dogiel, V.L. Ginzburg et al., Astrophysics of cosmic rays (1990)

M.L. Bernet, F. Miniati, S.J. Lilly, P.P. Kronberg et al., Strong magnetic fields in normal galaxies at high redshift. Nature **454**, 302–304 (2008). arXiv:0807.3347

R. Blandford, D. Eichler, Particle acceleration at astrophysical shocks—a theory of cosmic-ray origin. Phys. Rep. **154**, 1 (1987)

R. Blandford, S. Funk, The magnetic bootstrap, in *The First GLAST Symposium*, ed. by S. Ritz, P. Michelson, C.A. Meegan. American Institute of Physics Conference Series, vol. 921 (2007), pp. 62–64

A. Brandenburg, K. Subramanian, Astrophysical magnetic fields and nonlinear dynamo theory. Phys. Rep. **417**, 1–209 (2005). arXiv:astro-ph/0405052

Y.M. Butt, T.A. Porter, B. Katz, E. Waxman, X-ray hotspot flares and implications for cosmic ray acceleration and magnetic field amplification in supernova remnants. Mon. Not. R. Astron. Soc. Lett. **386**, L20–L22 (2008). arXiv:0801.4954

A.M. Bykov, R.A. Chevalier, D.C. Ellison et al., Nonthermal emission from a supernova remnant in a molecular cloud. Astrophys. J. **538**, 203–216 (2000). arXiv:astro-ph/0003235

A.M. Bykov, K. Dolag, F. Durret, Cosmological shock waves. Space Sci. Rev. **134**, 119–140 (2008a). arXiv:0801.0995

A.M. Bykov, S.M. Osipov, D.C. Ellison, Cosmic ray current driven turbulence in shocks with efficient particle acceleration: the oblique, long-wavelength mode instability. Mon. Not. R. Astron. Soc. Lett. **410**, 39–52 (2011). arXiv:1010.0408

A.M. Bykov, I.N. Toptygin, Particle kinetics in highly turbulent plasmas (renormalization and self-consistent field methods). Phys. Usp. **36**, 1020–1052 (1993)

A.M. Bykov, I.N. Toptygin, A model of particle acceleration to high energies by multiple supernova explosions in OB associations. Astron. Lett. **27**, 625–633 (2001)

A.M. Bykov, I.N. Toptygin, Generation of magnetic fluctuations near a shock front in a partially ionized medium. Astron. Lett. **31**, 748–754 (2005)

A.M. Bykov, Y.A. Uvarov, J.B.G.M. Bloemen, J.W. den Herder et al., A model of polarized X-ray emission from twinkling synchrotron supernova shells. Mon. Not. R. Astron. Soc. Lett. **399**, 1119–1125 (2009). arXiv:0907.2521

A.M. Bykov, Y.A. Uvarov, D.C. Ellison, Dots, clumps, and filaments: the intermittent images of synchrotron emission in random magnetic fields of young supernova remnants. Astrophys. J. **689**, L133–L136 (2008b). arXiv:0811.2498

G. Cassam-Chenaï, J.P. Hughes, J. Ballet, A. Decourchelle, The blast wave of Tycho's supernova remnant. Astrophys. J. **665**, 315–340 (2007). arXiv:astro-ph/0703239

G. Cassam-Chenaï, J.P. Hughes, E.M. Reynoso, C. Badenes et al., Morphological evidence for Azimuthal variations of the cosmic-ray ion acceleration at the blast wave of SN 1006. Astrophys. J. **680**, 1180–1197 (2008)

D. Castro, P. Slane, Fermi large area telescope observations of supernova remnants interacting with molecular clouds. Astrophys. J. **717**, 372–378 (2010). arXiv:1002.2738

S.V. Chalov, Instability of the structure of strong oblique MHD cosmic-ray shocks. Astrophys. Space Sci. **148**, 175–187 (1988)

R.A. Chevalier, Magnetic field amplification in interstellar collisionless shock waves. Nature **266**, 701 (1977)

K.T. Chyzy, Magnetic fields in dwarfs versus early-type galaxies. Highlights Astron. **15**, 454–455 (2010)

E. Costa, R. Bellazzini, G. Tagliaferri, G. Matt et al. POLARIX: a pathfinder mission of X-ray polarimetry. Exp. Astron. **9** (2010)

J.H. Croston, R.P. Kraft, M.J. Hardcastle, M. Birkinshaw et al., High-energy particle acceleration at the radio-lobe shock of Centaurus A. Mon. Not. R. Astron. Soc. Lett. **395**, 1999–2012 (2009). arXiv:0901.1346

A. Crusius, R. Schlickeiser, Synchrotron radiation in random magnetic fields. Astron. Astrophys. **164**, L16–L18 (1986)

E.A. Dorfi, L.O. Drury, A cosmic ray driven instability, in *International cosmic ray conference*, ed. by F.C. Jones. International Cosmic Ray Conference, vol. 3 (1985), pp. 121–123

L.O. Drury, S.A.E.G. Falle, On the stability of shocks modified by particle acceleration. Mon. Not. R. Astron. Soc. Lett. **223**, 353 (1986)

Y. Dubois, R. Teyssier, Magnetised winds in dwarf galaxies. Astron. Astrophys. **523**, A72 (2010). arXiv:0908.3862

D.C. Ellison, M.G. Baring, F.C. Jones, Nonlinear particle acceleration in oblique shocks. Astrophys. J. **473**, 1029 (1996). arXiv:astro-ph/9609182

D.C. Ellison, E. Moebius, G. Paschmann, Particle injection and acceleration at earth's bow shock—comparison of upstream and downstream events. Astrophys. J. **352**, 376–394 (1990)

D.C. Ellison, D.J. Patnaude, P. Slane, J. Raymond, Efficient cosmic ray acceleration, hydrodynamics, and self-consistent thermal X-ray emission applied to supernova remnant RX J1713.7-3946. Astrophys. J. **712**, 287–293 (2010). arXiv:1001.1932

K. Ferriere, Effect of the explosion of supernovae and superbubbles on the Galactic dynamo. Astrophys. J. **391**, 188–198 (1992)

B. Fryxell, C.C. Kuranz, R.P. Drake, M.J. Grosskopf et al., The possible effects of magnetic fields on laser experiments of Rayleigh-Taylor instabilities. High Energy Density Phys. **6**, 162–165 (2010)

V.L. Ginzburg, S.I. Syrovatskii, *The Origin of Cosmic Rays* (Macmillan, New York, 1964)

V.L. Ginzburg, S.I. Syrovatskii, Cosmic Magnetobremsstrahlung (synchrotron Radiation). Ann. Rev. Astron. Astrophys. **3**, 297 (1965)

H. Hanayama, K. Takahashi, K. Tomisaka, Generation of seed magnetic fields in primordial supernova remnants. arXiv:0912.2686 (2009)

L. Harvey-Smith, B.M. Gaensler, R. Kothes, R. Townsend et al., Faraday rotation of the supernova remnant G296.5+10.0: evidence for a magnetized progenitor wind. Astrophys. J. **712**, 1157–1165 (2010). arXiv:1001.3462

C. Heiles, Clustered supernovae versus the gaseous disk and halo. Astrophys. J. **354**, 483–491 (1990)

E.A. Helder, J. Vink, C.G. Bassa, A. Bamba et al., Measuring the cosmic-ray acceleration efficiency of a supernova remnant. Science **325**, 719 (2009). arXiv:0906.4553

T.S. Horbury, M. Forman, S. Oughton, Anisotropic scaling of magnetohydrodynamic turbulence. Phys. Rev. Lett. **101**(17), 175 005 (2008). arXiv:0807.3713

G.G. Howes, Inertial range turbulence in kinetic plasmas. Phys. Plasmas **15**(5), 055 904 (2008). arXiv:0711.4358

F.C. Jones, D.C. Ellison, The plasma physics of shock acceleration. Space Sci. Rev. **58**, 259–346 (1991)

B. Jun, M.L. Norman, On the origin of radial magnetic fields in Young supernova remnants. Astrophys. J. **472**, 245 (1996). arXiv:astro-ph/9606096

T.R. Kallman, J.H. Swank, G.E.M.S. Team, The gravity and extreme magnetism small explorer (GEMS). Bull. Am. Astron. Soc. **41**, 737 (2010).

T.N. Kato, H. Takabe, Nonrelativistic collisionless shocks in weakly magnetized electron-ion plasmas: two-dimensional particle-in-cell simulation of perpendicular shock. Astrophys. J. **721**, 828–842 (2010). arXiv:1008.0265

R.M. Kulsrud, *Plasma Physics for Astrophysics* (Princeton University Press, Princeton, 2005)

R.M. Kulsrud, E.G. Zweibel, On the origin of cosmic magnetic fields. Rep. Prog. Phys. **71**(4), 046 901 (2008). arXiv:0707.2783
Q. Luo, D. Melrose, Saturated magnetic field amplification at supernova shocks. Mon. Not. R. Astron. Soc. **397**, 1402–1409 (2009). arXiv:0904.1038
M.A. Malkov, L. Drury, Nonlinear theory of diffusive acceleration of particles by shock waves. Rep. Prog. Phys. **64**, 429–481 (2001)
M.A. Malkov, P.H. Diamond, Nonlinear dynamics of acoustic instability in a cosmic ray shock precursor and its impact on particle acceleration. Astrophys. J. **692**, 1571–1581 (2009)
M.A. Malkov, P.H. Diamond, R.Z. Sagdeev, On the structure and scale of cosmic ray modified shocks. arXiv:1007.3042 (2010)
A. Marcowith, F. Casse, Postshock turbulence and diffusive shock acceleration in young supernova remnants. Astron. Astrophys. **515**, A90 (2010). arXiv:1001.2111
A. Marcowith, M. Lemoine, G. Pelletier, Turbulence and particle acceleration in collisionless supernovae remnant shocks. II. Cosmic-ray transport. Astron. Astrophys. **453**, 193–202 (2006). arXiv:astro-ph/0603462
W.H. Matthaeus, S. Ghosh, S. Oughton, D.A. Roberts, Anisotropic three-dimensional MHD turbulence. J. Geophys. Res. **101**, 7619–7630 (1996)
D.K. Milne, Polarization and magnetic fields in supernova remnants, in *Galactic and Intergalactic Magnetic Fields*, ed. by R. Beck, R. Wielebinski, P.P. Kronberg. IAU Symposium, vol. 140 (1990), pp. 67–72
F. Miniati, A.R. Bell, Resistive magnetic field generation at cosmic dawn. arXiv:1001.2011 (2010)
Y. Ohira, T. Terasawa, F. Takahara, Plasma instabilities as a result of charge exchange in the downstream region of supernova remnant shocks. Astrophys. J. **703**, L59–L62 (2009). arXiv:0908.3369
D.J. Patnaude, RA Fesen, Proper motions and brightness variations of nonthermal X-ray filaments in the Cassiopeia a supernova remnant. Astrophys. J. **697**, 535–543 (2009). arXiv:0808.0692
G. Pelletier, M. Lemoine, A. Marcowith, Turbulence and particle acceleration in collisionless supernovae remnant shocks. I. Anisotropic spectra solutions. Astron. Astrophys. **453**, 181–191 (2006). arXiv:astro-ph/0603461
O. Petruk, G. Dubner, G. Castelletti et al., Aspect angle for interstellar magnetic field in SN 1006. Mon. Not. R. Astron. Soc. **393**, 1034–1040 (2009). arXiv:0811.2319
J.J. Podesta, Dependence of solar-wind power spectra on the direction of the local mean magnetic field. Astrophys. J. **698**, 986–999 (2009). arXiv:0901.4940
M. Pohl, H. Yan, A. Lazarian, Magnetically limited X-ray filaments in young supernova remnants. Astrophys. J. **626**, L101–L104 (2005)
V. Ptuskin, V. Zirakashvili, E. Seo, Spectrum of galactic cosmic rays accelerated in supernova remnants. Astrophys. J. **718**, 31–36 (2010). arXiv:1006.0034
E. Quataert, A. Gruzinov, Turbulence and particle heating in advection-dominated accretion flows. Astrophys. J. **520**, 248–255 (1999). arXiv:astro-ph/9803112
M.J. Rees, Origin of cosmic magnetic fields. Astron. Nachr. **327**, 395 (2006)
B. Reville, J.G. Kirk, P. Duffy, S. O'Sullivan, A cosmic ray current-driven instability in partially ionised media. Astron. Astrophys. **475**, 435–439 (2007). arXiv:0707.3743
S.P. Reynolds, Supernova remnants at high energy. Annu. Rev. Astron. Astrophys. **46**, 89–126 (2008)
M.A. Riquelme, A. Spitkovsky, Nonlinear study of bell's cosmic ray current-driven instability. Astrophys. J. **694**, 626–642 (2009). arXiv:0810.4565
M.A. Riquelme, A. Spitkovsky, Magnetic amplification by magnetized cosmic rays in supernova remnant shocks. Astrophys. J. **717**, 1054–1066 (2010). arXiv:0912.4990
A.A. Ruzmaikin, D.D. Sokolov, A.M. Shukurov (eds.), *Magnetic Fields of Galaxies*, Astrophysics and Space Science Library, vol. 133 (Springer, Berlin, 1988)
D. Ryu, H. Kang, J. Cho, S. Das, Turbulence and magnetic fields in the large-scale structure of the universe. Science **320**, 909 (2008). arXiv:0805.2466
S. Samui, K. Subramanian, R. Srianand, Cosmic ray driven outflows from high-redshift galaxies. Mon. Not. R. Astron. Soc. **402**, 2778–2791 (2010). arXiv:0909.3854
D.R.G. Schleicher, R. Banerjee, S. Sur, T.G. Arshakian et al., Small-scale dynamo action during the formation of the first stars and galaxies. I. The ideal MHD limit. Astron. Astrophys. **522**, A115 (2010). arXiv:1003.1135
R. Schlickeiser, *Cosmic Ray Astrophysics* (Springer, Berlin, 2002)
K.M. Schure, J. Vink, A. Achterberg, R. Keppens, Evolution of magnetic fields in supernova remnants, in *Revista Mexicana de Astronomia y Astrofisica Conference Series*, vol. 36 (2009), p. 350. arXiv:0810.5150
H. Siejkowski, M. Soida, K. Otmianowska-Mazur, M. Hanasz et al., Cosmic-ray driven dynamo in the interstellar medium of irregular galaxies. Astron. Astrophys. **510**, A97 (2010). arXiv:0909.0926
J.M. Stone, T. Gardiner, The magnetic Rayleigh-Taylor instability in three dimensions. Astrophys. J. **671**, 1726–1735 (2007). arXiv:0709.0452

W. Stroman, M. Pohl, Radio polarimetry signatures of strong magnetic turbulence in supernova remnants. arXiv:0902.1701 (2009)

I.N. Toptygin, *Cosmic Rays in Interplanetary Magnetic Fields* (1985)

R.A. Treumann, W. Baumjohann, *Advanced Space Plasma Physics* (1997)

Y. Uchiyama, F.A. Aharonian, T. Tanaka et al., Extremely fast acceleration of cosmic rays in a supernova remnant. Nature **449**, 576–578 (2007)

Y. Uchiyama, R. Blandford, S. Funk, H. Tajima et al., Gamma-ray emission from crushed clouds in supernova remnants. arXiv:1008.1840 (2010)

J. Vink, Multiwavelength signatures of cosmic ray acceleration by young supernova remnants, in *AIP Conference Series*, vol. 1085. ed. by F.A. Aharonian et al. (2008), p. 169

J. Vink, J.M. Laming, On the magnetic fields and particle acceleration in Cassiopeia A. Astrophys. J. **584**, 758–769 (2003). arXiv:astro-ph/0210669

A. Vladimirov, D.C. Ellison, A. Bykov, Nonlinear diffusive shock acceleration with magnetic field amplification. Astrophys. J. **652**, 1246–1258 (2006). arXiv:astro-ph/0606433

A.E. Vladimirov, A.M. Bykov, D.C. Ellison, Turbulence dissipation and particle injection in nonlinear diffusive shock acceleration with magnetic field amplification. Astrophys. J. **688**, 1084–1101 (2008). arXiv:0807.1321

A.E. Vladimirov, A.M. Bykov, D.C. Ellison, Spectra of magnetic fluctuations and relativistic particles produced by a nonresonant wave instability in supernova remnant shocks. Astrophys. J. **703**, L29–L32 (2009). arXiv:0908.2602

D.G. Wentzel, Cosmic-ray propagation in the Galaxy—Collective effects. Annu. Rev. Astron. Astrophys. **12**, 71–96 (1974)

K.C. Westfold, The polarization of synchrotron radiation. Astrophys. J. **130**, 241 (1959)

L.M. Widrow, Origin of galactic and extragalactic magnetic fields. Rev. Mod. Phys. **74**, 775–823 (2002). arXiv:astro-ph/0207240

A.M. Wolfe, R.A. Jorgenson, T. Robishaw, C. Heiles et al., An 84-μG magnetic field in a galaxy at redshift $z = 0.692$. Nature **455**, 638–640 (2008). arXiv:0811.2408

H. Xu, B.W. O'Shea, D.C. Collins, M.L. Norman et al., The Biermann battery in cosmological MHD simulations of population III star formation. Astrophys. J. **688**, L57–L60 (2008). arXiv:0807.2647

G.P. Zank, W.I. Axford, J.F. McKenzie, Instabilities in energetic particle modified shocks. Astron. Astrophys. **233**, 275–284 (1990)

Y.B. Zel'dovich, S.A. Molchanov, A.A. Ruzmaikin, D.D. Sokoloff, Intermittency in random media. Sov. Phys. Usp. **30**, 353–369 (1987)

V.N. Zirakashvili, V.S. Ptuskin, Diffusive shock acceleration with magnetic amplification by nonresonant streaming instability in supernova remnants. Astrophys. J. **678**, 939–949 (2008). arXiv:0801.4488

V.N. Zirakashvili, V.S. Ptuskin, H.J. Völk, Modeling Bell's nonresonant cosmic-ray instability. Astrophys. J. **678**, 255–261 (2008). arXiv:0801.4486

E.G. Zweibel, Cosmic-ray history and its implications for galactic magnetic fields. Astrophys. J. **587**, 625–637 (2003). arXiv:astro-ph/0212559

E.G. Zweibel, J.E. Everett, Environments for magnetic field amplification by cosmic rays. Astrophys. J. **709**, 1412–1419 (2010). arXiv:0912.3511

Space Sci Rev (2012) 166:97–132
DOI 10.1007/s11214-011-9770-3

Cosmic Rays in Galactic and Extragalactic Magnetic Fields

Felix Aharonian · Andrei Bykov · Etienne Parizot · Vladimir Ptuskin · Alan Watson

Received: 20 October 2010 / Accepted: 15 March 2011 / Published online: 12 April 2011

Abstract We briefly review sources of cosmic rays, their composition and spectra as well as their propagation in the galactic and extragalactic magnetic fields, both regular and fluctuating. A special attention is paid to the recent results of the X-ray and gamma-ray observations that shed light on the origin of the galactic cosmic rays and the challenging results of Pierre Auger Observatory on the ultra high energy cosmic rays. The perspectives of both high energy astrophysics and cosmic-ray astronomy to identify the sources of ultra high energy cosmic rays, the mechanisms of particle acceleration, to measure the intergalactic radiation fields and to reveal the structure of magnetic fields of very different scales are outlined.

Keywords Cosmic rays, ISM · (ISM:) supernova remnants · Clusters of galaxies · Shock waves · Magnetic fields

F. Aharonian
Center for Astroparticle Physics and Astrophysics, DIAS, Dublin, Ireland

F. Aharonian
MPIK, Heidelberg, Germany
e-mail: Felix.Aharonian@mpi-hd.mpg.de

A. Bykov (✉)
Ioffe Institute for Physics and Technology, 194021 St. Petersburg, Russia
e-mail: byk@astro.ioffe.ru

E. Parizot
APC, Université Paris Diderot 10, rue Alice Domon et Léonie Duquet, 75205 Paris Cedex 13, France
e-mail: parizot@apc.univ-paris7.fr

V. Ptuskin
Pushkov Institute of Terrestrial Magnetism, Ionosphere and Radio Wave Propagation of the Russian Academy of Science (IZMIRAN), Troitsk, Moscow Region 142190, Russia
e-mail: vptuskin@izmiran.ru

A. Watson
School of Physics and Astronomy University of Leeds, Leeds LS2 9JT, UK
e-mail: a.a.watson@leeds.ac.uk

1 Introduction

To a first approximation, the all-particle spectrum of cosmic rays (CR)—ionized atomic nuclei with energies extending from the MeV to more than 10^{20} eV, can be described by a power law over 11 decades on particle energy, so that the dependence of cosmic ray intensity on particle energy is close to $E^{-2.7}$ above 10 GeV/nucleon see Fig. 1. Spectra of different CR species are different however. Closer examination reveals some structures in the galactic cosmic ray spectrum that includes more flat low energy part at $E < 10$ GeV/nucleon (mainly due to the modulation of the Galactic cosmic ray intensity by the solar wind flow), the knee at about 4×10^{15} eV, the possible second knee at $\sim 4 \times 10^{17}$ eV, the ankle at $\sim 4 \times 10^{18}$ eV where the spectrum flattens again, and finally the suppression of flux above $\sim 5 \times 10^{19}$ eV. The latter can be interpreted as the so-called Greisen-Zatsepin-Kuzmin (GZK) cutoff (Greisen 1966; Zatsepin and Kuzmin 1966) caused by the interaction of the ultra-high-energy cosmic rays (UHECRs) with the cosmological microwave background (CMB). Therefore, the suppression of the UHECR flux, taken together with the detection of an anisotropy in their ar-

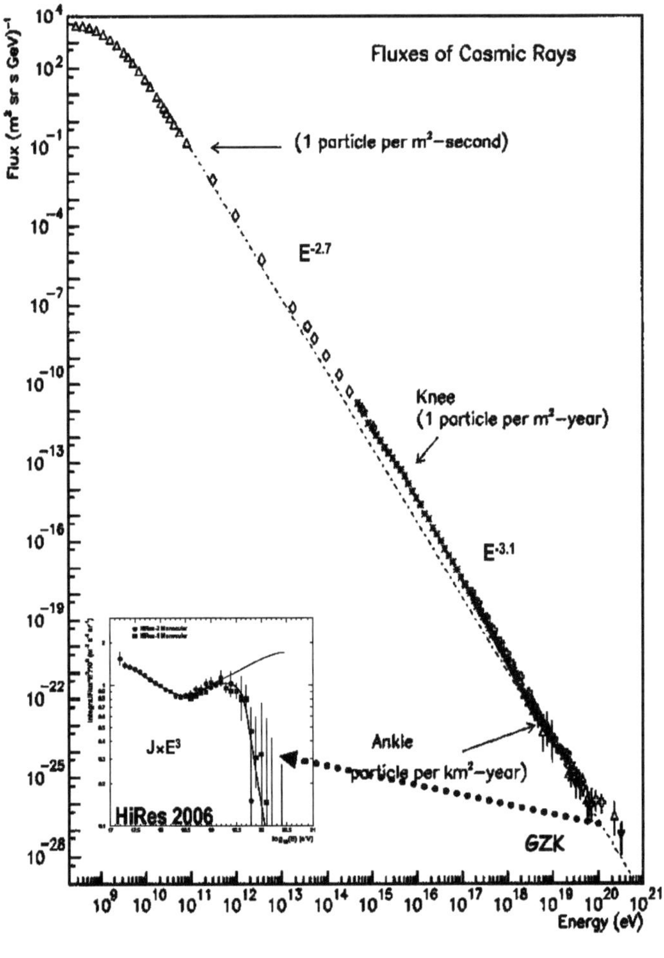

Fig. 1 All particle spectrum of cosmic rays (after S. Swordy and from Abbasi et al. (2008a)—*insert*)

rival direction (Abraham et al. 2007) and the absence of any anisotropy associated with galactic structure is often taken as evidence of the extragalactic origin of these particles. However, it should be noted that the reported cutoff at 10^{20} eV could also be related to the limited capability of even extreme cosmic accelerators to boost the energy of individual particles up to 10^{20} eV. In any case, it is widely believed that the highest energy cosmic rays above 10^{19} eV are of extragalactic origin (see however Blasi et al. 2000; Calvez et al. 2010).

On the other hand, the energy region below 10^{15} eV is dominated by particles produced by Galactic sources. A strong argument for this is provided by gamma-ray observations of the Galactic disk, and the comparison with the corresponding observations in the Magellanic clouds. Furthermore, it is likely that even low-intensity magnetic fields in the intergalactic space prevent low-energy CRs from reaching us from distant galaxies on cosmological timescales. Therefore, a transition from galactic to extragalactic cosmic rays is expected to occur somewhere between 10^{17} eV and 10^{19} eV, while it is tempting to associate the flattening of the spectrum at the so-called ankle in the energy spectrum as marking the transition region. However, different interpretations have been proposed, with a transition at lower energy (Berezinsky et al. 2004; Lemoine 2005; Aloisio et al. 2007), and this important part of high-energy CR phenomenology is still much debated (Allard et al. 2007; Berezinsky 2009). It is interesting to note that this question is strongly related to that of the source composition of extragalactic CRs (Allard et al. 2005), and of course, the energy scale of the transition depends on the capacity of the Galaxy to confine cosmic rays up to high energies, and thus on the intensity and structure of the Galactic magnetic field.

The spectra of both Galactic and extragalactic cosmic rays are shaped by two basic processes—the acceleration in the sources and the subsequent propagation in cosmic magnetic fields and radiation fields. We concentrate here only on the CR propagation and acceleration while the extensive review of the galactic and extragalactic magnetic fields is presented in the other chapters of the volume (see also Ferrière 2001; Beck 2008; Kulsrud and Zweibel 2008, and the references therein).

2 Potential Sources

If one uses the cosmic ray energy requirements and the nonthermal radiation as a guideline, then the most powerful accelerators of relativistic particles in the Galaxy should be supernovae and supernova remnants, pulsars, neutron stars in close binary systems, winds of young massive stars, and possibly gamma-ray bursts or hypernovæ. The total power L_{cr} needed to maintain the observed energy density of cosmic rays is estimated as 10^{41} erg/s. For the acceleration by a supernovae, this estimate implies the release of energy in the form of cosmic rays of approximately 1.5×10^{50} erg per supernova if the supernova rate in the Galaxy is 1 every 50 years. This value comes to about 10–20% of the kinetic energy of the ejecta, which is compatible with the expectations of the theory of diffusive shock acceleration. This assumes the acceleration of cosmic rays by the outward propagating shock, which results from the supernova explosion and propagates in the interstellar medium or in the wind of the progenitor star.

The rotational energy of a young pulsar with period P that remains after the supernova explosion is estimated to be $2 \times 10^{50}(10\ \mathrm{ms}/P)^2$ erg. It is an additional energy reservoir for particle acceleration and in particular it could be the source of very high energy electron-positron pairs. The production of relativistic positrons by pulsars may explain their presence in cosmic rays, see e.g. Profumo (2008). A general dimensional analysis by Hillas (1984,

2005) provided a useful tool to estimate the maximal CR energies that can be achieved by electromagnetic acceleration of charged particles in a source of given size and magnetic field magnitude. It follows that the highest energy CR particles of energy $\gtrsim 10^{20}$ eV can occur either in the compact sources with very large magnetic fields or in the most extended cosmological objects, though the particle energy losses may affect the maximal CR energy estimations.

2.1 Supernova Remnants as CR Accelerators and Gamma-Ray Astronomy

The observations of non-thermal radio emission of supernova remnants have established the presence of effective acceleration of cosmic ray electrons in these objects, e.g. Lozinskaya (1992), Drury et al. (2001). Shell-type supernova remnants exhibit a broad range of spectral indices, centered roughly on $\alpha = -0.5$. This implies a power-law distribution of electrons with an average index $\gamma = 2\alpha + 1 \approx 2.0$, but it varies in a broad range from about 1.4 to 2.6. The analysis of the synchrotron emission in the young supernova remnant Cas A (Jones et al. 2003) showed the presence of electrons with energies up to 200 GeV with a magnetic field strength of about 500 μG. The interpretation of nonthermal radio emission from external galaxies proved that supernova remnants are the sites of acceleration of relativistic electrons with an efficiency similar to the one needed to provide the observed intensity of Galactic cosmic-ray electrons (Duric et al. 1995).

In the case of young SNRs, the spectra of nonthermal synchrotron radiation continues into the X-ray region. In particular, bright nonthermal X-ray structures have been observed from SN 1006, RX J1713.7-3946, RCW 86, and Cas A. The only viable explanation of this emission is the synchrotron radiation of ultrarelativistic electrons accelerated to energies as high as 100 TeV (for a review, see Vink 2008; Reynolds 2008). Inverse Compton scattering of background photons by these electrons, together with gamma rays produced via π^0 decays are two major mechanisms of gamma-ray production related to the forward and reverse shocks.

It is generally believed that gamma-ray astronomy has a key role to play in solving the problem of origin of galactic cosmic rays. The basic concept is simple and concerns both the acceleration and propagation aspects of cosmic rays. Namely, while the localized gamma-ray sources exhibit the sites of production/acceleration of cosmic rays, the angular and spectral distributions of the diffuse galactic gamma-ray emission provide information about the character of propagation of relativistic particles in galactic magnetic fields. The crucial outcome of current efforts involving both the ground-based and space-borne gamma-ray detectors is hoped to be a decisive test of the popular hypothesis that supernova remnants (SNRs) are responsible for the bulk of the observed cosmic ray flux. Three factors, (i) high efficiency of transformation of the available kinetic energy of bulk motion into nonthermal particles, (ii) the predicted, within the DSA paradigm, hard energy spectra of protons extending to multi-TeV energies and (iii) relatively high density of the ambient gas—make the young supernova remnants viable sources of gamma-rays resulting from production and prompt decay of secondary π^0-mesons. Thus, the most straightforward test of acceleration of proton and nuclei in SNRs can be performed via search for π^0-decay gamma-rays—either directly from shells of young SNRs (Drury et al. 1994) or from nearby clouds interacting with an expanding SNR shell (Aharonian et al. 1994).

One may argue that the TeV gamma-ray domain is the best energy band to explore this possibility—from the point of view of the superior performance of the detection technique and because of the key information about particle acceleration carried by TeV gamma-rays. On the other hand, TeV gamma-rays are expected only from young SNRs, when the shock speed can be as high as 3000 km/s—a key condition for acceleration of protons to energies 100 TeV and beyond. With the age, typically $t \geq 2000$ yr, the shock speed and the magnetic field in the expanding shell are decreased to the point when the particle acceleration significantly slows down. Moreover, the highest energy particles accelerated at the early epochs, effectively leave the remnant. This results in dramatic reduction of gamma-ray emissivity at TeV energies. Meanwhile the production of GeV gamma-rays continues. Thus we expect MeV and GeV gamma-rays not only from young, but, to a large extent, also from mid-age SNRs. In this regard we expect significantly more GeV emitting SNRs compared to the TeV emitting SNRs. Unfortunately, due to relatively low gamma-ray fluxes, as well as very high gamma-ray background in the crowded regions of the galactic disk, the detection of GeV gamma-rays from SNRs is quite difficult. In the case of mid-age SNRs we expect also "delayed" gamma-ray emission produced by cosmic rays which have left the source and interact with nearby massive gas molecular clouds. The spectrum of cosmic rays, and, consequently, gamma-rays produced in p-p interactions depends on several factors, including the spectrum of protons that escaped the supernova remnant, the character of diffusion, the age of the source and the distance between the source and cloud. In this regard one should expect gamma-ray emission with GeV-to-TeV flux ratio which can vary significantly from a source to source.

Over the last several years TeV gamma-ray emission has been reported from several SNRs. Presently six young shell type SNRs—Cas A, RX J1713.7-3946, RX J0852-4622 (Vela Jr), RCW 86, SN 1006, Tycho—are identified as TeV gamma-ray emitters. Several others, in particular IC 433, W28 and CTB 37B are, most likely, SNR/molecular-cloud interacting systems. Also, several galactic TeV gamma-ray sources spatially coincide with the so-called composite supernova remnants, objects with characteristic features of both standard shell-type SNRs and Pulsar Wind Nebulae (see e.g. Aharonian et al. 2008).

In Fig. 2 we shown the TeV and X-ray images of the strongest and best studied gamma-ray emitting SNR—RX J1713.7-3946.

The age of this SNR RX J1713.7-3946 at a distance of about 1 kpc is estimated between 1 to 3 thousand years. Therefore, it can be formally teated as a representative of young galactic SNRs like SN 1006, Vela Junior, Tycho, Kepler, Cas A. However, RX J1713.7-3946 seems to be a unique object with quite unusual characteristics. This circumstance makes the identification of the gamma-ray mechanism, and hence the nature of the parent particles a nontrivial problem. Meanwhile the question of the origin of gamma-rays has a fundamental implication related to the role of SNRs to the production of galactic cosmic rays.

Remarkably, both the hadronic (gamma-rays produced in pp interactions) and leptonic models (gamma-ray produced via inverse Compton scattering) provide, with more or less success, but yet acceptable fits to the measured gamma-ray spectra and explain also the spatial distributions of gamma-rays and X-rays Zirakashvili and Aharonian (2010). So the test of these models relies essentially on multiwavelength data.

One of the most puzzling features of RX J1713.7-3946 is the lack of thermal X-ray emission which is a serious argument against the hadronic models of this source. The tight upper limit on the thermal X-ray flux of RX J1713.7-3946 is explained by the supernova explosion inside a wind-blown bubble with a very low gas density $n_{\rm gas} \ll 1\ {\rm cm}^{-3}$. This allows to suppress the free-free component of thermal emission but not the X-ray lines (Ellison et al. 2010; Zirakashvili and Aharonian 2010), thus an additional assumption about the peculiar composition with reduced content of heavy elements is required to avoid the conflict with X-ray data (Zirakashvili and Aharonian 2010). The hadronic models demand quite large magnetic field of order of 100 μG or larger. This, for the given flux of synchrotron radiation, requires only $\approx 10^{46}$ erg in electrons. On the other hand, the total energy in protons

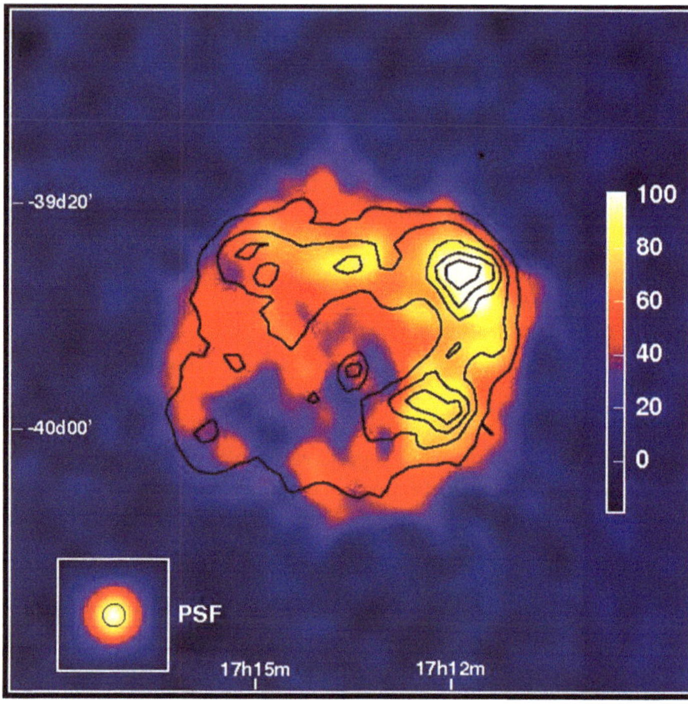

Fig. 2 The gamma-ray image of RX J1713.7-3946 obtained with the H.E.S.S. telescope array (Aharonian et al. 2007b). In the *left-hand side*, the overlaid *light-gray contours* illustrate the significance of the different gamma-ray features at 8, 18, and 24σ levels. In the *right-hand side*, the *black-lines* show the X-ray contours

should be 10^{50} erg or significantly larger if the background gas density $n \ll 1\ \text{cm}^{-3}$. Thus, in the hadronic scenario the electron-to-proton ratio is close to $K_{\text{ep}} = 10^{-4}$. This is a two orders of magnitude smaller than the "standard" electron/proton ratio observed in local cosmic rays.

A major challenge for the leptonic models is the demand for a small magnetic field in the shocked shell of the remnant. Within the framework of a one-zone model in which the electron acceleration and gamma-ray emission take place in the same region, the magnetic field cannot significantly exceed 10 μG, otherwise it leads, for the given X-ray flux, to significant suppression of IC gamma-rays.

A possible solution to the problems related to the pure (oversimplified) one-zone leptonic and hadronic models could be effective acceleration of electrons and protons in both forward and reverse shocks with an additional assumption of existence of dense condensations embedded in very low density shell. In this scenario the overall VHE gamma-ray emission is dominated by the IC scattering of electrons, but in addition to the leptonic component one should expect non-negligible contribution from hadronic gamma-rays produced at interactions of protons with dense gas condensations. Moreover, in this scenario both the low energy, $E \leq 100$ GeV and highest, $E \geq 10$ GeV gamma-rays should arrive from specific regions occupied by dense gas condensation (see Fig. 3). The "composite" scenario allows a quite relaxed parameter space compared to the "pure" hadronic model and leptonic models, and at the same time, provides better fits to the broad-band spectral energy distribution of the source.

Independent of the ability of different models to describe the spectral and morphological features of gamma-ray emission of RX J1713.7-3946, it is obvious that we deal with a source

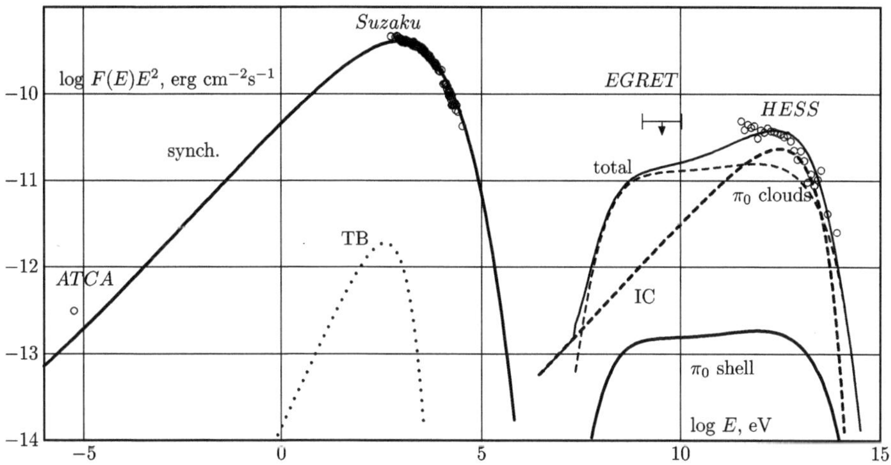

Fig. 3 Broad-band emission of RX J1713.7-3946 for the "composite" scenario of gamma-rays with non-modified forward and reverse shocks and dense clouds in the shell Zirakashvili and Aharonian (2010)

that effectively accelerates electrons and protons to energy of 100 TeV and beyond. On the other hand, this object has many unique features, and in this regard it could be misleading if we treat this source as a representative of the whole population of young SNRs. Therefore, the studies of other TeV emitting SNRs is crucial for definite conclusions concerning the role of SNRs in production of galactic cosmic rays.

The realization of different possible scenarios of particle acceleration and radiation in general, and the uncertainties in relevant model parameters like the strength of the magnetic field and the density of ambient gas, prevent us from unambiguous conclusions concerning the fraction of the hadronic component in the flux of observed gamma-rays, and consequently do not allow model-independent estimates of thew total energy released in accelerated protons and nuclei. The best hope in this regard is related to the detection of high energy neutrinos produced in interactions of accelerated protons and nuclei with the ambient plasma. Although the fluxes of neutrinos are quite similar to the gamma-ray fluxes, because of limited sensitivity of the neutrino detectors, this extremely important channel of information unfortunately remains unexplored. Even for the largest neutrino telescopes like the recently completed *IceCube* and the planned several km-cube underwater detector in the Mediterranean Sea, the high energy neutrino signal from RXJ 1713.7-3946 is expected to be marginal even under the assumption that the flux of detected TeV gamma-rays is fully consists of photons of hadronic origin (see e.g. Vissani et al. 2011; Morlino et al. 2009, and the references therein).

2.2 OB-Associations, Starburst, Superbubbles

Collective stellar winds and supernovae in clusters and associations of massive stars in galaxies can be potentially efficient cosmic ray accelerators (see e.g. Cesarsky and Montmerle 1982; Bykov and Fleishman 1992; Axford 1994; Higdon et al. 1998; Bykov 2001; Bykov and Toptygin 2001; Binns et al. 2007).

The young rich open cluster Westerlund 2 is spatially coincident with TeV photon source HESS J1023-575 (Aharonian et al. 2007a). The reported source extension excludes a single star origin of the observed VHE emission and argues in favor of the production of cosmic rays by collective stellar winds from the ensemble of at least a dozen of hot massive O-type and a few WR-type stars in the stellar cluster Westerlund 2 (of an estimated age of about a few million years).

Recently, TeV gamma-ray emission has be reported from two star-burst galaxies, NGC 253 and M82, by H.E.S.S. (Acero et al. 2009) and VERITAS (Acciari et al. 2009), respectively. Starburst galaxies exhibit in their central regions a highly increased rate of supernovae that are likely to be efficient producers of cosmic rays. As they are extended, with radii up to $\sim$100 pc, the superbubbles (SB) containing dozens of supernova remnants and winds of massive stars are difficult to identify in the Galaxy, and are thus currently best studied in the Magellanic Clouds. Nonthermal X-rays were detected by Bamba et al. (2004) from the shells of the superbubble 30 Dor C in the LMC. The X-ray observations revealed the SB morphology as a nearly circular shell with a radius of 40 pc, which is bright on the northern and western sides. The non-thermal X-ray shell traces the outer boundary of the radio shell. These features of thin thermal and non-thermal X-rays are similar to those of SN 1006, a prototype of a synchrotron X-ray shell, but the non-thermal component of 30 Dor C is about 10 times brighter than that of SN 1006. The source is also much older than that of SN 1006, so the particle acceleration time in this superbubble may be longer than those in normal shell-like supernova remnants.

Nonlinear modeling of CR acceleration in superbubbles by Bykov (2001) predicted a high efficiency of the kinetic energy conversion to the CRs that can be $\sim$20% and even higher with a significant temporal evolution of particle spectra inside the superbubble. Magnetic field inside the superbubble can be amplified up to about 30 μG providing the maximal energies of CR protons to be about 10^8 GeV and higher for CR nuclei that dominate the accelerated CR composition at highest energies in the model (see Bykov and Toptygin 2001). To model the CR composition in superbubbles a number of components including the ^{22}Ne-rich Wolf-Rayet and dust material must be accounted for (Binns et al. 2007; Meyer et al. 1997; Ellison et al. 1997).

2.3 CR Abundance Constraints on Potential CR Sources

The correction of the cosmic ray composition observed at the Earth for the effects of nuclear fragmentation in the interstellar medium makes it possible to determine the initial elemental and isotopic composition of accelerated particles, to clarify the process of cosmic ray acceleration and the nature of cosmic ray sources.

The relative abundance of chemical elements in cosmic ray sources is in general similar to the solar and local galactic abundances but with some interesting deviations. The popular scenarios which explain the cosmic ray source composition include the acceleration of grains together with relatively less abundant volatile ions by shocks in the interstellar medium (Meyer et al. 1997; Ellison et al. 1997), the acceleration of freshly formed material, particularly grains in young supernova remnants (Lingenfelter et al. 1998), and the acceleration in hot superbubbles with multiple supernova remnants (see e.g. Binns et al. 2007).

The isotopic composition of cosmic rays is now measured for all stable isotopes for elements from H to Ni. The isotopic composition of cosmic ray sources is strikingly similar to the composition found in the solar system (Wiedenbeck et al. 2001). However, a well-established anomaly in the isotopic composition of galactic cosmic rays is the excess of ^{22}Ne. The ratio ^{22}Ne/^{20}Ne is enhanced by a factor of 4 compared with the solar reference value (Duvernois et al. 1996a). It can be explained only by the special conditions of nucleosynthesis. The enhancement of neutron rich isotopes would be expected in the highly evolved very massive stars in their Wolf-Rayet stage when their surfaces contain large excesses of the products of core helium burning, including ^{22}Ne (Casse and Paul 1982). An increased cosmic ray ratio C/O by a factor of 2 is also in favor of Wolf-Rayet stars.

Using the data from *Cosmic Ray Isotope Spectrometer* aboard the *ACE* spacecraft Binns et al. (2007) measured the isotopic abundances of neon and a number of other species in the galactic cosmic rays. The authors found that the three largest deviations of galactic cosmic ray isotope ratios from solar-system ratios, namely ^{12}C/^{16}O, ^{22}Ne/^{20}Ne, and ^{58}Fe/^{56}Fe, are consistent with a model of CR source consisting of about 20% of Wolf-Rayet wind material mixed with about 80% material of the solar-system composition. Therefore Binns et al. (2007) concluded that OB associations within superbubbles are the likely source of at least a substantial fraction of GCRs.

The analysis of the ^{59}Co/^{56}Fe abundance ratio also provides important information. Indeed, the relatively high, close to the solar value, of this ratio (Mewaldt 1999) testifies that the major part of the originally synthesized ^{59}Ni, which is stable when it is ionized (i.e. when it is accelerated among the cosmic rays), has decayed by the K-capture of an orbital electron into ^{59}Co before the acceleration started. Thus, the delay between the synthesis of this material and its acceleration has to be larger than 10^5 yr severely constraining the models with the acceleration of freshly ejected matter in SNRs.

The production of cosmic rays with power $L_{\rm cr} = 10^{41}$ erg/s per normal galaxy can provide a cosmic-ray energy density of $w_{\rm Mg} = L_{\rm cr} N_{\rm g} T_H = 3 \times 10^{-17}$ erg cm^{-3}, where $N_{\rm g} = 2 \times 10^{-2}$ Mpc^{-3} is the number density of galaxies and $T_H = 1.4 \times 10^{10}$ yr is the Hubble time. This is almost 5 orders of magnitude smaller than the energy density of cosmic rays observed in the solar system. The contribution of active galaxies can reduce this discrepancy by only a factor of 10. All this is in agreement with the picture described in the Introduction, where the main part of the cosmic rays we observe at the Earth is produced by galactic sources (e.g. supernova remnants) and slowly diffuse in galactic magnetic fields out of the Galaxy to the intergalactic space, where the cosmic ray density is much smaller. The efficiency of the confinement of energetic particles is decreasing with energy, and because of that the difference between the cosmic ray densities inside and outside the Galaxy is also decreasing. The densities become equal at some energy between 10^{17} and 10^{19} eV and the extragalactic sources dominate at higher energies. The list of potential extragalactic sources includes in particular the active galactic nuclei, the progenitors of gamma-ray bursts, the accretion shocks of the size of galactic clusters and larger (see Torres and Anchordoqui 2004; Sigl 2009, for a review). These objects are probably able to provide the required power $q_{\rm Mg} = 3 \times 10^{36}$ erg s^{-1} Mpc^{-3} in cosmic rays with energies $>10^{19.5}$ eV. This estimate of $q_{\rm Mg}$ is based on the cosmic ray flux determined using the Auger Observatory (Pierre Auger Collaboration et al. 2010a) and it takes into account the energy loss time due to the GZK effect. It is notable that the entire population of supernovae in the Universe also meet the required power condition (Aublin and Parizot 2006), but as far as we know the highest energy particles can not be accelerated in such sources. The origin of the highest energy CRs in extragalactic sources remains a mystery, whose solution will require enhanced detections capabilities (because of the very low fluxes involved, see below), but will have a strong impact on the whole field of high-energy astrophysics.

3 Transport in the Galaxy

The interaction of relativistic charged particles with galactic magnetic fields explains the high isotropy and relatively large confinement time of the cosmic rays in the Galaxy. It is accepted that the diffusive approximation gives an adequate description of the cosmic ray propagation in the interstellar medium at energies up to $\sim 10^{17}$ eV. The diffusion model forms a basis for interpretation of the cosmic ray data and the related radio-astronomical, X-ray and gamma-ray observations.

3.1 Empirical Model of Cosmic Ray Propagation

The procedure to model the cosmic ray diffusion in the Galaxy can be summarized as follows. One must first specify the cosmic ray sources, define the shape of the cosmic ray halo and the conditions at its boundaries (it is usually assumed that energetic particles exit freely into the intergalactic space where the cosmic ray density is negligible). The basic diffusion-convection equations for different cosmic ray species describe the diffusion of relativistic charged particles and, if needed, their convective transport in the models including a galactic wind. The equations should incorporate all possible energy loss and gain processes in the interstellar medium, nuclear fragmentation, and the radioactive decay of unstable nuclei. One can then calculate the distribution functions of protons, electrons and the different types of nuclei. The transport coefficients of cosmic rays (diffusion coefficient and convection velocity), the properties of the cosmic ray sources (total power, energy spectra of the different components, elemental and isotopic composition), and the size of the confinement region of cosmic rays in the Galaxy can be inferred by fitting all the available data on cosmic rays. Hundreds of stable and radioactive isotopes are included in the calculations of nuclear fragmentation and the transformation of the energetic nuclei as they interact with the interstellar gas. The most advanced code developed for the numerical modeling of cosmic ray propagation in the Galaxy is the GALPROP code which uses a Crank-Nicholson implicit second-order scheme (Strong and Moskalenko 1998; Strong et al. 2007). It incorporates as much realistic astrophysical input as possible, together with the latest theoretical development, and numerically solves the transport diffusion-convection equations for all cosmic-ray species.

One of the key channels of information about cosmic-ray propagation is the abundance of the secondary energetic nuclei Li, Be, B, Sc, V, Ti, ^{2}H, ^{3}He and others produced as a result of the spallation of heavier primary nuclei interacting with the interstellar gas. The observed flux ratios between secondary and primary nuclei, for example the boron to carbon ratio, show a maximum at about $E = 1$ GeV/nucleon (e.g., Engelmann et al. 1990; Yanasak et al. 2001; Ahn et al. 2008). The mass distribution of secondary nuclei can be understood in terms of their propagation through about 10 g/cm^2 of interstellar material. This quantity, referred to as the escape length, can be approximately expressed through the parameters of the plain diffusion model with a flat cosmic ray halo of total thickness $2H \approx 8$ kpc as $X \approx v\mu_g H/2D$, where v is the particle velocity, $\mu_g \approx 2.4$ mg/cm^2 is the surface gas density of galactic disk, and D is the cosmic ray diffusion coefficient. This allows an estimate of the diffusion coefficient: $D \sim 3 \times 10^{28}$ cm^2/s at $E = 1$ GeV/nucleon.

The primary-to-secondary ratios decrease at higher energies (at least up to a few hundred GeV/nucleon where the measurements exist) as well as at lower energies. The understanding of this behavior is provided by the kinetic theory of particle interaction with the interstellar magnetic fields.

3.2 Kinetic Theory of Cosmic Ray Diffusion in Galactic Magnetic Fields

The theory of energetic particle transport in the galactic magnetic fields is constructed in much the same way as in the better studied case of particle transport in the interplanetary magnetic fields. The detailed treatment of cosmic ray diffusion in magnetic fields can be found in monographs (Toptygin 1985; Berezinskii et al. 1990; Schlickeiser 2002) (see also Ptuskin et al. 2006).

On the "microscopic level" the spatial and momentum diffusion of cosmic rays results from the particle scattering on random MHD waves and discontinuities. The effective "collision integral" for energetic charged particles moving in small-amplitude random fields $\delta B \ll B$ can be taken from the standard quasi-linear theory of plasma turbulence, see e.g. Kennel and Engelmann (1966). The wave-particle interaction is of resonant character and depends on the particle gyroradius, which is equal to $r_g = pc/ZeB = 10^{12} R_{GV}$ cm in a typical interstellar magnetic field $B = 3$ μG (here R_{GV} is the particle magnetic rigidity in GV and μ is the cosine of particle pitch angle). The particle is predominantly scattered by the irregularities of magnetic field δB that have the projection of the wave vector on the magnetic field direction equal to $k_\parallel = \pm s/(r_g \mu)$. The integers $s = 0, 1, 2, \ldots$ correspond to the cyclotron resonances of different orders. The efficiency of particle scattering depends on the polarization of the waves and on their distribution in **k**-space. The first-order resonance $s = 1$ is the most important for the isotropic and also for the one-dimensional distribution of random MHD waves along the average magnetic field, $\mathbf{k} \parallel \mathbf{B}$. In some cases—for calculation of scattering at small μ and for calculation of perpendicular diffusion—the broadening of resonances and magnetic mirroring effects should be taken into account. Locally, the cosmic ray diffusion is anisotropic and goes along the local magnetic field. The isotropization of diffusion is accounted for by the presence of strong large-scale (~100 pc) fluctuations of the galactic magnetic field. The problem of calculation of the average diffusion tensor is not trivial even in the case of relatively weak large scale random fields, since the field is almost static and the strictly one-dimensional diffusion along the magnetic field lines does not lead to non-zero diffusion perpendicular to **B**, see Casse et al. (2002), Webb et al. (2009).

The following simple equations for the diffusion coefficient are useful for estimates:

$$D = v r_g B^2 / [12\pi k_{res} W(k_{res})] = \frac{v r_g^a}{3(1-a) k_L^{1-a}} \frac{B^2}{\delta B_L^2}, \tag{1}$$

here $k_{res} = 1/r_g$ is the resonant wave number, and $W(k)$ is the spectral energy density of waves normalized as $\int dk W(k) = \delta B^2/4\pi$. The random field at the resonance scale is assumed to be weak, $\delta B_{res} \ll B$. The principle wave number of the turbulence is k_L and the amplitude of random field at this scale is δB_L. The second equation in (1) is valid for particles with $r_g < k_L^{-1}$ under the assumption of a power law spectrum of turbulence $W(k) \propto 1/k^{2-a}$, $k > k_L$. The diffusion coefficient has then a power law scaling on momentum $D \propto v(p/Z)^a$.

Equation (1) can be used to describe the turbulence which consists of random isotropic distribution of Alfvén and fast magnetosonic waves and it gives correct order of magnitude estimates for the wave distribution concentrated around the direction of average magnetic field. It should be pointed out however that the isotropization of the diffusion tensor does not occur in the case of a pure parallel wave propagation ($\mathbf{k} \parallel \mathbf{B}$). Another special case is 2D turbulence with perpendicular propagation of waves ($\mathbf{k} \perp \mathbf{B}$). In this case, the scattering occurs only for magnetosonic waves through the resonance $s = 0$ which leads to a very large diffusion coefficient, about a factor $(v/V_a)^2$ larger than given by (1).

The diffusion in momentum is approximately described by the equation:

$$D_{pp} = p^2 V_a^2 / (9D). \tag{2}$$

Here equal intensities of waves moving along the magnetic field in opposite directions are assumed. The momentum diffusion is vanished ($D_{pp} = 0$) if all waves move in one direction along the magnetic field with the same phase velocity. The flow of waves carry particles along the field in this case and the convection with velocity determined by the phase velocity of waves appears in the transport equation for cosmic rays, see Skilling (1975a), Berezinskii et al. (1990) for detail. Clearly the cosmic ray convection also arises as a result of the large scale motion of the interstellar gas with frozen magnetic field.

The random component of interstellar magnetic field with an extended spectrum of inhomogeneities can give the resonant particle scattering and spatial diffusion of cosmic rays. Information on the interstellar turbulence spectrum have been obtained from radio scintillation and refraction observations (sensitive to fluctuations of thermal electron density), measurements of the differential Faraday rotation angles from distant sources (mainly produced by fluctuations in the interstellar magnetic field), and the observations of random motions in the interstellar gas. These data are consistent with the assumption that a single close-to-Kolmogorov spectrum extends from scales 10^8 to 3×10^{20} cm, see Armstrong et al. (1995) and references therein. The Kolmogorov spectrum is of the form $W(k) \propto k^{-5/3}$. Other types of spectra frequently used to describe the interstellar turbulence are $W(k) \propto k^{-2}$ for the shock-dominated turbulence, see e.g. the model by Bykov and Toptygin (1987), and spectrum $W(k) \propto k^{-3/2}$ suggested by Iroshnikov (1964) and Kraichnan (1965) in their phenomenological theory of MHD turbulence. Comprehensive reviews of MHD turbulence with application to the interstellar conditions have been given by Zhou et al. (2004), Elmegreen and Scalo (2004), and Scalo and Elmegreen (2004).

The estimate based on the empirical value of the diffusion coefficient for GeV particles (see Sect. 3.1) gives the level of turbulence at the principal scale $k_L = 10^{-21}$ cm^{-1} of the order $\delta B_{tot}/B \sim 0.2$ for an Iroshnikov-Kraichnan spectrum $W(k) \propto k^{-3/2}$ ($a = 1/2$) and $\delta B_{tot}/B \sim 1$ for a Kolmogorov-type spectrum $W(k) \propto k^{-5/3}$ ($a = 1/3$). At the same time, the data on Faraday rotation angles favor the Kolmogorov spectrum with $\delta B_{tot}/B \sim 1$ and $k_L = 10^{-21}$ cm^{-1}. The cascades of Alfvén waves (with the scaling $k^{-5/3}$) and the fast magnetosonic waves ($k^{-3/2}$) are independent and may coexist in the Goldreich and Sridhar (1995) model of MHD turbulence. The amplitude of Alfvén wave cascade may dominate at the principle scale.

There are two physical explanations of the observed secondary-to-primary nuclei ratios in the diffusion model with a static cosmic ray halo.

The first explanation (Simon et al. 1986; Seo and Ptuskin 1994) refers to the diffusion model with distributed reacceleration of cosmic rays by the interstellar MHD turbulence which scatters particles and provides their spatial diffusion. The Kolmogorov-type spectrum is assumed that leads to the rising with rigidity of the diffusion coefficient $D \propto v(p/Z)^{1/3}$. For typical value of Alfvén velocity $V_a \sim 30$ km s^{-1}, the reacceleration is not essential for nuclei with energies $E > 40$ GeV/nucleon and the abundance of secondary nuclei is a decreasing function of rigidity. The impact of reacceleration on spectra of primary and secondary nuclei becomes stronger at smaller energies so that the characteristic time of distributed acceleration in the Galaxy becomes equal to the time of diffusion from the Galaxy at about 1 GeV/nucleon. In consequence the pronounced peak in the secondary-to-primary ratio arises, see Fig. 4. The asymptotic behavior of the escape length is $X \propto v(p/Z)^{-1/3}$ at $E > 40$ GeV/nucleon.

The second explanation (Ptuskin et al. 2006) assumes that the Iroshnikov-Kraichnan spectrum describes the interstellar MHD turbulence. It is characterized in particular by a relatively slow nonlinear cascade of wave from small to large wave numbers. The resonant wave-particle interaction results in the significant wave damping on cosmic rays and termination of the cascade at $k \sim 10^{-12}$ cm^{-1} (in contrast to the case of Kolmogorov cascade which is fast and not noticeably affected by cosmic rays). The amplitude of short waves is

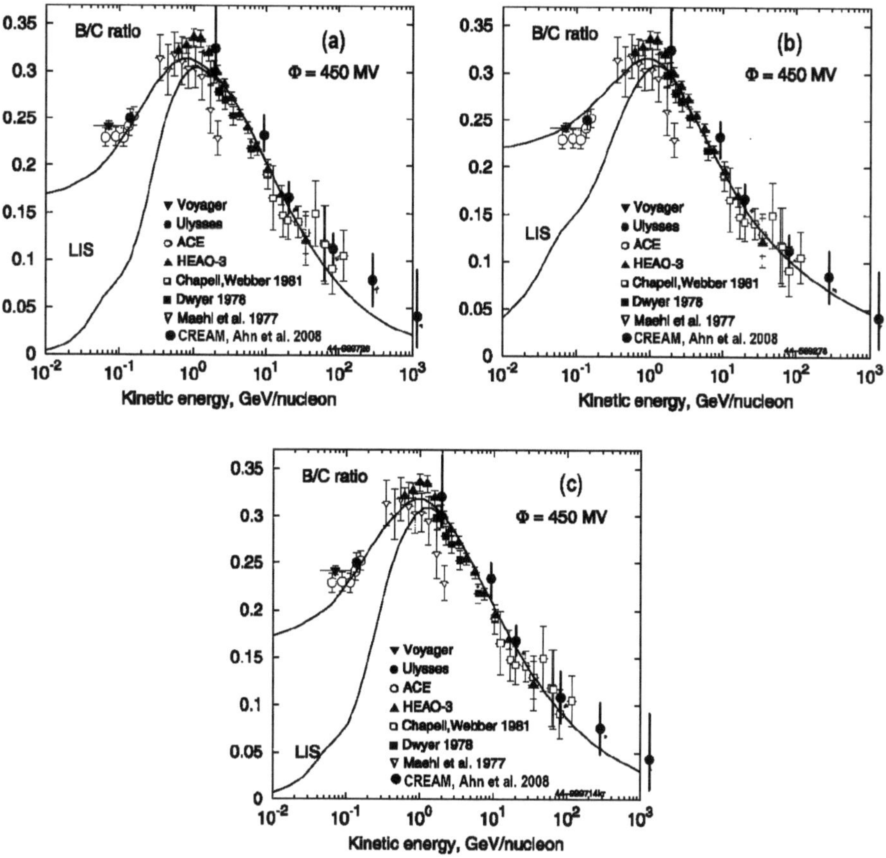

Fig. 4 B/C ratio as calculated in plain diffusion model with unphysical break of the diffusion coefficient (**a**), reacceleration model (**a**), and diffusive reacceleration with damping model (**c**). *Lower curve*—local interstellar spectra, *upper*—modulated (with modulation parameter $\Phi = 450$ MV). Data below 200 MeV/nucleon: ACE (Davis et al. 2000), Ulysses (Duvernois et al. 1996b), Voyager (Lukasiak 1999); high energy data: HEAO-3 (Engelmann et al. 1990), CREAM (Ahn et al. 2008) for other references see Stephens and Streitmatter (1998)

suppressed and the low energy particles rapidly exit the Galaxy without producing many secondaries. It explains the peaks in secondary/primary nuclei ratios at about 1 GeV/nucleon, see Fig. 4. The asymptotic behavior of the scape length in this case is $X \propto v(p/Z)^{-1/2}$ at $E > 3$ GeV/nucleon.

It is clear from Fig. 4 that the experimental data on the abundance of secondary nuclei are well described by both physical models. Also illustrated in Fig. 4 are the results of calculations in the plain diffusion model with the empirical diffusion coefficient of the form $D \propto \beta^{-2}$ and $\beta^{-2}(p/Z)^{0.6}$ below/above 3 GV chosen to fit the B/C data (the "unphysical" model).

It must be emphasized that the theoretical description of MHD turbulence is a complicated and not completely solved problem even in the case of small-amplitude random fields. Since the mid 1990s, there has been a renewed interest in understanding of magnetohydrodynamic turbulence as it applies to the interstellar magnetic field and density fluctuations (Goldreich and Sridhar 1995, 1997; Galtier et al. 2000). Goldreich and Sridhar (1995) exploited anisotropy in MHD turbulence and obtained Kolmogorov-like spectrum for the energy density of Alfvén waves. The main part of the energy density in this turbulence is concentrated perpendicular to the local magnetic field wave vectors $k_\perp \approx k$, while the parallel wave numbers are small: $k_\parallel \sim [kW(k)/(B_0^2/4\pi)]^{1/2} k_\perp$. The cascade is anisotropic with energy confined within the cone $k_\parallel \propto k_\perp^{2/3}$. This turbulence does not significantly scatter cosmic rays. The distribution of slow magnetosonic waves passively follows that of Alfvén waves (Lithwick and Goldreich 2001). They are not damped on cosmic rays (because of the property $k_\perp \gg k_\parallel$) and are probably responsible for the observed interstellar electron density fluctuations. The fast magnetosonic waves with the Iroshnikov-Kraichnan spectrum $W(k) \propto k^{-3/2}$ may have an independent nonlinear cascade which is isotropic and can efficiently scatter cosmic rays. These conclusions were supported by numerical simulations by Cho and Lazarian (2002). This concept of the MHD turbulence favors the second of the discussed above scenarios where cosmic rays are scattered by fast magnetosonic waves with the Iroshnikov-Kraichnan spectrum and suppress the turbulence at large wave numbers $k \geq 10^{-12}$ cm^{-1}.

The knowledge of the diffusion coefficient is absolutely essential for understanding the nature of the spectrum of galactic cosmic rays that is determined by the processes of acceleration in the sources (supernova remnants) and propagation in galactic magnetic fields. Two specific asymptotic power laws of the diffusion coefficient $D \propto (p/Z)^{1/3}$ and $D \propto (p/Z)^{1/2}$ at very high energies together with the observed spectrum approximated by the power law $J \propto E^{-2.7}$ imply the cosmic ray source spectra close to $q \propto E^{2.2}$ and $q \propto E^{2.4}$ respectively.

The contemporary modelling of cosmic ray production by supernova remnants was achieved by Ptuskin et al. (2010). The spectra of high-energy protons and nuclei accelerated by supernova remnant shocks were calculated taking into account magnetic field amplification and Alfvénic drift both upstream and downstream of the shock for different types of supernova remnants during their evolution. Four different types of SNRs (Ia, Ib/c, IIP and IIb) with corresponding burst rates were included in the calculations. The action of cosmic ray pressure on the shock structure was taken into account in the calculations. It was found that the maximum energy of accelerated particles may reach 5×10^{18} eV for Fe ions in Type IIb SNRs. The steady state spectrum of cosmic rays produced by SNRs in the Galaxy was calculated with the deduced source spectra of protons and different kind of nuclei up to Iron. The escape length with a high energy asymptotic $X \propto (p/Z)^{-0.54}$ determined by Jones et al. (2001) by the accurate fit to B/C data was used in the calculations. The derived energy spectrum of cosmic rays including the knee structure around 3×10^{15} eV is in good agreement with the spectrum measured at the Earth from low energies to about 5×10^{18} eV. This result is strongly in favor of the cosmic ray galactic diffusion on the MHD turbulence with spectrum close to $k^{-3/2}$. The Kolmogorov type spectrum $k^{-5/4}$ is unlikely.

The diffusion approximation cannot be used when the diffusion mean free path $l = 3D/v$ exceeds the size of cosmic ray halo $H \approx 4$ kpc that occurs at $E > 2 \times 10^{16} Z$ eV if $D \propto (p/Z)^{1/2}$. Calculations of particle trajectories in galactic magnetic field are needed to study cosmic ray propagation at these ultra high energies. At weaker dependence, $D \propto (p/Z)^{1/3}$, the diffusion approximation breaks at higher energies. The main limitation in this case is the inefficiency of particle scattering at $r_g k_L > 1$ when the diffusion coefficient along the average magnetic field rapidly rises with energy as $D \sim v r_g^2 k_L$, see e.g. Toptygin (1985). This regime is realized at $E > Z \times 10^{17}$ eV. The direct numerical modelling of particle motion in the regular and random galactic magnetic fields (Hörandel et al. 2007) confirmed the presented order of magnitude estimates. The most essential feature of cosmic ray transport at ultra-high energies above about 10^{17}eV is the prevalence of the drift motion across predominantly azimuthal average galactic magnetic field (Zirakashvili et al. 1998). It leads to the inverse dependence of the exit time from the Galaxy on particle momentum $T \propto (p/Z)^{-1}$.

3.3 Instabilities and Plasma Effects

Cosmic rays not always can be treated as test particles moving in given regular and random magnetic fields (Ginzburg 1965; Wentzel 1974; Cesarsky 1980; Kulsrud 2005). The energy density of cosmic rays estimated as $w_{cr} = 1$–$2\ \mathrm{eV\,cm^{-3}}$ is approximately equal to the energy density of the magnetic field and to the energy density of turbulent motions of the interstellar gas. The presence of relativistic charged particles in the interstellar medium leads to collective (plasma) effects during cosmic ray propagation.

A notable example is the cosmic ray streaming instability which develops when the bulk velocity of cosmic rays exceeds the Alfvén velocity V_a. The growth rate of MHD waves amplified by relativistic charged particles with number density $n_{cr}(E) \propto E^{-\gamma+1}$ at the resonance wave number $k_{res} = r_g^{-1}(E)$ is

$$\Gamma_{cr} \approx \Omega_p \frac{n_{cr}}{n}\left(\frac{v\delta_{cr}}{(\gamma+2)V_a} - 1\right), \tag{3}$$

where Ω_p is the gyrofrequency of thermal protons, δ_{cr} is the amplitude of cosmic ray anisotropy. Even for small anisotropy, $\delta_{cr} \approx 10^{-3}$, the instability for galactic cosmic rays with energies ~100 GeV develops in about 10^5 years, i.e. rather rapidly for the galactic timescale. The development of the instability leads to isotropization of the angular distribution of particles and turbulence amplification, see Zweibel (2003), Farmer and Goldreich (2004) and references cited therein. Plasma effects make the overall picture of cosmic ray diffusion in the Galaxy more complicated than it was discussed in Sect. 3. In principle, the cosmic ray diffusion coefficient should be calculated selfconsistently with the account taken for the generation of turbulence by streaming cosmic rays. The examples of such approach and the corresponding transport equations can be found in the papers by Skilling (1975a, 1975b) and Ptuskin et al. (2008). The necessity of considering the variety of dissipation processes of linear and nonlinear wave dissipation in the interstellar medium adds complexity to the investigation and makes the results strongly dependent on the interstellar gas parameters: the density, the state of ionization, the temperature. The wave dissipation makes the effect of streaming instability not efficient at high energies since the cosmic ray density is going down with particle energy and the growth rate (3) is decreasing correspondingly. One of the exceptions is the model of galactic wind driven by cosmic rays in a rotating galaxy developed by Zirakashvili et al. (1996) (it is briefly described at the end of the present section). Moving predominantly along the very long spiral magnetic field lines, the cosmic rays have enough time to amplify the resonant waves even at high energies of the order 10^{15} eV.

The effect of the streaming instability is more significant in the vicinity of the sources. The amplification of magnetic field by the cosmic-ray streaming instability at the shock in supernova remnant is an integral part of cosmic ray acceleration (Bell 1978, 2004; Blandford and Eichler 1987; Malkov and Drury 2001; Zirakashvili and Ptuskin 2008; Marcowith and Casse 2010; Bykov et al. 2011). The cosmic rays produce turbulence which selfconsistently determines the cosmic ray diffusion coefficient in the shock vicinity.

Cosmic rays may induce, in addition to kinetic effects, significant hydrodynamic effects in the Galaxy. Accounting for cosmic ray pressure is principally important for the formation of a halo filled with gas, magnetic field, and relativistic particles. The equilibrium distribution of the interstellar medium above the galactic plane in the gravitational field of stars is subject to the Parker instability (Parker 1966). Cosmic rays play a significant role in the development of this instability. The instability gives rise to large-scale turbulence and helps sustain an almost equipartition energy distribution among cosmic rays, magnetic fields, and turbulent gas motions. The characteristic time for instability development is ~10^7 years in the gaseous galactic disk, and $\sim 10^8$ years in the gas halo. Parker (1992) showed that magnetic arches and loops appearing above the galactic disk due to the action of cosmic rays are necessary for $\alpha\omega$ dynamo to operate, which is primary mechanism of magnetic field generation in the Galaxy. A numerical model of the magneto-hydrodynamical dynamo, driven by cosmic rays in the interstellar medium was developed by Hanasz et al. (2004). The cosmic rays in the model were accelerated in randomly occurring supernova remnants. The cosmic ray propagation was accounted with the diffusion-advection equation supplementing the MHD equations (see also Hanasz et al. 2006). The other essential elements of the model are: vertical gravity of the disk, differential rotation and resistivity leading to reconnection of magnetic field lines. They obtain amplification of the large-scale magnetic field on a timescale of galactic rotation. The model includes the ideas of a fast, cosmic ray driven, galactic dynamo proposed by Parker (1992). The authors find that both the resistivity and the introduced SN activity enhance the efficiency of the cosmic-ray dynamo. The timescale of magnetic field amplification in this model was as short as 140 Myr (Hanasz et al. 2006). The models assumed some CR diffusion prescriptions. A linear study of the effect of the anisotropic CR diffusion with very different but finite parallel ($\kappa_\parallel$) and transverse diffusion coefficients was performed by Ryu et al. (2003). It has been shown that the finiteness of parallel diffusion slows down the development of the Parker instability. However, with a realistic value of $\kappa_\parallel = 3 \times 10^{28}$ cm^2 s^{-1} in the ISM, the maximum growth rate is smaller by only a couple percent than that for $\kappa_\parallel \to \infty$, and the range of unstable wavenumbers remains the same. The inclusion of perpendicular diffusion with $\kappa_\parallel = 0.02\kappa_\perp$ does not change the growth rate noticeably. It should be noted however, that the turbulent advection due to magnetic fluctuations of scales larger than the CR mean free path would make the diffusion coefficient nearly isotropic on scales interesting for the Parker instability (see Bykov and Toptygin 1993). A non-linear development of Parker instability was simulated by Kuwabara et al. (2004). They stated that the growth rate of the instability is larger in the models with smaller diffusion coefficient only in the early linear stage. On the later stages the growth rate becomes smaller when compared to that of the large diffusion coefficient model. The growth of instability is impeded by the CR pressure gradient force interfering with the falling motion of the matter in the small-$\kappa_\parallel$ model, while the magnetic loop can grow up to larger scales in the models with large $\kappa_\parallel$.

It is possible that the gas in the galactic halo is not in static equilibrium but is involved in large-scale convective motion—the galactic wind. The data on galactic soft X-ray emission suggest that a wind exists in our Galaxy, see Everett et al. (2007). It can be supported by cosmic ray pressure (Ipavich 1975; Breitschwerdt et al. 1991). A model was constructed by Zirakashvili et al. (1996) where cosmic rays, after leaving the sources (supernova remnants), determine the wind outflow in the rotating Galaxy with a frozen magnetic field. See Fig. 5. Here, the streaming instability of cosmic rays exiting the Galaxy along the spiral magnetic field lines leads to the MHD turbulence generation, that self-consistently determines the diffusion-convection transport of relativistic particles. The level of turbulence is regulated by the nonlinear Landau damping on thermal ions. The outflow velocity is ~ 30 km s^{-1} at a distance ~ 20 kpc, and the speeds up to a velocity of ~ 400 km s^{-1} several hundred kpc away. The external pressure of the intergalactic gas produces a termination shock at a distance of ~ 300 kpc. In this model, the diffusion coefficient of cosmic ray is not given independently and is self-consistently calculated, being dependent on the power of sources and the spectrum of accelerated particles (Ptuskin et al. 1997). Remarkably, the obtained transport coefficients and other model parameters are consistent with the empirical diffusion-convection model, e.g. Bloemen et al. (1993), for cosmic ray propagation in the model with galactic wind. It should be emphasized that the existence of a galactic wind, either thermal or driven

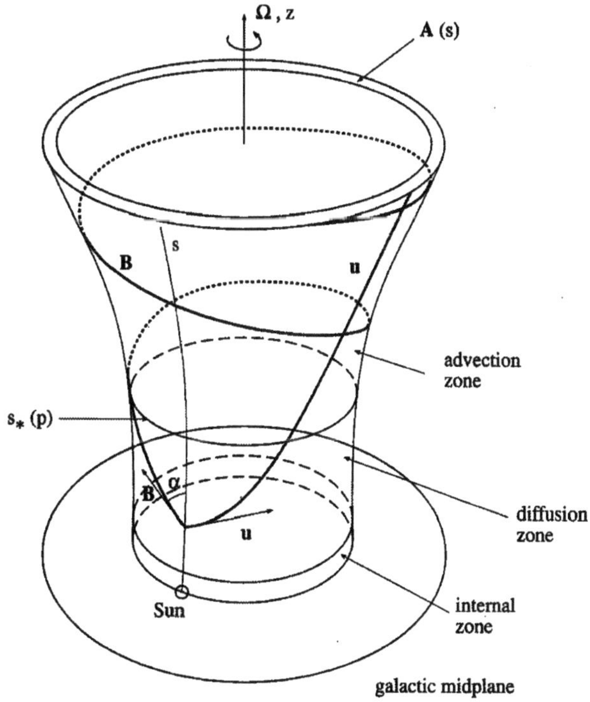

Fig. 5 The structure of the galactic wind flow driven by cosmic rays (Ptuskin et al. 1997). The large-scale magnetic field lines, the flow streamlines and the trajectories of cosmic ray bulk motion from the Galaxy are confined in the same flux tubes. The cylindrical flux tube with the cross section $\mathbf{A}(s)$ originating at the Sun location is shown (here s is the coordinate along the flux tube). Diffusion mainly determines the cosmic-ray transport in the diffusion zone (the analog of the boundary layer), whereas convection mainly makes it in the advection zone. The boundary between two zones $s_*(p)$ is moving up with energy of cosmic-ray particles

by cosmic rays, remains hypothetical. However, Everett et al. (2008) found that the observed galactic soft X-ray emission can be better explained by a wind than by the static gas models and the cosmic ray pressure is essential to drive the wind.

4 The Highest Energy Cosmic Rays

The "Hillas diagram" discussed above predicts that the compact sources like gamma ray bursts (GRB) (see e.g. Mészáros 2006, for a review) are potential sources of UHECRs (Vietri 1995; Waxman 1995; Milgrom and Usov 1995; Waxman 2004), as well as the cosmological sources like active galactic nuclei (AGN), radio galaxies (see e.g. Longair 2010; Blasi 2006), and clusters of galaxies (e.g. Norman et al. 1995). The observed widespread warm-hot intergalactic gas was likely heated by large scale cosmological shocks. The shocks in the Large Scale Structure formation scenario are driven by gravitationally accelerated flows (e.g. Kang et al. 2005). Non-thermal emission observed from clusters of galaxies indicate the presence of accelerated particles in scales larger the galactic halo sizes (e.g. Ferrari et al. 2008). Large-scale accretion shocks in the cluster of galaxies and more generally cosmological shocks in Large Scale Structure can accelerate particles to energies above 10^{19} eV. Namely, Norman et al. (1995) argued that cosmological shocks can be good sites for UHECR acceleration if there is an intergalactic field of order 10^{-9} G and even more important if microgauss regime fields can be self-generated in shocks accelerating particles. The maximal energies cosmic rays accelerated in the large scale shocks like the cluster accretion shocks or hot spots in radio galaxies can reach $\sim 10^{19}$ eV regime only if indeed microgauss fields can be produced (Norman et al. 1995). Current cosmological shock simulations (e.g. Bykov et al. 2008; Dolag et al. 2008) and the non-thermal emission models (e.g. Kushnir et al. 2009) are in favor of the reality of microgauss regime magnetic fields in clusters of galaxies. On the other hand the losses caused by interactions of the UHECRs with photons of the Cosmic Microwave Background Radiation could limit the maximal energies of protons accelerated in clusters of galaxies (see e.g. Vannoni et al. 2009).

4.1 Ultra-high Energy Cosmic Rays as Probes of Magnetic Fields

At the highest energies, the gyroradius of the cosmic rays becomes extremely large, especially in the intergalactic space where the magnetic fields are expected to be much lower than in the Galaxy. Their mode of propagation thus changes from being diffusive to being asymptotically rectilinear, so at extreme energies it is realistic to think of the possibility of a cosmic ray astronomy, with the sources of ultra-high energy cosmic rays (UHECRs) being identifiable directly. For example a proton of energy 6×10^{19} eV will be deflected by 1° to 5° in the galactic magnetic field depending upon the direction and length of the trajectory.

On the one hand, being able to actually *see* the UHECR sources would be a remarkable breakthrough: (i) it would (at least partially) solve the mystery of their origin, (ii) it would put strong constraints on the acceleration mechanism and its efficiency, teaching us a lot about the physics of particle acceleration in the universe in its most extreme manifestations, and (iii) it would offer an important piece of information, complementary to that gathered from multi-wavelengths observations of photons, to guide us in the astrophysical modeling of the sources. This would indeed enhance the power of current multi-messenger analysis, which are mostly relying on upper limits.

On the other hand, if we could be sure to know the source of (at least a set of) UHECRs, we could use these particles as messengers, not only to gather new information about the sources, but also to probe the intervening magnetic fields, by studying the deflections they suffered along their journey to the Earth.

In the following, we review the current status of UHECR observations, and discuss some of their implications. We also discuss why a new generation of detectors can be expected to lead to the above-mentioned breakthrough, and what can already be learned from the current data, including from the fact that its statistics is still insufficient. But to understand the data, we first need to understand the transport of UHECRs, and see how it can modify their energy spectrum, their composition and their arrival directions.

4.2 UHECR Transport: Energy, Mass and Direction

The key ingredient of UHECR transport in the extragalactic space is the so-called GZK effect, which refers to the interactions of the high-energy cosmic rays with the background radiation fields in the universe (Greisen 1966; Zatsepin and Kuzmin 1966). In the case of high-energy protons, the dominant interactions are with the photons of the cosmological microwave background (CMB), producing electron-positron pairs and/or pions as soon as the proton energy is high enough for the CMB photons to exceed the reaction threshold, when boosted in the proton rest frame. These interactions lead to strong energy losses, especially above $\sim 6 \times 10^{19}$ eV, when photopion production becomes important, which significantly limits the distance UHE protons can propagate in the universe (see e.g. Bhattacharjee and Sigl 2000, for a review). This translates into an effective horizon, from beyond which extragalactic sources cannot contribute to the observed UHE protons, and this in turn implies an expected strong suppression of the flux, referred to as the GZK cutoff. At lower energy, energy losses due to photopair-production on the CMB photons also modify the spectrum of UHE protons, producing a dip around 3×10^{18} eV, where the ankle is observed in the overall CR spectrum (Blumenthal 1970;

Berezinskii and Grigor'eva 1988). Another simple interpretation of the ankle, however, is that it marks the transition from galactic to extragalactic cosmic rays.

As already noted in the original work of Greisen (1966) and Zatsepin and Kuzmin (1966), a similar effect holds for heavier nuclei, which can be photodissociated by CMB and/or infrared photons as they propagate through the intergalactic space. The propagation of ultra-high-energy nuclei in the intergalactic space was studied in detail by Puget et al. (1976), and recently revisited with updated and additional cross-sections (Khan et al. 2005; Allard et al. 2005). A key result is that photodissociation is very efficient at ultra-high-energies, and lead to: (i) a modification of the composition of UHECRs as they propagate from their sources to the Earth, as well as (ii) energy losses associated with the loss of nucleons (although pair production also plays a role at lower energy, to a first approximation the nuclei essentially keep a constant Lorentz factor).

The energy threshold for the photodissociation of UHE nuclei depends on the energy of the background photons, as well as on the mass of the nuclei. At the threshold, the dominant process is the excitation of the giant dipole resonance, involving photons around 10 MeV in the nucleus rest frame. Photons in the tail of the CMB distribution reach this energy for nuclei with a Lorentz factor of the order of $\Gamma \gtrsim 10^9$, and energies $E \sim A \times \Gamma m_p c^2$, where A is the mass number of the nucleus. Thus, heavier nuclei can survive photodissociation up to higher energies. By a striking coincidence of Nature, although the processes involved are very different, and the cross section thresholds and sizes are also different, the energy loss length for UHE protons and for Fe nuclei are essentially the same in the intergalactic space, so both species have a very similar horizon structure. Therefore, their elemental GZK cutoff leads to the same spectral shape, and indeed the energy distribution of UHECRs can be understood just as well in terms of pure proton sources or of pure Fe sources (or a combination of both, provided the source spectral index and/or the evolution of the source luminosity as a function of redshift is adjusted correspondingly). In principle, intermediate mass nuclei may also be accelerated in the same UHE sources. However, at a given energy, their Lorentz factor is smaller than the Lorentz factor of Fe nuclei, so they see higher energy photons and get dissociated more rapidly. As a consequence, the very highest energy cosmic rays are likely to be mostly protons and/or Fe nuclei (heavier nuclei would also survive, but *a priori* they are much less abundant in the sources).

The studies on the propagation of UHE nuclei by Allard et al. (2005, 2007) also drew a renewed attention to the question of the galactic/extragalactic transition, whose energy scale turns out to depend on the assumed UHECR source composition. Schematically, a pure proton composition (at the sources) would allow one to fit the high energy spectrum down to below 10^{18} eV, assuming a steep source spectrum (in $E^{-\alpha}$, with $\alpha \sim 2.6$ or 2.7, depending on the assumptions on the cosmological source evolution), and would then imply a galactic/extragalactic transition at a few times 10^{17} eV. Conversely, a mixed source composition (e.g. similar to that of low energy cosmic-rays in the Galaxy) would allow one to fit the energy spectrum with a source spectral index $\alpha \sim 2.3$, and imply that the extragalactic component overcomes the galactic one at the ankle (Allard et al. 2005). We note in passing that this question is also related to the question of the galactic magnetic fields, which must be able to confine the cosmic rays up to the transition energy.

Any information about the UHECR composition would thus have important implications for the general phenomenology of cosmic rays, including low energy ones. Unfortunately, a corollary of the above remarks is that the overall UHE spectrum cannot be used, on its own, to constrain the source composition of the UHECRs. A direct measurement of the mass of UHE particles is thus very important, and although this is experimentally difficult, noticeable progress has been made recently (see below). However, it is fair to say that the experimental situation is not clear yet, and since the measurements also rely on hadronic physics at energies well beyond the energy accessible through direct experiments, it is not clear how conclusively this kind of measurements can resolve the issue in the near future.

In this context, two other aspects of UHECR transport can play a role, which are worth mentioning here since they also involve magnetic fields. First, the interactions associated with the above-mentioned GZK effect(s) produce secondary particles, including the so-called cosmogenic neutrinos and the electromagnetic cascades initiated by secondary gamma-rays (from pion decay) or electrons and positrons. These cascades involve the usual mechanisms of synchrotron radiation, pair production and inverse Compton interactions (see e.g., in the context of UHECR propagation, Bhattacharjee and Sigl 2000). These secondary particles can in principle be detected (or their fluxes constrained). In the case of secondary neutrinos, the expected flux and energy distribution strongly depends on assumptions about the phenomenology of the UHECR sources (for a recent update, including scenarios with a mixed composition, see Allard et al. 2006; Kotera et al. 2010, and references therein). It should also be noted that secondary particles produced in the acceleration site may be detected or constrained, and thus provide additional, multi-messenger information about the UHECR sources and their environment, as well as about the acceleration mechanism itself. Finally, extragalactic magnetic fields can also be constrained by the observation of TeV gamma-ray halos around potential acceleration sites. Recently Taylor et al. (2011) discussed the constraints on the strength of extragalactic magnetic field derived from the simultaneous detections of TeV emitting Blazars in GeV photons by the *Fermi* telescope. The measured GeV flux was found to be lower than that calculated in the cascade model under the assumption of zero magnetic field. Assuming that the reason for the suppression of the cascade component is the extended nature of the cascade emission, the authors concluded that the extragalactic magnetic field above 10^{-15} G of correlation length of 1 Mpc is consistent with the data.

The second aspect, directly associated with magnetic fields, is the deflection of UHE particles between their sources and the Earth. If the source of each individual cosmic ray was known, the angular distance between the source and the arrival direction would give a very precious measurement of the deflection integrated along the path of the cosmic ray. Since the deflection, in a ballistic or semi ballistic regime, is proportional to the particle charge, an analysis of the deflection patterns would provide information about the UHECR charges and/or about the intervening magnetic fields. Although the identification of sources apparently remains out of reach of the current detectors, we shall mention below what the current observational situation can nevertheless teach us, and why this situation is expected to change with a ten times larger statistics at the highest energies. It should also be mentioned that additional hints about the composition of UHECRs might come from the study of their anisotropies, comparing patterns in the arrival directions at different energies and/or angular scale. The basic idea is that protons of energy E have the same gyroradius as heavier nuclei with energy $Z \times E$. Now, in a regime where energy losses can be neglected in a first approximation (say, below the GZK energy scale), the angular transport of the different UHE particles only depend on their gyroradii. Therefore Fe nuclei from a given source follow exactly the same path as protons at an energy 26 times lower. This has recently been turned into a very interesting argument by Lemoine and Waxman (2009), relating anisotropy studies at different energies to composition assertions. Extensive studies with larger statistics should provide important clues about this crucial issue in the future. Likewise, at the highest energies, if the number of sources is low and a few hot spots appear in the UHECR sky, halos of different angular sizes around the centroid of the source should appear, corresponding to different nuclear species (essentially protons and Fe or sub-Fe nuclei, if both are present at the source).

More generally, it is clear that the study of the UHECR angular deflections have a lot to teach us, both about the cosmic rays and about the intervening magnetic fields. The main aspects of this transport can be summarized as follows. We may distinguish deflections occurring: (i) at or around the sources, (ii) in the intergalactic space, and (iii) in the Milky Way. Even though the sources remain unknown, it is not unlikely that they are embedded in environments where large scale magnetic fields exist, e.g. in galactic clusters. These magnetic fields may be able to deflect even the highest energy CRs significantly, or even confine them for some time before they escape from the source environment. This can have an effect of the energy spectrum reaching out in the intergalactic space, if the confinement time becomes comparable to the energy loss time. It may also affect the composition of the UHECR leaving the sources, since, at a given energy, heavier nuclei are more easily confined. However, this would require effective lengths of several tens of Mpc at $\sim 10^{20}$ eV. While such a confinement would increase the angular size of the source on the sky, turning it, say, from a localized source to the size of a whole cluster of galaxies for instance, this should not affect our ability to identify clusters of events from individual sources on the sky, as long as the number of sources is limited (as expected at the highest energies) and their typical angular separation is larger than the apparent size of the confinement halo. Finally, it is worth noting that a specificity of UHE cosmic rays, compared to the low-energy, galactic ones, is that the constancy of the fluxes cannot be assumed anymore, since the dynamical timescales of the sources can be shorter than the confinement timescales around them and/or the differential propagation timescales to the Earth. This may result in time dependent effects (on time scales of typically thousands to tens of thousands of years) as to which sources contribute to the observed flux at different energies, as well as in different mass ranges, depending on the elapsed time since their activity as UHECR sources.

Once they leave their sources, UHECRs may also be deflected by magnetic fields in the intergalactic voids, if their strength is large enough. This, however, is highly uncertain. Among the interesting effects that may be associated with such magnetic fields, the existence of a magnetic horizon has received particular attention, as it was realized that it could act as an anti-GZK effect, i.e. essentially a high-pass filter for energetic particles, preventing lower energy cosmic rays from reaching us from distant sources, if they wander around in the extragalactic space (due to magnetic diffusion) for a time longer than the energy loss timescale, or even the age of the universe (Parizot 2004; Lemoine 2005; Aloisio and Berezinsky 2005). This can have an impact on the matching between the galactic and extragalactic components, and also on the cosmic-ray composition in the corresponding energy range: at a given energy, heavier nuclei diffuse for a longer time than light ones, and thus extragalactic Fe nuclei should not be able to reach us from distant galaxies at a energy as low as protons do. It should be noted, however, that the actual outcome of such propagation effects on the UHECR spectrum and composition depend on the granularity of the sources (Deligny et al. 2004; Aloisio and Berezinsky 2005). If the sources were uniformly distributed in the universe, no effect at all would be expected, so no definite prediction can be made at our current stage of ignorance about the sources, the source density and of course the intergalactic magnetic fields. More recent works have focused on the propagation of UHECRs in strongly inhomogeneous extragalactic magnetic (Kotera and Lemoine 2008a, 2008c). This is indeed a much more plausible assumption, under which UHECRs appear to propagate essentially in straight lines between more localized interactions with large scale magnetized structures, acting as scattering centers. As a consequence, potential hot spots in the UHECR sky may reveal the position of the most nearby scattering centers rather than that of the most nearby sources (Kotera and Lemoine 2008a).

Different aspects of propagation of cosmic rays in IGMF recently have been comprehensively studied Dolag et al. (2005), Globus et al. (2008), Kotera and Lemoine (2008b).

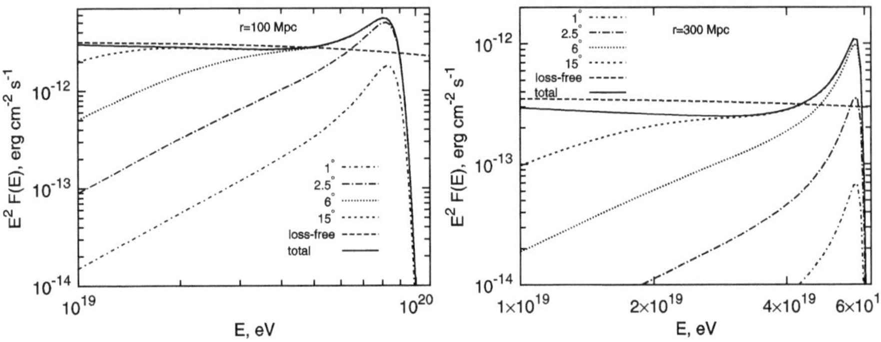

Fig. 6 Energy distributions of protons observed within different angles for the source at distances $r = 100$ Mpc (*left panel*) and $r = 300$ Mpc (*right panel*). The energy spectrum of protons is assumed in the form of power-law with exponential cutoff, with $\alpha = 2$ and $E_0 = 3 \times 10^{20}$ eV; the rate $L = 10^{44}$ erg/s

The conclusions of these studies are generally different, basically because of different assumptions and approaches in the modeling of the IGMF. The propagations of protons and accompanying them gamma-rays and neutrinos recently have been studied by Aharonian et al. (2010) in small angle approximation limit. In Fig. 6 we show the energy distributions of protons within different angles arriving from a point source located at distances 100 and 300 Mpc. It is assumed that the source is injects protons into the intergalactic medium with magnetic field 1 nG and correlation length $\lambda = 1$ Mpc. The upper dashed lines in Fig. 6 correspond to the case when protons propagate in empty space; flux is determined by the geometrical factor $1/r^2$. The solid line present the case when the deflections in the magnetic field are ignored. Comparison of these two curves demonstrates strong dependence of the energy distribution of protons on the solid angle within which they are detected; the flux of protons at highest energies is concentrated along the direction to the source. This is the result of selective deflection of protons depending on their energy. In addition to this effect there are additional two spectral features—a bump and a sagging at lower energies. The bump preceding the cutoff is due to strong growth of energy losses at the threshold of photomeson production that makes particles to be accumulated in this energy region; the sagging is a consequence of the energy losses due to the electron-positron pair production. However, as it is seen in Fig. 6, the maximum due to the deflection is more distinct than the "photoproduction energy loss bump".

Finally, at the other end of the propagation track, the galactic magnetic fields also deflect the UHECRs, with deflections of the order of a degree times the charge of the particle, or possibly much more if an extended magnetized wind surrounds our Galaxy. However, the exact deflection patterns (both size and direction) depend on the structure of the magnetic field, which remains highly uncertain. The regular component should provide a global shift on the sky of the image of the actual UHECR sources, while the turbulent component essentially broadens the apparent cosmic-ray source around that image. Finally, the deflections are expected to be inversely proportional to energy, which should lead to characteristic structures in the arrival direction/energy space. Chains of UHECR events, ordered in decreasing energy away from what would be a common source, have been sought in the various data sets, with no conclusive result so far. Obviously, while such structures would be extremely informative about the source positions as well as about the magnetic field lines responsible, they can be detected only if the sources do not overlap (i.e., at the highest energies, but then, unfortunately, with a limited lever arm in energy), or with much larger statistics, to allow the detection of these patterns on top of a background mixing UHECRs from overlapping sources, with a large enough significance.

In conclusion, the study of UHECR transport in the energy space, mass space and angular space, makes it clear that the road to a better understanding of the sources and phenomenology of UHECRs passes through the joint study of the three fundamental observables (energy spectrum, composition and arrival directions). It should also be stressed that the GZK cut-off, while implying a limitation of the UHECR energies in the universe (independently of the sources) and strongly reducing the expected flux, is also the best chance we have to answer the central questions about UHECRs. As the current data show (see below), no obvious source and/or pattern has appeared in the sky yet. This implies either that the sources are numerous, and we have not seen multiplets (i.e. several events coming from the same astrophysical source), or not strong enough for angular associations in the sky to be significant, or that the overall deflections of UHECRs are large, and the sources largely overlap. The obvious way out is by reducing the number of sources as well as the deflections. This is precisely what the GZK allows (and actually imposes). By reducing the horizon scale as the energy increases, the GZK effect guarantees that fewer and fewer sources contribute at the highest energies. In addition, the number of sources capable of accelerating cosmic rays up to a given energy is likely to be a decreasing function of that energy. In addition, the deflections are inversely proportional to energy, so by concentrating on the UHECR sky at 10^{20} eV or so, we are bound to face a situation where a (very) limited number of sources contribute to the observed flux, with limited deflections and well separated "centroids" of the corresponding hot spots in the UHECR sky. The price to pay for such a situation, of course, is that the corresponding cosmic ray fluxes are very low, so further progress is conditioned by the resolution on an important observational challenge. The target has been set at the level of a total exposure of the sky of several 10^5 up to 10^6 km sr yr (see the recent white papers on UHECRs: Olinto et al. 2009), i.e. more than an order of magnitude larger than what is currently available.

4.3 UHECR Observations

The highest-energy particles are so rare that they are detectable only by means of the giant cascades or extensive air showers they create in the atmosphere. Details of how these extensive air showers are observed and of how the parameters of importance are measured can be found in reviews (e.g. Nagano and Watson 2000). The results that are most relevant to the topic of this chapter have been obtained with the Pierre Auger Observatory (Abraham and Aglietta 2004; Dawson and for the Auger Collaboration 2008) in Argentina and with the HiRes instrument (Boyer et al. 2002) operated in the Dugway Desert, Utah, USA. The HiRes device uses telescopes comprising mirrors and photomultipliers to detect the fluorescence radiation emitted from the excitation of atmospheric nitrogen by the shower particles passing through the atmosphere. The Auger Observatory has similar telescopes but in addition has an array of water-Cherenkov detectors deployed over 3000 km^2 to measure the particles in the showers at ground level. The fluorescence telescopes are used to obtain a measure of how the cascade grows and decays in the atmosphere while the water-Cherenkov detectors are used to obtain the pattern of the distribution of shower particles on the ground from which a shower size, closely proportional to the primary energy, is obtained. With both devices the direction of the incoming cosmic ray is measured with an accuracy of $\sim 1^\circ$ using GPS-based timing.

A 21st century innovation has been to build instruments combining the fluorescence and the surface detector techniques. This combination has allowed what is known as the hybrid method to be developed in which the time of the shower front traversing one of the detectors on the ground is combined with the times of arrival of light at the photomultipliers of the fluorescence telescopes, to enhance significantly the precision with which the direction, energy and depth of maximum of the showers can be determined. The fluorescence technique has the unique advantage of enabling the energy of a shower to be found without resorting to assumptions about hadronic physics. This makes it possible to calibrate the shower size found with the surface array, and so exploit the near 100% on-time of a surface array. Two implementations of the hybrid approach are now operating: the Pierre Auger Observatory in Argentina and the Telescope Array in Utah (Kawai et al. 2005). Data from the Telescope Array are not yet available and the exposure from the Auger Observatory (early 2010) is about 10 times greater.

4.4 Status of Measurements of the Energy Spectrum

An energy spectrum based on an exposure of 12,790 km^2 sr yr has recently been reported by the Auger Collaboration (Pierre Auger Collaboration et al. 2010b). This exposure is a factor 4 higher than achieved with the HiRes detector at the highest energies and even greater than this at lower energies. When deriving the primary energy from fluorescence measurements, an estimate of the missing energy carried into the ground by hadrons, muons and neutrinos must be made based on assumptions about the mass of cosmic rays and of the hadronic model. For a primary beam that is a 50/50 mixture of protons and iron, it has been found from simulations of showers with the QGSJET01 model of hadronic interactions that the energy found from the fluorescence signal should be increased by 10% (Barbosa et al. 2004). The systematic uncertainties on the energy scale sum to 22% with the largest coming from the absolute fluorescence yield (14%), the absolute calibration of the fluorescence telescopes (9%) and that due to the reconstruction method of the longitudinal shower profile (10%). Efforts are underway to reduce these uncertainties.

The Auger spectrum reported in Pierre Auger Collaboration et al. (2010b), and based on over 36,000 events, is shown in Fig. 7 together with the final measurement from the HiRes instrument (Abbasi et al. 2009). It is customary to represent parts of the spectrum with a differential form, $J = kE^{-\gamma}$. It is seen that there is clear evidence of a flattening of the slope at an energy of $\log E(\text{eV}) = (18.61 \pm 0.01)$ where γ changes from (3.26 ± 0.04) to

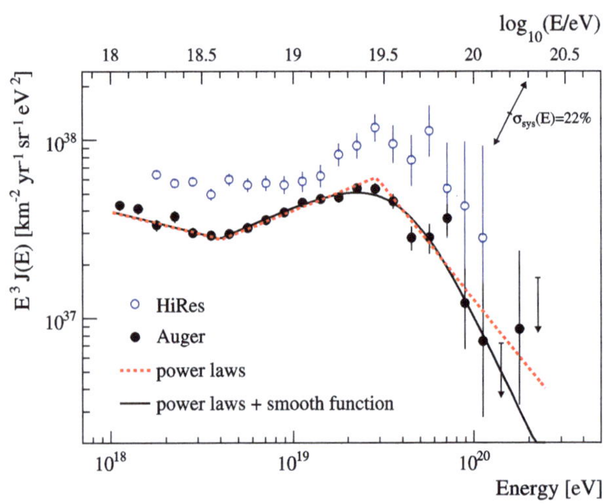

Fig. 7 The Auger energy spectrum. The spectrum is fitted with two functions [see Pierre Auger Collaboration et al. (2010b) for details] and the data are compared with the stereo spectrum of HiRes (Abbasi et al. 2009). The uncertainty of the flux scaled by E^3 arising from the systematic uncertainty in the energy scale of 22% is indicated

(2.59 ± 0.02). At $\log E(\text{eV}) = (19.46 \pm 0.03)$ the slope increases very sharply to (4.3 ± 0.2). This suppression of the flux is what would be expected because of the GZK-effect and is significant at the level of 20 standard deviations.

An intriguing question is whether the slopes in the region of the suppression are different in the Northern and Southern Hemispheres. This would be expected if the source distribution was different in the two hemispheres with the region with sources at a closer average distance being expected to show a less suppressed slope, assuming that the mean injection spectra are the same. However the present measurements of (5.5 ± 1.8, HiRes) and (4.3 ± 0.2, Auger) are not significantly different and as the HiRes instrument is not longer operating the possibility of any difference will need to be explored with future instruments.

In any case, one may expect that the flux at the very highest energies, namely *in the GZK cutoff region* itself, will be dominated by only a few sources within the horizon. For this reason, the assumption of a uniform source distribution usually made to produce the synthetic spectra used to fit the observed spectrum should not be valid anymore. Even with infinite statistics, the overall spectrum around 10^{20} eV will not be of much interest, as it will be shaped mostly by the most luminous sources that just happen to be here or there in the local sky—in other words, *cosmic variance* is the dominant factor in the region of the spectrum.

The flux above 10^{20} eV, as estimated from the 3 Auger events and 1 HiRes event, is $(2.4^{+1.9}_{-1.1}) \times 10^{-4}\ \text{km}^{-2}\,\text{sr}^{-1}\,\text{yr}^{-1}$ or about 1 per square kilometre per millennium. It is really quite remarkable that the energy of such rare events can be measured to within about 20% and indeed estimates of the energies of particle initiating showers at the highest energies are more certain than at 10^{15} eV where there has to be greater reliance on models of hadronic physics.

4.5 Current Status of Measurements of the Arrival Direction Distribution

From a quick look at the first data recorded with the Auger Observatory at the highest energies it was evident that no strong sources stood out. It is not clear what classes of objects are likely to provide the necessary acceleration sites so a well-defined search strategy to look for signals was adopted. Active galactic nuclei, radio-galaxy lobes and gamma-ray bursts have all been suggested as possibilities and if these are anisotropically distributed and are not too numerous within the distance associated with the GZK-suppression then an anisotropy might be expected. However there is no prediction for source features as compelling as was that of the GZK-effect for the spectral shape. Faced with this dilemma the Auger Collaboration made the decision to search for correlations on different angular scales with objects in a particular catalogue using energy and distance as additional parameters. The catalogue chosen was the Véron-Cetty and Véron (2006) catalogue (VCV) of quasars and active galactic nuclei. Using data recorded between 1 January 2004 and 26 May 2006 a scan was made for a minimum in the probability P for a set of N events expected from an isotropic flux to contain k or more events at a maximum angular distance Ψ from any object in the VCV catalogue. P is given by the cumulative binomial distribution $\sum_{j=k}^{N} C_j^N p^j (1-p)^{N-j}$, where p is the fraction of the sky (weighted by exposure) defined by the region at angular separation less than ψ from a selected source. The parameter space was defined by the angular separation ψ, the maximum red-shift $z_{\max}$ and the threshold energy E_{th}. A minimum in P was found with $\psi = 3.1°$, $z_{\max} = 0.018$ ($\sim$120 Mpc) and $E_{\text{th}} = 5.6 \times 10^{19}$ eV. With these values, 12 events among 15 correlated with AGNs in the catalogue, whereas only 3.2 would have been expected by chance. These numbers were found after a search and so no probability could be attached to the significance of the association. However the observation led to the

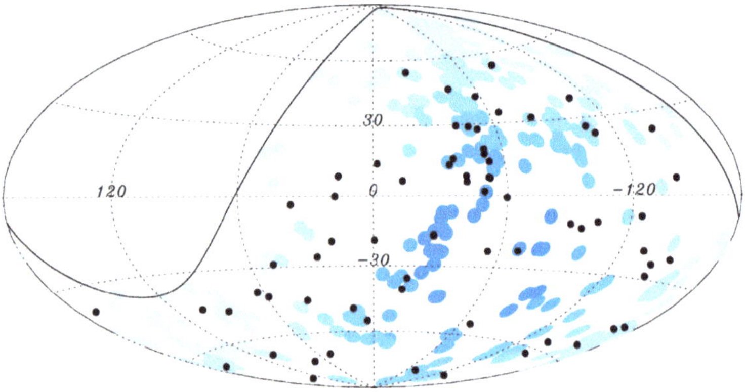

Fig. 8 The directions, in galactic coordinates, of 69 events above 5×10^{19} eV recorded at the Pierre Auger Observatory up to 31 December 2009 (Abreu et al. 2010). The events are shown as *black dots* on an Aitoff-Hammer projection with the galactic plane across the centre of the diagram. The *shaded circles* of 3.1° are centered at the positions of 318 AGNs in the VCV catalogue that lie within 75 Mpc and are within the field of view of the Observatory. The *darker shading* corresponds to greater relative exposure: the exposure weighted-fraction covers 21% of the sky

definition of a test to validate the result with an independent data set using these parameters a priori. Details of the test are given in Abraham et al. (2007, 2008). The test was applied to data collected between 27 May 2006 and 31 August 2007 using exactly the same reconstruction algorithms as for the initial period. In this independent data set, which had almost exactly the same exposure, 8 of 13 events above the same energy threshold were found to correlate with events in the VCV catalogue. The probability of this occurring by chance is 1.7×10^{-3} and the result was reported (Abraham et al. 2007). This correlation corresponds to an association of $(69^{+11}_{-13})\%$.

The Auger Collaboration has continued to take data and as at 31 December 2009 a total of 69 events above the threshold energy have been recorded (Abreu et al. 2010). The correlation with objects in the VCV catalogue, excluding the 15 events used to define the correlation parameters, is now $(38 \pm 6)\%$ with 21 events correlated. The fraction expected from an isotropic distribution of event directions is 0.21. From an isotropic flux the cumulative probability of such a correlation is $P = 0.003$. A sky map using an Aitoff-Hammer projection is shown in Fig. 8. With the current estimate of the correlation at 38%, a 5 sigma signal ($P < 3 \times 10^{-5}$) will require 110 events which should be obtained with about 2 further years of operation.

Using a cross-correlation study the Auger Collaboration have also noted that there is an excess of events from the region around Centaurus A (Abraham et al. 2007, 2008; Abreu et al. 2010) with 13 of the arrival directions within 18° of Centaurus A forming 6 pairs separated by less than 4° and 28 pairs separated by 11°. Clearly this is a region of sky that will be studied carefully but as the search was made *a posteriori* no statistical significance can be attached to the observation. No similar object lies as close to us in the Northern Hemisphere.

The HiRes collaboration does not find a similarly correlation of their events with AGNs in the VCV catalogue (Abbasi et al. 2008b). They find that 2 events out of 13 are correlated with VCV objects: this compares with 5 that would be expected assuming that the correlation is really 38%. The difference between 2 and 5 does not rule out a 38% correlation in the northern hemisphere, the part of the sky dominantly observed with the HiRes detector. It may be that the source distributions are different in the two parts of the sky and, with the steeply-falling spectrum, there may be a small difference in the definition of the energy threshold. The situation is inconclusive and will not be clarified until new Observatories are built.

However that may be, keeping in mind that the excess of correlation observed by Auger does not necessarily imply that the AGNs of the catalog are related with the sources, and may simply reflect the accidental correlation between these AGNs and the actual sources in that part of the sky (convolved with the deflections in the intervening magnetic fields), one must be prepared to a situation where other sources in the Northern hemisphere lead to a different level of correlation with the AGNs of that hemisphere. In other words, there is no direct conflict between the observations currently available, and a systematic study of the entire sky with comparable data sets should tell us more about the origin of the detected anisotropy, and what it implies concerning the UHECR sources and deflections.

In the mean time, it is interesting to note that the current data sets do not reveal obvious multiplets or close correlations with known sources. This may be due to a very large number of sources, from which only one event at most has typically been observed so far, or to relatively large deflections, causing the sources to overlap in the sky. In both cases, the observational solution consists in increasing the statistics at high energy, where the deflections are reduced and the horizon scale becomes smaller, so that only a few sources can significantly contribute. The fact that some anisotropy is already detectable gives strong confidence that the expected breakthrough is within reach of the next generation detectors, provided they can gather a statistics of the order of a few 10^5 km^2 sr yr or more (Olinto et al. 2009).

From the moment when the first few significant hot spots are observed in the UHECR sky, a new era will begin, with the study of individual sources. As indicated above, this should lead to key parameters such as the source power and spectral shape. The current statistics does not allow one to confirm whether the excess of events possibly seen around the location of Centaurus A is related to the first of such sources, but if it is the case, then the next generation of detectors should accumulate tens or even hundreds of events from that or similar sources.

4.6 Measurement of the Chemical Composition of the Primaries

To interpret the information that we have on the energy spectrum and arrival direction pattern of ultra high-energy cosmic rays we need to know the charge, Z, of the particles, ideally on an event-by-event basis. Then, for example, if one had identified a point source, the bending of the particles from it as a function of energy could be used to distinguish between different models of the magnetic fields lying between the source and earth. However it is very difficult to be sure even of the atomic mass, A, as the only methods that are practical for estimating it require assumptions about the hadronic interactions at centre-of-mass energies well above what will be reached with the LHC.

The concept of the most promising method is to compare the average depth of shower maximum (called $X_{\max}$) as a function of energy for protons, iron nuclei and photons with what is predicted for different hadronic models. The depth of maximum can be determined to $\sim$(20 to 30) g/cm^2 using the fluorescence technique. For a mass composition of a single species, $X_{\max}$ is expected to increase with energy as γ-rays from π^0 decay become more energetic and so the position of shower maximum is pushed deeper into the atmosphere. The average $X_{\max}$ for iron nuclei lies below that for protons because the energy per nucleon of an incoming primary is smaller for heavier nuclei at a given energy. The change of $X_{\max}$ with energy is known as the elongation rate. In addition to the elongation rate being a useful

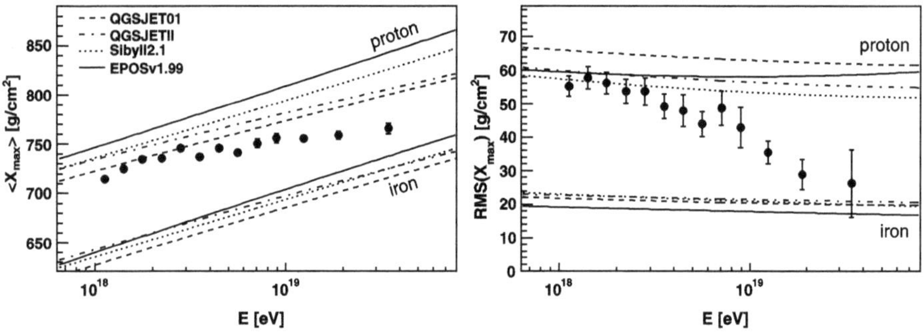

Fig. 9 *Left*: The average measurements of $X_{\rm max}$ as a function of energy reported by the Auger Collaboration based on 3754 events compared with the predictions of 4 different calculations (see Abraham et al. 2010, for details). *Right*: The measurements of the rms of $X_{\rm max}$ as a function of energy for the data of the figure on the left, compared with calculations (Abraham et al. 2010)

indicator of primary mass, a key parameter is the fluctuation in the depth of shower maximum at a particular energy. It is intuitively obvious that as an iron nucleus is larger than a proton, the cross-section for interaction will be greater. Thus fluctuations in the position of shower maximum also contain information about the primary mass. However, although these parameters can be measured rather accurately, the interpretation of the measurements must rely on models of shower development.

Work on this problem has recently been reported by the Auger Collaboration (Abraham et al. 2010) and by the HiRes Collaboration (Abbasi et al. 2010). With the Auger Observatory the depth of maximum has been determined for over 3700 events of energy $>10^{18}$ eV. Each event has been reconstructed using the hybrid method: 2% of them are stereo events for which two independent estimates of $X_{\rm max}$ are made. Using stereo information it can be demonstrated that the accuracy of estimation of $X_{\rm max}$ in a non-stereo event is $\sim$20 g/cm^{-2} (about 200 m at the height of the maximum of a typical shower, 750 g/cm^{-2}, or 2.5 km above sea-level in the vertical direction).

The Auger data on the depth and width of the fluctuation of shower maximum are shown in Fig. 9. The predictions for proton and iron nuclei of $X_{\rm max}$ and rms($X_{\rm max}$) are given for different models of the hadronic interactions (QGSJET01 etc.). It is clear that if these models are approximately correct then the mean mass of the cosmic rays is becoming heavier as the energy increases. However the results from the HiRes collaboration seem to tell a different story. They conclude that their data, totaling 815 events, is entirely consistent with proton primaries from $\sim 3 \times 10^{18}$ eV. The reasons for the differences between the two measurements are not understood. The HiRes result is highly dependent on Monte Carlo calculations whereas the Auger measurement makes very little use of such calculations except, as is true for both measurements, when a comparison is made with the predictions. The Auger Collaboration have taken precautions in their analysis to test that the effects observed are not dependent on zenith angle, demonstrating that deeply penetrating events are not missed, and it is hard to see how distributions can be made too narrow as a consequence of the analysis procedures.

5 Some Conclusions and Perspectives

Magnetic fields play a key role in the acceleration, propagation and radiation of galactic cosmic rays. Current generation of X-ray *Chandra*, *XMM-Newton* and *Suzaku* and gamma ray telescopes *H.E.S.S.*, *MAGIC*, *MILAGRO*, *VERITAS* and *Fermi* have provided conclusive evidences for CR particle acceleration up to the energies at least about 100 TeV in young supernova remnants, starburst regions, radio galaxies and AGNs. With the coming generation of the focusing hard X-ray telescopes *Astro-H* (Takahashi et al. 2010) and *NuSTAR* (Harrison et al. 2010) new possibilities will be opened to study in details hard X-rays images and spectra that of cosmic ray sources providing unique information on both particle spectra and fluctuating magnetic fields in the sources. A project of very large X-ray telescope *IXO* that is currently under discussion in addition to the unprecedent sensitivity will be likely supplied with X-ray imaging polarimetry (Barcons et al. 2011). Imaging polarimetry is sensitive to details of magnetic field amplification mechanism in the particle acceleration sources like supernova remnants and radio galaxies (Bykov et al. 2009). The Cherenkov Telescope Array *CTA* project under development is the next generation ground-based gamma-ray instrument, which is supposed to provide at least 5 times better sensitivity in the current energy domain of about 100 GeV to about 10 TeV and an extension of the accessible energy range well below 100 GeV and to above 100 TeV.

At extremely high energies, $E \sim 10^{20}$ eV, the impact of galactic and extragalactic magnetic fields on the propagation of cosmic rays becomes less dramatic, which should result in large and small scale anisotropies of the cosmic rays. Although the corresponding fluxes are very low, considerable progress has been made in the recent years, with the detection of the giant cascades (so-called extensive air showers) induced by these cosmic rays in the atmosphere, by large detectors consisting of ground arrays (AGASA), fluorescence telescopes (HiRes) or hybrid detectors (Auger, currently in operation).

The extension of cosmic ray studies up to energies of the order of 10^{20} eV is likely to give rise to what is referred to as "proton astronomy" (or more generally "charged particle astronomy"), where charged cosmic rays are used as astronomical messengers pointing roughly back to their sources, because of the reduced deflections in the intervening magnetic fields. Another key aspect of the cosmic rays phenomenology in this energy range is the existence of a physical horizon, associated with the energy losses suffered by the cosmic rays as they interact with the background radiation fields (notably the cosmological microwave background).

Whether protons or heavier nuclei, particles of such high energies can arrive only from relatively nearby accelerators, say within 100 Mpc, which decreases dramatically the number of potential sources capable of accelerating cosmic rays up to $\geq 10^{20}$ eV and contributing to the observed flux. In principle, many weak sources may be at work, which may prevent one from detecting them individually, by the accumulation of events in a given direction. However, even with large source densities, a few most nearby and brightest sources may be expected to contribute a significant fraction of the total flux, in a similar way as what is observed at all photon wavelengths, where a few sources dominate the sky. Alternatively, there may be only a few sources within the horizon, especially if one considers the tough requirements associated with 10^{20} eV proton accelerators (Aharonian et al. 2002). This excludes, in particular, astrophysical objects like ordinary galaxies, unless these galaxies accelerate the highest energy cosmic rays through transient events related to compact objects like Gamma Ray Bursts. Whether we can identify the accelerators of extragalactic cosmic rays using the highest energy protons is a question which largely depends on the strength and structure of the large scale IGMF. However, if the latter are not too large, as seems to indicate the current UHECR data showing anisotropy above 55 EeV or so, the cosmic rays from identified sources may be used in turn to probe the intergalactic magnetic fields in the nearby universe, as well as the structure and strength of the galactic magnetic field, from the observed (energy- and charge-dependent) deflections.

The identification of the first cosmic ray sources by astronomical means, i.e. through hot spots or elongated multiplets with a characteristic rigidity ordering, may be provided by the next generation detectors, either on the ground, with the Northern site of the Pierre Auger Observatory (Blümer and the Pierre Auger Collaboration 2010), proposed for construction over 20,000 km^2, with the proven technique of Auger South and similar performance, or in space, with the *JEM-EUSO* mission (Ebisuzaki et al. 2010), proposed to be installed on the Japanese module of the International Space Station, for an even larger aperture at the highest energies. This is expected to lead to a new era of cosmic ray studies, where nearby sources may be studied individually, and particle acceleration may be explored in the most challenging regime, up to above 10^{20}eV.

Finally, we stress that the strong coupling between cosmic ray studies and the understanding of the magnetic field generation, amplification and structure in the universe calls for an increased exchange of knowledge between the different communities involved. Important progress is expected in both fields in the coming years, and any progress in one of the various aspects of the problem will benefit to the others, and to the whole field of high-energy astrophysics, where energetic particles are responsible for non-thermal emission at all energies, as well as to the physics of the interstellar medium in general, in which the magnetic field is a key component.

Acknowledgements We thank the referee for the constructive comments. A.M.B. was supported in part by RBRF grants 09-02-12080, 11-02-00429 by the RAS Presidium Programm, and the Russian government grant 11.G34.31.0001 to Sankt-Petersburg State Politechnical University. He performed some of the simulations at the Joint Supercomputing Centre (JSCC RAS) and the Supercomputing Centre at Ioffe Institute, St. Petersburg. V.S.P. was supported by RFBR grant 10-02-00110.

References

R.U. Abbasi, T. Abu-Zayyad, M. Al-Seady, M. Allen et al., Measurement of the flux of ultra high energy cosmic rays by the stereo technique. Astropart. Phys. **32**, 53–60 (2009). arXiv:0904.4500

R.U. Abbasi, T. Abu-Zayyad, M. Al-Seady, M. Allen et al., Indications of proton-dominated cosmic-ray composition above 1.6 EeV. Phys. Rev. Lett. **104**(16), 161101 (2010). arXiv:0910.4184

R.U. Abbasi, T. Abu-Zayyad, M. Allen, J.F. Amman et al., First observation of the Greisen-Zatsepin-Kuzmin suppression. Phys. Rev. Lett. **100**(10), 101101 (2008a). arXiv:astro-ph/0703099

R.U. Abbasi, T. Abu-Zayyad, M. Allen, J.F. Amman et al., Search for correlations between HiRes stereo events and active galactic nuclei. Astropart. Phys. **30**, 175–179 (2008b). arXiv:0804.0382

J. Abraham, P. Abreu, M. Aglietta, C. Aguirre et al., Correlation of the highest-energy cosmic rays with nearby extragalactic objects. Science **318**, 938 (2007). arXiv:0711.2256

J. Abraham, P. Abreu, M. Aglietta, C. Aguirre et al., Correlation of the highest-energy cosmic rays with the positions of nearby active galactic nuclei. Astropart. Phys. **29**, 188–204 (2008). arXiv:0712.2843

J. Abraham, P. Abreu, M. Aglietta, E.J. Ahn et al., Measurement of the depth of maximum of extensive air showers above 10^{18} eV. Phys. Rev. Lett. **104**(9), 091101 (2010). arXiv:1002.0699

J. Abraham, M. Aglietta, I.C. Aguirre, M. Albrow et al., Properties and performance of the prototype instrument for the Pierre Auger Observatory. Nucl. Instrum. Methods Phys. Res. A **523**, 50–95 (2004)

P. Abreu, M. Aglietta, E.J. Ahn, D. Allard et al., Update on the correlation of the highest energy cosmic rays with nearby extragalactic matter. arXiv:1009.1855 (2010)

V.A. Acciari, E. Aliu, T. Arlen, T. Aune et al., A connection between star formation activity and cosmic rays in the starburst galaxy M82. Nature **462**, 770–772 (2009). arXiv:0911.0873

F. Acero, F. Aharonian, A.G. Akhperjanian, G. Anton et al., Detection of gamma rays from a Starburst Galaxy. Science **326**, 1080 (2009). arXiv:0909.4651

F. Aharonian, A.G. Akhperjanian, A.R. Bazer-Bachi, M. Beilicke et al., Detection of extended very-high-energy γ-ray emission towards the young stellar cluster Westerlund 2. Astron. Astrophys. **467**, 1075–1080 (2007a). arXiv:astro-ph/0703427

F. Aharonian, A.G. Akhperjanian, A.R. Bazer-Bachi, M. Beilicke et al., Primary particle acceleration above 100 TeV in the shell-type supernova remnant RX J1713.7-3946 with deep HESS observations. Astron. Astrophys. **464**, 235–243 (2007a). arXiv:astro-ph/0611813

F. Aharonian, J. Buckley, T. Kifune, G. Sinnis, High energy astrophysics with ground-based gamma ray detectors. Rep. Prog. Phys. **71**(9), 096901 (2008)
F.A. Aharonian, A.A. Belyanin, E.V. Derishev, V.V. Kocharovsky et al., Constraints on the extremely high-energy cosmic ray accelerators from classical electrodynamics. Phys. Rev. D **66**(2), 023005 (2002). arXiv:astro-ph/0202229
F.A. Aharonian, L.O. Drury, H.J. Voelk, GeV/TeV gamma-ray emission from dense molecular clouds overtaken by supernova shells. Astron. Astrophys. **285**, 645–647 (1994)
F.A. Aharonian, S.R. Kelner, A.Y. Prosekin, Angular, spectral, and time distributions of highest energy protons and associated secondary gamma rays and neutrinos propagating through extragalactic magnetic and radiation fields. Phys. Rev. D **82**(4), 043002 (2010). arXiv:1006.1045
H.S. Ahn, P.S. Allison, M.G. Bagliesi, J.J. Beatty et al., Measurements of cosmic-ray secondary nuclei at high energies with the first flight of the CREAM balloon-borne experiment. Astropart. Phys. **30**, 133–141 (2008). arXiv:0808.1718
D. Allard, M. Ave, N. Busca, M.A. Malkan et al., Cosmogenic neutrinos from the propagation of ultrahigh energy nuclei. J. Cosmol. Astropart. Phys. **9**, 5 (2006). arXiv:astro-ph/0605327
D. Allard, E. Parizot, A.V. Olinto, On the transition from galactic to extragalactic cosmic-rays: spectral and composition features from two opposite scenarios. Astropart. Phys. **27**, 61–75 (2007). arXiv: astro-ph/0512345
D. Allard, E. Parizot, A.V. Olinto, E. Khan et al., UHE nuclei propagation and the interpretation of the ankle in the cosmic-ray spectrum. Astron. Astrophys. **443**, L29–L32 (2005). arXiv:astro-ph/0505566
R. Aloisio, V. Berezinsky, P. Blasi, A. Gazizov et al., A dip in the UHECR spectrum and the transition from galactic to extragalactic cosmic rays. Astropart. Phys. **27**, 76–91 (2007). arXiv:astro-ph/0608219
R. Aloisio, V.S. Berezinsky, Anti-GZK effect in ultra-high-energy cosmic ray diffusive propagation. Astrophys. J. **625**, 249–255 (2005). arXiv:astro-ph/0412578
J.W. Armstrong, B.J. Rickett, S.R. Spangler, Electron density power spectrum in the local interstellar medium. Astrophys. J. **443**, 209–221 (1995)
J. Aublin, E. Parizot, On the viability of holistic cosmic-ray source models. Astron. Astrophys. **452**, L19–L22 (2006). arXiv:astro-ph/0605046
W.I. Axford, The origins of high-energy cosmic rays. Astrophys. J. Suppl. Ser. **90**, 937–944 (1994)
A. Bamba, M. Ueno, H. Nakajima, K. Koyama, Thermal and nonthermal X-Rays from the large Magellanic cloud superbubble 30 Doradus C. Astrophys. J. **602**, 257–263 (2004). arXiv:astro-ph/0310713
H.M.J. Barbosa, F. Catalani, J.A. Chinellato, C. Dobrigkeit, Determination of the calorimetric energy in extensive air showers. Astropart. Phys. **22**, 159–166 (2004). arXiv:astro-ph/0310234
X. Barcons, D. Barret, M. Bautz, J. Bookbinder et al., International X-ray Observatory (IXO) assessment study report for the ESA cosmic vision 2015–2025 (2011). arXiv:1102.2845
R. Beck, Galactic and extragalactic magnetic fields, in *American Institute of Physics Conference Series*, ed. by F.A. Aharonian, W. Hofmann, F. Rieger, vol. 1085 (2008), pp. 83–96. arXiv:0810.2923
A.R. Bell, The acceleration of cosmic rays in shock fronts. I. Mon. Not. R. Astron. Soc. **182**, 147–156 (1978)
A.R. Bell, Turbulent amplification of magnetic field and diffusive shock acceleration of cosmic rays. Mon. Not. R. Astron. Soc. **353**, 550–558 (2004)
V.S. Berezinskii, S.V. Bulanov, V.A. Dogiel, V.L. Ginzburg, V.S. Ptuskin, *Astrophysics of Cosmic Rays* (North-Holland, Amsterdam, 1990)
V.S. Berezinskii, S.I. Grigor'eva, A bump in the ultra-high energy cosmic ray spectrum. Astron. Astrophys. **199**, 1–12 (1988)
V. Berezinsky, Ultra high energy cosmic ray protons: signatures and observations. Nucl. Phys. B, Proc. Suppl. **188**, 227–232 (2009). arXiv:0901.0254
V. Berezinsky, A. Gazizov, S. Grigorieva, Propagation and signatures of ultra high energy cosmic rays. Nucl. Phys. B, Proc. Suppl. **136**, 147–158 (2004). arXiv:astro-ph/0410650
P. Bhattacharjee, G. Sigl, Origin and propagation of extremely high energy cosmic rays. Phys. Rep. **327**, 109–247 (2000). arXiv:astro-ph/9811011
W.R. Binns, M.E. Wiedenbeck, M. Arnould, A.C. Cummings et al., OB Associations, Wolf Rayet Stars, and the origin of galactic cosmic rays. Space Sci. Rev. **130**, 439–449 (2007). arXiv:0707.4645
R. Blandford, D. Eichler, Particle acceleration at astrophysical shocks—a theory of cosmic-ray origin. Phys. Rep. **154**, 1 (1987)
P. Blasi, The origin of ultra high energy cosmic rays. J. Phys. Conf. Ser. **39**, 372–378 (2006). arXiv:astro-ph/0512438
P. Blasi, R.I. Epstein, A.V. Olinto, Ultra-high-energy cosmic rays from young neutron star winds. Astrophys. J. **533**, L123–L126 (2000). arXiv:astro-ph/9912240
J.B.G.M. Bloemen, V.A. Dogiel, V.L. Dorman, V.S. Ptuskin Galactic diffusion and wind models of cosmic-ray transport. I—Insight from CR composition studies and gamma-ray observations. Astron. Astrophys. **267**, 372–387 (1993)

G.R. Blumenthal, Energy loss of high-energy cosmic rays in pair-producing collisions with ambient photons. Phys. Rev. D **1**, 1596–1602 (1970)

J. Blümer (The Pierre Auger Collaboration), The northern site of the Pierre Auger Observatory. New J. Phys. **12**(3), 035001 (2010)

J.H. Boyer, B.C. Knapp, E.J. Mannel, M. Seman, FADC-based DAQ for HiRes Fly's eye. Nucl. Instrum. Methods Phys. Res. A **482**, 457–474 (2002)

D. Breitschwerdt, J.F. McKenzie, H.J. Voelk, Galactic winds. I—Cosmic ray and wave-driven winds from the Galaxy. Astron. Astrophys. **245**, 79–98 (1991)

A.M. Bykov, Particle acceleration and nonthermal phenomena in superbubbles. Space Sci. Rev. **99**, 317–326 (2001)

A.M. Bykov, K. Dolag, F. Durret, Cosmological shock waves. Space Sci. Rev. **134**, 119–140 (2008). arXiv:0801.0995

A.M. Bykov, G.D. Fleishman, On non-thermal particle generation in superbubbles. Mon. Not. R. Astron. Soc. **255**, 269–275 (1992)

A.M. Bykov, S.M. Osipov, D.C. Ellison, Cosmic ray current driven turbulence in shocks with efficient particle acceleration: the oblique, long-wavelength mode instability. Mon. Not. R. Astron. Soc. **410**, 39–52 (2011). arXiv:1010.0408

A.M. Bykov, I.N. Toptygin, Effect of shocks on interstellar turbulence and cosmic-ray dynamics. Astrophys. Space Sci. **138**, 341–354 (1987)

A.M. Bykov, I.N. Toptygin, Particle kinetics in highly turbulent plasmas (renormalization and self-consistent field methods). Phys. Usp. **36**, 1020–1052 (1993)

A.M. Bykov, I.N. Toptygin, A model of particle acceleration to high energies by multiple supernova explosions in OB associations. Astron. Lett. **27**, 625–633 (2001)

A.M. Bykov, Y.A. Uvarov, J.B.G.M. Bloemen, J.W. den Herder et al., A model of polarized X-ray emission from twinkling synchrotron supernova shells. Mon. Not. R. Astron. Soc. **399**, 1119–1125 (2009). arXiv:0907.2521

A. Calvez, A. Kusenko, S. Nagataki, Role of galactic sources and magnetic fields in forming the observed energy-dependent composition of ultrahigh-energy cosmic rays. Phys. Rev. Lett. **105**(9), 091101 (2010). arXiv:1004.2535

F. Casse, M. Lemoine, G. Pelletier, Transport of cosmic rays in chaotic magnetic fields. Phys. Rev. D **65**(2), 023002 (2002). arXiv:astro-ph/0109223

M. Casse, J.A. Paul, On the stellar origin of the Ne-22 excess in cosmic rays. Astrophys. J. **258**, 860–863 (1982)

C.J. Cesarsky, Cosmic-ray confinement in the galaxy. Annu. Rev. Astron. Astrophys. **18**, 289–319 (1980)

C.J. Cesarsky, T.M. Montmerle, Cosmic rays from OB associations and supernovae—anti-protons and the origin of local cosmic rays, in *International Cosmic Ray Conference*, vol. 9 (1982), pp. 207–210

J. Cho, A. Lazarian, Compressible sub-Alfvénic MHD turbulence in low-β plasmas. Phys. Rev. Lett. **88**(24), 245001 (2002). arXiv:astro-ph/0205282

A.J. Davis, R.A. Mewaldt, W.R. Binns, E.R. Christian et al., On the low energy decrease in galactic cosmic ray secondary/primary ratios, in *Acceleration and Transport of Energetic Particles Observed in the Heliosphere*, ed. by R.A. Mewaldt, J.R. Jokipii, M.A. Lee, E. Möbius, T.H. Zurbuchen. American Institute of Physics Conference Series, vol. 528 (2000), pp. 421–424

B.R. Dawson (for The Auger Collaboration), Hybrid performance of the Pierre Auger Observatory, in *International Cosmic Ray Conference*, vol. 4 (2008), pp. 425–428. arXiv:0706.1105

O. Deligny, A. Letessier-Selvon, E. Parizot, Magnetic horizons of UHECR sources and the GZK feature. Astropart. Phys. **21**, 609–615 (2004). arXiv:astro-ph/0303624

K. Dolag, A.M. Bykov, A. Diaferio, Non-thermal processes in cosmological simulations. Space Sci. Rev. **134**, 311–335 (2008). arXiv:0801.1048

K. Dolag, D. Grasso, V. Springel, I. Tkachev, Constrained simulations of the magnetic field in the local Universe and the propagation of ultrahigh energy cosmic rays. J. Cosmol. Astropart. Phys. **1**, 9 (2005) arXiv:astro-ph/0410419

L.O. Drury, F.A. Aharonian, H.J. Voelk, The gamma-ray visibility of supernova remnants. A test of cosmic ray origin. Astron. Astrophys. **287**, 959–971 (1994). arXiv:astro-ph/9305037

L.O. Drury, D.E. Ellison, F.A. Aharonian, E. Berezhko et al., Test of galactic cosmic-ray source models—Working Group Report. Space Sci. Rev. **99**, 329–352 (2001)

N. Duric, S.M. Gordon, W.M. Goss, F. Viallefond et al., The relativistic ISM in M33: role of the supernova remnants. Astrophys. J. **445**, 173–181 (1995)

M.A. Duvernois, M. Garcia-Munoz, K.R. Pyle, J.A. Simpson et al., The isotopic composition of galactic cosmic-ray elements from carbon to silicon: the combined release and radiation effects satellite investigation. Astrophys. J. **466**, 457 (1996a)

M.A. Duvernois, J.A. Simpson, M.R. Thayer, Interstellar propagation of cosmic rays: analysis of the ULYSSES primary and secondary elemental abundances. Astron. Astrophys. **316**, 555–563 (1996b)

T. Ebisuzaki, Y. Takahashi, F. Kajino, H. Mase et al., The JEM-EUSO mission to explore the extreme universe, in *American Institute of Physics Conference Series*, ed. by H. Susa, M. Arnould, S. Gales, T. Motobayashi, C. Scheidenberger, H. Utsunomiya, vol. 1238 (2010), pp. 369–376

D.C. Ellison, L.O. Drury, J. Meyer, Galactic cosmic rays from supernova remnants. II. Shock acceleration of gas and dust. Astrophys. J. **487**, 197 (1997). arXiv:astro-ph/9704293

D.C. Ellison, D.J. Patnaude, P. Slane, J. Raymond, Efficient cosmic ray acceleration, hydrodynamics, and self-consistent thermal X-ray emission applied to supernova remnant RX J1713.7-3946. Astrophys. J. **712**, 287–293 (2010). arXiv:1001.1932

B.G. Elmegreen, J. Scalo, Interstellar turbulence I: Observations and processes. Annu. Rev. Astron. Astrophys. **42**, 211–273 (2004). arXiv:astro-ph/0404451

J.J. Engelmann, P. Ferrando, A. Soutoul, P. Goret et al., Charge composition and energy spectra of cosmic-ray nuclei for elements from Be to NI—results from HEAO-3-C2. Astron. Astrophys. **233**, 96–111 (1990)

J.E. Everett, E.G. Zweibel, R.A. Benjamin, D. McCammon et al., Does the Milky Way launch a large-scale wind? Astrophys. Space Sci. **311**, 105–110 (2007)

J.E. Everett, E.G. Zweibel, R.A. Benjamin, D. McCammon et al., The Milky Way's kiloparsec-scale wind: a hybrid cosmic-ray and thermally driven outflow. Astrophys. J. **674**, 258–270 (2008). arXiv:0710.3712

A.J. Farmer, P. Goldreich Wave damping by magnetohydrodynamic turbulence and its effect on cosmic-ray propagation in the interstellar medium. Astrophys. J. **604**, 671–674 (2004). arXiv:astro-ph/0311400

C. Ferrari, F. Govoni, S. Schindler, A.M. Bykov et al., Observations of extended radio emission in clusters. Space Sci. Rev. **134**, 93–118 (2008). arXiv:0801.0985

K.M. Ferrière, The interstellar environment of our galaxy. Rev. Mod. Phys. **73**, 1031–1066 (2001). arXiv:astro-ph/0106359

S. Galtier, S.V. Nazarenko, A.C. Newell, A. Pouquet, A weak turbulence theory for incompressible magnetohydrodynamics. J. Plasma Phys. **63**, 447–488 (2000). arXiv:astro-ph/0008148

V.L. Ginzburg, Cosmic Rays and Plasma Phenomena in the Galaxy and Metagalaxy. Sov. Astron. **9**, 877 (1965)

N. Globus, D. Allard, E. Parizot, Propagation of high-energy cosmic rays in extragalactic turbulent magnetic fields: resulting energy spectrum and composition. Astron. Astrophys. **479**, 97–110 (2008). arXiv:0709.1541

P. Goldreich, S. Sridhar, Toward a theory of interstellar turbulence. 2: Strong Alfvénic turbulence. Astrophys. J. **438**, 763–775 (1995)

P. Goldreich, S. Sridhar, Magnetohydrodynamic turbulence revisited. Astrophys. J. **485**, 680 (1997). arXiv:astro-ph/9612243

K. Greisen, End to the cosmic-ray spectrum? Phys. Rev. Lett. **16**, 748–750 (1966)

M. Hanasz, G. Kowal, K. Otmianowska-Mazur, H. Lesch, Amplification of galactic magnetic fields by the cosmic-ray-driven dynamo. Astrophys. J. **605**, L33–L36 (2004). arXiv:astro-ph/0402662

M. Hanasz, K. Otmianowska-Mazur, G. Kowal, H. Lesch, Cosmic ray driven dynamo in galactic disks: effects of resistivity, SN rate and spiral arms. Astron. Nachr. **327**, 469 (2006)

F.A. Harrison, S. Boggs, F. Christensen, W. Craig et al., The nuclear spectroscopic telescope array (NuSTAR), in *Society of Photo-Optical Instrumentation Engineers (SPIE) Conference Series*, vol. 7732 (2010). arXiv:1008.1362

J.C. Higdon, R.E. Lingenfelter, R. Ramaty, Cosmic-ray acceleration from supernova ejecta in superbubbles. Astrophys. J. **509**, L33–L36 (1998)

A.M. Hillas, The origin of ultra-high-energy cosmic rays. Annu. Rev. Astron. Astrophys. **22**, 425–444 (1984)

A.M. Hillas, TOPICAL REVIEW: can diffusive shock acceleration in supernova remnants account for high-energy galactic cosmic rays? J. Phys. G, Nucl. Part. Phys. **31**, 95 (2005)

J.R. Hörandel, N.N. Kalmykov, A.V. Timokhin, Propagation of super-high-energy cosmic rays in the Galaxy. Astropart. Phys. **27**, 119–126 (2007). arXiv:astro-ph/0609490

F.M. Ipavich, Galactic winds driven by cosmic rays. Astrophys. J. **196**, 107–120 (1975)

P.S. Iroshnikov, Turbulence of a conducting fluid in a strong magnetic field. Sov. Astron. **7**, 566 (1964)

F.C. Jones, A. Lukasiak, V. Ptuskin, W. Webber, The modified weighted slab technique: models and results. Astrophys. J. **547**, 264–271 (2001). arXiv:astro-ph/0007293

T.J. Jones, L. Rudnick, T. DeLaney, J. Bowden, The identification of infrared synchrotron radiation from Cassiopeia A. Astrophys. J. **587**, 227–234 (2003). arXiv:astro-ph/0212544

H. Kang, D. Ryu, R. Cen, D. Song, Shock-heated gas in the large-scale structure of the universe. Astrophys. J. **620**, 21–30 (2005). arXiv:astro-ph/0410477

H. Kawai, T. Nunomura, N. Sakurai, S. Yoshida et al., Telescope array; progress of surface array, in *International Cosmic Ray Conference*, vol. 8 (2005), p. 181

C.F. Kennel, F. Engelmann, Velocity space diffusion from weak plasma turbulence in a magnetic field. Phys. Fluids **9**, 2377–2388 (1966)
E. Khan, S. Goriely, D. Allard, E. Parizot et al., Photodisintegration of ultra-high-energy cosmic rays revisited. Astropart. Phys. **23**, 191–201 (2005). arXiv:astro-ph/0412109
K. Kotera, D. Allard, A.V. Olinto, Cosmogenic neutrinos: parameter space and detectabilty from PeV to ZeV. arXiv:1009.1382 (2010)
K. Kotera, M. Lemoine, Inhomogeneous extragalactic magnetic fields and the second knee in the cosmic ray spectrum. Phys. Rev. D **77**(2), 023005 (2008a). arXiv:0706.1891
K. Kotera, M. Lemoine, Inhomogeneous extragalactic magnetic fields and the second knee in the cosmic ray spectrum. Phys. Rev. D **77**(2), 023005 (2008b). arXiv:0706.1891
K. Kotera, M. Lemoine, Optical depth of the Universe to ultrahigh energy cosmic ray scattering in the magnetized large scale structure. Phys. Rev. D **77**(12), 123003 (2008c). arXiv:0801.1450
R.H. Kraichnan, Inertial-range spectrum of hydromagnetic turbulence. Phys. Fluids **8**, 1385–1387 (1965)
R.M. Kulsrud, *Plasma Physics for Astrophysics* (Princeton University Press, Princeton, 2005)
R.M. Kulsrud, E.G. Zweibel, On the origin of cosmic magnetic fields. Rep. Prog. Phys. **71**(4), 046901 (2008). arXiv:0707.2783
D. Kushnir, B. Katz, E. Waxman, Magnetic fields and cosmic rays in clusters of galaxies. J. Cosmol. Astropart. Phys. **9**, 24 (2009). arXiv:0903.2275
T. Kuwabara, K. Nakamura, C.M. Ko, Nonlinear Parker instability with the effect of cosmic-ray diffusion. Astrophys. J. **607**, 828–839 (2004). arXiv:astro-ph/0402350
M. Lemoine, Extragalactic magnetic fields and the second knee in the cosmic-ray spectrum. Phys. Rev. D **71**(8), 083007 (2005). arXiv:astro-ph/0411173
M. Lemoine, E. Waxman, Anisotropy vs chemical composition at ultra-high energies. J. Cosmol. Astropart. Phys. **11**, 9 (2009). arXiv:0907.1354
R.E. Lingenfelter, R. Ramaty, B. Kozlovsky, Supernova grains: the source of cosmic-ray metals. Astrophys. J. **500**, L153 (1998)
Y. Lithwick, P. Goldreich, Compressible magnetohydrodynamic turbulence in interstellar plasmas. Astrophys. J. **562**, 279–296 (2001). arXiv:astro-ph/0106425
M.S. Longair, *High Energy Astrophysics* (Cambridge University Press, Cambridge, 2010)
T.A. Lozinskaya, *Supernovae and Stellar Wind in the Interstellar Medium* (American Institute of Physics, New York, 1992)
A. Lukasiak, Voyager measurements of the charge and isotopic composition of cosmic ray Li, Be and B nuclei and implications for their production in the galaxy, in *International Cosmic Ray Conference*, vol. 3 (1999), p. 41
M.A. Malkov, L. Drury, Nonlinear theory of diffusive acceleration of particles by shock waves. Rep. Prog. Phys. **64**, 429–481 (2001)
A. Marcowith, F. Casse, Postshock turbulence and diffusive shock acceleration in young supernova remnants. Astron. Astrophys. **515**, A90 (2010). arXiv:1001.2111
P. Mészáros, Gamma-ray bursts. Rep. Prog. Phys. **69**, 2259–2321 (2006). arXiv:astro-ph/0605208
R.A. Mewaldt, The time delay between nucleosynthesis and acceleration based on ACE measurements of primary electron-capture nuclides, in *International Cosmic Ray Conference*, vol. 3 (1999), p. 1
J. Meyer, L.O. Drury, D.C. Ellison Galactic cosmic rays from supernova remnants. I. A cosmic-ray composition controlled by volatility and mass-to-charge ratio. Astrophys. J. **487**, 182 (1997). arXiv:astro-ph/9704267
M. Milgrom, V. Usov, Possible association of ultra-high-energy cosmic-ray events with strong gamma-ray bursts. Astrophys. J. **449**, L37 (1995). arXiv:astro-ph/9505009
G. Morlino, P. Blasi, E. Amato, Gamma rays and neutrinos from SNR RX J1713.7-3946. Astropart. Phys. **31**, 376–382 (2009). arXiv:0903.4565
M. Nagano, A.A. Watson, Observations and implications of the ultrahigh-energy cosmic rays. Rev. Mod. Phys. **72**, 689–732 (2000)
C.A. Norman, D.B. Melrose, A. Achterberg, The origin of cosmic rays above 10 18.5 eV. Astrophys. J. **454**, 60 (1995)
A.V. Olinto, J.H. Adams, C.D. Dermer, J.F. Krizmanic et al., White paper on ultra-high energy cosmic rays, in *astro2010: The Astronomy and Astrophysics Decadal Survey*. ArXiv Astrophysics e-prints, vol. 2010 (2009), p. 225. arXiv:0903.0205
E. Parizot, GZK horizon and magnetic fields. Nucl. Phys. B, Proc. Suppl. **136**, 169–178 (2004). arXiv: astro-ph/0409191
E.N. Parker, The dynamical state of the interstellar gas and field. Astrophys. J. **145**, 811 (1966)
E.N. Parker, Fast dynamos, cosmic rays, and the galactic magnetic field. Astrophys. J. **401**, 137–145 (1992)
Pierre Auger Collaboration, J. Abraham, P. Abreu, M. Aglietta et al., Measurement of the energy spectrum of cosmic rays above 10^{18} eV using the Pierre Auger Observatory. Phys. Lett. B **685**, 239–246 (2010a). arXiv:1002.1975

Pierre Auger Collaboration, J. Abraham, P. Abreu, M. Aglietta et al., Measurement of the energy spectrum of cosmic rays above 10^{18} eV using the Pierre Auger Observatory. Phys. Lett. B **685**, 239–246 (2010b). arXiv:1002.1975

S. Profumo, Dissecting cosmic-ray electron-positron data with Occam's Razor: the role of known Pulsars (2008). arXiv:0812.4457

V. Ptuskin, V. Zirakashvili, E. Seo, Spectrum of galactic cosmic rays accelerated in supernova remnants. Astrophys. J. **718**, 31–36 (2010). arXiv:1006.0034

V.S. Ptuskin, I.V. Moskalenko, F.C. Jones, A.W. Strong et al., Dissipation of magnetohydrodynamic waves on energetic particles: impact on interstellar turbulence and cosmic-ray transport. Astrophys. J. **642**, 902–916 (2006). arXiv:astro-ph/0510335

V.S. Ptuskin, H.J. Voelk, V.N. Zirakashvili, D. Breitschwerdt, Transport of relativistic nucleons in a galactic wind driven by cosmic rays. Astron. Astrophys. **321**, 434–443 (1997)

V.S. Ptuskin, V.N. Zirakashvili, A.A. Plesser, Non-linear diffusion of cosmic rays. Adv. Space Res. **42**, 486–490 (2008)

J.L. Puget, F.W. Stecker, J.H. Bredekamp, Photonuclear interactions of ultrahigh energy cosmic rays and their astrophysical consequences. Astrophys. J. **205**, 638–654 (1976)

S.P. Reynolds, Supernova remnants at high energy. Annu. Rev. Astron. Astrophys. **46**, 89–126 (2008)

D. Ryu, J. Kim, S.S. Hong, T.W. Jones, The effect of cosmic-ray diffusion on the Parker instability. Astrophys. J. **589**, 338–346 (2003). arXiv:astro-ph/0301625

J. Scalo, B.G. Elmegreen, Interstellar turbulence II: Implications and effects. Annu. Rev. Astron. Astrophys. **42**, 275–316 (2004). arXiv:astro-ph/0404452

R. Schlickeiser, *Cosmic Ray Astrophysics* (Springer, Berlin, 2002)

E.S. Seo, V.S. Ptuskin, Stochastic reacceleration of cosmic rays in the interstellar medium. Astrophys. J. **431**, 705–714 (1994)

G. Sigl, Time structure and multi-messenger signatures of ultra-high energy cosmic ray sources. New J. Phys. **11**(6), 065014 (2009)

M. Simon, W. Heinrich, K.D. Mathis, Propagation of injected cosmic rays under distributed reacceleration. Astrophys. J. **300**, 32–40 (1986)

J. Skilling, Cosmic ray streaming. I—Effect of Alfvén waves on particles. Mon. Not. R. Astron. Soc. Lett. **172**, 557–566 (1975a)

J. Skilling, Cosmic ray streaming. III—Self-consistent solutions. Mon. Not. R. Astron. Soc. Lett. **173**, 255–269 (1975b)

S.A. Stephens, R.E. Streitmatter, Cosmic-ray propagation in the galaxy: techniques and the mean matter traversal. Astrophys. J. **505**, 266–277 (1998)

A.W. Strong, I.V. Moskalenko, Propagation of cosmic-ray nucleons in the galaxy. Astrophys. J. **509**, 212–228 (1998). arXiv:astro-ph/9807150

A.W. Strong, I.V. Moskalenko, V.S. Ptuskin, Cosmic-ray propagation and interactions in the galaxy. Annu. Rev. Nucl. Part. Sci. **57**, 285–327 (2007). arXiv:astro-ph/0701517

T. Takahashi, K. Mitsuda, R. Kelley, F. Aharonian et al., The ASTRO-H mission, in *Society of Photo-Optical Instrumentation Engineers (SPIE) Conference Series*, vol. 7732 (2010). arXiv:1010.4972

A.M. Taylor, I. Vovk, A. Neronov, EGMF constraints from simultaneous GeV-TeV observations of Blazars (2011). arXiv:1101.0932

I.N. Toptygin, *Cosmic Rays in Interplanetary Magnetic Fields* (Reidel, Dordrecht, 1985)

D.F. Torres, L.A. Anchordoqui, Astrophysical origins of ultrahigh energy cosmic rays. Rep. Prog. Phys. **67**, 1663–1730 (2004). arXiv:astro-ph/0402371

G. Vannoni, F.A. Aharonian, S. Gabici, S.R. Kelner et al., Acceleration and radiation of ultra-high energy protons in galaxy clusters (2009). arXiv:0910.5715

M. Véron-Cetty, P. Véron, A catalogue of quasars and active nuclei: 12th edition. Astron. Astrophys. **455**, 773–777 (2006)

M. Vietri, The acceleration of ultra-high-energy cosmic rays in gamma-ray bursts. Astrophys. J. **453**, 883 (1995). arXiv:astro-ph/9506081

J. Vink, Multiwavelength signatures of cosmic ray acceleration by young supernova remnants. in *American Institute of Physics Conference Series*, ed. by F.A. Aharonian, W. Hofmann, F. Rieger, vol. 1085 (2008), pp. 169–180

F. Vissani, F. Aharonian, N. Sahakyan, On the detectability of high-energy galactic neutrino sources (2011). arXiv:1101.4842

E. Waxman, Cosmological gamma-ray bursts and the highest energy cosmic rays. Phys. Rev. Lett. **75**, 386–389 (1995). arXiv:astro-ph/9505082

E. Waxman, High-energy cosmic rays from gamma-ray burst sources: a stronger case. Astrophys. J. **606**, 988–993 (2004). arXiv:astro-ph/0210638

G.M. Webb, E.K. Kaghashvili, J.A. le Roux, A. Shalchi et al., Compound and perpendicular diffusion of cosmic rays and random walk of the field lines: II. Non-parallel particle transport and drifts. J. Phys. A, Math. Gen. **42**(23), 235502 (2009)
D.G. Wentzel, Cosmic-ray propagation in the Galaxy—collective effects. Annu. Rev. Astron. Astrophys. **12**, 71–96 (1974)
M.E. Wiedenbeck, N.E. Yanasak, A.C. Cummings, A.J. Davis et al., The origin of primary cosmic rays: constraints from ACE elemental and isotopic composition observations. Space Sci. Rev. **99**, 15–26 (2001)
N.E. Yanasak, M.E. Wiedenbeck, R.A. Mewaldt, A.J. Davis et al., Measurement of the secondary radionuclides and implications for the galactic cosmic-ray age. Astrophys. J. **563**, 768–792 (2001)
G.T. Zatsepin, V.A. Kuzmin, Upper limit of the spectrum of cosmic rays. JETP Lett. **4**, 78–80 (1966)
Y. Zhou, W.H. Matthaeus, P. Dmitruk, Colloquium: magnetohydrodynamic turbulence and time scales in astrophysical and space plasmas. Rev. Mod. Phys. **76**, 1015–1035 (2004)
V.N. Zirakashvili, F.A. Aharonian, Nonthermal radiation of young supernova remnants: the case of RX J1713.7-3946. Astrophys. J. **708**, 965–980 (2010). arXiv:0909.2285
V.N. Zirakashvili, D. Breitschwerdt, V.S. Ptuskin, H.J. Voelk, Magnetohydrodynamic wind driven by cosmic rays in a rotating galaxy. Astron. Astrophys. **311**, 113–126 (1996)
V.N. Zirakashvili, D.N. Pochepkin, V.S. Ptuskin, S.I. Rogovaya, Propagation of ultra-high-energy cosmic rays in Galactic magnetic fields. Astron. Lett. **24**, 139–143 (1998)
V.N. Zirakashvili, V.S. Ptuskin, Diffusive shock acceleration with magnetic amplification by nonresonant streaming instability in supernova remnants. Astrophys. J. **678**, 939–949 (2008). arXiv:0801.4488
E.G. Zweibel, Cosmic-ray history and its implications for galactic magnetic fields. Astrophys. J. **587**, 625–637 (2003). arXiv:astro-ph/0212559

Space Sci Rev (2012) 166:133–144
DOI 10.1007/s11214-011-9757-0

Magnetic Fields in Galactic Haloes

M. Haverkorn · V. Heesen

Received: 7 December 2010 / Accepted: 2 February 2011 / Published online: 5 March 2011

Abstract Magnetic fields on a range of scales play a large role in the ecosystems of galaxies, both in the galactic disk and in the extended layers of gas away from the plane. Observing magnetic field strength, structure and orientation is complex, and necessarily indirect. Observational data of magnetic fields in the halo of the Milky Way are scarce, and non-conclusive about the large-scale structure of the field. In external galaxies, various large-scale configurations of magnetic fields are measured, but many uncertainties about exact configurations and their origin remain. There is a strong interaction between magnetic fields and other components in the interstellar medium such as ionized and neutral gas and cosmic rays. The energy densities of these components are comparable on large scales, indicating that magnetic fields are not passive tracers but that magnetic field feedback on the other interstellar medium components needs to be taken into account.

Keywords Cosmic magnetism · Spiral galaxies · Milky Way · Magnetic fields · Interstellar medium · Galactic haloes

1 Introduction

Significant magnetic fields are present not only in galaxy disks, but also in the gaseous galactic halo.[1] Magnetic fields in these galactic haloes can provide a significant pressure

[1] The term 'galactic halo' is used here to describe the medium above the galactic disk, also sometimes referred to as the thick disk. In this definition, the gaseous halo of a galaxy has not necessarily a direct relationship to the spherical stellar halo.

M. Haverkorn (✉)
ASTRON, PO Box 2, 7990 AA Dwingeloo, the Netherlands
e-mail: haverkorn@astron.nl

M. Haverkorn
Leiden Observatory, Leiden University, PO Box 9513, 2300 RA Leiden, the Netherlands

V. Heesen
Centre for Astrophysics Research, University of Hertfordshire, Hatfield AL 10 9AB, UK

component to counteract the gravity of the halo gas (Boulares and Cox 1990). Also, they play a major role in the disk-halo interaction, influencing superbubble break-out and funneling particles out to intergalactic space. Finally, the determination of the morphology of halo magnetic fields can shed light on the validity of various dynamo models to explain the origin and evolution of galactic magnetism.

We will first give a brief summary of the conditions in the haloes of the Milky Way and nearby spirals in Sects. 2 and 3. Sections 4 to 6 describe magnetic field strengths, scale heights, and structure in the Milky Way and external spiral galaxies. Finally, in Sect. 7 we briefly discuss the interaction of magnetic fields with other interstellar medium components.

2 Gaseous Components in Galactic Haloes

2.1 The Galaxy

The interstellar medium (ISM) in the halo of the Milky Way consists of the three-phase medium as in the disk, but with varying filling factors and scale heights. While there is a warm neutral hydrogen layer and cold neutral hydrogen clouds at large distances from the Galactic plane (Dickey and Lockman 1990), coupled to magnetic fields through charge exchange, the most direct influence on magnetic field structure comes from the halo plasma.

Estimates for the scale height of the Warm Ionized Medium (WIM) in the Milky Way vary between $\sim$1000 and $\sim$1800 pc (Reynolds 1991; Gaensler et al. 2008; Savage and Wakker 2009). The volume filling factor of the WIM is believed to increase from $f \sim 0.1$ at the Galactic mid-plane to $f > 0.3$ at $|z| = 1000$ pc (Berkhuijsen et al. 2006). Whereas H II regions dominate the WIM distribution in the Galactic plane, at high latitudes Hα emission from the WIM is more diffuse and pervasive (Haffner et al. 2003).

Hot, $\sim 10^6$ K gas existing inside supernova remnants and superbubbles is predicted to have a filling factor of about $\sim$20% (Ferrière 1998). This estimate is confirmed by the filling factor of large neutral hydrogen holes (Heiles 1980) and observations of O VI UV absorption lines (Shelton and Cox 1994). A patchy distribution of hot gas at high latitudes is seen in fluctuations in the projected O VI column densities (Hurwitz and Bowyer 1996) and from X-ray shadows of interstellar clouds. Estimates of the hot gas scale height, from absorption lines of high-ionization ions like O VI and N V, range from about 3 to 5 kpc (Ferrière 2001).

2.2 External Galaxies

Rossa and Dettmar (2003a, 2003b) reported a mean vertical extent of 1–2 kpc of the diffuse ionized gas (DIG)[2] in external spiral galaxies, if they possess warm dust as indicated by the FIR $\lambda_{60}/\lambda_{100}$ ratio. This shows that gaseous haloes are connected to the star formation activity in the underlying disk. External galaxies also possess hot X-ray emitting gas with a scale height in the range 4–8 kpc (Strickland et al. 2004). Tüllmann et al. (2000) showed that this gas is again correlated with the presence of warm dust, as for the extraplanar DIG.

[2]Following convention, we call the layer of ionized gas in spiral galaxies the Warm Ionized Medium (WIM) in the Milky Way, and Diffuse Ionized Gas (DIG) in external galaxies.

3 Cosmic Rays

3.1 Scale Height

The Galaxy: From estimates of the gamma-ray radiation from the Compton Gamma-Ray Observatory in the 1–30 MeV band, and isolating the inverse Compton component, Strong et al. (1996) concluded that the cosmic ray scale height is probably much larger than that of the gas. An alternative method to estimate the vertical scale height of cosmic rays is based on counts of cosmic ray nuclei such as Li, Be, and B, created in collisions of primary cosmic ray nuclei with interstellar hydrogen. Using this method, Bloemen et al. (1993) derived a cosmic ray scale height ≤ 3 kpc at the solar radius, consistent with values found by Webber et al. (1992). The cosmic ray equivalent scale height derived from observed synchrotron emissivities in the Galaxy is about 2 kpc, in agreement with the estimates above (Ferrière 2001).

Modeling of the Galactic synchrotron distribution from an all-sky survey at 408 MHz (Haslam et al. 1982) shows that the synchrotron emissivity in the Milky Way consists of a thick and a thin disk (Beuermann et al. 1985). The scale heights of these components are $\sim$150 pc and $\sim$1500 pc, respectively (after rescaling to a Galactocentric radius of 8.5 kpc), and 90% of the total power is emitted in the thick disk. The equipartition[3] value for the cosmic ray scale height derived from this again roughly agrees with the above estimates.

External Galaxies: Similarly, in external galaxies vertical profiles of the synchrotron emission can be characterized by two components, a thin disk with a scale height of 300 pc and a thick disk with a scale height of 1.8 kpc (at 4.8 GHz). These numbers are remarkably independent of the star-formation rate (Krause 2009). If we assume equipartition, the scale height of the electrons would thus be 3.6 kpc. Not much is known if the cosmic-ray nuclei and the electrons are transported together but this is generally assumed. Therefore, for electrons with an energy of a few GeV, the scale height in external galaxies agrees roughly with the ones quoted above for the Galaxy.

3.2 Cosmic-Ray Transport

The luminosity of X-ray emission in the center of the Milky Way can only be explained if the Galaxy possesses a wind that is hybridly driven by cosmic rays and the thermal gas (Everett et al. 2008). Due to the interaction with the ions and electron in the so-called streaming instability the cosmic rays can push out the ionized gas by momentum transfer, and—if there are sufficient ion-neutrals collisions—also the neutral gas. In a study of the resolved emission in NGC 253 it was shown that the scale height of the electrons is proportional to their lifetime leading to an estimate of 300 $\mathrm{km\,s^{-1}}$ as the cosmic-ray bulk speed. This is similar to the escape velocity in this galaxy and rather constant across the full extent of the disk, so that a "disk wind" was proposed (Heesen et al. 2009a).

[3] The equipartition assumption indicates equal energy densities in the magnetic field and cosmic rays. It is often used to estimate magnetic field strengths from measured synchrotron emissivities, and generally holds well on galaxy-size scales.

4 Magnetic Field Scale Height and Strength

4.1 Scale Heights

The Galaxy: From hydrostatic equilibrium considerations, Boulares and Cox (1990) determined that cosmic rays, magnetic fields, and a significant fraction of the ISM should have scale heights in excess of 1 kpc. In particular, the magnetic field strength only decreases by ~30% out to a distance of a kpc from the Galactic plane. A measure for the scale height of the Galactic magnetic field can also be found from the synchrotron scale height. Beuermann et al. (1985) modeled synchrotron emissivity from measurements of Haslam et al. (1981, 1982), which indicates a magnetic field scale height at the location of the Sun of about 4.5 kpc (Ferrière 2001). This scale height agrees with estimates considering hydrostatic equilibrium in the halo (Kalberla and Kerp 1998). The discrepancy with the smaller scale height of the regular magnetic field component found from Faraday rotation measurements was explained by Boulares and Cox (1990) by stating that the Galactic magnetic field becomes less regular away from the plane. Alternatively, these measurements may be explained by two separate magnetic field components with different scale heights (Andreasyan and Makarov 1988; Han and Qiao 1994), consistent with observations of a two-layer structure of the Galactic non-thermal synchrotron emission (Beuermann et al. 1985).

External Galaxies: The scale heights in external galaxies can be estimated from the synchrotron scale height. If we assume equipartition, we obtain that the magnetic field scale height is a factor of 4 larger than the synchrotron scale height and thus about 6 kpc.

4.2 Field Strengths

The Galaxy: Much is still unknown about the magnetic field strength and structure in the Milky Way at large distances from the Galactic plane. Sun et al. (2008) modeled the magnetic field in the Galactic halo as a plane parallel field superimposed on a turbulent component, based on synchrotron emissivity and Faraday rotation data. They found that the halo field is unrealistically high with a 1 kpc thermal electron scale height, but when using a scale height of 1.8 kpc, their estimate of the regular magnetic field strength in the Galactic halo goes down to reasonable values of 2 μG.

External Galaxies: Using the energy equipartition condition the total magnetic field strength in galaxies is 9 ± 3 μG (Niklas and Beck 1997). From the linear polarization one can derive the regular magnetic field strength as typically 1–5 μG in the disk (Krause 2009). In the halo, the magnetic field is dominated by the regular component as opposed to the disk. The regular component in the halo is of the order of the disk component, if we neglect stronger fields that are found in the interarm regions in the disk such as in NGC 6946 (Beck 2007, 2011, this issue).

5 Magnetic Field Structure in the Galaxy

Galactic magnetic fields are maintained and amplified by dynamo action. The simplest model is that of the $\alpha - \omega$ dynamo, which amplifies the radial magnetic field component through differential rotation, and amplifies the azimuthal and poloidal components of the field by turbulent loops twisted by the Coriolis force. Although the $\alpha - \omega$ dynamo

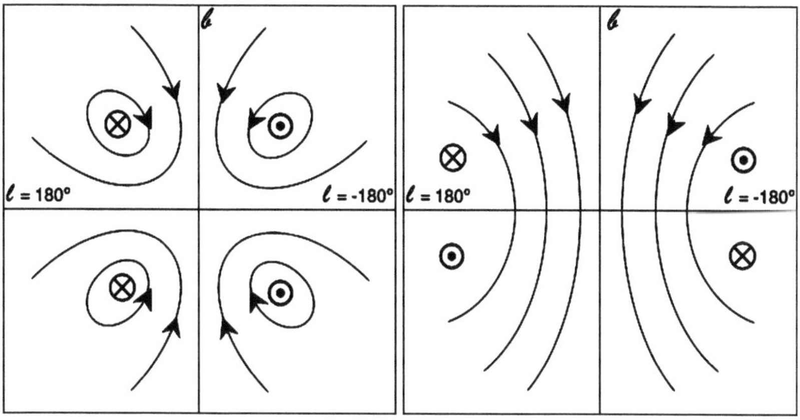

Fig. 1 Quadrupole (*left*) and dipole (*right*) large-scale magnetic field configurations of the Milky Way as a function of Galactic longitude and latitude. Magnetic field towards the viewer is denoted by a *dot*, field away from the observer as a *cross*

is believed to act in the Sun (Ossendrijver 2003), it is considered unable to sufficiently amplify galactic magnetic fields to observed values (eddy diffusion time scales are much shorter than time scales for amplification of the regular field, thereby suppressing turbulent motions, and quenching the dynamo). Various solutions to this problem are discussed in Widrow (2002). For a flat disk-like galaxy which is differentially rotating, mean-field dynamo theory predicts a quadrupolar magnetic field configuration, where the direction of the azimuthal magnetic field is the same above and below the plane, but the direction of the vertical magnetic field component reverses with respect to the plane. However, for a spherical, weakly rotating structure—such as a Galactic halo—the dipolar configuration is more easily excited, i.e. an azimuthal magnetic field with reversing direction across the Galactic plane, while the vertical field is directed in the same way above and below the plane (see Fig. 1).

These two configurations can be observationally distinguished by the directions of the azimuthal magnetic field component above and below the Galactic plane, and by the directions of the vertical field components above and below the plane. Firstly, measurements of Faraday rotation of extragalactic sources show a pronounced anti-symmetric "butterfly pattern" with respect to the Galactic plane and the meridian through the Galactic center (Simard-Normandin and Kronberg 1980), only visible in the inner Galaxy (first and fourth Galactic quadrants). This has been interpreted as a reversal of the direction of the azimuthal magnetic field with respect to the Galactic plane indicative of a dipolar magnetic field configuration and an A0 dynamo (Han et al. 1997), but can possibly also be due to the influence of local structures on large angular scales (Wolleben et al. 2010; Stil et al. 2011). Secondly, a vertical component of the Galactic magnetic field is predicted by Boulares and Cox (1990) to provide a magnetic tension term that can balance other pressure components in the Galactic disk and halo (see also Andreasyan and Makarov 1988). Mao et al. (2010) tried to measure this vertical magnetic field component at high latitudes. At the solar radius, they found a small vertical magnetic field component of $B_z = 0.31 \pm 0.03$ μG towards negative Galactic latitudes but not towards positive latitudes, indicating smaller-scale structure in the halo magnetic field. These conclusions agree with the structure in the X-ray halo (Ferrière 2001) and with star light polarization measurements. Taylor et al. (2009) include the North Polar Spur region in their determination of vertical magnetic field

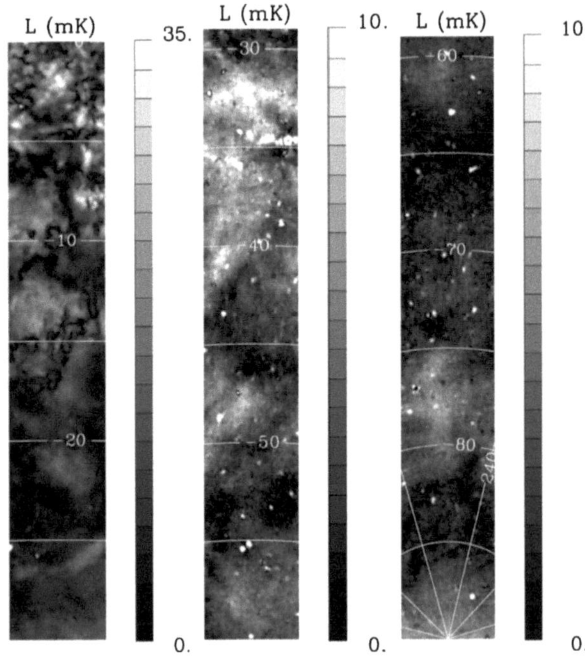

Fig. 2 Linearly polarized intensity from the Parkes Galactic Meridian Survey (Carretti et al. 2010) at 2.3 GHz shows small-scale structure in linear polarization at all Galactic latitudes, indicating small-scale structure in the electron-density weighted magnetic field

strengths and do find a small vertical magnetic field component $B_z = -0.14 \pm 0.02$ μG, while their estimates towards the southern sky agree with Mao et al. (2010).

Schnitzeler et al. (2007) also find a small, varying vertical magnetic field component at intermediate positive latitudes in the outer Galaxy. De Bruyn et al. (2006) conclude from radio polarimetric measurements at high Galactic latitudes that there must be significant variations in the vertical component of the Galactic magnetic field on the level of 1 μG, similar to the strength of the random magnetic field component at high Galactic latitudes (Mao et al. 2010). However, the studies by Schnitzeler et al. (2007) and De Bruyn et al. (2006) were performed at low frequencies, which indicates that they may be probing mostly nearby structure. The higher-frequency Parkes Galactic Meridian Survey at 2.3 GHz detects polarized emission from a larger path length and shows that even towards the southern Galactic pole, there is still small-scale structure in radio polarization, indicating structure in the density-weighted Galactic magnetic field on scales of degrees (Carretti et al. 2010, see Fig. 2).

Small-scale structure in the high latitude magnetic field is confirmed by measurements of linear polarization of starlight. Starlight which propagates through asymmetric dust grains aligned in a magnetic field will become partially polarized with a polarization angle in the average direction of the magnetic field perpendicular to the line of sight in the dusty regions between the star and the observer. Berdyugin and co-workers published a series of papers (e.g. Berdyugin and Teerikorpi 2001; Berdyugin et al. 2001, 2004) on polarization directions of stars at high latitude, which probe mostly the direction of the field component parallel to the Galactic plane. They concluded (1) that this magnetic field towards the southern Galactic pole is oriented towards $l \approx 80°$, in agreement with a spiral magnetic field, and (2) that smaller-scale structure towards the northern Galactic pole precluded any conclusions about the ordered field component there. As the polarization percentages keep increasing towards higher distances from the plane, out to their furthest stars at about 500 pc, they concluded that Galactic magnetic fields reach at least to 500 pc without significant decrease.

6 Magnetic Field Structure in External Galaxies

6.1 Field Orientation

The magnetic field is mainly aligned parallel to the disk in regions close to the mid plane. In the halo, the vertical component becomes stronger with increasing height above the disk. Also the magnetic field is more ordered as indicated by the higher polarization degree in the halo. Galaxies that have a radio halo possess also vertical fields.

Several disk galaxies viewed edge-on show X-shaped magnetic fields in their haloes as shown in Fig. 3 for the case of NGC 891. The orientation of the magnetic field aligns with the X in the four quadrants and the polarized emission is enhanced there. Some galaxies have distinctive sites with significant vertical magnetic field components that are referred to as "radio spurs". They correlate in position with filamentary Hα emission sticking out from the disk into the halo. Examples for this are NGC 4666 (Dahlem et al. 1997) and NGC 5775 (Tüllmann et al. 2000). This is illustrated in Figs. 4 and 5.

The reason for this structure is not yet clear. It could be related to the flow of gas in the halo that has X-shaped structure as well (DallaVecchia and Schaye 2008). The energy densities of the magnetic field and the gas are comparable, so that interaction between the two is possible. Heesen et al. (2009b) showed that NGC 253 also has an X-shaped field. This is in so far remarkable as the halo field can only be revealed after subtraction of a model field for the disk, because this galaxy has an inclination angle of 78° and is therefore not exactly edge-on. In the case of NGC 253, Heesen et al. (2009b) suggested that the halo magnetic field may be enhanced by compression due to expanding hot gas in the halo that is visible as a conical outflow in X-ray emission (see Fig. 6). Consequently, radio spurs should align with the boundary of the outflow cone and the magnetic field be roughly tangential to it.

An example of non-X-shaped fields is NGC 4631 with a rather more dipolar field structure, where the vertical component of the magnetic field dominates as shown in Fig. 7. This may be a case where the field is dominated by an interaction with a nearby companion or

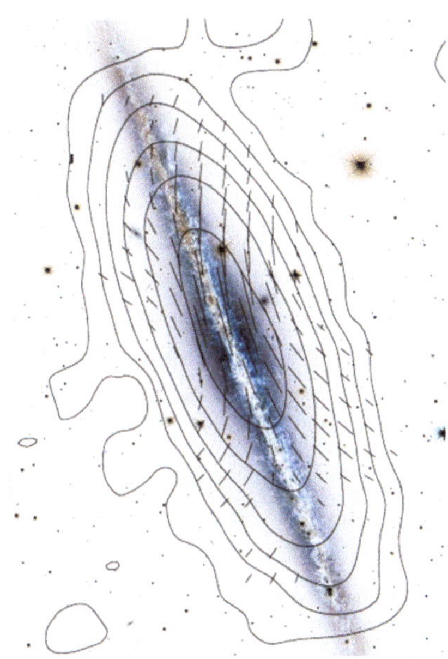

Fig. 3 Radio continuum emission of the edge-on spiral galaxy NGC 891 at λ3.6 cm (8.35 GHz) with the 100 m Effelsberg telescope with a resolution of 84″ HPBW. The contours give the total intensity, the vectors the intrinsic magnetic field orientation (Copyright: MPIfR Bonn). The radio map is overlaid on an optical image of NGC 891 from the Canada-France-Hawaii Telescope/(c)1999 CFHT/Coelum. From Krause (2009)

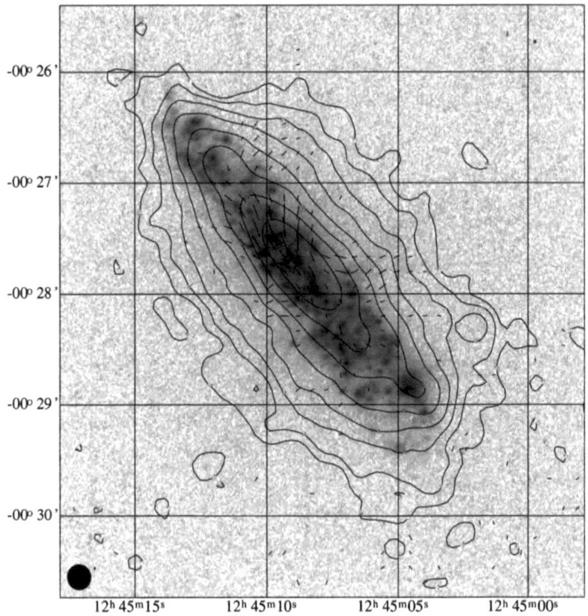

Fig. 4 Radio continuum emission of NGC 4666 at λ6 cm from VLA observations with 14″ HPBW. The vectors show the magnetic field orientation where the length is proportional to the polarized emission. The background shows Hα + N[II] emission, showing the spatial correlation of optical emission line and (polarized) radio synchrotron emission from the halo. From Dahlem et al. (1997)

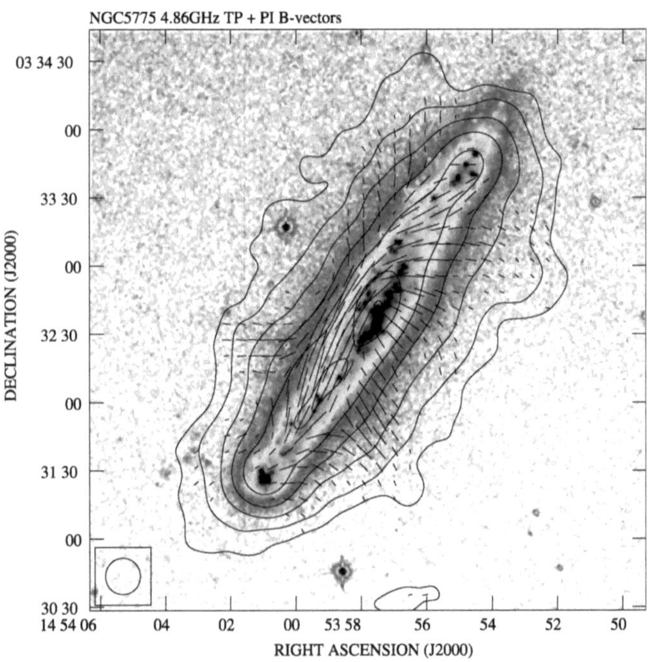

Fig. 5 Radio continuum emission of NGC 5775 at λ6 cm from VLA observations with 14″ HPBW. The vectors show the magnetic field orientation where the length is proportional to the polarized emission. The background shows Hα emission. From Tüllmann et al. (2000)

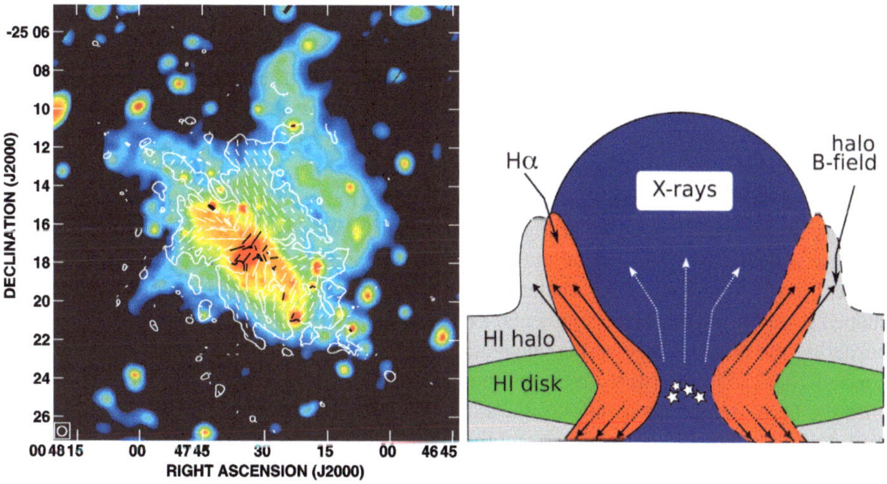

Fig. 6 *Left*: halo magnetic field in NGC 253 overlaid on diffuse X-ray emission. Contours show the intensity of the polarized emission of the halo magnetic field. Vectors show the orientation of the halo magnetic field where the length is proportional to the polarized intensity. *Right*: proposed halo structure of NGC 253. The superbubble, filled with soft X-ray emitting gas, expands into the surrounding medium (indicated by *dotted lines with arrows*). The halo magnetic field is aligned with the walls of the superbubble. *Dashed lines* denote components that are not (or only weakly) detected in the southwestern half of NGC 253. Figures from Heesen et al. (2009b)

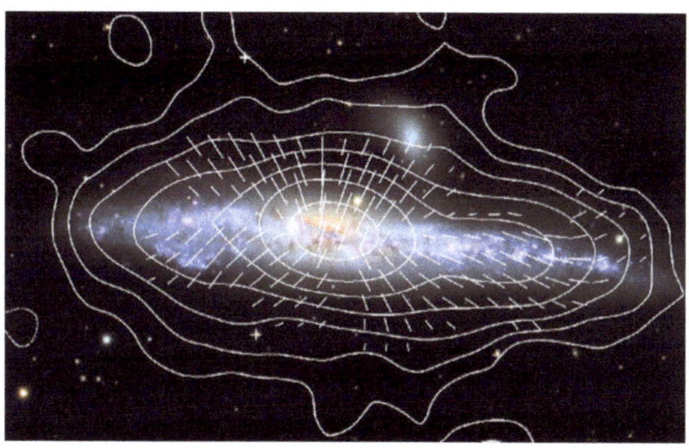

Fig. 7 Radio continuum emission of NGC 4631 at λ3.6 cm with 84″ HPBW from observations with the 100 m Effelsberg telescope. The vectors show the magnetic field orientation where the length is proportional to the polarized emission. The background is an optical image taken at the Misti Mountain Observatory (Copyright: MPIfR Bonn). From Krause (2009)

where differential rotation is weak and thus the A0 dynamo mode is excited (Golla and Hummel 1994). Deviations from the X-shaped field structure are also found for galaxies which do not have star formation driven outflows. NGC 4258 (Krause and Löhr 1994) and NGC 4569 (Chyży et al. 2006) are examples for jet magnetic fields. There the magnetic field is vertically aligned to the disk and parallel to the outflow in the jet.

Vertical magnetic fields are also indirectly detected. M 82 has vertical filaments and gaps in the radio continuum emission at λ20 cm with a width of 100 pc (Reuter et al. 1992; Wills et al. 1999). Vertical fields were detected with radio continuum polarimetry in this galaxy albeit at a lower resolution, which likely prevented the detection of filaments (Reuter et al. 1994). Wills et al. (1999) tried to connect the filaments to chimney structures, i.e. blow outs by clustered supernovae, but the outflow visible in CO line emission was only related to a weaker minimum in the continuum emission. Duric et al. (1998) found vertical "tentacles" of a flat spectral index component protruding from the disk into the halo in NGC 5775. If they are interpreted as outflows of young cosmic-ray electrons, there needs to be some vertical alignments for which magnetic fields are the favored mechanism. There are other indicators of vertical fields like vertical dust filaments in NGC 253 (Sofue et al. 1994). These filaments are thin ($\leq$50 pc) compared to the resolution of the available radio maps and are thus far not detected in the continuum or polarized emission.

Nuclear outflows are a good site to study filaments in the radio continuum, because they are very bright and allow us to use high resolution. Although they are not located in the halo, they are at the base of the conical outflows discussed above. They may be also a source for vertical fields. An example for nuclear outflow is NGC 3079 which has a prominent bowl-shaped structure visible as Hα filaments. The polarized emission is concentrated in filaments bordering the outflow cone and is tangential to it (Cecil et al. 2001). Similarly, the outflow cone in NGC 253 is bordered by continuum filaments with a scale height of 150 ± 20 pc with the field orientation parallel to the filaments (Heesen et al. 2011). The magnetic field may collimate the nuclear outflows and thus explain the small opening angles $\sim$30° of the cones.

Whereas interaction in the interstellar medium can explain the distribution of the fields in filaments, it does not *generate* vertical magnetic fields. For this the galactic dynamo is a favorite process. Vertical magnetic fields may also be generated by the Parker instability where cosmic-ray gas inflates magnetic field lines that reconnect and give rise to the fast Parker dynamo (Parker 1992). These magnetic fields would be expected to be anisotropic with frequent field reversals. This configuration is expected to have reversals in the line-of-sight component which can be detected in Faraday rotation.

6.2 Field Direction

The halo rotation measures (RMs) are difficult to measure, because they can overlap with the RMs in the disk which dominate the polarized emission. Recent papers have tried to separate the contributions from the disk and halo by using models of the magnetic field structure. Heesen et al. (2009b) found in NGC 253 a magnetic field that points outward on the near side of the halo, but the data is not conclusive for the halo behind the bright polarized emission in the disk. The field direction has some consequence for the dynamo models because in the presence of a galactic wind the parity of the disk and halo field are expected to be identical (Moss et al. 2010).

Braun et al. (2010) explained the asymmetry of the polarized emission with respect to the receding and approaching side of the galaxy with halo fields. In their sample of 21 galaxies detected in polarization (Heald et al. 2008) they found directions that resemble both the quadrupolar type of fields and dipolar radially directed fields. The magnetic field structure in the halo is also of interest for angular momentum transport. If the field lines are stiff and we do not measure any azimuthal component then we know that we are below the Alfvénic point. The higher up the Alfvénic point the more angular momentum is transported from the disk into the halo (Zirakashvili et al. 1996).

The Parker fast dynamo would see RM reversals as part of Parker loops. Such a reversal was observed in the nuclear outflow in NGC 3079 (Cecil et al. 2001). In NGC 1569 the vertical magnetic field borders an outflow bubble and is also accompanied with a field reversal (Kepley et al. 2010). In principle it is possible to distinguish between the anisotropic and the isotropic component of the halo field by comparing the field strengths from the RM and equipartition. This was attempted by Heesen et al. (2009b) where they found the field component from the RM to be only 50% lower than that from equipartition. This would mean that the halo field has only a small anisotropic component generated by Parker loops.

7 Interaction of the Magnetic Field with the Interstellar Medium

Interstellar magnetic fields are in constant interaction with the other components in the interstellar medium. Magnetic fields are mostly directly frozen into the ionized gas component, but also closely interact with the neutral gas through charge exchange (for an example, see McClure-Griffiths et al. 2006). On large scales, the magnetic field energy density is comparable to the cosmic ray energy density and the turbulent gas energy density (see Heiles and Haverkorn 2011, this issue, and Beck 2011, this issue). This means that any of these components can only be fully understood if the interaction with the other components is taken into account.

Therefore, gas dynamics is greatly influenced by magnetic fields. Numerical simulations show that expansion of supernova remnants in a magnetized medium will be anisotropic (Stil et al. 2009) and slower (Ferrière et al. 1991) than in the non-magnetized case. Magnetohydrodynamic turbulence can be a significant source of heating in the ISM (e.g. Scalo and Elmegreen 2004), as can magnetic reconnection (Zimmer et al. 1997).

Magnetic fields play a large role in the hydrostatic equilibrium of the Milky Way. Thick gas disks, as observed in the Milky Way, with purely azimuthal magnetic fields, have a tendency to become unstable (e.g. Parker 1969). Boulares and Cox (1990) show that magnetic tension in a small vertical component of the magnetic field may stabilize this configuration and so help to keep the thick gas layer from collapsing onto the disk.

Cosmic rays are tightly coupled to galactic magnetic fields. These charged particles gyrate around magnetic field lines with a Larmor radius depending on their energy and the magnetic field strength. Therefore, magnetic fields effectively trap all except the most energetic cosmic rays in galaxies and fully determine their trajectories. Also, magnetic fields are a major factor in acceleration of these charged particles to their relativistic velocities.

Acknowledgements The authors want to thank the International Space Science Institute for support and organisation of an excellent meeting. We are also indebted to Rainer Beck and Elias Brinks for fruitful discussions and useful comments which improved the paper.

References

R.R. Andreasyan, A.N. Makarov, Astrophysics **28**, 247 (1988)
R. Beck, Astron. Astrophys. **470**, 539 (2007)
R. Beck, Space Sci. Rev. (2011, this issue)
A. Berdyugin, P. Teerikorpi, Astron. Astrophys. **368**, 635 (2001)
A. Berdyugin, P. Teerikorpi, L. Haikala, M. Hanski, J. Knude, T. Markkanen, Astron. Astrophys. **372**, 276 (2001)
A. Berdyugin, V. Piirola, P. Teerikorpi, Astron. Astrophys. **424**, 873 (2004)
E.M. Berkhuijsen, D. Mitra, P. Mueller, Astron. Nachr. **327**, 82 (2006)
K. Beuermann, G. Kanbach, E.M. Berkhuijsen, Astron. Astrophys. **153**, 17 (1985)
J.B.G.M. Bloemen, V.A. Dogiel, V.L. Dorman, V.S. Ptuskin, Astron. Astrophys. **267**, 372 (1993)
A. Boulares, D.P. Cox, Astrophys. J. **365**, 544 (1990)
R. Braun, G. Heald, R. Beck, Astron. Astrophys. **514**, A42 (2010)
E. Carretti, M. Haverkorn, D. McConnell, G. Bernardi, N.M. McClure-Griffiths, S. Cortiglioni, S. Poppi, Mon. Not. R. Astron. Soc. **405**, 1670 (2010)
G. Cecil, J. Bland-Hawthorn, S. Veilleux, A.V. Filippenko, Astrophys. J. **555**, 338 (2001)
K.T. Chyży, M. Soida, D.J. Bomans et al., Astron. Astrophys. **447**, 465 (2006)
M. Dahlem, M.G. Petr, M.D. Lehnert et al., Astron. Astrophys. **320**, 731 (1997)
C. DallaVecchia, J. Schaye, Mon. Not. R. Astron. Soc. **387**, 1431 (2008)
A.G. De Bruyn, P. Katgert, M. Haverkorn, D.H.F.M. Schnitzeler, Astron. Nachr. **327**, 487 (2006)
J.M. Dickey, F.J. Lockman, Annu. Rev. Astron. Astrophys. **28**, 215 (1990)
N. Duric, J. Irwin, H. Bloemen, Astron. Astrophys. **331**, 428 (1998)

J.E. Everett, E.G. Zweibel, R.A. Benjamin et al., Astrophys. J. **674**, 258 (2008)
K.M. Ferrière, Astrophys. J. **503**, 700 (1998)
K.M. Ferrière, Rev. Mod. Phys. **73**, 1031 (2001)
K.M. Ferrière, M.-M. Mac Low, E.G. Zweibel, Astrophys. J. **375**, 239 (1991)
B.M. Gaensler, G.J. Madsen, S. Chatterjee, S.A. Mao, Publ. Astron. Soc. Aust. **25**, 184 (2008)
G. Golla, E. Hummel, Astron. Astrophys. **284**, 777 (1994)
L.M. Haffner, R.J. Reynolds, S.L. Tufte, G.J. Madsen, K.P. Jaehnig, J.W. Percival, Astrophys. J. Suppl. Ser. **149**, 405 (2003)
J.L. Han, G.J. Qiao, Astron. Astrophys. **288**, 759 (1994)
J.L. Han, R.N. Manchester, E.M. Berkhuijsen, R. Beck, Astron. Astrophys. **322**, 98 (1997)
C.G.T. Haslam, U. Klein, C.J. Salter, H. Stoffel, W.E. Wilson, M.N. Cleary, D.J. Cooke, P. Thomasson, Astron. Astrophys. **100**, 209 (1981)
C.G.T. Haslam, H. Stoffel, C.J. Salter, W.E. Wilson, Astron. Astrophys. Suppl. Ser. **47**, 1 (1982)
G. Heald, R. Braun, R. Edmonds, Astron. Astrophys. **503**, 409 (2008)
V. Heesen, R. Beck, M. Krause, R.-J. Dettmar, Astron. Astrophys. **494**, 563 (2009a)
V. Heesen, M. Krause, R. Beck, R.-J. Dettmar, Astron. Astrophys. **506**, 1123 (2009b)
V. Heesen, R. Beck, M. Krause, R.-J. Dettmar, U. Klein (2011, in prep.)
C. Heiles, Astrophys. J. **235**, 833 (1980)
C. Heiles, M. Haverkorn, Space Sci. Rev. (2011, this issue)
M. Hurwitz, S. Bowyer, Astrophys. J. **465**, 296 (1996)
P.M.W. Kalberla, J. Kerp, Astron. Astrophys. **339**, 745 (1998)
A.A. Kepley, S. Mühle, J. Everett, E.G. Zweibel, E.M. Wilcots, U. Klein, Astrophys. J. **712**, 536 (2010)
M. Krause, in *Magnetic Fields in the Universe II*, ed. by A. Esquivel, RevMexAA, vol. 36 (2009), p. 25
M. Krause, A. Löhr, Astron. Astrophys. **420**, 115 (1994)
S.A. Mao, B.M. Gaensler, M. Haverkorn, E.G. Zweibel, G.J. Madsen, N.M. McClure-Griffiths, A. Shukurov, P.P. Kronberg, Astrophys. J. **714**, 1170 (2010)
N.M. McClure-Griffiths, J.M. Dickey, B.M. Gaensler, A.J. Green, M. Haverkorn, Astrophys. J. **652**, 1339 (2006)
D. Moss, D. Sokoloff, R. Beck, M. Krause, Astron. Astrophys. **512**, A61 (2010)
S. Niklas, R. Beck, Astron. Astrophys. **320**, 54 (1997)
M. Ossendrijver, Astron. Astrophys. Rev. **11**, 287 (2003)
E.N. Parker, Space Sci. Rev. **9**, 651 (1969)
E.N. Parker, Astrophys. J. **401**, 137 (1992)
H.-P. Reuter, U. Klein, H. Lesch, R. Wielebinski, P.P. Kronberg, Astron. Astrophys. **256**, 10 (1992)
H.-P. Reuter, U. Klein, H. Lesch, R. Wielebinski, P.P. Kronberg, Astron. Astrophys. **282**, 724 (1994)
R.J. Reynolds, Astrophys. J. Lett. **372**, 17 (1991)
J. Rossa, R.-J. Dettmar, Astron. Astrophys. **406**, 493 (2003a)
J. Rossa, R.-J. Dettmar, Astron. Astrophys. **406**, 505 (2003b)
B.D. Savage, B.P. Wakker, Astrophys. J. **702**, 1472 (2009)
J. Scalo, B.G. Elmegreen, Annu. Rev. Astron. Astrophys. **42**, 275 (2004)
D.H.F.M. Schnitzeler, P. Katgert, M. Haverkorn, A.G. de Bruyn, Astron. Astrophys. **461**, 963 (2007)
R.L. Shelton, D.P. Cox, Astrophys. J. **434**, 599 (1994)
M. Simard-Normandin, P.P. Kronberg, Astrophys. J. **242**, 74 (1980)
Y. Sofue, K.-I. Wakamatsu, D.F. Malin, Astron. J. **108**, 2102 (1994)
J. Stil, N. Wityk, R. Ouyed, A.R. Taylor, Astrophys. J. **701**, 330 (2009)
J. Stil, A.R. Taylor, C. Sunstrum, Astrophys. J. **726**, 4 (2011)
D.K. Strickland, T.M. Heckman, E.J.M. Colbert, C.G. Hoopes, K.A. Weaver, Astrophys. J. Suppl. Ser. **151**, 193 (2004)
A.W. Strong, K. Bennett, H. Bloemen et al., Astron. Astrophys. Suppl. Ser. **120**, 381 (1996)
X.H. Sun, W. Reich, A. Waelkens, T.A. Ensslin, Astron. Astrophys., 573 (2008)
A.R. Taylor, J.M. Stil, C. Sunstrum, Astrophys. J. **702**, 1230 (2009)
R. Tüllmann, R.-J. Dettmar, M. Soida, M. Urbanik, J. Rossa, Astron. Astrophys. **364**, L36 (2000)
W.R. Webber, M.A. Lee, M. Gupta, Astrophys. J. **390**, 96 (1992)
L.M. Widrow, Rev. Mod. Phys. **74**, 775 (2002)
K.A. Wills, M.P. Redman, T.W.B. Muxlow, A. Pedlar, Mon. Not. R. Astron. Soc. **309**, 395 (1999)
M. Wolleben, A. Fletcher, T.L. Landecker, M. Haverkorn, E. Carretti, J.M. Dickey, B.M. Gaensler, J.L. Han, J.P. Leahy, N.M. McClure-Griffiths, D. McConnel, W. Reich, A.R. Taylor, Astrophys. J. **724**, 48 (2010)
F. Zimmer, H. Lesch, G.T. Birk, Astron. Astrophys. **320**, 746 (1997)
V.N. Zirakashvili, D. Breitschwerdt, V.S. Ptuskin, Astron. Astrophys. **311**, 113 (1996)

Space Sci Rev (2012) 166:145–185
DOI 10.1007/s11214-011-9771-2

Magnetic Fields in Massive Stars, Their Winds, and Their Nebulae

Rolf Walder · Doris Folini · Georges Meynet

Received: 9 September 2010 / Accepted: 13 March 2011 / Published online: 13 April 2011

Abstract Massive stars are crucial building blocks of galaxies and the universe, as production sites of heavy elements and as stirring agents and energy providers through stellar winds and supernovae. The field of *magnetic* massive stars has seen tremendous progress in recent years. Different perspectives—ranging from direct field measurements over dynamo theory and stellar evolution to colliding winds and the stellar environment—fruitfully combine into a most interesting and still evolving overall picture, which we attempt to review here. Zeeman signatures leave no doubt that at least some O- and early B-type stars have a surface magnetic field. Indirect evidence, especially non-thermal radio emission from colliding winds, suggests many more. The emerging picture for massive stars shows similarities with results from intermediate mass stars, for which much more data are available. Observations are often compatible with a dipole or low order multi-pole field of about 1 kG (O-stars) or 300 G to 30 kG (Ap/Bp stars). Weak and unordered fields have been detected in the O-star ζ Ori A and in Vega, the first normal A-type star with a magnetic field. Theory offers essentially two explanations for the origin of the observed surface fields: fossil fields, particularly for strong and ordered fields, or different dynamo mechanisms, preferentially for less ordered fields. Numerical simulations yield the first concrete stable (fossil) field configuration, but give contradictory results as to whether dynamo action in the radiative envelope of massive main sequence stars is possible. Internal magnetic fields, which may not even show up at the stellar surface, affect stellar evolution as they lead to a more uniform rotation, with

R. Walder (✉) · D. Folini
CRAL: Centre de Recherche Astrophysique de Lyon, ENS-Lyon, 46, allée d'Italie, 69364 Lyon Cedex 07, France
e-mail: Rolf.Walder@ens-lyon.fr

D. Folini
e-mail: Doris.Folini@ens-lyon.fr

R. Walder · D. Folini
UMR CNRS 5574, Université de Lyon, Lyon, France

G. Meynet
Faculté des Sciences, Observatoire de Genève, Chemin des Maillettes 51, 1290 Versoix, Switzerland
e-mail: Georges.Meynet@unige.ch

more slowly rotating cores and faster surface rotation. Surface metallicities may become enhanced, thus affecting the mass-loss rates.

Keywords Massive stars · Magnetic fields · Dynamos · Fossil field · Stellar evolution · Binaries · Colliding winds · Non-thermal emission

1 Introduction

One may wonder why a review on magnetic fields in massive stars enters a book dealing with large scale magnetic fields in the universe. Several reasons may be given.

A first line of argument may stress the relevance of massive stars as cosmic engines for the chemical and dynamical evolution of galaxies and the universe, their role as production sites of heavy elements and as stirring agents and energy providers in the form of stellar winds, supernova explosions, gamma ray bursts, and galactic super-bubbles. Also, if the stars are magnetic, their wind driven structures will be magnetized as well. One may further argue with the magnetic dipole field observed at the surface of some massive stars, whose large scale ordering not trivial to achieve or maintain. Or with the conditions under which magnetic fields in massive stars exist or form, which represent relative exotic locations of the parameter space. Massive stars may be viewed as laboratories to study and observe magnetic fields under such conditions. Better understanding of the role of magnetic fields in massive stars then will not only help to better understand the role of massive stars in the universe but will also broaden our knowledge on the existence and generation of magnetic fields in the universe. Like the field of magnetism in massive stars has profited from combining different points of view, from direct field measurements to modeling of colliding winds, the reader of this book may profit from the presented largely different perspectives on magnetism in the universe.

The term massive star is typically used for stars with initial masses above 8 $M_\odot$, which will ultimately end their lives as supernovae. Early type massive stars refer to massive stars of spectral type O or B, thus stars on or close to the main sequence but not, for example, massive stars having entered the red super-giant phase. The first definite magnetic field detections in massive stars, by means of spectropolarimetry, date back only about a decade, to around the year 2000. Since then, observational techniques have been further refined and field detections, although still difficult for physical reasons, have multiplied. We stress that all these detections refer only to *surface magnetic fields*, and that any *magnetic fields in the interior* of stars need not even show up at the surface. A 'snapshot' overview of the rapidly evolving observational results is given within the frame of this review, Sect. 2.

Crucial and interesting with regard to magnetic fields in massive stars is the radiative envelope of these stars, which prevents the existence of a solar type dynamo. Any magnetic field thus must be due to another cause. Theories on the origin of magnetic fields in early type massive stars may be roughly divided into two categories: dynamos working differently than the solar dynamo or fossil fields of debated origin that are present before the star reaches the main sequence. The dynamo aspects may link to another review in this book on turbulent dynamos. Similarly, the fossil fields may link to the review by Beck et al., which touches on magnetic fields in the ISM. Part of this review, Sect. 3, is devoted to a more detailed presentation of these theories, their analytical argumentation as well as numerical simulations which allow to access the non-linear and time-dependent regime.

Equally important as the cause of magnetic fields in massive stars are the consequences of such fields. The issue may again be roughly subdivided into two parts, consequences for the stellar evolution and consequences for the ambient medium. Even weak magnetic fields in the stellar interior, which may not even show up as surface fields, are expected to affect the stellar evolution from 'cradle to graveyard', by modifying the internal transport of angular momentum. Subsequent changes in the transport of chemical yields, surface abundances, or mass loss, may even influence the ultimate fate of the star as a neutron star, gamma-ray burst, or black hole. Sufficiently strong surface fields can, by contrast, convincingly explain a number of observational signatures originating at the surface of or at some distance from the star. Termination shocks of magnetized winds or wind collision in binaries or open clusters can contain indirect evidence on the presence of a stellar magnetic field. Moreover, such structures could potentially serve as a laboratory to test non-thermal shock models, super-enhancement of magnetic fields in shocks, or particle acceleration, aspects which may link to two other contributions in this volume, one on supernova remnants and pulsar nebulae, the other on particle acceleration. The consequences of stellar magnetic fields add another two parts to this review, Sects. 4 and 5.

The presented material and the rapid speed at which this knowledge, observational and theoretical, currently grows leads us to believe that some of the above questions will have a more definite answer in the not too far future. Some more room to these considerations will be given in the closing section of this review, Sect. 6.

2 Observations of Surface Fields in Intermediate and High Mass Stars

Stellar magnetism is ubiquitous in most parts of the HR diagram, along the main sequence as well as during pre- and post-main-sequence evolution (Berdyugina 2009; Donati and Landstreet 2009). With the possible exception of a narrow mass range between 1.5 and 1.6 $M_\odot$ (Donati and Landstreet 2009) stellar magnetism has been found throughout the main sequence. Widely different are, however, the concrete manifestations of stellar magnetism, such as the strength of the field, the field topology, or just the percentage of stars of a given spectral type which have a detectable magnetic field. Our knowledge on these aspects has been rapidly evolving in recent years, as better instruments became available and data analysis techniques were refined. This development is still ongoing and, consequently, some of the current conclusions may need revision in the future, when both the quality and quantity of the observational data will have increased even further. We include a brief overview of current observational techniques in Sect. 2.1 in the hope that this will help the reader to better rate the reviewed observational results.

Current observational data suggests the following rules of thumb, which may be subject to change as new data becomes available. Magnetic fields are common among low-mass stars ($M < 1.5\ M_\odot$), but only a minority of intermediate- and high-mass stars ($M > 1.5\ M_\odot$) show a magnetic field. The field topology in intermediate- and high-mass stars is frequently dominated by a simple dipole, while low-mass stars typically display a much richer topology. For intermediate- and high-mass stars, the presence of a magnetic field often goes hand in hand with a comparatively slow rotation of the star and with a chemically peculiar photosphere.

These rules of thumb demand for physical explanation, thus inspiring a lively and ongoing debate on the detailed physical origins and consequences of magnetic fields in intermediate- and high-mass stars. Some aspects of this debate deserve mentioning here as they directly feed back in new observational campaigns. Fossil fields (see Sect. 3.3), for example, are the currently favored explanation for the dipole magnetic fields in intermediate- and high-mass stars, in contrast to convection in low-mass stars. This explanation promoted recent observational surveys to search for magnetic fields in intermediate and high-mass pre-main-sequence stars. Magnetic braking as the proposed agent to explain the on average slower rotation of magnetic stars inspired observational efforts to find a possible age—magnetic field strength relation ship. The argument that magnetic dipole fields are stable only above a certain threshold field strength led to an intense search for stars with field strengths below this threshold value (see Sect. 2.2).

2.1 Measuring Stellar Magnetic Fields

Babcock (1947) showed that the circular polarization observed in some absorption lines of the spectrum of 78 Vir can be interpreted in terms of the Zeeman effect and thus give access to the measurement of magnetic fields in stars. Since then, observational techniques have been greatly refined (Mathys 1989; Kochukhov 2006b; Donati and Landstreet 2009). A recent review on current instruments and techniques, given at the IAU symposium 272 on active OB stars, can be found in Petit (2010).

The basic idea behind Zeeman-measurements is simple. The component of a magnetic field parallel to the line of sight of an observer, B_l, manifests itself in a split of spectral lines into two components of opposite circular polarization (Landstreet 2009). For relatively weak magnetic fields the resulting Zeeman shift $\Delta\lambda$ between the two components is proportional to B_l, the magnetic splitting sensitivity or effective Landé factor g_{eff}, and the wave length squared,

$$\Delta\lambda \propto B_l g_{\text{eff}} \lambda^2. \tag{1}$$

The Zeeman shift thus is particularly pronounced for strong fields (large B_l) and large wave lengths (infra red).

In real observations, complications of this simple picture arise from mainly two sides. First, thermal and stellar rotation broadening are superimposed on the $\Delta\lambda$ of the Zeeman shift. The Zeeman effect thus may only result in some additional line broadening and not in any clear line splitting. Second, the observed spectral signal is an average signal from the entire stellar disk, thus not due to one particular value of B_l but rather due to a composite of magnetic field strengths and orientations. If polarity of the field changes over the stellar disk, significant cancellation of the observable signal results. Taken together these effects demand for a measurement accuracy on the order of 10^{-3} to 10^{-4} if stellar magnetic fields are to be detected (Donati and Landstreet 2009; Berdyugina 2009). Only for strong dipole fields of slowly rotating stars (as in some Ap stars) the Zeeman-effect may be directly visible as a splitting of spectral lines in intensity spectra (Stokes I). In these cases, or if at least a substantial line broadening exists, one may determine the mean value of the field strength averaged over the stellar disk, $\langle|B|\rangle$, also known as the mean field modulus or mean surface field $\langle B\rangle$ without detailed line profile modeling (Mathys 1995; Mathys and Hubrig 2006; Landstreet 2009).

The magnetic field observations reviewed in this article address these difficulties in mainly two ways. First, on the 'hardware' side, by measuring circular polarization profiles (Stokes V) with high frequency resolution. Second, on the 'software' side, by combining Zeeman signatures from several spectral lines to increase the signal to noise ratio and by combining observations taken at different times, thus different rotational phases of the star, to learn more about the 3D field topology.

The currently most advanced instruments for magnetic field observations in massive stars are two high-resolution spectropolarimeters. One of them is ESPaDOnS (Donati 2003), installed in 2004 on the Canada-France-Hawaii Telescope on Mauna Kea, Hawaii. The other

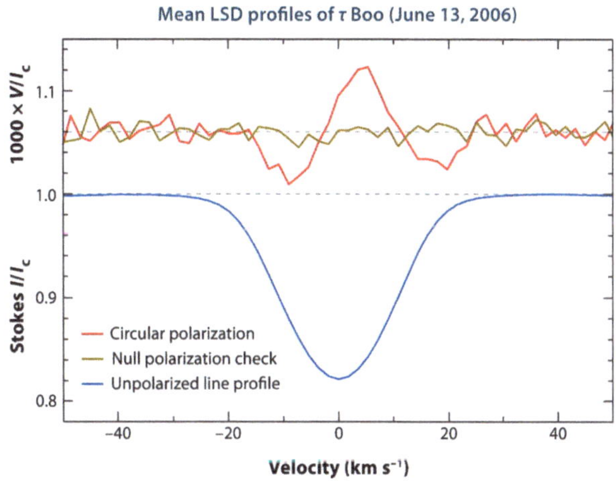

Fig. 1 LSD circular polarization (Stokes V) Zeeman signature (*red line*), null polarization check (*dark yellow line*, both expanded by 1000 and shifted vertically by 1.06 for graphical purposes) and unpolarized (Stokes I) profile (*blue line*) from the photospheric lines of τ Boo, as derived from ESPaDOnS data. A clear Zeeman signature is detected. The figure is taken from Donati and Landstreet (2009), their Fig. 1

is, NARVAL (Aurière 2003), a clone of ESPaDOnS, installed at the Télescope Bernard Lyot on Pic du Midi, France. Both instruments are regularly used by the MiMeS (Magnetism in Massive Stars) project (Wade et al. 2009; Wade and the MiMeS Collaboration 2010). Two other instruments that were frequently used in recent years for magnetic field observations in massive stars are the low-resolution spectropolarimeter FORS1 on the ESO Very Large Telescope (VLT) on Cerro Paranal, Chile, and MUSICOS (Donati et al. 1999), also on Pic du Midi, France.

The basic idea for increasing the signal to noise ratio is to combine the information from numerous spectral lines instead of analyzing only a single line. Different approaches exist to achieve this combination of multiple lines and Semel et al. (2009) recently suggested to summarize all these different methods under the term multi-line Zeeman signature (MZS). One very basic approach is known as line addition technique (Semel 1989; Semel and Li 1996). Here, several observed lines are just added up. Another technique is the Least Squares Deconvolution (LSD) introduced by Donati et al. (1997) and illustrated in Fig. 1. Comparing the two techniques by applying them to a synthetic test, case Semel et al. (2009) obtain very similar results. A detailed investigation of the capabilities and limitations of LSD can be found in Kochukhov et al. (2010). More recently, yet another technique was proposed for de-noising the Zeeman signatures, based on principle component analysis (PCA) (Semel et al. 2006; Martínez González et al. 2008).

To obtain information not only on the line of sight magnetic field component but on its 3D structure it would be necessary to measure also Stokes Q and U, in addition to Stokes V. However, these components carry an even weaker signature than Stokes V. Therefore, they are hardly explored so far in the context of massive stars. An alternative approach is followed instead, by exploring the time dependence of the Zeeman signatures. To illustrate the basic idea, assume a star with a bipolar magnetic field, the dipole not being aligned with the rotation axis of the star. A distant observer will see the Zeeman signature to vary with time as the dipole axis processes about the rotation axis of the star. Note that the interpretation of such a signal requires the inclination i of the rotation axis of the star to be known. The time dependent Zeeman signature carries the imprint of the magnetic dipole field. Information on the dipole (or higher moments) can be retrieved again from this signal by decomposing the observational data in terms of spherical harmonics. This technique today is known under the name Zeeman Doppler imaging (ZDI). An example of ZDI is shown in Fig. 2, derived field lines using the field-extrapolation technique by Jardine et al. (1999) are shown in Fig. 3.

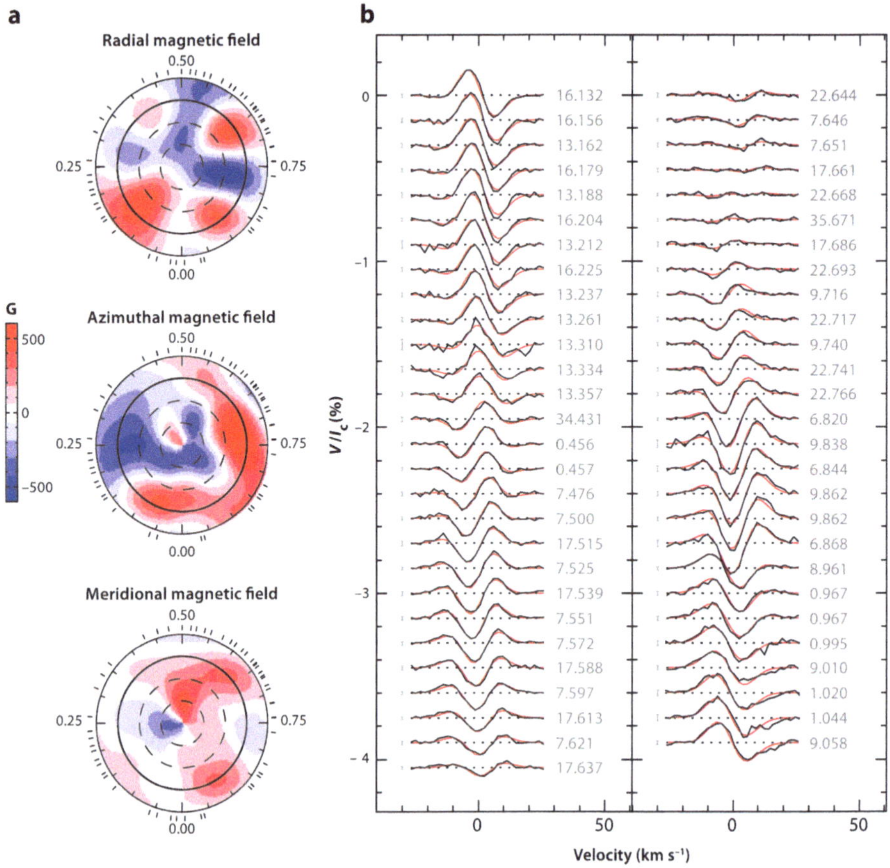

Fig. 2 Large-scale magnetic topology of the young early B-star τ Sco (*right*) derived with ZDI from a time series of circular polarization (Stokes V) Zeeman signatures covering the whole rotation cycle (*left*). The figure is taken from Donati and Landstreet (2009), their Fig. 5

Note that the term ZDI originally stood for the detection of a Zeeman signature thanks to the Doppler effect (Semel 1989; Semel et al. 2009). In fast rotating stars, the later can help to disentangle the contributions from opposite magnetic polarities and their respective opposite polarization which otherwise may cancel. The combination of Zeeman and Doppler effects is one of the reasons why the detection of the Zeeman effect in fast rotating active solar type stars became possible.

2.2 A and Late B Stars

A and B stars account for around 0.6% and 0.1%, respectively, of all main sequence stars in the HR diagram. Their masses range between 1.4–2.1 $M_\odot$ (A stars) and 2.1–16 $M_\odot$ (B stars). A small fraction of these stars distinguishes themselves from normal A- and B-type stars by chemically peculiar atmospheric compositions, for example high abundances of Si, Cr, Sr, or Eu (Preston 1974; Landstreet et al. 2007). These stars are commonly referred to as CP stars. A subset of these stars are the Ap/Bp stars. In the following, *we will collectively refer to these stars as Ap stars*.

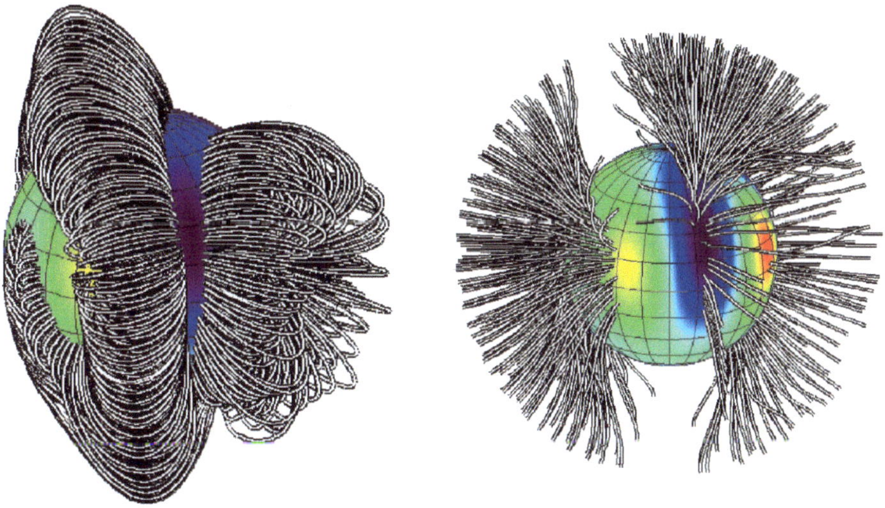

Fig. 3 Closed (*left*) and open (*right*) magnetic-field lines of the extended magnetic field configuration of τ Sco, extrapolated from photospheric maps using the technique by Jardine et al. (1999). The star is shown at phase 0.83. The figure is taken from Donati et al. (2006b), their Figs. 11 and 12

Typically, Ap stars display periodic variability on time scales that range from about half a day up to several decades and that are inversely correlated with $v \cdot \sin i$, where i denotes the inclination of the rotational axis of the star and v the rotation velocity. The variability is commonly attributed to a magnetic dipole field, whose axis is not aligned with the rotation axis of the star. In fact, the strongest magnetic fields observed to date in non-degenerated stars belong Ap stars: HD 215441 (34 kG) (Babcock 1960), HD 75049 (30 kG) (Freyhammer et al. 2008; Elkin et al. 2010), HD 137509 (29 kG) (Mathys 1995; Kochukhov 2006a), HD 154708 (24.5 kG) (Hubrig et al. 2005).

Assuming the field topology to be bipolar generally gives reasonable agreement with the observational data (Preston 1967; Landstreet 1992; Mestel and Landstreet 2005; Aurière et al. 2007). Evidence for additional higher-order multi-polar contributions is, however, found in most cases when searched for (Preston 1969; Landstreet 1988; Landstreet and Mathys 2000; Aurière et al. 2007). Nevertheless, magnetic fields of Ap stars are often specified in terms of the surface strength of the dipole field. The relative frequency of magnetic Ap stars compared to the number of A- and B-type stars of similar mass is below 10% and may be decreasing with mass from about 10% at 3 $M_\odot$ to around 0% at about 1.6 $M_\odot$ (Power et al. 2008; Donati and Landstreet 2009).

Whether all Ap stars are magnetic remains controversial in the sense that no direct field measurements are available for the majority of the known Ap stars. A recently published catalog by Renson and Manfroid (2009) lists 3652 Ap stars. By contrast, an equally recent catalog by Bychkov et al. (2009) of direct line of sight magnetic field measurements contains 1223 stars of spectral type O to M, 610 of which are chemically peculiar and only 410 belong to the classical Ap class as defined by Preston (1974). To clarify the question, Bagnulo et al. (2006) used the low-resolution FORS1 instrument to monitor 97 Ap stars for magnetic fields. A clear detection was obtained for only 41 stars. Meanwhile, it seems reasonable to assume that the many non-detections in this study were due to the low-resolution.

Aurière et al. (2007) employed the high-resolution spectropolarimeter ESPaDOnS to monitor 28 firmly established Ap stars. They found magnetic fields well above the detection limit in all of them. The result is all the more remarkable as the 28 stars were deliberately selected to have a weak magnetic field, if one at all. Nevertheless, for all but two stars the inferred surface dipole field is larger than 300 G. The authors conclude that not only all Ap stars have a detectable magnetic field but also that this field is always larger than some magnetic threshold value of about 300 G. For a possible explanation for the existence of such a threshold value they refer to Spruit (1999). The basic argument is that a too weak magnetic field will be wound up due to differential rotation, a pinch-type instability will set in, and the magnetic field will be destroyed. In more formal terms, the order of magnitude estimate for the critical field B_c given by the authors is

$$\frac{B_c}{B_{eq}} \simeq 2\left(\frac{P_{rot}}{5_{day}}\right)^{-1}\left(\frac{r}{3R_\odot}\right)\left(\frac{T}{10^4\ \mathrm{K}}\right)^{-1/2} \tag{2}$$

with B_{eq} the equipartition field ($B_{eq}^2 = 8\pi P$, P the gas pressure) at the surface ($\tau_{5000} = 2/3$), and P_{rot}, r, and T the stellar rotation period, radius, and temperature, respectively. For a typical Ap star, $B_{eq} = 170$ G and $B_c \approx 300$ G.

The authors stress that such a mechanism would not only explain the apparent lower bound in the strength of magnetic fields in Ap stars. If a large enough field strength is required in order that the magnetic field does not decay within a short time, this may explain why only a small fraction of A- and B-type stars have a magnetic field. Moreover, that the critical field strength increases with initial stellar mass (thus increasing temperature T and stellar radius r) would naturally explain the even greater scarcity of magnetic field detections in more massive stars.

Additional high-resolution observations have further refined the picture of magnetic Ap stars. Power et al. (2008) identified 57 Ap stars within 100 pc from the Sun, by means of the Hypparcos Catalog and additional sources. They find for the distribution of surface dipole field strengths a plateau at 2.5 ± 0.5 kG, dropping off to higher and lower field strengths. The mass distribution of these stars peaks around 2.1 to 2.3 $M_\odot$ with a strong decrease towards lower and higher masses. The lowest mass stars in the sample are 1.6 ± 0.1 $M_\odot$. The findings for the magnetic fields are consistent with the weak field results obtained by Aurière et al. (2007). The picture is further supported by Kholtygin et al. (2010), who carried out detailed statistical analysis of the stars catalogued by Bychkov et al. (2009).

To detect a potential dependence of the magnetic field strength on the main sequence age of the star, a number of observations of Ap stars in open clusters where performed by Bagnulo et al. (2006) and Landstreet et al. (2007, 2008). The observational data suggests that the magnetic field of an Ap star indeed decays with time, but that the time scale for the decay depends strongly on the mass of the star. The time over which an initially strong (about 1 kG) magnetic field essentially disappears ranges from about 250 Myr for stars of 2–3 $M_\odot$ over 40 Myr for stars of 3–4 $M_\odot$ to 15 Myr for stars of 4–5 $M_\odot$.

Further peculiarities of magnetic Ap stars may or may not be related to their magnetic field. Probably related is their comparatively slow rotation, with periods on the order of a few days instead of a few hours to one day. Landstreet and Mathys (2000) find that most slow rotators (periods longer than 25 days) have small inclination angles of less than 20 degrees between the rotation axis and the axis of the magnetic field, whereas stars with short rotation periods have larger inclination angles. Maybe also related to the presence of magnetic fields are chemical abundance variations in the observed spectra. Their period is mostly identical to the rotation period, implying that the variations stem from chemical inhomogeneities on the surface. Not clear is the reason for the lack of short period binaries among magnetic Ap stars. Based on a sample of 95 Ap stars, North et al. (1998) find 27% to be spectroscopic binaries, as compared to (47 $\pm$ 5)% for normal A- and B-type stars, particularly apparent is the lack of tight orbits as compared to normal stars. Renson and Manfroid (2009) report 5 definite detections of binary Ap stars with orbital periods below 3 days. Tutukov and Fedorova (2010) ascribe this lack of close binaries to mergers. Effective temperatures of Ap stars are reported to lie in a range between 7000 K and 23000 K (North and Debernardi 2004). The authors argue that for temperatures outside this range all chemical peculiarities will be erased by either strong radiatively driven winds (higher temperatures) or convection (lower temperatures).

The only types of Ap/Bp stars for which no large scale magnetic fields have been detected yet are the helium weak Bp stars of type PGaS, and the Bp HgMn stars (North and Debernardi 2004). More recent observations of the HgMn star in AR Aur, the only eclipsing binary known to contain a HgMn star, place an upper limit of 100 G and 400 G on any longitudinal or dipole field, respectively (Folsom et al. 2010). For the same star, Savanov et al. (2009) report the detection of patchy chemical inhomogeneities at the surface. As an explanation they suggest a weak magnetic field, caused by differential rotation in the wake of a depth dependent tidal torque. For another HgMn star, α And, Wade et al. (2006) find an upper limit on the order of 100 G as well, depending on the assumed field geometry. Studying a larger sample of 12 Am and 3 HgMn stars with NARVAL, Auriere et al. (2010) confirm these upper limits. Together, these observations support the existence of a dichotomy between spectroscopically defined Ap stars and any other A-type stars, with only the former harboring substantial magnetic fields.

Turning finally back to normal A- and B-type stars it is interesting to note that Bagnulo et al. (2006) examined 138 normal A- and B-type stars but found no evidence for any magnetic field. It has to be stressed, however, that this does not necessarily imply that normal A- and B-stars do not have a magnetic field. The field may just be too weak to be measured or its topology may be too complicated, leading to cancellation effects. In fact, Lignières et al. (2009) recently reported the detection of a weak magnetic field for the young (a few hundred Myr) normal A-type star Vega. Petit et al. (2010) meanwhile confirmed the detection using the NARVAL and ESPaDOnS instruments. The field distinguishes itself, however, in several aspects from the field of Ap stars. Averaged over the visible part of the stellar surface the field strength does not exceed 1 G. The field topology is comparatively complex, putting it at odds with the fossil field hypothesis for which more simple, bipolar like geometries are expected.

2.3 Herbig Ae/Be Stars

Herbig Ae/Be stars (HAeBe) are the progenitors of main sequence A/B stars. A fraction of 5–10% of the later are known to have a magnetic field, the magnetic Ap stars. Their field typically is well organized with a dominant dipole component of around 300 G or more at the surface. As the envelope of Ap stars is not convective but radiative, two hypothesis have been brought forward to explain the presence of the observed magnetic fields. One idea is that the field is generated below the stellar surface, in the stellar core or envelope, from where it is transported to the surface of the star (see Sect. 3.2). The other suggestion is that the magnetic field is of fossil origin, that it existed already during the formation of the Ap star (see Sect. 3.3). This second hypothesis lead to intense measurement campaigns in recent years to search for magnetic fields in HAeBe stars, the progenitors of Ap/Bp stars (Hubrig et al. 2004; Wade et al. 2007; Hubrig et al. 2007; Alecian et al. 2008a, 2008b; Power et al. 2008; Hubrig et al. 2009; Wade et al. 2009a).

A synthesis of the results from the above campaigns can be found in Wade et al. (2009a). Based on observational data obtained with the ESPaDOnS instrument on the Canada-France-Hawaii Telescope, this survey is the largest and most sensitive published survey for magnetic fields in HAeBe stars we are aware of. The survey covers about 130 HAeBe stars with masses from about 2 to 13 $M_\odot$. Those stars in which magnetic fields are observed correspond to about 7% of the observed stars. The detected magnetic fields all show organization on large scales. Those three stars which so far have been studied in greater detail, namely V380 Ori, HD 72106A, and HD 200775, display an important dipole component with characteristic polar strength on the order of 1 kG, have ages between 0.1 and 10 Myr, and rotation periods between half a day and a few days (see Table 1 in Alecian et al. (2007) for details).

The 7% are remarkably similar to the 5–10% of magnetized A/B stars. Concerning the field strength, Wade et al. (2009a) argue that the dipole intensity derived by Alecian et al. (2008a, 2008b) and Folsom et al. (2008) for the detected magnetic HAeBe stars is compatible with the typical field strengths of Ap stars (few kG) if flux conservation as described in Alecian et al. (2008b) applies when the HAeBe stars evolve towards the main sequence.

One may wonder whether the fraction of 7% is detection limited or physical in origin. Using Monte Carlo simulations to study synthetic populations of HAeBe stars and their observational signatures in FORS1 data, Wade et al. (2007) place the following limits on the non-detections of magnetic fields in HAeBe stars: within statistical uncertainty, the observational data is consistent with a distribution of non-magnetic stars; the observations are inconsistent with a *uniform* population of magnetic stars with dipole intensities above 300 G or 500 G in case the dipole axis is aligned or perpendicular to the rotational axis of the star, respectively.

While the above Monte Carlo study was carried out for the FORS1 instrument, Wade et al. (2009a) point out that the results essentially apply to ESPaDOnS as well, where they have a 90% chance of detecting even weak fields of only 300 G. The authors conclude that, on the one hand, it is likely that some magnetic HAeBe stars still went undetected in their survey, because of a too weak or too unorganized magnetic field. On the other hand, it is highly probable that they captured the majority of magnetic HAeBe stars in their sample. Consequently, they do not expect the fraction of 7% magnetic stars among all HAeBe stars to change substantially with future surveys.

Somewhat a by product of the above surveys is the finding that only two of the detected magnetic HAeBe stars, HD 72106A and NGC 6611-601, display chemical peculiarities similar to those of Ap stars (Wade et al. 2009a). According to the authors, this aspect of the data has, however, not yet been investigated in detail.

The above results seem compatible with the idea of the fossil origin of the magnetic fields observed in Ap stars. Arguments in favor of this point of view are, in particular, the definite detection of magnetic fields in the HAeBe stars, the progenitors of Ap stars, and the observation that magnetic fields of Ap stars on the main sequence apparently weaken with the age of the star. Whether the surface field observed in HAeBe stars are directly related to the surface fields of Ap stars remains, however, debated. Hubrig et al. (2009) observe 21 HAeBe stars and detect a magnetic field in 6 of them, for which they further find indications of a decline of surface field strength with age. The age range spanned by the sample is 2–14 Myr. On this basis, the authors speculate that the surface fields of HAeBe may ultimately vanish and thus need not be direct progenitors of surface fields in Ap stars. Instead, they suggest that the surface fields in both Ap stars and HAeBe stars are due to emergence of interior fields in the wake of slow stellar rotation and that, consequently, the fields observed in Ap and HAeBe stars need not be directly related.

Also enigmatic remains the origin of the field in the HAeBe stars. At least three hypothesis have been brought forward. One idea is that the magnetic fields of HAeBe stars are generated by a convective dynamo during the formation process of the star. A second suggestion is that the magnetic field results in the course of the merger of two stars during pre-main-sequence or early main-sequence evolution. Yet another idea is that the magnetic fields stem from the molecular cloud in which the stars are born. All ideas will have to explain the fact that probably only a small number of HAeBe stars have a (detectable) magnetic field. A possible mechanism might again be the critical field strength introduced earlier on, below which the magnetic field is wound up and destroyed by instabilities.

2.4 O and Early B Stars

Up to now, less than 20 O-type and early B-type stars have definite detections of surface magnetic fields. Analysis thus is mostly restricted to case studies, unlike for intermediate mass magnetic stars where several hundred detections (Bychkov et al. 2009) enable statistical data analysis. Much more magnetic OB-stars might be expected, based on indirect indicators. For example, cyclic wind variability in OB stars has long been related to the potential presence of magnetic fields, in particular dipole fields (Moffat and Michaud 1981; Kaper et al. 1997; Rauw et al. 2001; Fullerton 2003). Searching the IUE archive for corresponding signatures, Henrichs et al. (2005) identify 100 O-stars with statistically significant variability, 60 of which show periodic variability. This large fraction of stars showing periodic variability contrasts with the fraction of stars with clear magnetic field detections. On the other hand, a systematic search by Petit et al. (2009) for a direct link between field strength and X-ray emission, as suggested by Stelzer et al. (2005), yielded no confirmation of this hypothesis. Another indication for a large number of magnetic OB stars comes from non-thermal radio emission of colliding wind binaries (see Sect. 5.3).

A potential explanation for the comparatively few detections lies certainly in the difficulty of direct field measurements in O-type stars as they have rather few suitable spectral lines in spectral ranges amenable by today's instruments. Schnerr et al. (2008) use cyclic wind variability to select 25 OB candidate stars, for which they then carried out observations with the MUSICOS instrument. No clear field detection was obtained for any of these stars.

Definite field detection in early B-type stars (estimated field in parentheses) exist for β Cep (360 G), ζ Cas (335 G), τ Sco (500 G), ξ^1 CMa (300 G, line of sight), Par 1772 (1150 G), and ν Ori (620 G) (Henrichs et al. 2000; Neiner et al. 2003; Donati et al. 2006b; Hubrig et al. 2006; Petit et al. 2008).

For O-type stars, five reliable detections of a magnetic field are known: θ^1 Ori C (Donati et al. 2002), HD 191612 (Donati et al. 2006a), ζ Ori A (Bouret et al. 2008), HD 57682 (Grunhut et al. 2009), and HD 108 (Martins et al. 2010). Additional detections that are generally regarded as tentative, as they are based on FORS1 observations only, exist for HD 36879, HD 148937, HD 152408, and HD 164794 (Hubrig et al. 2008). See Silvester et al. (2009) for a more detailed discussion.

θ^1 Ori C is a very young star, upper age limits ranging between 0.2 and 0.6 Myr (Donati et al. 2002, 2006a). Observed Stokes V signatures are consistent with a dipole field of 1.1 $\pm$ 0.1 kG (Donati et al. 2002). The magnetic wind confinement parameter η_* (see Sect. 5.1) is about 20. The star has a rotational period of 15.4 days, a mass of about 40 $M_\odot$, and its spectral type varies between O4 and O6. θ^1 Ori C is a binary with a period of 11–26 yr (Nazé et al. 2008).

The Of?p star HD 191612 varies between spectral types O6 and O8 (Walborn et al. 2004). The dipole strength inferred from ESPaDOnS data is 1.5 kG, the magnetic confinement parameter is $\eta_* \sim 10$ (Donati et al. 2006a). The star has a rotation period of 538 days. The estimated age is between 3 and 4 Myr. Its mass is similar to that of θ^1 Ori C, about 40 $M_\odot$. Like θ^1 Ori C, HD 191612 is a binary with a period of 1542 days (Nazé et al. 2008).

ζ Ori A, an O9.7 super-giant, has by far the weakest magnetic field detected so far in a hot massive star. Observations were done by Bouret et al. (2008) using NARVAL. Their data analysis indicates a much more intricate field topology than a simple dipole field. Local surface magnetic fluxes are estimated at only a few tens of Gauss, certainly not exceeding 100 G. Values for the magnetic confinement parameter η_* are between 0.03 and 0.07. A rotation period of 7 days is found, as well as a mass of 40 $M_\odot$ and an age of 5–6 Myr.

HD 57682 is an O9 sub-giant. Based on ESPaDOnS data, Grunhut et al. (2009) inferred a magnetic dipole field of about 1680 G. The authors point out that a single inclined dipole field is equally consistent with the observational data as individual dipole configurations. The rotational period is estimated at about 31.5 days, the mass at about 17 $M_\odot$. The wind confinement parameter η_* is estimated to lie in a range 4×10^3 to 2×10^4.

For HD 108, the second Of?p star with a clear field detection, Martins et al. (2010) infer a bipolar large-scale field of at least 0.5 kG and most likely on the order of 1–2 kG, based on longitudinal field measurements of 100–150 G obtained with ESPaDOnS and NARVAL. For the magnetic confinement parameter $\eta_* \geq 100$ is obtained. The rotation period is several decades, probably between 50–60 years, the age is about 4 Myr. Estimates for the stellar mass range of 35–43$M_\odot$ (Nazé et al. 2008; Martins et al. 2010).

Some properties of the above five O-type stars fit nicely with theoretical expectations. For example, older stars are expected to rotate more slowly as magnetic braking has had more time to act. The young and fast rotating star θ^1 Ori C, as well as the older and more slowly rotating stars HD 191612 and HD 108 seem to confirm this idea (Martins et al. 2010). Why ζ Ori A does not fit into this scheme one can only speculate (Bouret et al. 2008). Support for theoretical models of rotational braking due to a magnetized line-driven wind came recently from Mikulášek et al. (2008). Analyzing 31 years of observational data of the helium-strong star HD 37776, they found a lengthening of the 1.5387 d period by 17.7±0.7 s. Similarly, Townsend et al. (2010) find for the Bp star σ Ori E a linear decrease of the rotational period of 77 ms per year, based on 30 years of observational data. The corresponding spin-down time of 1.34 Myr corresponds to theoretical expectations.

Yet another star, HD 148937, deserves mentioning here although field detection is only tentative so far: -276 ± 88 G, based on FORS1 data (Hubrig et al. 2008). Spectropolarimetric observations are not yet available. HD 148937 is interesting as it is the third of only three known galactic Of?p stars, the other two stars having definitive field detections (HD 191612 and HD 108). The Of?p class was originally introduced by Walborn (1972) and recently slightly refined (Nazé et al. 2008; Walborn et al. 2010). A particularity of HD 148937 is the bipolar circumstellar nebula around it, NGC 6164-6165. As the nebula displays similar chemical anomalies as HD 148937 (nitrogen 4 times overabundant as compared to solar) it was suggested to have been formed by an eruption of the Of?p central star (Leitherer and Chavarria-K 1987; Dufour et al. 1988). All three galactic Of?p stars show nitrogen overabundances as compared to solar which, together with other spectral features, lead Walborn et al. (2003) to suggest that there may exist a similarity between Of?p stars and WN9 objects.

HD 148937 has a rotational period of 7 days (Nazé et al. 2008, 2010). Based on its position in the HR diagram, its mass and age are estimated at 55 $M_\odot$ and 2–4 Myr, respectively (Nazé et al. 2008). With these properties—magnetic field, old age, yet short rotation period—HD 148937 would be much more similar to ζ Ori A than to the other two galactic Of?p stars HD 191612 and HD 108.

2.5 Late Stages of Massive Star Evolution

Late-type super giants (spectral type F and later) represent the late evolutionary stages of most massive stars. Characteristics of these stars are extended radii, helium-burning core, a convective hydrogen-burning envelope, and slow rotation. A recent systematic survey of more than 30 late-type super-giants using ESPaDOns revealed clear detection of magnetic fields in one third of all stars (Grunhut et al. 2010). Clear detections were obtained for the following stars (Grunhut et al. 2010, Table 1): α Lep, α Per, η Aql, β Dra, ξ Pup, ϵ Gem, c Pup, 32 Cyg, λ Vel. The most massive and also the hottest star among this sample is α Lep (15 $M_\odot$, $T_{\rm eff} \approx 7200$ K). The lowest mass star is β Dra (5 $M_\odot$), the coolest c Pup ($T_{\rm eff} \approx 3700$ K). The observations suggest topologically complex fields, longitudinal magnetic fields are generally below 1 G. The authors speculate that probably even a larger fraction of late-type super-giants or even all of them may have a magnetic field, but that the signal to noise ratio of their measurements is insufficient to detect these fields. They see their point confirmed by the detection of a magnetic field in Betelgeuse (Aurière et al. 2010) where a signal to noise (S/N) about twice as high as in their study was crucial.

The case of Betelgeuse and other late-type super-giants is of interest as these stars occupy an extreme position in parameter space. Their rotation is too slow that a solar like dynamo could be at work. An alternative idea for the origin of the observed magnetic fields are large convection cells (Schwarzschild 1975). The idea obtains further support from 3D convection simulations (Freytag et al. 2002; Dorch 2004).

Very high mass stars with initial masses between 25 $M_\odot$ and 40 $M_\odot$ evolve into Wolf-Rayet (WR) stars towards the end of their life (Hirschi et al. 2010). To date, no direct magnetic field detection for any WR exists. However, WR 142 recently gained attention because of the detection of weak ($L_X \approx 7 \cdot 10^{30}$ ergs s^{-1} or $L_X/L_{\rm bol} \leq 10^{-8}$) but hard X-rays with XMM-Newton (Oskinova et al. 2009) and Chandra (Sokal et al. 2010). The inferred plasma temperature is on the order of 10^8 K (Oskinova et al. 2009), which is difficult to explain in terms of internal shocks in the WR wind. The authors suggest that the X-rays may be due to the presence of a stellar magnetic field that deviates the stellar wind and creates magnetically confined wind shocks (MCWS, see Sect. 5.1). The authors estimate that a surface field strength of some 10 kG would be required, a value they consider compatible with the order 1 kG measured magnetic fields in the much larger O-type stars. Alternative explanations have been suggested as well, most notably inverse Compton scattering in colliding winds in a (yet unresolved) binary (Sokal et al. 2010). In fact, the 'colliding winds community' has been assembling indirect evidence for the existence of magnetic fields in WR + OB and O + OB binaries for several decades already, mostly in the form of non-thermal radio emission (see Sect. 5.3).

2.6 Summary

Measurements of Zeeman signatures clearly demonstrate the presence of dipole fields of about 1 kG strength in four O-type stars. One more O-type star, ζ Ori A, has a clear field detection as well, but the field is much weaker (tens of Gauss) and less ordered. A clear field detection also exists for the late-type supergiant Betelgeuse. For WR stars only indirect evidence for a magnetic field exists, in the form of hard X-ray emission and non-thermal radio emission from WR + O binaries.

Much more data is available for intermediate mass Ap/Bp stars. Evidence is growing that probably all Ap/Bp stars have a magnetic field. The field of these stars often is dipole like with strength between about 300 G and 30 kG. The lower limit of 300 G is probably not detection limited but constitutes a physical lower limit instead. Only one normal A-type star, Vega, is currently known to have a magnetic field. The field is weak and unordered, resembling in this sense the field of ζ Ori A.

Observations from open clusters suggest the field strength in Ap/Bp stars to decrease with the age of the stars. Apart from field detections in main sequence Ap/Bp stars, there exist a few field detections in pre-main-sequence Herbig Ae/Be stars. Measured field strenghts are compatible with those of Ap/Bp stars if stellar evolution is taken into account.

3 Theoretical Models for Magnetic Fields in Massive Stars

3.1 Introduction

Even strong—magnetic fields on their surfaces demands for explanation in terms of analytical and numerical models. Beyond the interpretation of observational data, a better physical understanding of magnetic fields in massive stars will allow a more consistent inclusion of such fields in stellar evolution models.[1] For the potentially far reaching consequences of magnetic fields for stellar evolution we refer to Sect. 4. Here we only stress that magnetic fields relevant for stellar evolution need not necessarily emerge on the stellar surface. It is sufficient that they dynamically link the (convective) core and (radiative) envelope of the star. This poses some difficulty as current observations of magnetic fields in massive stars (see Sect. 2) deal for the most part with surface fields only.

The basic challenge associated with magnetic fields in early type massive stars is that the envelope of such stars is for the most part radiative. This excludes a solar type dynamo as the origin of the magnetic field. Two groups of alternative explanations are favored instead: some other kind of dynamo, the precise nature of which is under debate, or fossil fields, which have been preserved since the formation time of the star and whose precise origin is under debate as well. Both groups of mechanisms may be required to explain observed surface magnetism in the upper main sequence: fossil fields for the kG dipole fields typical of Ap-stars, dynamos for weak (1G) and unordered fields as in the normal A-type star Vega (Lignières et al. 2009; Petit et al. 2010).

A common issue in both explanations, dynamo or fossil field, is the stability or instability of the involved magnetic field. Some dynamo theories rely on instabilities for closing the dynamo loop, while fossil field theories rely on a sufficiently stable field configuration for the field to last over the lifetime of the star. The relevance of the question is reflected in a quite exhaustive literature. We sketch here only the basic picture. Further details will be given later on as needed.

A first point to note is that Ohmic dissipation alone is not efficient in destroying a stellar magnetic field. The corresponding destruction time scale is on the order of the life time of the star. Turning to instabilities instead, it has long been known that the field topology plays a decisive role. Only mixed poloidal-toroidal topologies may be sufficiently stable. This hypothesis got recent support from 3D simulations by Braithwaite and Spruit (2004) and Braithwaite and Nordlund (2006) which show that initially unordered fields often relax to a long-lasting mixed topology. Several decades earlier, Prendergast (1956, 1958) showed already that an exact solution of the hydromagnetic equilibrium exists if the field is of mixed

[1] A very good presentation of the physics of rotating stars, including an elusive chapter on the role of magnetism, can be found in the book by Maeder (2009). We also refer to the book by Mestel (1999), which is perhaps the most complete compilation of stellar magnetism.

topology, has a toroidal and a poloidal component. Probably one of the earliest results in this direction is the Ferraro-law (Ferraro 1937): "*The magnetic field of a star can only remain steady if it is symmetrical about the axis of rotation and each line of force lies wholly in a surface which is symmetrical about the axis and rotates with uniform angular velocity.*"

By contrast, purely poloidal fields were shown to be unstable if some (Markey and Tayler 1973, 1974; Wright 1973) or all (Flowers and Ruderman 1977) field lines are closed outside of the star. A purely toroidal field is unstable to non-axisymmetric perturbations of low azimuthal order, with a strong dominance of the $m = 1$ mode (Tayler 1973; Wright 1973; Goossens et al. 1981). Rotation (Pitts and Tayler 1985) or the addition of thermal and Ohmic diffusion (Acheson 1978) help to stabilize such a situation, but cannot suppress the $m = 1$ instability. Note that the above considerations refer to the full field, inside and outside the star. A bipolar field outside of the star can still exist, provided that some toroidal component exists within the star.

In the rest of this section, we review dynamo models proposed for the different layers of massive stars (in Sect. 3.2). This includes the question whether such fields can raise to the surface, for example by buoyancy or due to meridional circulation. In Sect. 3.3 we review fossil field models, which propose that the field observed in massive stars was enclosed during the process of their formation. A summary concludes the section (Sect. 3.4).

3.2 Dynamo Generated Fields and their Appearance at the Surface

3.2.1 Convective Core

An obvious candidate for a place in a massive star where magnetic fields may be created is its convective core. By mean field dynamo theory, Charbonneau and MacGregor (2001) found that $\alpha 2$-, $\alpha 2\Omega$- and $\alpha\Omega$ dynamos can operate in such cores. In addition, the authors investigate whether meridional circulation could bring the field to the surface. They conclude, however, that this is not the case. The main argument is: *In all models with strong core-to-envelope magnetic diffusivity contrast (presumably closest to reality) whenever circulation is strong enough to carry a significant magnetic flux, it is also strong enough to prevent dynamo action.*

In a follow-up paper, MacGregor and Cassinelli (2003) looked at alternative transport mechanisms—buoyant, centrifugal, Coriolis, magnetic tension, and aerodynamic drag forces—whether they may be able to bring the field generated in the convective core to the surface. They found that thin, axisymmetric, toroidal flux tubes can be advected from the outer boundary of the convective core to the atmosphere within a life-time of a 9 solar mass star. The trajectories of the raising ring are more or less parallel to the rotation axis of the star. They also found that the rise-time for smaller rings is faster. Mullan and MacDonald (2005) argue, however, that correct inclusion of yield gradients may suppress the buoyancy.

The idea of substantial field generation in the convective cores of A-type stars is further supported by more recent 3D simulations by Browning et al. (2004) and Brun et al. (2005). Mega-Gauss fields are found in the core, at its boundary layer to the radiative envelope. Featherstone et al. (2009) find, by means of 3D simulations of the innermost 30% of a 2 $M_\odot$ A-type star, that the presence of a fossil field in the radiative envelope which threads into the convective core can considerably augment the field generated by a core dynamo. Apart from pointing out that stronger fields generally have a better chance of reaching the surface, the authors do not comment on whether or not the core dynamo generated fields may emerge at the stellar surface.

Taken together, the above papers demonstrate that dynamo action in the convective cores of upper main sequence stars is likely to exist, but that it is unclear whether the generated fields can reach the stellar surface.

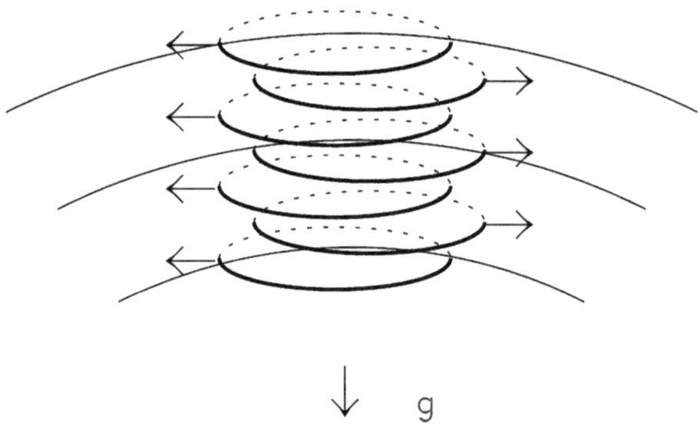

Fig. 4 Unstable displacements in an azimuthal field near the pole. Shown is the $m = 1$ mode, which occurs under the widest range of conditions. The displacements are along horizontal surfaces (indicated by *arcs*). The figure is taken from Spruit (1999), his Fig. 1

3.2.2 Radiative Envelope

Dynamos that can operate without a convection-related α-mechanism have been known for several years. An example is the dynamo generated by differential rotation and the magneto-rotational instability in accretion-disks (Hawley et al. 1996). The recognition that a similar dynamo mechanism, one without a convection-related α-mechanism, may also take place in the stably stratified radiative regions of stars had far reaching consequences. The two corresponding papers (Spruit 1999, 2002) are seminal for two reasons. Firstly, because they re-shape the theory on stably stratified radiative regions and, secondly, because of their impact on stellar evolution models, the topic of Sect. 4. Whether such a dynamo can indeed operate in real stars of the upper main sequence has been questioned (Zahn et al. 2007) and is the topic of a still ongoing debate.

One starting point of the above developments were results of helioseismology, which found that the radiative solar core is uniformly rotating. At the tachocline, this uniform rotation switches rather abruptly, in only 0.05 $R_\odot$, to the differential rotation of the envelope, which itself shows a 30 percent contrast in its rotation between equator and poles. The observed uniform rotation of the solar core was puzzling. A potential agent for the necessary efficient transport of angular momentum could be magnetic fields, but under the prevailing radiative conditions a conventional 'convective dynamo'—and thus magnetic fields—would not exist. The connection of this problem with the radiative envelopes of upper main sequence stars is obvious.

Spruit (1999) re-investigated the stability of magnetic fields in stably stratified, but differentially rotating stars. He reviews the processes which contribute to the evolution of an initially weak magnetic field in a differentially rotating star: rotational smoothing and five instabilities, among them magnetorotational instability, buoyancy instability, and pinch-type instabilities. The Pitts-Taylor instability, a pinch-instability driven by the free energy of the field itself, is found to set in first. It generically occurs in a region near the pole, in the form of an $m = 1$ displacement of the field lines along horizontal surfaces (see Fig. 4). The analysis is complemented by an investigation of the relevance of thermal and magnetic diffusion on these instabilities. Based on heuristic arguments, concrete numbers are provided that characterize the behavior of the instabilities.

In a follow up paper, Spruit (2002) argues that on the basis of the Pitts-Taylor instability a dynamo can operate in the stably stratified radiative regions of stars. He states *'the generation of a magnetic field in a star requires only one essential ingredient: a sufficiently powerful differential rotation. The recreation of poloidal field components which is needed to close the dynamo loop can be achieved by an instability in the toroidal field.'*. The paper gives concrete formulas for the field strength in radial and azimuthal direction produced by a dynamo of this type. The crucial parameter is the differential rotation, which of course will be affected by the presence of a field.

Numerical confirmation of the dominance of the Pitts-Taylor instability and the associated growth rate predicted by Spruit (2002) is reported by Braithwaite (2006b). The $m = 1$ kink-mode is indeed found to be the dominant instability in a toroidal field, where the field strength is proportional to the distance from the axis, such as the field formed by the winding up of a weak field by differential rotation. The growth rate of the instability for initially weak fields is found to agree with the analytic predictions.

Braithwaite (2006a) further report that he finds the dynamo mechanism to work: an initial small poloidal field is wound up. The resulting toroidal field is subject to the Pitts-Taylor instability, thus producing a poloidal field again, which is then wound up again. This process continues up to a certain saturation level, at which *'the field is being wound up by differential rotation at the same rate as it is decaying through its inherent magneto-hydrodynamic instability.'*

Doubts on the above results were raised by Zahn et al. (2007). While numerically investigating the solar tachocline they found inconsistencies between their results and results reported by Spruit (1999, 2002) and Braithwaite (2006a). In their paper, Zahn et al. (2007) first provide a more rigorous analysis than the more heuristic argumentation by Spruit (1999). They confirmed the scaling of Spruit but give better bounds and factors (differences are of order 1). Comparing their analytical results with numerical results from the ASH code Zahn et al. (2007) find excellent agreement. In strong contrast to Braithwaite (2006a) they observe: *'Although the instability generated field reaches an energy comparable to that of the mean poloidal field, that field seems unaffected by the instability: it undergoes Ohmic decline, and is neither eroded nor regenerated by the instability.'* They conclude that *'In our simulations we observe no sign of dynamo action, of either mean field or fluctuation type, up to a magnetic Reynolds number of* 10^5.' The authors also doubt that a large scale dynamo-loop can be closed in the way suggested by Spruit (2002) and Braithwaite (2006b), but suggest an alternative instead.

The contradictory numerical results demand for some details, given the importance of the question. The two simulation codes employed differ in several respects, including magnetic boundary conditions and the way differential rotation is enforced (see Zahn et al. (2007) for details). Maybe the most important difference is the way the equations are solved. The ASH code used by Zahn et al. (2007) is of pseudo-spectral type and should allow to reach a magnetic Reynolds number of 10^5 in the simulations under debate, which the authors consider enough to detect any dynamo action if existent. Braithwaite (2006b) use a 6th order finite difference code on a Cartesian domain. Whether or not dynamo action sets in the model by Braithwaite (2006b) crucially depends on the artificially enforced differential rotation. Adding a uniform rotation does not suppress the dynamo but the field strength shows an oscillatory behavior. Further investigations are clearly needed to disentangle the issue.

Apart from generating a magnetic field by dynamo in the radiative envelope—if possible at all—there is again the question of whether such a field may eventually become visible at the surface. Two possibilities come to mind: the field is generated near the surface or the field is generated further below and transported to the surface. As to the first possibility, Zahn et al. (2007) agree with Maeder and Meynet (2005) who pointed out that there is too little differential rotation in the layers immediately below the surface to operate the Taylor-Spruit-dynamo (see Sect. 4). The second possibility is rejected as well, as meridional circulation very likely does not reach the surface layers and as the field generated would be too weak for the magnetic buoyancy instability. The raising fields reported in numerical simulations by Mullan and MacDonald (2005) rely on field strengths on the order of 1 MG. These field strengths, in turn stem from their stellar structure model (MacDonald and Mullan 2004), which neglects the back-coupling of the magnetic field on the rotation rate, thus resulting in a very high differential rotation. In fact, as will be discussed in Sect. 4, even a weak field will suppress differential rotation to a high degree and the consistent rotation rate is nearly flat. Since the performance of the dynamo depends on the root of $d\Omega/dr$, this makes a huge difference.

3.2.3 Giant Convection Cells and Surface Dynamos

Yet another form of dynamo action has been proposed to explain observed magnetic fields in red super giants (see Sect. 2.5), the late stages of stellar evolution of massive stars. Freytag et al. (2002) performed detailed 3D radiation hydrodynamic simulations of the late-type supergiant Betelgeuse. The simulations, in comparison with observations, convincingly demonstrated the existence of giant convection cells. In a subsequent paper, Dorch (2004) presented magneto-hydrodynamical simulations of non-linear dynamo action in Betelgeuse and found that the giant convection cells can indeed result in the formation of surface magnetic fields of up to 500 G. More recent simulation results further support the existence of giant convection cells (Chiavassa et al. 2010, 2009).

3.3 Fossil Fields

An alternative to dynamo generated fields are field configurations which—once formed—are stable over an essential part of the life-time of the star. Such fields are generally called *fossil fields*. They are particularly attractive to explain the existence of magnetic fields in Ap-stars and the like: stars with a static, strong, large-scale magnetic field. Although, as discussed in the last section, there are competing ideas which ascribe such fields to dynamo action in the convective core with subsequent raise to the surface (Charbonneau and MacGregor 2001; MacGregor and Cassinelli 2003), a fossil origin of these fields seems more attractive. One main reason in favor of a fossil origin are that no correlation is observed between field strength and stellar rotation, as would be expected for a dynamo dragging its force from the differential rotation of the star. Also, the field strengths are so large that dynamo action is suppressed (Spruit 1999). The two main questions associated with fossil fields are their origin and their stability.

We already briefly touched the question of the stability of a magnetic field over essentially the life time of a star at the beginning of Sect. 3. There we pointed out that a mixed poloidal-toroidal topology is a necessary prerequisite. The results quoted there have been further refined in recent years. For an overview see for example Mestel and Landstreet (2005).

More recently, Braithwaite and Spruit (2004) and Braithwaite and Nordlund (2006) identified by means of numerical simulations a first concrete field topology that is stable over long time, illustrated in Fig. 5. They follow the evolution of a random field initially concentrated in a sphere of radius $r_m < R^*$ within a star of radius R^* and using a polytrope with index $n = 3$. For $r_m \lesssim 0.5R^*$, the initial configuration rapidly changes, within a few Alvén times the field strength decays strongly before stabilizing into a configuration which is always similar, independent of the concrete initialization: *'the poloidal field component is very similar to that which would be produced by an azimuthal current loop near the equator of the star. The toroidal component then threads along this loop.'* The field subsequently

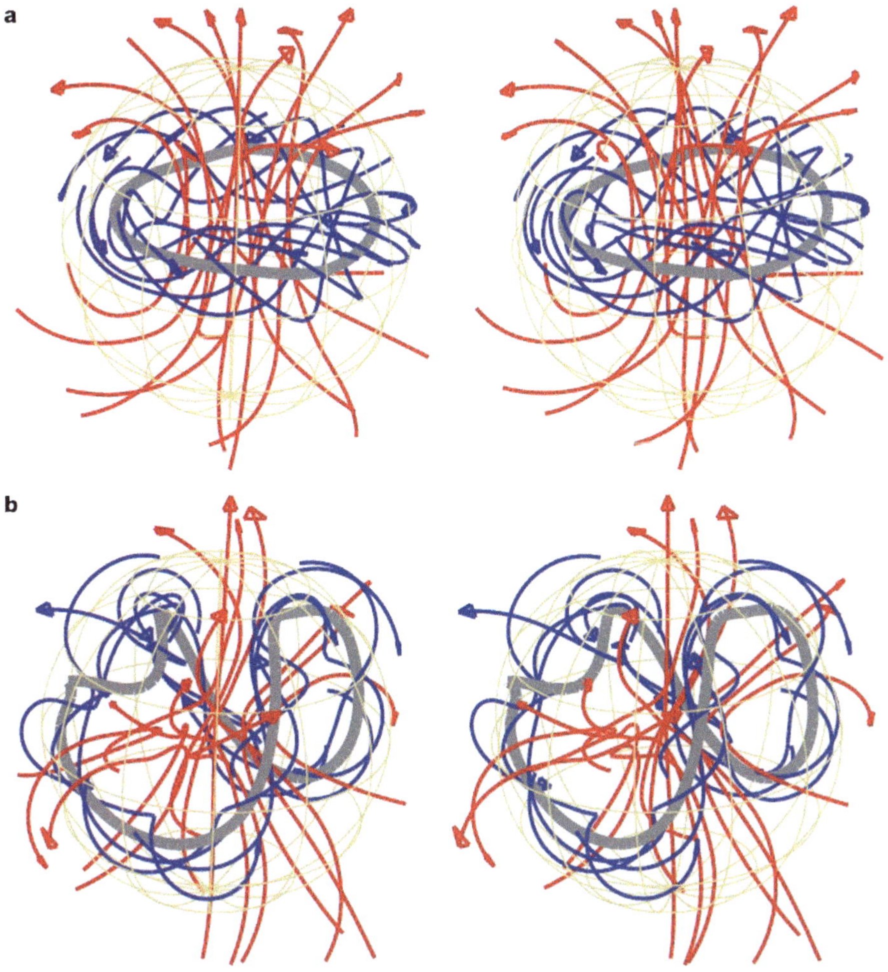

Fig. 5 Structure of stable magnetic fields, as found with three-dimensional numerical simulations. (**a**) Stereographic view of the long-lived magnetic field configuration as it evolves from a random initial condition. The stable core of the configuration is formed by a torus of twisted field lines inside the star (*blue*, with axis of torus shown in *grey*). Field lines that pass through the stellar surface (*red*) are stabilized by the torus. The configuration slowly evolves outwards by magnetic diffusion. When the torus reaches the surface it becomes unstable (stereographic view (**b**)). From Braithwaite and Spruit (2004), their Fig. 1

diffuses outwards, on a much longer time scale. While roughly preserving the overall configuration, the poloidal component gets stronger compared to the toroidal component. Once the imaginary current associated with the toroidal field component reaches the surface, the entire field rapidly decays. For initial field realizations of $r_m \gtrsim 0.5R^*$, the field directly enters this final distorted state and decays rapidly.

The authors predict that the first appearance of the field on the surface is after about $1/3$ of the life-time of the star, that the field strength subsequently should increase, and that contributions from lower order modes should gradually add to the initially bipolar surface field. Comparison of these predictions with observations shows a mixed picture. Hubrig et al. (2000) find that magnetic Ap stars are typically older than 30% of their total main sequence life span. By contrast, Landstreet et al. (2008) find no such age barrier. In addition, these authors find a decrease of magnetic field strength with the main sequence age of the star. As possible explanations for the discrepancy they suggest the non-consideration of both, rotation and the changing stellar radius and structure with age, in the simulations by Braithwaite and Spruit (2004).

The above picture essentially still holds today, although it has been refined in details since (Braithwaite et al. 2010). Using numerical simulations, Braithwaite (2009) found that for a stable field topology the poloidal component must not be significantly stronger than the toroidal component, while the toroidal may well be much stronger than the poloidal component. Reisenegger (2009) investigated the effect of a non-barotropic stratification. Mestel and Moss (2010) constructed a simple analytical model of tori as those found in numerical models and show that radiative heat transfer, Archimedes' principle, Lorentz force, and Ohmic decay all play a significant role. Duez and Mathis (2010) largely generalized the Prendergast models (Prendergast 1956, 1958) to barotropic magneto-hydrostatic equilibrium states of realistic stellar interiors. The analytic solutions they obtain strongly resemble the numerical results by Braithwaite and Spruit (2004) and Braithwaite and Nordlund (2006). Although stability has not yet been proofed for the analytical model, this combined effort between numerical and analytical work is exemplary and has already proofed to be seminal. Note, however, that Duez et al. (2010) tested the model numerically and found no significant change of the configuration over 10 Alfvénic times.

Turning now to the second corner stone of the fossil field hypothesis, namely the physical origin of these fields, we find a much less settled debate. At least three ideas have been put forward and are being investigated today: field generation during the convective pre-main sequence evolution, field generation in the wake of early binary mergers (Ferrario et al. 2009), and fields inherited from the molecular cloud in which the stars are born (Tout et al. 2004; Ferrario and Wickramasinghe 2006). Not surprisingly, each idea has its pros and cons.

So far, only the most simple model, the flux-conservation model put forward by Tout et al. (2004) and Ferrario and Wickramasinghe (2006), is worked out to a stage which allows confrontation with observations. The model relies on flux conservation and assumes that the field of a star is built by the magnetic flux trapped in the collapsing pre-stellar cloud. It then assumes that the magnetic flux is given as the sum of two Gaussians in the logarithm with dispersions σ_i with appropriate weightings. The model builds up on the remarkable fact that Ap- and Bp-stars as well as magnetic white dwarfs have magnetic fluxes of the same order. Moreover, when fields of magnetic white dwarfs are appropriately rescaled, the distribution of their strengths matches to a good degree the distribution of the same quantity of neutron stars. They also point out that the flux of θ Orion C (1.1×10^{27} G cm^2) and HD191512 (7.5×10^{27} G cm^2) is remarkably similar to the flux of the highest-field magnetar SGR 1806-20 (5.7×10^{27} G cm^2).

For massive main sequence stars, this model predicts a distribution which peaks at 46 G with 5 percent of the stars having a field in excess of 1 kG (originally 8 percent, but see the footnote in Petit et al. (2008)). The still very small sample of massive stars in the Orion Nebular Cluster (ONC) allows a first test of this prediction. As reported by Petit et al. (2008), the ONC contains nine massive OB stars, ranging from B3 V ($\sim$8 $M_\odot$) to O7 V ($\sim$40 $M_\odot$), three of them showing a very strong field. These authors find that, according to the multinomial distribution of Ferrario and Wickramasinghe (2006), the probability of obtaining the distribution of magnetic field strengths observed in the ONC is about 1 per cent. As pointed out by Petit et al. (2008), the sample is much too small to be representative. Also, the ONC may just be particularly magnetized, thus biasing the sample. Nevertheless, this demonstrates that observations now are at a level which starts to allow discrimination between different models.

3.4 Summary

Theory basically offers two hypothesis for the origin of magnetic fields in intermediate mass and massive stars: fossil fields, which existed already before the star reached the main sequence, and several kinds of dynamo generated fields. Fossil fields are the preferred explanation for strong and ordered (dipole) fields, whereas dynamo generated fields are believed to be less ordered and weaker. In fact, dynamo generated fields may not even emerge on the surface.

The details of both hypothesis are not well known so far. The origin of the fossil fields—from the parent molecular cloud, an early merger, or pre-main-sequence convection—are debated. Simulation results disagree on whether or not a dynamo can operate in the radiative envelope of a massive star. Similarly, it is unclear whether fields from the convective core can reach the surface of the star.

A particular case are surface fields in late-type supergiants like Betelgeuze, which numerical simulations indicate to be due to a surface dynamo operating in the outer convection layers of the star.

4 Magnetic Fields in Massive Stars: Implications for Stellar Evolution

The role of magnetic fields in stellar evolution is an area of research which largely remains to be explored and represents a top priority challenge in astrophysics. This is illustrated by the fact that the topic of magnetic field has been identified with rotation as one of the areas of discovery potential in the report "New Worlds, New Horizons in Astronomy and Astrophysics" written by the Committee for a Decadal Survey of Astronomy and Astrophysics (2010).

Magnetic fields have probably a very crucial impact at the two extreme points of the lifetime of a star, namely during their formation and when they disappear in a supernova explosion. But the magnetic field has also likely an important impact, at least when strong enough, during the hydrostatic phases of stellar evolution.

At the moment, stellar evolution computations taking into account the effects of magnetic fields remain rare. The available models have focused on exploring the consequences of the Tayler-Spruit dynamo mentioned in Sect. 3.2.2 above. In the present section we shall briefly describe the main results of these computations.

Let us recall that in the Tayler-Spruit dynamo (Spruit 1999, 2002), a pristine poloidal magnetic field is amplified through a dynamo effect. In a first step, the poloidal field is sheared by differential rotation. This produces a toroidal component which is sensitive to a pinch-type instability (the Pitts & Tayler instability). This generates an amplified poloidal component. The process can then begin again and the dynamo loop is closed. The energy needed for this amplification is obtained from the excess energy contained in differentially rotating layers.

The amplification will of course not be pursued indefinitely. The regulating mechanism is due to two counteracting effects: on one hand the source of the instability, which is differential rotation, will be progressively eroded by the strong radial coupling exerted by the increasing poloidal magnetic field, on the other hand meridional currents, which become stronger when the angular velocity gradient flattens, will rebuild the differential rotation. As a result of these two counteracting effects an "equilibrium profile" of the angular velocity in the star is reached. This angular profile presents the right degree of shear to sustain both the dynamo that continually erodes the shear and the meridional currents which continually rebuild it. Such a dynamo will produce internal magnetic fields of the order of 10^4 G in massive stars with an initial rotation of 300 km s^{-1} (Maeder and Meynet 2005).

The theory and the equations describing these effects are presented in (Spruit 1999, 2002). In this theoretical frame, two modifications of the equations have been brought by further works: (1) using the condition that the energy of the magnetic field created by the Tayler-Spruit dynamo cannot be larger than the energy excess present in the differential rotation, Maeder and Meynet (2003) propose a criterion for the existence of the magnetic field in stellar interior; (2) Maeder and Meynet (2004) avoid the simplifying assumptions that either the mean molecular weight- or the temperature-gradient dominates, but they treat the general case and also account for the nonadiabatic effects, which favor the growth of the magnetic field.

An interesting property of the Tayler-Spruit dynamo is that it can explain the internal near solid body rotation of the Sun (Eggenberger et al. 2005). However some critics of the theory have been presented: Zahn et al. (2007), using the 3-dimensional ASH code conclude that, although the Pitts & Tayler instability is present in their simulations, the mean poloidal field remains unaffected by it and thus that the dynamo loop is not closed. Denissenkov and Pinsonneault (2007) critic the fact that an important approximation used in the derivation of the equations is only valid for small length scales while, for obtaining the equations presented in Spruit (2002), it is applied for scales of the order of the radius of the star.

Despite these difficulties, the effects of the Spruit-Tayler dynamo are still explored in stellar models. A justification for using this theory is that, as discussed in Maeder et al. (2009), except for the expression of the growth rate of the instability, the equations presented in Spruit (2002) are quite general and can be derived by simply imposing that the rate of dissipation of the excess energy of the shear is equal to the production rate of magnetic energy. Said in other words, everything can be computed once the growth rate is known. As long as this growth rate is short enough, magnetic field will produce solid body rotation during the Main-Sequence phase.

So, to first order, we can say that models accounting for the Tayler-Spruit dynamo will differ from models with rotation only by the fact they have a solid body rotation all along the MS phase while the models with rotation only (and no magnetic fields) show a moderate contrast (by a factor of a few) between the rotation rate of the core and that of the surface (see Fig. 6).

An immediate consequence of the Tayler-Spruit dynamo is that mixing of the chemical species by the shear is no longer efficient. However, as shown by Maeder and Meynet (2005), meridional currents are more rapid in the rigid rotating models than in those with differential rotation. As a result, the deficit of mixing from the shear mechanism is more than compensated by the increasing efficiency of the mixing due to the meridional currents. These authors thus obtain more important surface enrichment (at equivalent age) in magnetic models with respect to non-magnetic ones.

Another important consequence of the magnetic models is that they predict for a given value of the initial rotation, higher surface velocities during the MS phase. This is because the magnetic field imposes a strong coupling between the core and the envelope. It allows to extract angular momentum from the contracting still faster rotating convective core and to bring it to the expanding and slower rotating radiative envelope.

So from what has been said above, magnetic fields will produce, for given initial conditions and a given age, faster rotating and more surface enriched stars. Also, the angular momentum of the core will be decreased in magnetic models. Heger et al. (2005) find that

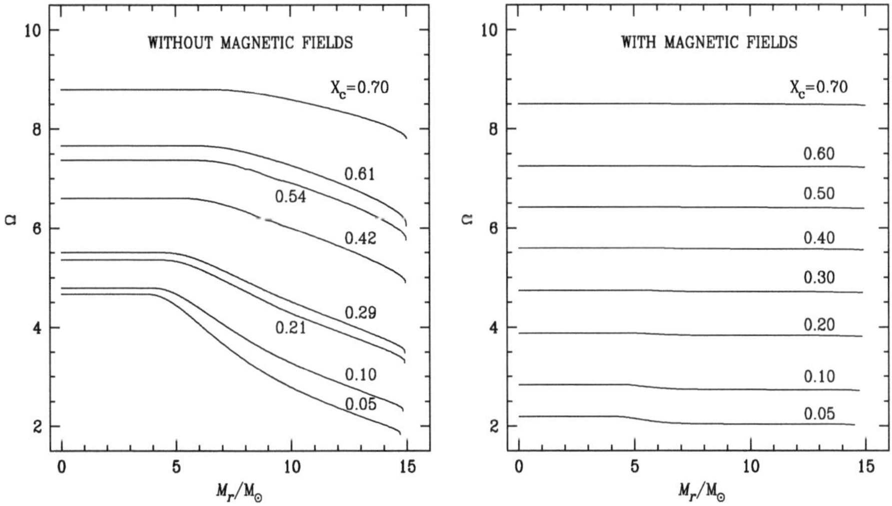

Fig. 6 *Left panel*: Internal distribution of the angular velocity $\Omega(r)$ as a function of the Lagrangian mass in solar units in a 15 $M_{\odot}$ model, without magnetic fields, at various stages of the model evolution indicated by the central H-content X_c during the Main-Sequence phase. The initial velocity is 300 km s^{-1}. *Right panel*: Same a *left panel* but with magnetic fields. Figures taken from Maeder and Meynet (2005)

magnetic torques decrease the final rotation rate of the collapsing iron core by about a factor of 30–50 when compared with the nonmagnetic models. From their 15 $M_{\odot}$ model, they predict pulsar periods at birth near 15 ms which is in agreement with the observed periods of the most rapidly rotating pulsars (see detailed discussion in the above reference). Thus we see that magnetic models can reasonably well account for the observed rotation rate of young pulsars, while models with rotation only would predict much too fast rotating pulsars. Let us however mention that such arguments in favor of magnetic models should be taken with some caution because extraction of angular momentum from the core can also occur at the time of the explosion and/or during the early life of the pulsars.

For very massive stars (above 40 $M_{\odot}$ at solar metallicity), strong stellar winds during the Main-Sequence phase remove the outer layers and thus angular momentum from the star. Since magnetic field, by tightly coupling the core to the envelope increases the angular momentum content of the outer layers, models with magnetic fields lose more angular momentum by mass loss than their non-magnetic counterparts (everything else being kept the same). This feature of magnetic models makes the production of a very fast spinning core more difficult. As just seen above, this is a nice feature in the frame of explaining the rotation rate of young pulsars, but it becomes a difficulty when one wants to explain the progenitors of collapsars proposed by Woosley (1993) to be at the origin of the long soft Gamma Ray Bursts (GRB). Let us briefly recall here that collapsars are core collapsing stars producing a fast rotating black hole. The fast rotation allows the formation of an accretion disk around the black hole. Gravitational energy extracted from the accretion disk is used (at least in part) to power polar jets. The gamma ray burst comes from shocks in these jets characterized by very high Lorentz factors. In order for the gamma ray burst to be observable, the star must have shed away its H-rich envelope. Thus models producing collapsars should present two properties which at first sight do appear very difficult to reconcile: on the one hand, a core with a high angular momentum content,[2] and at the same time no H-rich envelope, which means that the model had to suffer strong mass loss and hence strong angular momentum loss.

Non-magnetic models for high mass stars at low metallicity can easily reach the needed conditions as discussed by Hirschi et al. (2005). However such models would predict a rate of long soft gamma ray burst which is at the upper level of the observed range. Magnetic models can reproduce the required condition for having a collapsar but for a more restrained range of initial velocities. Only stars initially very rapidly rotating have a chance to give birth to a collapsar and then to a long soft gamma ray burst. Magnetic fileds thus reduce the predicted rate of GRBs.

The initial velocity needed to produce a collapsar is such that the star will follow a homogeneous evolution during the Main-Sequence phase, i.e. the stellar composition of the star will be nearly equal from the core to the surface at every time. Such an evolution has many advantages for producing a collapsar: quasi homogeneity during the MS phase allows the star to remove its H-rich envelope by nuclear processing rather than by mass loss, moreover the tracks of such homogeneous evolution in the Hertzsprung-Russel diagram remain in the blue, reducing the mass loss rates during the pre-WR phases. Evolution of these very fast rotating stars has been computed by Hirschi et al. (2005). These authors estimate that such stars might comprise roughly 1% of all stars above 10 $M_\odot$ and that they can, under certain circumstances, retain enough angular momentum to make GRBs. They underline the fact that the possibility to make GRBs is very sensitive to mass loss and is favored in regions of low metallicity. A similar scenario has been explored by Yoon and Langer (2005).

The rapid rotation needed in such scenario may be obtained either as a result of the initial very rapid rotation of the star or acquired by mass transfer in a close binary system (Petrovic et al. 2005; Detmers et al. 2008; Cantiello et al. 2007).

Let us end this section by saying a few words about the possibility for massive stars to suffer magnetic braking. At the moment of writing this review, there are no evolutionary computations accounting for this effect in the range of massive stars.[3] We expect that such computations will soon appear. Recently Ud-Doula et al. (2009) have proposed a numerical recipe for accounting for the angular momentum loss and associated rotational spin-down for magnetic hot stars with a line-driven stellar wind and a rotation-aligned dipole magnetic field. The numerical scaling relation that they obtain gives typical spin-down times of the order of 1 Myr for several known magnetic massive stars. This is quite in line with the recent spin down time scale for σ Ori E estimated to be 1.34 Myr by Townsend et al. (2010). These authors thus conclude that the observations are consistent with σ Ori E undergoing rotational braking due to its magnetized line-driven wind. Other stars present wind and surface magnetic field characteristics which are compatible with the existence of a magnetized line-driven wind. This is the case for instance of the Of?p star HD108 (Martins et al. 2010) which has also a slow rotation rate (lower than 50 km s^{-1}). How the surface chemical enrichments are affected by such a magnetic braking remains to be seen and will probably constitute an interesting line of research for the near future.

[2]Typically, the massive stars that give rise to gamma-ray bursts must have an amount of angular momentum in their inner regions, 1–2 orders of magnitude greater than the ones that make common pulsars (Woosley and Heger 2006).

[3]This effect is taken into account in solar models on a standard basis.

5 Magnetic Massive Stars in Their Environment

Despite the comparatively strong wind of massive stars in all stages of their evolution, the dynamics of the mass-loss in the vicinity of the star will be affected by the magnetic field of the star if such a field is present at all and if it is sufficiently strong. This likely also affects photospheric emission and probably also leads to X-ray emission (Sect. 5.1). If observed and if properly understood, such effects may in turn be used to confine properties of the stellar magnetic fields.

Going to larger scales, due to the faster decrease of the magnetic energy density as compared to the kinetic energy density, the kinetic energy of the stellar wind will dominate again. Owocki (2009) estimates that for the star with the largest ratio of magnetic to kinetic energy density detected so far, σ Ori E, this transition from magnetically to kinetically dominated flow takes place within less than 100 stellar radii, assuming a dipole field. The transition radius is approximately the radius where the first open field lines appear. Note that although the magnetic field no longer dominates the flow, the flow remains magnetized as the magnetic field is advected with the flow.

Observing consequences of the magnetic field at these larger scales, like particle acceleration at shocks and non-thermal emission, could again give a clue on the magnetism of the star. The general situation is potentially similar to strong supernova-remnant shocks (see the contributions in this volume: 'Magnetic Fields in Supernova Remnants and Pulsar Wind Nebulae' and 'Magnetic Fields in Cosmic Particle Acceleration Sources'). Although we are not aware of many definitive results using this approach, we nevertheless will discuss some points in this respect. The discussion is split into two parts, which may roughly be described as single stars interacting with the interstellar medium (Sect. 5.2) and multiple stars interacting among themselves, in binaries, or open clusters (Sect. 5.3).

5.1 Magnetic Wind Structure and Associated Emission

An excellent review on stellar magnetospheres, their theoretical and observational aspects especially also in early-type stars, has recently been given by Owocki (2009). Much of the following is adopted from this source.

Typical wind speeds and mass losses in O-stars are on the order of 1000 km/s and 10^{-6} $M_\odot$/yr. The winds are driven by the radiation field of the stars, by photon scattering in metal lines. Analytical considerations and numerical simulations show this acceleration mechanism to be subject to an instability, leading to the formation of internal shocks in the O-star wind (Owocki and Rybicki 1984; Owocki et al. 1988; Feldmeier et al. 1997). The emission from these shocks is the generally accepted explanation for the observed soft (10^7 K), thermal, broad-lined X-ray emission ($L_X/L_{\rm bol}$ 10^{-7}) from O-type stars.

Magnetic fields affecting the stellar winds of early-type stars have been suggested as a possible explanation for a number of additional observational peculiarities of these stars. Among these are strong X-ray emission in combination with narrow spectral lines of some stars (Babel and Montmerle 1997a; Favata et al. 2009), occasional flares of very hard X-rays (Townsend and Owocki 2005; Mullan 2009), or the rotation modulated Balmer line emission of Be stars (Townsend and Owocki 2005; Oksala et al. 2010). The concrete theoretical and numerical models invoked to explain these observational data have become more and more elaborate over the years.

As a simple measure to estimate the relative importance of a magnetic field for the wind of an early type star, ud-Doula and Owocki (2002) introduced the wind magnetic confine-

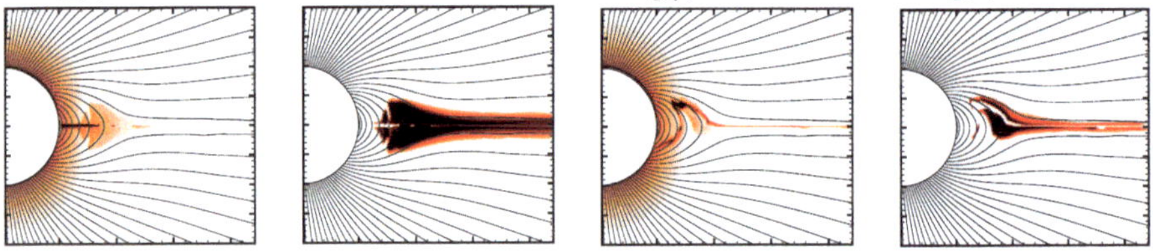

Fig. 7 MHD simulations of the MCWS for θ^1 Ori C, compiled by Owocki (2009) on the basis of ideas developed in Ud-Doula et al. (2008), his Fig. 5. Shown is the logarithmic density ρ (from -14 to -12 (gm/cm^3)) and the temperature T (from 4.7 to 7.7 (K)) in a meridional plane at 80 ksec (*left*) and 180 ksec (*right*) after initialization. At the earlier time, the magnetic field has channeled wind material into a compressed, hot disk at the magnetic equator. At the later time, the cooled equatorial material is falling back toward the star, along field lines. The *darkest areas* in temperature correspond to about 10^7 K

ment parameter

$$\eta_* = \frac{B_*^2 R_*^2}{\dot{M} v_\infty}. \tag{3}$$

Here, B_* denotes the surface magnetic field of the star, R_* the stellar radius, $\dot{M}$ the mass loss rate, and v_∞ the terminal wind speed. Basically, η_* measures the ratio of the magnetic to the kinetic energy densities (ud-Doula and Owocki 2002; Owocki 2009). Values of $\eta_* \ll 1$ indicate that the stellar wind dominates the magnetic field, and as the magnetic field is carried with the wind a roughly radial field configuration results. By contrast, if $\eta_* \gg 1$ the magnetic field dominates, an extreme example here being σ Ori E with $\eta_* \sim 10^7$. Closed magnetic field lines occur, at least in the vicinity of the star, guiding the stellar wind. With increasing distance from the star the stellar wind will generally dominate again, as the magnetic field energy decreases faster than the kinetic energy of the wind.

The basic effect of a strong enough dipole field on the wind structure is to deflect the stellar wind towards the equatorial plane separating the two magnetic hemispheres. There, collision flows results in strong shocks and associated X-ray emission. This magnetically confined wind shock model (MCWS) was first brought forward by Babel and Montmerle (1997b). In its semi-analytical formulation, the model provided a natural explanation of the periodic X-ray emission of θ^1 Ori C (Babel and Montmerle 1997a). Subsequent numerical simulations demonstrated that the X-ray emitting collision zone is not stable (ud-Doula and Owocki 2002; Ud-Doula et al. 2008). Shocked matter, once cooled, falls down along magnetic field lines, unless supported by rapid rotation of the star and associated centrifugal forces, as illustrated in Fig. 7. The simulations also demonstrated, however, that despite the much more complicated, unstable flow structure the X-ray emission predicted by Babel and Montmerle (1997a) is approximately recovered if time-averages are considered.

Considering the effect of a strong dipole field in rapidly rotating stars, Cassinelli et al. (2002) suggested that the coupling of matter and field could result in a spin-up of the stellar wind and the formation of a quasi-Keplerian magnetically torqued disk (MTD) at the magnetic equator. The idea was not supported by subsequent numerical simulations (Owocki and Ud-Doula 2003). For the field strengths required to spin up material to Keplerian velocities, the numerical models showed a tendency for centrifugal mass ejection instead of disk formation. Alternatively, a situation where material rotates at about Keplerian velocities can be obtained if much stronger fields ($\eta_* \to \infty$) are invoked, as assumed in the rigidly rotating magnetosphere (RRM) model (Townsend and Owocki 2005). Although the disk rotation in the RRM model is rigid body rather than Keplerian, approximate Keplerian velocities result as the material accumulates at the right distance from the star in the form of a thin disk or of clouds. Applying the RRM model to σ Ori E, Townsend et al. (2005) could well reproduce the periodic modulations observed in the light curve, Hα emission-line profile, and longitudinal field strength.

Numerical simulations confirm the general working mechanism of RRM but also reveal that the accumulation of matter cannot continue for infinite time. Eventually, the centrifugal force will overcome the magnetic tension forces and plasma ejection will result. During such a centrifugal break-out (Townsend et al. 2004; ud-Doula et al. 2006), magnetic field lines will first be drawn away from the star before they snap and reconnect again. The energy release associated with this reconnection may explain the occasional observed hard X-ray flares in some early-type stars.

5.2 Single Stars and Their Wind Blown Bubbles

We restrict ourselves to three kinds of wind blown bubbles and the potential effect stellar magnetic fields may have on their shape and emission: the case of an O-star wind blowing against the interstellar medium, a Wolf Rayet (WR) wind blowing into the remnant flow structures from earlier evolutionary phases, and planetary nebulae as a phase in the life of an intermediate-mass A-type or late B-type star.

As the wind of an O-type star runs highly supersonically against the interstellar medium, a strong wind-termination shock forms. There, particles will be accelerated and non-thermal emission is likely to take place. In principal, the observation of this shock and in particular of the associated non-thermal emission should give access to the field strength and topology in the wind. Unfortunately, if the O-star wind runs against the low-density ISM ($\sim$1 cm^{-3}) the termination shock is found at distances from the star, where wind densities are well below 1/100 cm^{-3}. This makes detection of non-thermal emission extremely difficult and we are actually not aware of any detection of such emission so far. One should add perhaps that—from an astronomical point of view—the observation of such a shock has hardly been of interest at all so far. However, with the increasing interest in the magnetism of massive stars this may change now.

For the roughly 300 galactic WR stars, van der Hucht, van der Hucht (2001, 2006) estimates that only about one quarter have a ring nebula. These nebulae are commonly attributed to the fast WR wind ploughing its way through the slow wind shed by the star when being in its super-giant phase. The WR wind compresses the material from the super-giant phase into a high density shell. Not all WR stars evolve, however, through a super-giant phase and about 40% of them live in binaries (van der Hucht 2001, 2006) with different evolutionary features. If the above wind collision scenario applies, theoretical considerations by Chevalier and Luo (1994) suggest that the field might even be able to affect the shape of the resulting ring nebula.

The work by Chevalier and Luo (1994) is, however, not particularly designed for WR ring nebulae but looks instead at the general situation where a fast, powerful wind catches up with a slow, massive wind. A thin interaction zone is produced, bounded by two shocks. The inner shock is the termination shock of the magnetic wind from the central star. As they point out, the field in the wind gets increasingly toroidal as this component drops in the free wind only linearly with the distance r, in contrast to the radial component of the field, which drops quadratically in r. Also in contrast to the radial field, the toroidal field is compressed at the termination shock. Thus, even if the magnetic field is not dynamically important in the free wind, it can become dynamically very important in the shocked wind bubble. A bipolar or elliptical nebula is a natural consequence of the fact that magnetic tension exists in the equatorial direction and the lack of such effects in the polar direction. As derived in this paper, the shape of the nebula depends on two parameters, namely, on λ, the ratio between the expansion velocity of the nebula and the velocity of the slow wind, and $\sigma = \eta_*(v_{\mathrm{rot}}/v_\infty)^2$. The paper provides the shaping of the nebulae for different combination of these parameters.

García-Segura et al. (1999) performed a numerical study of this effect for the case of planetary nebulae and also provided a series of different shapes of nebulae depending on two parameters: rotation and magnetic field strength. The authors point out that magnetic hook-stresses can collimate the flow along the rotation axis and even produce jet-like features and ansae which are indeed observed in planetary nebulae of low-mass stars. Zhekov and Myasnikov (2000) point out that heat-conduction by thermal electrons further contributes to the shape of such structures. Heat conduction, in turn, is largely influenced by the presence of magnetic fields, suppressing to a great deal conduction normal to the field.

5.3 Wind Collision in Binaries and Open Clusters

To test the magnetization of the outflow from a massive star, and thus the presence of a stellar magnetic field, wind collision regions in binary star systems and open clusters offer much better conditions than the wind blown bubbles of Sect. 5.2. While both scenarios, wind collisions and wind blown bubbles, are accompanied by strong shocks, densities are much higher in the wind collision case and accompanying emissions thus are stronger. In fact, Eichler and Usov (1993) predicted by analytical means that the wind collision zone in early type binaries may be a strong source of synchrotron generated, non-thermal radio emission. Key ingredients of the model are the strong shocks confining the interaction zone, where electrons can undergo Fermi acceleration and reach relativistic speeds, and the availability of a magnetic field, which causes the electrons to spin and emit synchrotron radiation.

A number of WR- and O-stars are indeed known to show non-thermal radio emission due to synchrotron radiation, indicating the presence of a magnetic field. Based on observational data, van der Hucht et al. (1992) noted that many WR-stars with non-thermal emission were long-period binaries. The hypothesis that binarity is even a pre-requisite for non-thermal emission of WR stars was put forward by Dougherty and Williams (2000), based on observations of 9 non-thermal WR emitters, 7 of which known binaries. In a recent review, De Becker (2007) lists 17 WR stars that show non-thermal radio-emission. Of these, 13 live in binaries, for one binarity is strongly suspected (as of November 2005), while for the remaining three stars no companion has been detected to date. Binarity remains, however, a possible explanation also for those non-thermal emitters without a detected companion, as excluding binarity is very difficult, especially if the companion is substantially less luminous. The situation is similar for O-stars. De Becker (2007) lists 16 O-stars with non-thermal emission, 14 of them are confirmed binaries or multiple systems.

That the inverse is not true, that binarity does not imply observable non-thermal emission, was already pointed out by Dougherty and Williams (2000). Of the 11 WR + OB binary star systems they observed, only those 7 systems with binary periods longer than about one year showed non-thermal emissions, while the 4 short period systems showed thermal emissions only. The authors argue that such short periods result in densities that are so high as to be opaque to radio emissions. Qualitatively the same conclusion is reached by Eichler and Usov (1993) on analytical grounds and by Pittard and Dougherty (2006) and Blomme (2010) using numerical simulations.

With binarity (or multiplicity in general) observationally established to accompany most if not all non-thermal radio emission in WR- and O-stars, the wind collision zone becomes

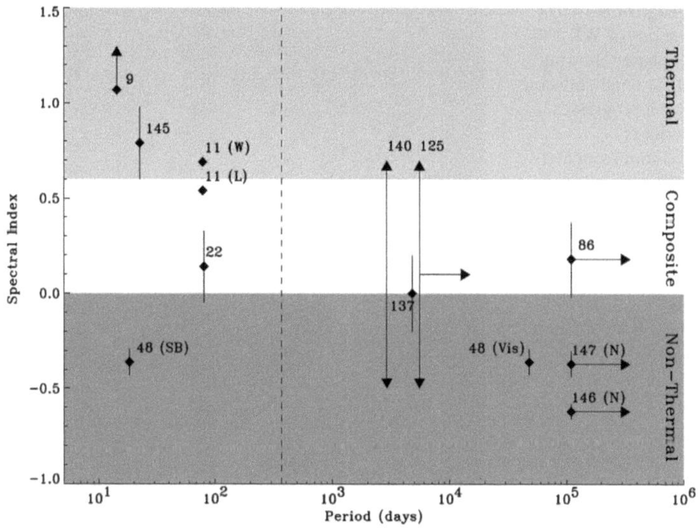

Fig. 8 Radio spectral index of WR + O binary systems against their binary period, taken from Dougherty and Williams (2000), their Fig. 1. While all long period binaries show some non-thermal emission, no such emission is found in short period binaries. WR 48 is a particular case, for details see the original work

the prime suspect for the source of the observed emission, as suggested by Usov (1991) and Eichler and Usov (1993). That the interaction zone of the two stellar winds indeed coincides with the source of the non-thermal emission was demonstrated by Williams et al. (1997) for the system WR147 (WN8 + B0.5V) by means of spatially resolved radio observations. From the radio flux they further derive a field strength between 1 and 9 mG in the WR wind, corresponding to a field on the stellar surface in the range of 30–300 G.

Internal shocks embedded in the stellar wind have been discussed as an alternative origin of the shocks needed for the Fermi acceleration of the electrons. However, recent numerical simulations indicate that the shocks are too weak when they reach the outer layers of the stellar wind, from where radio emission would be observable at all (van Loo et al. 2006; van Loo 2010; Blomme et al. 2010). This makes the colliding wind interaction zone not only one but possibly the source of non-thermal radio emission.

With regard to magnetism in massive stars, the above observational results are interesting in at least two ways. First, the detection of 33 non-thermal emitters as listed by De Becker (2007) indicates the presence of a stellar magnetic surface field in at least 33 and maybe up to 66 WR and O stars, thus outnumbering by far the direct detection of fields in such stars (Sects. 2.4, 2.5). From the point of view that the observation of non-thermal radio emission requires a long enough binary period, the number of 33 stars even constitutes a lower limit.

Second, and more speculatively, one might argue that Dougherty and Williams (2000) found indications for the presence of a magnetic field (non-thermal radio emission) in all those WR stars which fulfilled the two necessary conditions for such a field to leave an observable signature: binarity to have a wind collision zone with strong shocks, and a long enough binary period to avoid complete absorption of the non-thermal emission within the system (see Fig. 8). Simply speaking, the limiting factor may have been detectability and not the presence of a field as such. One may then further speculate whether all WR + OB and O + OB star binaries have a magnetic field. Or, if the existence of a stellar magnetic field does not depend on the star living in a binary system or not, whether all WR- and O-stars

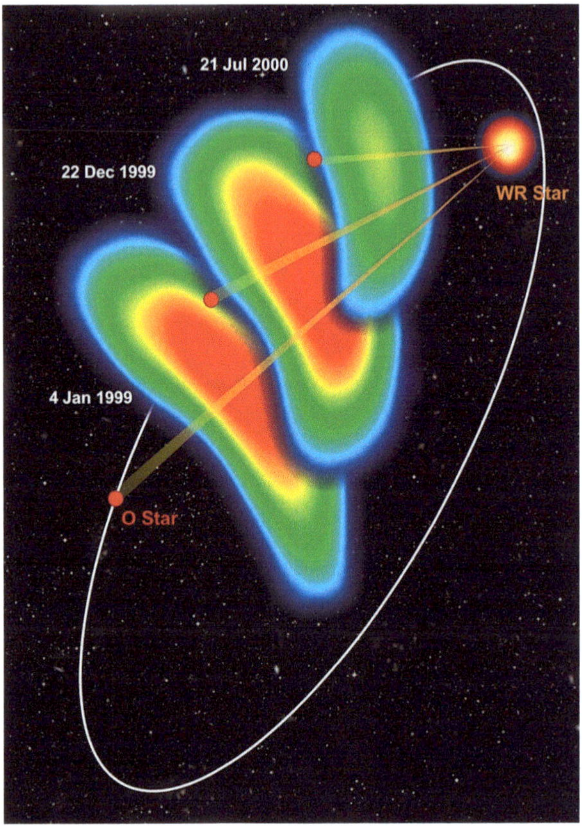

Fig. 9 A montage of 8.4-GHz VLBA observations of WR 140 at three orbital phases showing the rotation of the wind collision region as the orbit progresses. The deduced orbit is superimposed. From Dougherty and Pittard (2006), their Fig. 4

have a magnetic field. While it is unlikely that these speculations hold true in this rigid form, they indicate that it would be most interesting to search for non-thermal emission in additional WR + O star binaries. This would allow to see to what degree the above findings, based on 11 WR + O binaries only, carry over to a larger sample of the around 300 known galactic WR-stars, about 40% of which are binaries (van der Hucht 2001, 2006).

Equally desirable is a better theoretical understanding of the wind collision zone. The physics of the this zone is very complex, in particular if the conditions allow for radiative cooling, and has been studied extensively by analytical and numerical means (Usov 1991; Stevens et al. 1992; Nussbaumer and Walder 1993; Stevens and Pollock 1994; Owocki and Gayley 1995; Walder 1995; Walder and Folini 1996, 1998, 2003, Gayley et al. 1997; Folini and Walder 2000a, 2000b, 2002, 2006; Zhekov and Myasnikov 2000; Pittard 2009, 2010a; Pittard and Parkin 2010). Positively speaking, shocks in colliding wind binaries are an ideal laboratory to test the physics of particle acceleration and other kinetic processes related to collisionless shock waves. With comparatively high densities and large magnetic fields as well as a very intense UV-photon field from the involved stars, the parameter space of colliding winds in massive binaries complements the region covered by shocks in supernova remnants (see the contributions in this volume: 'Magnetic Fields in Supernova Remnants and Pulsar Wind Nebulae' and 'Magnetic Fields in Cosmic Particle Acceleration Sources').

The currently most detailed numerical studies of colliding winds in WR binaries and associated non-thermal emission in radio, X-rays, and γ-rays are the series of papers by Dougherty et al. (2003), Pittard et al. (2006), and Pittard and Dougherty (2006). Apart from hydrodynamical modeling of the collision zone, the authors developed numerical tools to fit synthetic spectra in radio, X-ray, and γ-rays to observed ones and applied these mainly to the two systems WR147 and WR140. A composite of radio observations of the latter is shown in Fig. 9. The model assumes diffuse shock acceleration of particles at the confining shocks of the wind collision zone. Corresponding particle aspects are described in terms of bulk properties, like the total energy and the energy distribution of the particles. Details on the models may be taken from a recent review of one of the authors (Pittard 2010b) or from corresponding, more recent work on O + O star binaries (Pittard 2009, 2010a; Pittard and Parkin 2010).

Noteworthy in the context of the present review, the results by Pittard and Dougherty (2006) underline again that any non-thermal emission in short-period binaries is unlikely to be observed. It is absorbed already in the stellar winds, by inverse Compton cooling, free-free absorption, and the Razin effect. Due to the strong radiation field, inverse Compton cooling is found to be the by far most effective cooling mechanism. Consequently, the authors expect non-thermal X ray and γ-ray emission to outpower the synchrotron radio emission by orders of magnitude. The numerical results even predict the emission of TeV photons due to π^0-decay, the π^0 being produced by the collision of shock-accelerated non-thermal hadrons (see also Eichler and Usov 1993).

Related to the 'classical' colliding wind scenario are processes taking place in dense open clusters, where massive stars often live. Such clusters blow strong galactic wind bubbles. Presumably, such bubbles play a significant role in the chemical enrichment of galaxies. By their energy impact, they furnish the turbulence of the interstellar medium and they can induce secondary star formation. Large bubbles cover a significant part of the galaxy and reach far out of the galactic plane. This huge space is magnetized by the massive star winds, which drive the bubble. As massive stars live very close to each other (roughly 1000 AU), such a bubble is created by some combined effect of single star nebulae and colliding winds. A network of strong, magnetic shocks is established. Very likely, galactic super bubbles are also a strong source of cosmic rays and another field to study particle acceleration. We are not aware of a comprehensive study of the role of magnetic fields in such bubbles. Pittard and Dougherty (2006) and Butt (2009) discuss some points, however.

5.4 Summary

In the immediate vicinity of a magnetic massive star, a magnetic dipole field has a strong effect on the stellar wind, if the ratio of the magnetic to the kinetic energy density, the wind magnetic confinement parameter η_*, is much larger than one. A shock compressed disk forms at the magnetic equator. Apart from periodic variabilities of different kind, a prominent observational signature expected from such a disk are hard X-rays.

For larger scales, further away from the star, much less is known on the effect of a magnetic field on the stellar environment. The only notable exception is the case of colliding winds in massive binary star systems. Non-thermal radio emission from the wind collision zone has been predicted and observed. The emission is considered to be indicative of a magnetic field, which must be the advected, in the wind frozen, magnetic field of the star (unless a dynamo is operating in the wind-collision zone). While the numerical models so far do not explicitly account for the stellar magnetic field, its strength and geometry, the available numerical and observational results suggest that many more massive stars (perhaps even a significant fraction) have a magnetic field than is known from direct field detections through Zeeman signatures. Performing further radio observations of massive binaries thus seems worthwhile.

6 Conclusion

The research on magnetic fields in massive and intermediate mass stars has seen tremendous progress in recent years. Reviewing corresponding developments and results is an intimidating task, all the more so as the author's backgrounds in modeling of astrophysical flows or stellar evolution touch only partly on the subject. Impressing also the range of perspectives taken on the issue: direct surface field measurements by means of the Zeeman effect, from pre-main sequence to post-main sequence stars; indirect indications of surface magnetic fields from observation and numerical modeling of colliding winds and other circumstellar phenomena; stellar evolution models exploring the effects of internal magnetic fields; dynamo theories and considerations on the stability of fossil fields to explain the origin of magnetic fields. Concluding this review by trying to sketch a general, emerging picture in this rapidly developing field is notoriously difficult and prone to misjudgments.

Starting on firm grounds, current Zeeman observations leave no doubt that at least some massive main-sequence stars poses a magnetic field. Indirect evidence, especially the observed non-thermal radio emission from a number of WR + OB and O + OB star binaries, hints at an even much larger number of magnetic massive stars. More speculatively, the only five direct field observations in main sequence O-stars may suggest a dichotomy: relatively strong ($\sim$1 kG) dipole fields on the one hand, as for θ^1 Ori C, and rather weak (few tens of Gauss) and unordered fields on the other hand, as for ζ Ori A.

Observational data from intermediate mass main sequence stars seem to support and enlarge this general picture. The 100% detection rate in a sample of 28 Ap/Bp stars suggests that maybe all Ap/Bp stars have a relatively strong (up to 30 kG) magnetic dipole field and that there exists, in addition, a lower limit to the magnetic dipole field of about 300 G. The recent—and first—Zeeman detection of a weak ($\sim$1 G) and unordered magnetic field in the normal A-type star Vega may suggest a similar dichotomy as for the massive stars. Observations of Ap/Bp stars in open clusters further suggest that the strengths of their surface fields decrease with the age of the star.

Theory offers essentially two hypothesis on the origin of the observed magnetic fields, but few firm conclusions. Compared to observations, much fewer publications are available.

The fossil field hypothesis assumes that the field was present already before the star reached the main sequence. It requires a field configuration that is stable enough for the field survive for long enough without being replenished by a dynamo. Analytical results showed long ago that such a configuration may exist in the form of a mixed poloidal-toroidal field configuration. But it was only recently that numerical simulations provided the first concrete realization of such a stable field configuration. The origin of fossil fields generally is still speculative. Suggestions range from the field being a relict of the parent molecular cloud, being generated by convection in the forming star, or being due to the merger of a pre-main-sequence binary. Whether the recently detected magnetic fields in some pre-main sequence Herbig Ae/Be stars constitute fossil fields that will eventually evolve into the field of the main-sequence Ap/Bp stars is debated.

The other explanation for the origin of observed surface magnetic fields is some sort of dynamo mechanism. There exists a lively debate on whether the radiative envelope of massive and intermediate mass stars can host a dynamo and how this would work. The only two numerical simulations addressing the topic give contradictory answers. Equally undecided is the debate on whether fields generated in the convective core can reach the surface. For late stages of massive star evolution, during the super giant phase, surface dynamos may explain the observed magnetic field. One may speculate whether the observed dichotomy between a 'strong and ordered field' and a 'weak and unordered field', if such a dichotomy exists at all, is due to two different origins, fossil or dynamo, of the field.

Rather detailed numerical models exist for the consequences of magnetic fields on the stellar evolution and environment. These models are of particular interest as they often link the magnetism in massive stars with a wider range of phenomena and, obviously, observation with theory. Stellar evolution models link magnetism in massive stars with the rotation rates of pulsars or the origin of gamma ray bursts. Simulations of how stellar winds are affected by a stellar magnetic field lead to the identification of indirect observational evidence of the later. Colliding wind models fulfill a similar task and, in addition, allow to gain insight into the physics of particle acceleration at collision-less magnetic shocks. The dependencies are often mutual and illustrate that advances in the field of magnetism in massive stars go hand in hand with advances in some other fields. In the future, one may want to exploit these perspectives and mutual dependence's even further.

To improve, for example, the statistics on what fraction of massive stars have a magnetic field, it would be interesting to have an observation based estimate of the fraction of WR + O star binaries showing non-thermal emission. The interpretation of such data would greatly benefit from improved colliding wind models that relate the magnetic fields in the interaction zone and at the surface of the star. Such a program would offer a complementary view on the certainly anyway growing observational data base of direct magnetic surface field detections in intermediate and high mass stars.

Given that stellar evolution is probably particularly sensitive to the presence of a magnetic field when a star is formed or dies, another interesting combination could be the observation of early stages of stellar evolution or even molecular clouds with more elaborate models of star formation in the presence of a magnetic field. Also most promising in combination with stellar evolution models is the emerging field of astroseismology of intermediate and high mass stars. First data and proof of concept studies have been published, which use gravity modes that provide information on the stellar interior down to the convective core (Aerts 2008; Degroote et al. 2010a, 2010b; Balona et al. 2010).

The above considerations make clear the necessity of combined efforts of observation, analytical theory, and numerical simulations to unravel the mysteries of magnetic fields in upper main sequence stars.

Acknowledgements We would like to thank Maxime Viallet for his comments on the manuscript which helped to improve the text. RW and DF acknowledge support from the French Stellar Astrophysics Program PNPS. We are indebted to the people running the NASA's Astrophysics Data System Abstract Service and arXiv of Cornell University Library. Both data-bases have greatly facilitated our work.

References

D.J. Acheson, On the instability of toroidal magnetic fields and differential rotation in stars. Philos. Trans. R. Soc. Lond. Ser. A **289**, 459–500 (1978)

C. Aerts, Core overshoot and nonrigid internal rotation of massive stars: current status from asteroseismology, in *IAU Symposium*, vol. 250, ed. by F. Bresolin, P.A. Crowther, J. Puls (2008), pp. 237–244

E. Alecian, G.A. Wade, C. Catala, C. Folsom, J. Grunhut, J. Donati, P. Petit, S. Bagnulo, T. Boehm, J. Bouret, J.D. Landstreet, Magnetism, rotation and accretion in Herbig Ae-Be stars, in *IAU Symposium*, vol. 243, ed. by J. Bouvier, I. Appenzeller (2007), pp. 43–50

E. Alecian, C. Catala, G.A. Wade, J. Donati, P. Petit, J.D. Landstreet, T. Böhm, J. Bouret, S. Bagnulo, C. Folsom, J. Grunhut, J. Silvester, Characterization of the magnetic field of the Herbig Be star HD200775. Mon. Not. R. Astron. Soc. **385**, 391–403 (2008a)

E. Alecian, G.A. Wade, C. Catala, C. Folsom, J. Grunhut, J. Donati, P. Petit, S. Bagnulo, S.C. Marsden, J.C. Ramirez Velez, J.D. Landstreet, T. Boehm, J. Bouret, J. Silvester, Magnetism in pre-MS intermediate-mass stars and the fossil field hypothesis. Contrib. Astron. Obs. Skaln. Pleso, **38**, 235–244 (2008b)

M. Aurière, Stellar polarimetry with NARVAL, in *EAS Publications Series*, vol. 9, ed. by J. Arnaud, N. Meunier (2003), p. 105

M. Aurière, G.A. Wade, J. Silvester, F. Lignières, S. Bagnulo, K. Bale, B. Dintrans, J.F. Donati, C.P. Folsom, M. Gruberbauer, A. Hui Bon Hoa, S. Jeffers, N. Johnson, J.D. Landstreet, A. Lèbre, T. Lueftinger, S. Marsden, D. Mouillet, S. Naseri, F. Paletou, P. Petit, J. Power, F. Rincon, S. Strasser, N. Toqué, Weak magnetic fields in Ap/Bp stars. Evidence for a dipole field lower limit and a tentative interpretation of the magnetic dichotomy. Astron. Astrophys. **475**, 1053–1065 (2007)

M. Auriere, G.A. Wade, F. Lignieres, A. Hui-Bon-Hoa, J.D. Landstreet, I. Iliev, J. Donati, P. Petit, T. Roudier, S. Theado, No detection of large-scale magnetic fields at the surfaces of Am and HgMn stars (2010)

M. Aurière, J. Donati, R. Konstantinova-Antova, G. Perrin, P. Petit, T. Roudier, The magnetic field of Betelgeuse: a local dynamo from giant convection cells? Astron. Astrophys. **516**, 2 (2010)

H.W. Babcock, Zeeman effect in stellar spectra. Astrophys. J. **105**, 105 (1947)

H.W. Babcock, The 34-KILOGAUSS magnetic field of HD 215441. Astrophys. J. **132**, 521 (1960)

J. Babel, T. Montmerle, On the periodic X-ray emission from the O7 V star theta 1 Orionis C. Astrophys. J. Lett. **485**, 29 (1997a)

J. Babel, T. Montmerle, X-ray emission from Ap-Bp stars: a magnetically confined wind-shock model for IQ Aur. Astron. Astrophys. **323**, 121–138 (1997b)

S. Bagnulo, J.D. Landstreet, E. Mason, V. Andretta, J. Silaj, G.A. Wade, Searching for links between magnetic fields and stellar evolution. I. A survey of magnetic fields in open cluster A- and B-type stars with FORS1. Astron. Astrophys. **450**, 777–791 (2006)

L. Balona, M. Cunha, D. Kurtz, I.M. Brandao, M. Gruberbauer, H. Saio, R. Ostensen, V. Elkin, W. Borucki, J. Christensen-Dalsgaard, H. Kjeldsen, D. Koch, Kepler observations of a roAp star: delta Scuti and gamma Doradus pulsations in Ap stars (2010)

S.V. Berdyugina, Stellar magnetic fields across the H-R diagram: observational evidence, in *IAU Symposium*, vol. 259 (2009), pp. 323–332

R. Blomme, Non-thermal radio emission from colliding wind binaries, in *Astronomical Society of the Pacific Conference Series*, vol. 422, ed. by J. Martí, P.L. Luque-Escamilla, J.A. Combi (2010), p. 178

R. Blomme, M. De Becker, D. Volpi, G. Rauw, Non-thermal radio emission from O-type stars. IV. Cyg OB2 No. 8A (2010)

J. Bouret, J. Donati, F. Martins, C. Escolano, W. Marcolino, T. Lanz, I.D. Howarth, The weak magnetic field of the O9.7 supergiant ζ Orionis A. Mon. Not. R. Astron. Soc. **389**, 75–85 (2008)

J. Braithwaite, A differential rotation driven dynamo in a stably stratified star. Astron. Astrophys. **449**, 451–460 (2006a)

J. Braithwaite, The stability of toroidal fields in stars. Astron. Astrophys. **453**, 687–698 (2006b)

J. Braithwaite, Axisymmetric magnetic fields in stars: relative strengths of poloidal and toroidal components. Mon. Not. R. Astron. Soc. **397**, 763–774 (2009)

J. Braithwaite, Å. Nordlund, Stable magnetic fields in stellar interiors. Astron. Astrophys. **450**, 1077–1095 (2006)

J. Braithwaite, H.C. Spruit, A fossil origin for the magnetic field in A stars and white dwarfs. Nature **431**, 819–821 (2004)

J. Braithwaite, T. Akgün, E. Alecian, A.F. Kholtygin, J.D. Landstreet, S. Mathis, G. Michaud, J. Portnoy, G. Alecian, V.D. Bychkov, L.V. Bychkova, N. Drake, S.N. Fabrika, A. Reisenegger, R. Steinitz, M. Vick, CP and related phenomena in the context of Stellar evolution. Highlights Astron. **15**, 161–171 (2010). doi:10.1017/S1743921310008562

M.K. Browning, A.S. Brun, J. Toomre, Simulations of core convection in rotating A-type stars: differential rotation and overshooting. Astrophys. J. **601**, 512–529 (2004)

A.S. Brun, M.K. Browning, J. Toomre, Simulations of core convection in rotating A-type stars: magnetic dynamo action. Astrophys. J. **629**, 461–481 (2005)

Y. Butt, Beyond the myth of the supernova-remnant origin of cosmic rays. Nature **460**, 701–704 (2009). doi:10.1038/nature08127

V.D. Bychkov, L.V. Bychkova, J. Madej, Catalogue of averaged stellar effective magnetic fields—II. Re-discussion of chemically peculiar A and B stars. Mon. Not. R. Astron. Soc. **394**, 1338–1350 (2009)

M. Cantiello, S. Yoon, N. Langer, M. Livio, Binary star progenitors of long gamma-ray bursts. Astron. Astrophys. **465**, 29–33 (2007)

J.P. Cassinelli, J.C. Brown, M. Maheswaran, N.A. Miller, D.C. Telfer, A magnetically torqued disk model for Be stars. Astrophys. J. **578**, 951–966 (2002)

P. Charbonneau, K.B. MacGregor, Magnetic fields in massive stars. I. Dynamo models. Astrophys. J. **559**, 1094–1107 (2001)

R.A. Chevalier, D. Luo, Magnetic shaping of planetary nebulae and other stellar wind bubbles. Astrophys. J. **421**, 225–235 (1994)

A. Chiavassa, B. Plez, E. Josselin, B. Freytag, Radiative hydrodynamics simulations of red supergiant stars. I. Interpretation of interferometric observations. Astron. Astrophys. **506**, 1351–1365 (2009)

A. Chiavassa, X. Haubois, J.S. Young, B. Plez, E. Josselin, G. Perrin, B. Freytag, Radiative hydrodynamics simulations of red supergiant stars. II. Simulations of convection on Betelgeuse match interferometric observations. Astron. Astrophys. **515**, 12 (2010)

M. De Becker, Non-thermal emission processes in massive binaries. Astron. Astrophys. Rev. **14**, 171–216 (2007)

P. Degroote, M. Briquet, M. Auvergne, S. Simon-Diaz, C. Aerts, A. Noels, M. Rainer, M. Hareter, E. Poretti, L. Mahy, R. Oreiro, M. Vuckovic, K. Smolders, A. Baglin, F. Baudin, C. Catala, E. Michel, R. Samadi, Detection of frequency spacings in the young O-type binary HD 46149 from CoRoT photometry (2010a)

P. Degroote, C. Aerts, A. Baglin, A. Miglio, M. Briquet, A. Noels, E. Niemczura, J. Montalban, S. Bloemen, R. Oreiro, M. Vučković, K. Smolders, M. Auvergne, F. Baudin, C. Catala, E. Michel, Deviations from a uniform period spacing of gravity modes in a massive star. Nature **464**, 259–261 (2010b)

P.A. Denissenkov, M. Pinsonneault, A revised prescription for the Tayler-Spruit dynamo: magnetic angular momentum transport in stars. Astrophys. J. **655**, 1157–1165 (2007)

R.G. Detmers, N. Langer, P. Podsiadlowski, R.G. Izzard, Gamma-ray bursts from tidally spun-up Wolf-Rayet stars? Astron. Astrophys. **484**, 831–839 (2008)

J. Donati, ESPaDOnS: an Echelle SpectroPolarimetric device for the observation of stars at CFHT, in *Astronomical Society of the Pacific Conference Series*, vol. 307, ed. by J. Trujillo-Bueno, J. Sanchez Almeida (2003), p. 41

J. Donati, J.D. Landstreet, Magnetic fields of nondegenerate stars. Annu. Rev. Astron. Astrophys. **47**, 333–370 (2009)

J. Donati, M. Semel, B.D. Carter, D.E. Rees, A. Collier Cameron, Spectropolarimetric observations of active stars. Mon. Not. R. Astron. Soc. **291**, 658 (1997)

J. Donati, C. Catala, G.A. Wade, G. Gallou, G. Delaigue, P. Rabou, A dedicated polarimeter for the MuSiCoS échelle spectrograph. Astron. Astrophys. Suppl. Ser. **134**, 149–159 (1999)

J. Donati, J. Babel, T.J. Harries, I.D. Howarth, P. Petit, M. Semel, The magnetic field and wind confinement of θ^1 Orionis C. Mon. Not. R. Astron. Soc. **333**, 55–70 (2002)

J. Donati, I.D. Howarth, J. Bouret, P. Petit, C. Catala, J. Landstreet, Discovery of a strong magnetic field on the O star HD 191612: new clues to the future of θ^1 Orionis C*. Mon. Not. R. Astron. Soc. **365**, 6–10 (2006a)

J. Donati, I.D. Howarth, M.M. Jardine, P. Petit, C. Catala, J.D. Landstreet, J. Bouret, E. Alecian, J.R. Barnes, T. Forveille, F. Paletou, N. Manset, The surprising magnetic topology of τ Sco: fossil remnant or dynamo output? Mon. Not. R. Astron. Soc. **370**, 629–644 (2006b)

S.B.F. Dorch, Magnetic activity in late-type giant stars: numerical MHD simulations of non-linear dynamo action in Betelgeuse. Astron. Astrophys. **423**, 1101–1107 (2004)

S.M. Dougherty, J.M. Pittard, Winds in collision: high-energy particles in massive binary systems, in *Proceedings of the 8th European VLBI Network Symposium* (2006)

S.M. Dougherty, P.M. Williams, Non-thermal emission in Wolf-Rayet stars: are massive companions required? Mon. Not. R. Astron. Soc. **319**, 1005–1010 (2000)

S.M. Dougherty, J.M. Pittard, L. Kasian, R.F. Coker, P.M. Williams, H.M. Lloyd, Radio emission models of colliding-wind binary systems. Astron. Astrophys. **409**, 217–233 (2003)

V. Duez, S. Mathis, Relaxed equilibrium configurations to model fossil fields. I. A first family. Astron. Astrophys. **517**, 58 (2010). doi:10.1051/0004-6361/200913496

R.J. Dufour, R.A.R. Parker, K.G. Henize, The spectrophotometry and chemical composition of the oxygen-poor bipolar nebula NGC 6164-5. Astrophys. J. **327**, 859–869 (1988)

V. Duez, K. Braithwaite, S. Mathis, On the stability of non-force-free magnetic equilibria in stars. Astrophys. J. Lett. **724**, 34–38 (2010). doi:10.1088/2041-8205/724/1/L34

P. Eggenberger, A. Maeder, G. Meynet, Stellar evolution with rotation and magnetic fields. IV. The solar rotation profile. Astron. Astrophys. **440**, 9–12 (2005)

D. Eichler, V. Usov, Particle acceleration and nonthermal radio emission in binaries of early-type stars. Astrophys. J. **402**, 271–279 (1993)

V.G. Elkin, G. Mathys, D.W. Kurtz, S. Hubrig, L.M. Freyhammer, A rival for Babcock's star: the extreme 30-kG variable magnetic field in the Ap star HD75049. Mon. Not. R. Astron. Soc. **402**, 1883–1891 (2010)

F. Favata, C. Neiner, P. Testa, G. Hussain, J. Sanz-Forcada, Testing magnetically confined wind shock models for β Cephei using XMM-Newton and Chandra phase-resolved X-ray observations. Astron. Astrophys. **495**, 217–229 (2009)

N.A. Featherstone, M.K. Browning, A.S. Brun, J. Toomre, Effects of fossil magnetic fields on convective core dynamos in A-type stars. Astrophys. J. **705**, 1000–1018 (2009)

A. Feldmeier, J. Puls, A.W.A. Pauldrach, A possible origin for X-rays from O stars. Astron. Astrophys. **322**, 878–895 (1997)

L. Ferrario, D. Wickramasinghe, Modelling of isolated radio pulsars and magnetars on the fossil field hypothesis. Mon. Not. R. Astron. Soc. **367**, 1323–1328 (2006)
L. Ferrario, J.E. Pringle, C.A. Tout, D.T. Wickramasinghe, The origin of magnetism on the upper main sequence. Mon. Not. R. Astron. Soc. **400**, 71–74 (2009)
V.C.A. Ferraro, The non-uniform rotation of the Sun and its magnetic field. Mon. Not. R. Astron. Soc. **97**, 458 (1937)
E. Flowers, M.A. Ruderman, Evolution of pulsar magnetic fields. Astrophys. J. **215**, 302–310 (1977)
D. Folini, R. Walder, 3D hydrodynamical simulations of colliding wind binaries: theory confronts observations. Astrophys. Space Sci. **274**, 189–194 (2000a)
D. Folini, R. Walder, Theory of thermal and ionization effects in colliding winds of WR + O binaries, in *Thermal and Ionization Aspects of Flows from Hot Stars*, ed. by H. Lamers, A. Sapar. Astronomical Society of the Pacific Conference Series, vol. 204 (2000b), p. 267
D. Folini, R. Walder, Theoretical predictions for the cold part of the colliding-wind interaction-zone, in *Interacting Winds from Massive Stars*, ed. by A.F.J. Moffat, N. St-Louis. Astronomical Society of the Pacific Conference Series, vol. 260 (2002), p. 605
D. Folini, R. Walder, Supersonic turbulence in shock-bound interaction zones. I. Symmetric settings. Astron. Astrophys. **459**, 1–19 (2006)
C.P. Folsom, G.A. Wade, O. Kochukhov, E. Alecian, C. Catala, S. Bagnulo, T. Böhm, J. Bouret, J. Donati, J. Grunhut, D.A. Hanes, J.D. Landstreet, Magnetic fields and chemical peculiarities of the very young intermediate-mass binary system HD 72106. Mon. Not. R. Astron. Soc. **391**, 901–914 (2008)
C.P. Folsom, O. Kochukhov, G.A. Wade, J. Silvester, S. Bagnulo, Magnetic field, chemical composition and line profile variability of the peculiar eclipsing binary star AR Aur. Mon. Not. R. Astron. Soc. (2010)
L.M. Freyhammer, V.G. Elkin, D.W. Kurtz, G. Mathys, P. Martinez, Discovery of 17 new sharp-lined Ap stars with magnetically resolved lines. Mon. Not. R. Astron. Soc. **389**, 441–460 (2008)
B. Freytag, M. Steffen, B. Dorch, Spots on the surface of Betelgeuse—results from new 3D stellar convection models. Astron. Nachr., **323**, 213–219 (2002)
A.W. Fullerton, Cyclical wind variability from O-type stars, in *Astronomical Society of the Pacific Conference Series*, vol. 305, ed. by L.A. Balona, H.F. Henrichs, R. Medupe (2003), p. 333
G. García-Segura, N. Langer, M. Różyczka, J. Franco, Shaping bipolar and elliptical planetary nebulae: effects of stellar rotation, photoionization heating, and magnetic fields. Astrophys. J. **517**, 767–781 (1999)
K.G. Gayley, S.P. Owocki, S.R. Cranmer, Sudden radiative braking in colliding hot-star winds. Astrophys. J. **475**, 786 (1997)
M. Goossens, D. Biront, R.J. Tayler, Additional results for unstable stratified toroidal magnetic fields in stars. Astrophys. Space Sci. **75**, 521–526 (1981)
J.H. Grunhut, G.A. Wade, W.L.F. Marcolino, V. Petit, H.F. Henrichs, D.H. Cohen, E. Alecian, D. Bohlender, J. Bouret, O. Kochukhov, C. Neiner, N. St-Louis, R.H.D. Townsend, Discovery of a magnetic field in the O9 sub-giant star HD 57682 by the MiMeS Collaboration. Mon. Not. R. Astron. Soc. **400**, 94–98 (2009)
J.H. Grunhut, G.A. Wade, D.A. Hanes, E. Alecian, Systematic detection of magnetic fields in massive, late-type supergiants (2010)
J.F. Hawley, C.F. Gammie, S.A. Balbus, Local three-dimensional simulations of an accretion disk hydromagnetic dynamo. Astrophys. J. **464**, 690 (1996)
A. Heger, S.E. Woosley, H.C. Spruit, Presupernova evolution of differentially rotating massive stars including magnetic fields. Astrophys. J. **626**, 350–363 (2005)
H.F. Henrichs, R.S. Schnerr, E. ten Kulve, Observed magnetism in massive stars, in *The Nature and Evolution of Disks Around Hot Stars*, ed. by R. Ignace, K.G. Gayley. Astronomical Society of the Pacific Conference Series, vol. 337 (2005), p. 114
H.F. Henrichs, J.A. de Jong, J. Donati, C. Catala, G.A. Wade, S.L.S. Shorlin, P.M. Veen, J.S. Nichols, L. Kaper, The magnetic field of β Cep and the Be phenomenon, in *IAU Colloq. 175: The Be Phenomenon in Early-Type Stars*, ed. by M.A. Smith, H.F. Henrichs, J. Fabregat. Astronomical Society of the Pacific Conference Series, vol. 214 (2000), p. 324
R. Hirschi, G. Meynet, A. Maeder, Stellar evolution with rotation. XIII. Predicted GRB rates at various Z. Astron. Astrophys. **443**, 581–591 (2005)
R. Hirschi, G. Meynet, A. Maeder, S. Ekström, C. Georgy, Stellar evolution in the upper HR diagram, in *Astronomical Society of the Pacific Conference Series*, vol. 425, ed. by C. Leither, P. Bennet, P. Morris, J. van Loon (2010), p. 13
S. Hubrig, P. North, G. Mathys, Magnetic AP stars in the Hertzsprung-Russell diagram. Astrophys. J. **539**, 352–363 (2000)
S. Hubrig, M. Schöller, R.V. Yudin, Magnetic fields in Herbig Ae stars. Astron. Astrophys. **428**, 1–4 (2004)

S. Hubrig, N. Nesvacil, M. Schöller, P. North, G. Mathys, D.W. Kurtz, B. Wolff, T. Szeifert, M.S. Cunha, V.G. Elkin, Detection of an extraordinarily large magnetic field in the unique ultra-cool Ap star HD 154708. Astron. Astrophys. **440**, 37–40 (2005)

S. Hubrig, M. Briquet, M. Schöller, P. De Cat, G. Mathys, C. Aerts, Discovery of magnetic fields in the β Cephei Star ζ1 CMa and in several slowly pulsating B stars. Mon. Not. R. Astron. Soc. **369**, 61–65 (2006). doi:10.1111/j.1745-3933.2006.00175.x

S. Hubrig, M.A. Pogodin, R.V. Yudin, M. Schöller, R.S. Schnerr, The magnetic field in the photospheric and circumstellar components of Herbig Ae stars. Astron. Astrophys. **463**, 1039–1046 (2007)

S. Hubrig, M. Schöller, R.S. Schnerr, J.F. González, R. Ignace, H.F. Henrichs, Magnetic field measurements of O stars with FORS 1 at the VLT. Astron. Astrophys. **490**, 793–800 (2008)

S. Hubrig, B. Stelzer, M. Schöller, C. Grady, O. Schütz, M.A. Pogodin, M. Curé, K. Hamaguchi, R.V. Yudin, Searching for a link between the magnetic nature and other observed properties of Herbig Ae/Be stars and stars with debris disks. Astron. Astrophys. **502**, 283–301 (2009)

M. Jardine, J.R. Barnes, J. Donati, A. Collier Cameron, The potential magnetic field of AB Doradus: comparison with Zeeman-Doppler images. Mon. Not. R. Astron. Soc. **305**, 35–39 (1999). doi:10.1046/j.1365-8711.1999.02621.x

L. Kaper, H.F. Henrichs, A.W. Fullerton, H. Ando, K.S. Bjorkman, D.R. Gies, R. Hirata, E. Kambe, D. McDavid, J.S. Nichols, Coordinated ultraviolet and Hα spectroscopy of bright O-type stars. Astron. Astrophys. **327**, 281–298 (1997)

A.F. Kholtygin, S.N. Fabrika, N.A. Drake, V.D. Bychkov, L.V. Bychkova, G.A. Chountonov, T.E. Burlakova, G.G. Valyavin, Statistics of magnetic fields for OB stars. Astron. Lett. **36**, 370–379 (2010). doi:10.1134/S1063773710050087

O. Kochukhov, Remarkable non-dipolar magnetic field of the Bp star HD 137509. Astron. Astrophys. **454**, 321–325 (2006a)

O. Kochukhov, Spectro-polarimetry of magnetic hot stars, in *Astronomical Society of the Pacific Conference Series*, vol. 358, ed. by R. Casini, B.W. Lites (2006b), p. 345

O. Kochukhov, V. Makaganiuk, N. Piskunov, Least squares deconvolution of the stellar intensity and polarization spectra (2010)

J.D. Landstreet, The magnetic field and abundance distribution geometry of the peculiar A star 53 Camelopardalis. Astrophys. J. **326**, 967–987 (1988)

J.D. Landstreet, Magnetic fields at the surfaces of stars. Astron. Astrophys. Rev. **4**, 35–77 (1992)

J.D. Landstreet, Observing and modelling stellar magnetic fields 1. Basic physics and simple models, in *EAS Publications Series*, vol. 39, ed. by C. Neiner, J.-P. Zahn (2009), pp. 1–20

J.D. Landstreet, G. Mathys, Magnetic models of slowly rotating magnetic Ap stars: aligned magnetic and rotation axes. Astron. Astrophys. **359**, 213–226 (2000)

J.D. Landstreet, S. Bagnulo, V. Andretta, L. Fossati, E. Mason, J. Silaj, G.A. Wade, Searching for links between magnetic fields and stellar evolution: II. The evolution of magnetic fields as revealed by observations of Ap stars in open clusters and associations. Astron. Astrophys. **470**, 685–698 (2007)

J.D. Landstreet, J. Silaj, V. Andretta, S. Bagnulo, S.V. Berdyugina, J. Donati, L. Fossati, P. Petit, J. Silvester, G.A. Wade, Searching for links between magnetic fields and stellar evolution. III. Measurement of magnetic fields in open cluster Ap stars with ESPaDOnS. Astron. Astrophys. **481**, 465–480 (2008)

C. Leitherer, C. Chavarria-K, The O6.5f?p star HD 148937 and its interstellar environment. Astron. Astrophys. **175**, 208–218 (1987)

F. Lignières, P. Petit, T. Böhm, M. Aurière, First evidence of a magnetic field on Vega. Towards a new class of magnetic A-type stars. Astron. Astrophys. **500**, 41–44 (2009)

J. MacDonald, D.J. Mullan, Magnetic fields in massive stars: dynamics and origin. Mon. Not. R. Astron. Soc. **348**, 702–716 (2004)

K.B. MacGregor, J.P. Cassinelli, Magnetic fields in massive stars. II. The buoyant rise of magnetic flux tubes through the radiative interior. Astrophys. J. **586**, 480–494 (2003)

A. Maeder, *Physics, Formation and Evolution of Rotating Stars*. Astronomy and Astrophysics Library (Springer, Berlin, 2009). ISBN 978-3-540-76948-4

A. Maeder, G. Meynet, Stellar evolution with rotation and magnetic fields. I. The relative importance of rotational and magnetic effects. Astron. Astrophys. **411**, 543–552 (2003)

A. Maeder, G. Meynet, Stellar evolution with rotation and magnetic fields. II. General equations for the transport by Tayler-Spruit dynamo. Astron. Astrophys. **422**, 225–237 (2004)

A. Maeder, G. Meynet, Stellar evolution with rotation and magnetic fields. III. The interplay of circulation and dynamo. Astron. Astrophys. **440**, 1041–1049 (2005)

A. Maeder, G. Meynet, C. Georgy, S. Ekström, The basic role of magnetic fields in stellar evolution, in *IAU Symposium*, vol. 259 (2009), pp. 311–322

P. Markey, R.J. Tayler, The adiabatic stability of stars containing magnetic fields—II. Poloidal fields. Mon. Not. R. Astron. Soc. **163**, 77 (1973)

P. Markey, R.J. Tayler, The adiabatic stability of stars containing magnetic fields—III. Additional results for poloidal fields. Mon. Not. R. Astron. Soc. **168**, 505–514 (1974)

M.J. Martínez González, A. Asensio Ramos, T.A. Carroll, M. Kopf, J.C. Ramírez Vélez, M. Semel, PCA detection and denoising of Zeeman signatures in polarised stellar spectra. Astron. Astrophys. **486**, 637–646 (2008)

F. Martins, J. Donati, W. Marcolino, J. Bouret, G.A. Wade, C. Escolano, I. Howarth, Detection of a magnetic field on HD108: clues to extreme magnetic braking and the Of?p phenomenon. Mon. Not. R. Astron. Soc. (2010)

G. Mathys, The observation of magnetic fields in nondegenerate stars. Fundam. Cosm. Phys. **13**, 143–308 (1989)

G. Mathys, Spectropolarimetry of magnetic stars. V. The mean quadratic magnetic field. Astron. Astrophys. **293**, 746–763 (1995)

G. Mathys, S. Hubrig, The diagnosis of the mean quadratic magnetic field of Ap stars. Astron. Astrophys. **453**, 699–715 (2006)

L. Mestel, *Stellar Magnetism*, International Series of Monographs on Physics (Clarendon, Oxford, 1999)

L. Mestel, J.D. Landstreet, Stellar magnetic fields, in *Cosmic Magnetic Fields*, ed. by R. Wielebinski, R. Beck. Lecture Notes in Physics, vol. 664 (Springer, Berlin, 2005), p. 183

L. Mestel, D. Moss, The evolution of stable magnetic fields in stars: an analytical approach. Mon. Not. R. Astron. Soc. **405**, 1845–1853 (2010)

Z. Mikulášek, J. Krtička, G.W. Henry, J. Zverko, J. Žižåovský, D. Bohlender, I.I. Romanyuk, J. Janík, H. Božić, D. Korčáková, M. Zejda, I.K. Iliev, P. Škoda, M. Šlechta, T. Gráf, M. Netolický, M. Ceniga, The extremely rapid rotational braking of the magnetic helium-strong star HD 37776. Astron. Astrophys. **485**, 585–597 (2008)

A.F.J. Moffat, G. Michaud, Zeta Puppis—an O-type oblique rotator. Astrophys. J. **251**, 133–138 (1981)

D.J. Mullan, Flares on a Bp star. Astrophys. J. **702**, 759–766 (2009)

D.J. Mullan, J. MacDonald, Dynamo-generated magnetic fields at the surface of a massive star. Mon. Not. R. Astron. Soc. **356**, 1139–1148 (2005)

Y. Nazé, N.R. Walborn, F. Martins, The mysterious Of?p class and the magnetic O-star θ^1 Ori C: confronting observations. Rev. Mex. Astron. Astrofís. **44**, 331–340 (2008)

Y. Naze, A. ud-Doula, M. Spano, G. Rauw, M. De Becker, N.R. Walborn, New findings on the prototypical Of?p stars (2010)

C. Neiner, V.C. Geers, H.F. Henrichs, M. Floquet, Y. Frémat, A. Hubert, O. Preuss, K. Wiersema, Discovery of a magnetic field in the Slowly Pulsating B star zeta Cassiopeiae. Astron. Astrophys. **406**, 1019–1031 (2003)

P. North, Y. Debernardi, Chemically Peculiar stars in binaries, in *Spectroscopically and Spatially Resolving the Components of the Close Binary Stars*, ed. by R.W. Hilditch, H. Hensberge, K. Pavlovski. Astronomical Society of the Pacific Conference Series, vol. 318 (2004), pp. 297–305

P. North, N. Ginestet, J. Carquillat, F. Carrier, S. Udry, Binaries among AP and AM stars. Contrib. Astron. Obs. Skaln. Pleso **27**, 179–183 (1998)

H. Nussbaumer, R. Walder, Modification of the nebular environment in symbiotic systems due to colliding winds. Astron. Astrophys. **278**, 209–225 (1993)

M.E. Oksala, G.A. Wade, W.L.F. Marcolino, J. Grunhut, D. Bohlender, N. Manset, R.H.D. Townsend, Discovery of a strong magnetic field in the rapidly rotating B2Vn star HR 7355. Mon. Not. R. Astron. Soc. **405**, 51–55 (2010)

L.M. Oskinova, W. Hamann, A. Feldmeier, R. Ignace, Y. Chu, Discovery of X-ray emission from the Wolf-Rayet star WR 142 of oxygen subtype. Astrophys. J. Lett. **693**, 44–48 (2009)

S. Owocki, Stellar magnetospheres, in *EAS Publications Series*, vol. 39, ed. by C. Neiner, J.-P. Zahn (2009), pp. 223–254

S. Owocki, A. Ud-Doula, Magnetic spin-up of line-driven stellar winds, in *Astronomical Society of the Pacific Conference Series*, vol. 305, ed. by L.A. Balona, H.F. Henrichs, R. Medupe (2003), p. 350

S.P. Owocki, K.G. Gayley, The importance of radiative braking for the wind interaction in the close WR + O binary V444 cygni. Astrophys. J. Lett. **454**, 145 (1995)

S.P. Owocki, G.B. Rybicki, Instabilities in line-driven stellar winds. I—Dependence on perturbation wavelength. Astrophys. J. **284**, 337–350 (1984)

S.P. Owocki, J.I. Castor, G.B. Rybicki, Time-dependent models of radiatively driven stellar winds. I—Nonlinear evolution of instabilities for a pure absorption model. Astrophys. J. **335**, 914–930 (1988)

P. Petit, F. Lignières, G.A. Wade, M. Aurière, T. Böhm, S. Bagnulo, B. Dintrans, A. Fumel, J. Grunhut, J. Lanoux, A. Morgenthaler, V. Van Grootel, The rapid rotation and complex magnetic field geometry of Vega (2010)

V. Petit, Observations of magnetic fields in hot stars (2010)

V. Petit, G.A. Wade, L. Drissen, T. Montmerle, E. Alecian, Discovery of two magnetic massive stars in the Orion Nebula Cluster: a clue to the origin of neutron star magnetic fields? Mon. Not. R. Astron. Soc. **387**, 23–27 (2008)

V. Petit, G.A. Wade, L. Drissen, T. Montmerle, E. Alecian, Magnetic fields, winds and X-rays of massive stars in the Orion Nebula Cluster, in *IAU Symposium*, vol. 259 (2009), pp. 449–452

J. Petrovic, N. Langer, S. Yoon, A. Heger, Which massive stars are gamma-ray burst progenitors? Astron. Astrophys. **435**, 247–259 (2005)

J.M. Pittard, 3D models of radiatively driven colliding winds in massive O + O star binaries—I. Hydrodynamics. Mon. Not. R. Astron. Soc. **396**, 1743–1763 (2009)

J.M. Pittard, 3D models of radiatively driven colliding winds in massive O + O star binaries—II. Thermal radio to submillimetre emission. Mon. Not. R. Astron. Soc. **403**, 1633–1656 (2010a)

J.M. Pittard, Models of the non-thermal emission from early-type binaries, in *Astronomical Society of the Pacific Conference Series*, vol. 422, ed. by J. Martí, P.L. Luque-Escamilla, J.A. Combi (2010b), p. 145

J.M. Pittard, S.M. Dougherty, Radio, X-ray, and γ-ray emission models of the colliding-wind binary WR140. Mon. Not. R. Astron. Soc. **372**, 801–826 (2006)

J.M. Pittard, E.R. Parkin, 3D models of radiatively driven colliding winds in massive O + O star binaries—III. Thermal X-ray emission. Mon. Not. R. Astron. Soc. **403**, 1657–1683 (2010)

J.M. Pittard, S.M. Dougherty, R.F. Coker, E. O'Connor, N.J. Bolingbroke, Radio emission models of colliding-wind binary systems. Inclusion of IC cooling. Astron. Astrophys. **446**, 1001–1019 (2006)

E. Pitts, R.J. Tayler, The adiabatic stability of stars containing magnetic fields. IV—The influence of rotation. Mon. Not. R. Astron. Soc. **216**, 139–154 (1985)

J. Power, G.A. Wade, M. Aurière, J. Silvester, D. Hanes, Propertiesofavolume-limitedsampleofAp-stars. Contrib. Astron. Obs. Skaln. Pleso **38**, 443–444 (2008)

K.H. Prendergast, The equilibrium of a self-gravitating incompressible fluid sphere with a magnetic field. I. Astrophys. J. **123**, 498 (1956)

K.H. Prendergast, The equilibrium of a self-gravitating incompressible fluid sphere with a magnetic field. II. Astrophys. J. **128**, 361 (1958)

G.W. Preston, A statistical investigation of the orientation of magnetic axes in the periodic magnetic variables. Astrophys. J. **150**, 547 (1967)

G.W. Preston, The magnetic field of HD 215441. Astrophys. J. **156**, 967 (1969)

G.W. Preston, The chemically peculiar stars of the upper main sequence. Annu. Rev. Astron. Astrophys. **12**, 257–277 (1974)

G. Rauw, N.D. Morrison, J. Vreux, E. Gosset, C.L. Mulliss, The spectral variability of HD 192639 and its implications for the star's wind structure. Astron. Astrophys. **366**, 585–597 (2001)

A. Reisenegger, Stable magnetic equilibria and their evolution in the upper main sequence, white dwarfs, and neutron stars. Astron. Astrophys. **499**, 557–566 (2009)

P. Renson, J. Manfroid, Catalogue of Ap, HgMn and Am stars. Astron. Astrophys. **498**, 961–966 (2009)

I.S. Savanov, S. Hubrig, J.F. González, M. Schöller, Searching for a link between the presence of chemical spots on the surface of HgMn stars and their weak magnetic fields, in *IAU Symposium*, vol. 259 (2009), pp. 401–402

R.S. Schnerr, H.F. Henrichs, C. Neiner, E. Verdugo, J. de Jong, V.C. Geers, K. Wiersema, B. van Dalen, A. Tijani, B. Plaggenborg, K.L.J. Rygl, Magnetic field measurements and wind-line variability of OB-type stars. Astron. Astrophys. **483**, 857–867 (2008)

M. Schwarzschild, On the scale of photospheric convection in red giants and supergiants. Astrophys. J. **195**, 137–144 (1975)

M. Semel, Zeeman-Doppler imaging of active stars. I—Basic principles. Astron. Astrophys. **225**, 456–466 (1989)

M. Semel, J. Li, Zeeman-Doppler imaging of solar-type stars: multi line technique. Sol. Phys. **164**, 417–428 (1996)

M. Semel, D.E. Rees, J.C. Ramírez Vélez, M.J. Stift, F. Leone, Multi-line spectro-polarimetry of stellar magnetic fields using principal components analysis, in *Astronomical Society of the Pacific Conference Series*, vol. 358, ed. by R. Casini, B.W. Lites (2006), p. 355

M. Semel, J.C. Ramírez Vélez, M.J. Martínez González, A. Asensio Ramos, M.J. Stift, A. López Ariste, F. Leone, Multiline Zeeman signatures through line addition. Astron. Astrophys. **504**, 1003–1009 (2009)

J. Silvester, C. Neiner, H.F. Henrichs, G.A. Wade, V. Petit, E. Alecian, A. Huat, C. Martayan, J. Power, O. Thizy, On the incidence of magnetic fields in slowly pulsating B, β Cephei and B-type emission-line stars. Mon. Not. R. Astron. Soc. **398**, 1505–1511 (2009)

K.R. Sokal, S.L. Skinner, S.A. Zhekov, M. Güdel, W. Schmutz, Chandra detects the rare oxygen-type Wolf-Rayet star WR 142 and OB stars in Berkeley 87. Astrophys. J. **715**, 1327–1337 (2010)

H.C. Spruit, Differential rotation and magnetic fields in stellar interiors. Astron. Astrophys. **349**, 189–202 (1999)

H.C. Spruit, Dynamo action by differential rotation in a stably stratified stellar interior. Astron. Astrophys. **381**, 923–932 (2002)

B. Stelzer, E. Flaccomio, T. Montmerle, G. Micela, S. Sciortino, F. Favata, T. Preibisch, E.D. Feigelson, X-ray emission from early-type stars in the Orion Nebula cluster. Astrophys. J. Suppl. Ser. **160**, 557–581 (2005)

I.R. Stevens, A.M.T. Pollock, Stagnation-point flow in colliding wind binary systems. Mon. Not. R. Astron. Soc. (1994)

I.R. Stevens, J.M. Blondin, A.M.T. Pollock, Colliding winds from early-type stars in binary systems. Astrophys. J. **386**, 265–287 (1992)

R.J. Tayler, The adiabatic stability of stars containing magnetic fields—I. Toroidal fields. Mon. Not. R. Astron. Soc. **161**, 365 (1973)

C.A. Tout, D.T. Wickramasinghe, L. Ferrario, Magnetic fields in white dwarfs and stellar evolution. Mon. Not. R. Astron. Soc. **355**, 13–16 (2004)

R.H.D. Townsend, S.P. Owocki, A rigidly rotating magnetosphere model for circumstellar emission from magnetic OB stars. Mon. Not. R. Astron. Soc. **357**, 251–264 (2005)

R.H.D. Townsend, S.P. Owocki, D. Groote, The rigidly rotating magnetosphere of σ Orionis E. Astrophys. J. Lett. **630**, 81–84 (2005)

R.H.D. Townsend, S.P. Owocki, I.D. Howarth, Be-star rotation: how close to critical? Mon. Not. R. Astron. Soc. **350**, 189–195 (2004)

R.H.D. Townsend, M.E. Oksala, D.H. Cohen, S.P. Owocki, A. ud-Doula, Discovery of rotational braking in the magnetic helium-strong star Sigma Orionis E. Astrophys. J. Lett. **714**, 318–322 (2010)

A.V. Tutukov, A.V. Fedorova, Possible scenarios for the formation of Ap/Bp stars. Astron. Rep. **54**, 156–162 (2010)

A. ud-Doula, S.P. Owocki, Dynamical simulations of magnetically channeled line-driven stellar winds. I. Isothermal, nonrotating, radially driven flow. Astrophys. J. **576**, 413–428 (2002)

A. Ud-Doula, S.P. Owocki, R.H.D. Townsend, Dynamical simulations of magnetically channelled line-driven stellar winds—II. The effects of field-aligned rotati on. Mon. Not. R. Astron. Soc. **385**, 97–108 (2008)

A. Ud-Doula, S.P. Owocki, R.H.D. Townsend, Dynamical simulations of magnetically channelled line-driven stellar winds—III. Angular momentum loss and rotation al spin-down. Mon. Not. R. Astron. Soc. **392**, 1022–1033 (2009)

A. ud-Doula, R.H.D. Townsend, S.P. Owocki, Centrifugal breakout of magnetically confined line-driven stellar winds. Astrophys. J. Lett. **640**, 191–194 (2006)

V.V. Usov, Stellar wind collision and dust formation in long-period, heavily interacting Wolf-Rayet binaries. Mon. Not. R. Astron. Soc. **252**, 49–52 (1991)

K.A. van der Hucht, The VIIth catalogue of galactic Wolf-Rayet stars. New Astron. Rev. **45**, 135–232 (2001)

K.A. van der Hucht, New Galactic Wolf-Rayet stars, and candidates. An annex to the VIIth catalogue of galactic Wolf-Rayet stars. Astron. Astrophys. **458**, 453–459 (2006)

K.A. van der Hucht, P.M. Williams, T.A.T. Spoclstra, A.C. de Bruyn, Non-thermal radio observations of Wolf-Rayet stars: a case for long-period binaries, in *Nonisotropic and Variable Outflows from Stars*, ed. by L. Drissen, C. Leitherer, A. Nota. Astronomical Society of the Pacific Conference Series, vol. 22 (1992), p. 249 (Invited paper)

S. van Loo, Non-thermal radio emission from presumably single O stars, in *Astronomical Society of the Pacific Conference Series*, ed. by J. Martí, P.L. Luque-Escamilla, J.A. Combi. Astronomical Society of the Pacific Conference Series, vol. 422 (2010), p. 157

S. van Loo, M.C. Runacres, R. Blomme, Can single O stars produce non-thermal radio emission? Astron. Astrophys. **452**, 1011–1019 (2006)

G.A. Wade, the MiMeS Collaboration, The MiMeS Project: Current status and recent results (2010)

G.A. Wade, M. Aurière, S. Bagnulo, J. Donati, N. Johnson, J.D. Landstreet, F. Lignières, S. Marsden, D. Monin, D. Mouillet, F. Paletou, P. Petit, N. Toqué, E. Alecian, C. Folsom, A search for magnetic fields in the variable HgMn star α Andromedae. Astron. Astrophys. **451**, 293–302 (2006)

G.A. Wade, S. Bagnulo, D. Drouin, J.D. Landstreet, D. Monin, A search for strong, ordered magnetic fields in Herbig Ae/Be stars. Mon. Not. R. Astron. Soc. **376**, 1145–1161 (2007)

G.A. Wade, E. Alecian, J. Grunhut, C. Catala, S. Bagnulo, C.P. Folsom, J.D. Landstreet, Magnetism of Herbig Ae/Be stars (2009a)

G.A. Wade, E. Alecian, D.A. Bohlender, J. Bouret, J.H. Grunhut, H. Henrichs, C. Neiner, V. Petit, N.S. Louis, M. Aurière, O. Kochukhov, J. Silvester, A. ud-Doula, The MiMeS project: magnetism in massive stars, in *IAU Symposium*, vol. 259 (2009), pp. 333–338

N.R. Walborn, Spectral classification of OB stars in both hemispheres and the absolute-magnitude calibration. Astron. J. **77**, 312–318 (1972)

N.R. Walborn, I.D. Howarth, A. Herrero, D.J. Lennon, The remarkable alternating spectra of the Of?p star HD 191612. Astrophys. J. **588**, 1025–1038 (2003)

N.R. Walborn, I.D. Howarth, G. Rauw, D.J. Lennon, H.E. Bond, I. Negueruela, Y. Nazé, M.F. Corcoran, A. Herrero, A. Pellerin, A period and a prediction for the Of?p spectrum alternator HD 191612. Astrophys. J. Lett. **617**, 61–64 (2004)

N.R. Walborn, A. Sota, J. Maíz Apellániz, E.J. Alfaro, N.I. Morrell, R.H. Barbá, J.I. Arias, R.C. Gamen, Early results from the galactic O-star spectroscopic survey: C III emission lines in of spectra. Astrophys. J. Lett. **711**, 143–147 (2010)

R. Walder, Simulations of colliding winds in 3 dimensions, in *Wolf-Rayet Stars: Binaries; Colliding Winds; Evolution*, ed. by K.A. van der Hucht, P.M. Williams. IAU Symposium, vol. 163 (1995), p. 420

R. Walder, D. Folini, Radiative cooling instability in 1D colliding flows. Astron. Astrophys. **315**, 265–283 (1996)

R. Walder, D. Folini, Knots, filaments, and turbulence in radiative shocks. Astron. Astrophys. **330**, 21–24 (1998)

R. Walder, D. Folini, 3D-hydrodynamics of colliding winds in massive binaries, in *A Massive Star Odyssey: From Main Sequence to Supernova*, ed. by K. van der Hucht, A. Herrero, C. Esteban, IAU Symposium, vol. 212, (2003), p. 139

P.M. Williams, S.M. Dougherty, R.J. Davis, K.A. van der Hucht, M.F. Bode, D.Y.A. Setia Gunawan, Radio and infrared structure of the colliding-wind Wolf-Rayet system WR147. Mon. Not. R. Astron. Soc. **289**, 10–20 (1997)

S.E. Woosley, Gamma-ray bursts from stellar mass accretion disks around black holes. Astrophys. J. **405**, 273–277 (1993)

S.E. Woosley, A. Heger, The progenitor stars of gamma-ray bursts. Astrophys. J. **637**, 914–921 (2006)

G.A.E. Wright, Pinch instabilities in magnetic stars. Mon. Not. R. Astron. Soc. **162**, 339 (1973)

S. Yoon, N. Langer, Evolution of rapidly rotating metal-poor massive stars towards gamma-ray bursts. Astron. Astrophys. **443**, 643–648 (2005)

J. Zahn, A.S. Brun, S. Mathis, On magnetic instabilities and dynamo action in stellar radiation zones. Astron. Astrophys. **474**, 145–154 (2007)

S.A. Zhekov, A.V. Myasnikov, Colliding stellar winds: "Asymmetric" thermal conduction. Astrophys. J. Lett. **543**, 53–56 (2000)

Space Sci Rev (2012) 166:187–213
DOI 10.1007/s11214-011-9785-9

Magnetic Fields, Relativistic Particles, and Shock Waves in Cluster Outskirts

Marcus Brüggen · Andrei Bykov · Dongsu Ryu · Huub Röttgering

Received: 22 September 2010 / Accepted: 2 May 2011 / Published online: 9 June 2011

Abstract It is only now, with low-frequency radio telescopes, long exposures with high-resolution X-ray satellites and γ-ray telescopes, that we are beginning to learn about the physics in the periphery of galaxy clusters. In the coming years, Sunyaev-Zel'dovich telescopes are going to deliver further great insights into the plasma physics of these special regions in the Universe. The last years have already shown tremendous progress with detections of shocks, estimates of magnetic field strengths and constraints on the particle acceleration efficiency. X-ray observations have revealed shock fronts in cluster outskirts which have allowed inferences about the microphysical structure of shocks fronts in such extreme environments. The best indications for magnetic fields and relativistic particles in cluster outskirts come from observations of so-called radio relics, which are megaparsec-sized regions of radio emission from the edges of galaxy clusters. As these are difficult to detect due to their low surface brightness, only few of these objects are known. But they have provided unprecedented evidence for the acceleration of relativistic particles at shock fronts and the existence of μG strength fields as far out as the virial radius of clusters. In this review we summarise the observational and theoretical state of our knowledge of magnetic fields, relativistic particles and shocks in cluster outskirts.

Keywords Galaxy clusters · Magnetic fields · Cosmic rays

M. Brüggen (✉)
Jacobs University Bremen, Campus Ring 1, 28759 Bremen, Germany
e-mail: m.brueggen@jacobs-university.de

A. Bykov
Ioffe Institute for Physics and Technology, 194021 St. Petersburg, Russia
e-mail: byk@astro.ioffe.ru

D. Ryu
Department of Astronomy and Space Science, Chungnam National University, Daejeon 305-764, Korea
e-mail: ryu@canopus.cnu.ac.kr

H. Röttgering
Leiden Observatory, Leiden University, P.O. Box 9513, 2300 RA Leiden, The Netherlands
e-mail: rottgering@strw.leidenuniv.nl

1 Introduction

Clusters of galaxies are filled with a dilute ($n_e \leq 0.1$ cm^{-3}), hot ($T > 10^7$ K) gas, the so-called intracluster medium (ICM). In this review, we are primarily concerned with regions outside of cluster cores, close to the virial radius ($\sim$1 Mpc). The ICM contains the majority of baryons in galaxy clusters and has a temperature close to the virial temperature of the gravitational potential of the cluster.

Shocks play a fundamental role in the evolution of the ICM and affect clusters in two ways: First, shocks thermalize the incoming gas allowing it to virialise and providing much of the pressure support in baryons. Secondly, because these shocks are collisionless features whose interactions in the hot plasma are mediated by electromagnetic fields, it is possible for a fraction of the thermal distribution of particles to be accelerated and transformed into nonthermal populations of cosmic rays (CRs) through diffusive shock acceleration. This process results in a part of the kinetic energy to be converted to non-thermal components. Non-thermal populations of particles can in principle be observed in the hard X-rays through inverse Compton scattering of the Cosmic Microwave background, in the gamma-rays through hadronic interactions of non-thermal protons with the thermal component and in the radio band through synchrotron radiation of the electrons. The latter has been observed in galaxy clusters in the form of diffuse radio sources. These sources are usually subdivided into two classes, denoted as 'radio haloes' and 'radio relics' (see e.g. Ferrari et al. 2008, for a review).

Radio haloes are diffuse, low surface brightness ($\simeq 10^{-6}$ Jy/arcsec^2 at 1.4 GHz), steep-spectrum[1] ($\alpha > 1$) sources, permeating the central regions of clusters, produced by synchrotron radiation of relativistic electrons with energies of $\simeq$10 GeV in magnetic fields with $B \simeq 0.5$–1 μG. Radio haloes represent the best evidence for the presence of large-scale magnetic fields and relativistic particles throughout the intra-cluster medium (ICM). They are mostly unpolarized and have diffuse morphologies that are similar to those of the thermal X-ray emission of the cluster gas. Their origin and the role that shocks play in their formation are still subject to debate, (see e.g., Petrosian and Bykov 2008; Petrosian et al. 2008; Brunetti et al. 2009, for recent review). Unlike haloes, radio relics are typically located near the periphery of the cluster. They often exhibit sharp emission edges and many of them show strong radio polarization. For some relics there is now good evidence that these electrons are produced by shock waves and we will present this evidence in Sect. 4.

If a significant fraction of the pressure of the ICM is provided by CRs, this will affect gas mass fraction estimates. Since gas mass fractions in clusters are used to make inferences about dark energy, a knowledge about the non-thermal content of galaxy clusters is important for their use in precision cosmology (e.g. Vikhlinin et al. 2009).

There are three types of phenomena that create shocks in the ICM. The first class of shocks are accretion shocks: At very large off-center distances, typically several Mpc, cosmological simulations predict that intergalactic medium (IGM) continues to accrete onto the clusters through a system of shocks that separate the IGM from the hot, mostly virialized inner regions. The IGM is much cooler than the ICM, so these accretion (or infall) shocks should be strong, with Mach numbers $M \sim 10$–100 (e.g., Bertschinger 1984; Zeldovich 1972; Miniati et al. 2000). As such, they were suggested as sites of effective cosmic ray acceleration, with consequences for the cluster energy budget and the cosmic gamma-ray background.

The second class of shocks are merger shocks: as subhalos fall into the main clusters, they create moderate-strength shocks that allow the gas to virialise. If an infalling subcluster

[1] $S(\nu) \propto \nu^{-\alpha}$.

has a deep enough gravitational potential to retain at least some of its gas when it enters the dense, X-ray bright region of the cluster into which it is falling, a merger shock is produced. In Sect. 2, we will discuss some prominent observations of this type of shocks.

The third class of shocks are AGN shocks: Powerful AGN are observed to inflate large bubbles in the ICM, which may generate shocks within the central few hundred kiloparsecs (Brüggen et al. 2007; Simionescu et al. 2009). Like merger shocks, these shocks have moderate Mach numbers, and have been detected in cluster centres, such as in Virgo and Hydra A. These shocks are very important because they convert the energy produced by the central supermassive black hole into thermal energy and, thus, complete the feedback loop which is regarded as essential for galaxy evolution (Cattaneo et al. 2009). As these shocks are primarily found in cluster centres, they will not be the subject of this review.

The setup of this review is as follows: In Sect. 2, we introduce some phenomena that we will cover in this review and explain some basic terms. In Sect. 3, we describe our current knowledge of cluster shocks from X-ray and γ-ray observations. Then we summarise what we can learn from radio observations in Sect. 4, including estimates of magnetic fields in cluster outskirts from recent observations of radio relics. In Sect. 5, we review recent results concerning structure formation shocks from cosmological simulations. In Sect. 6, we then discuss the origin of magnetic fields and relativistic particles in shocks. Finally, in Sect. 7, we summarise insights gained in the last years on the physics of shocks in such dilute environments as found in cluster outskirts and identify open questions.

2 Fundamentals

Shock Structure Shock waves are usually regarded as a sharp transition between a supersonic (and super-Alfvénic) upstream flow and a subsonic downstream flow. In a rarefied hot cosmic plasma the Coulomb collisions are not sufficient to provide the viscous dissipation of the incoming flow, and collective effects due to the plasma flow instabilities play a major role. Hence, cosmological shocks in a rarefied highly ionised plasma are collisionless. Protons, whose thermal velocity is lower than the shock velocity, are heated dissipatively at the shock layer. However, the equilibration of electrons may differ substantially from that of the ions because of their smaller gyroradii. The amount of collisionless heating of electrons depends on the shock velocity and the Mach number (see e.g. Schwartz et al. 1988; Bykov and Uvarov 1999). Then the Coulomb equilibration eventually occurs on scales that are orders of magnitude longer than the collisionless heating scale length (see e.g. Fox and Loeb 1997; Bykov et al. 2008b). Under such circumstances, the shock structure can be more complex because electrons and ions can remain out of equilibrium for a while. Then, such a shock consists of an ion viscous jump and an electron-ion thermal relaxation zone. We will revisit the microphysics of cluster shocks in Sect. 6.

Shock Acceleration Shocks can also change the momentum distribution of a gas by accelerating a fraction of the particles to very high energies. This has been observed in astrophysical shocks on a wide range of scales. Examples are the shock waves that form when the solar wind collides with planetary magnetospheres, shocks surrounding supernova remnants in the interstellar medium and shocks in active, extragalactic sources such as quasars and radio galaxies. The resulting energies reach up to about 100 keV in the shock in the Earth's magnetosphere to about 100 TeV in supernova remnants. Particles gain energy whenever they cross the shock front in either direction. Scattering processes make the velocity distribution isotropic with respect to the local rest frame on either side of the shock front. Consequently, there is a certain probability that a particle that has just gone through the shock and been accelerated will diffuse back through the shock to where it came from. Every time the particle crosses the shock it will gain energy. Moreover, the increase in energy is the same no matter from which direction the particle crosses the (non-relativistic) shock. Particle passing back and forth through the shock can thus attain very high energies. This process is called first-order Fermi (Fermi-I) acceleration or diffusive shock acceleration (DSA; Krymskii 1977; Axford et al. 1977; Bell 1978a, 1978b; Blandford and Ostriker 1978; Drury 1983; Blandford and Eichler 1987; Jones and Ellison 1991; Malkov and O'C Drury 2001). The cosmic rays thus generated can penetrate far into the upstream flow, thus creating a shock precursor. The particles decelerate the flow and preheat the gas. If the cosmic ray pressure inside the shock becomes comparable to the thermal pressure, the cosmic rays will also modify the structure of the shock. Finally, shocks can efficiently generate strong fluctuating magnetic fields in the upstream region due to cosmic-ray driven instabilities (Bell 2004; Bykov et al. 2011).

Cluster Magnetic Fields The origin of magnetic fields in galaxy clusters remains unclear (see e.g. Dolag et al. 2008, for a review). It has been suggested that they are primordial; i.e., a seed field that has formed prior to recombination is subsequently amplified by compression and turbulence. Alternatively, it has been proposed that the magnetic field is of proto-galactic origin (Kulsrud et al. 1997; Ryu et al. 2008) or that it had been produced by a cluster dynamo (Roettiger et al. 1999). Others conclude that magnetic fields can be produced efficiently in shocks by the Weibel instability (Medvedev et al. 2004) or by small-scale plasma instabilities (Schekochihin et al. 2005). Finally, it has been suggested that the intergalactic medium has been magnetised by bubbles of radio plasma that are ejected by active galactic nuclei (Furlanetto and Loeb 2001). Cluster magnetic fields are the subject of a separate chapter in this review and in this review we will focus on magnetic fields in cluster outskirts only.

Turbulence Mergers, the motions of galaxies and the action of AGN produce turbulence in the ICM. Turbulence may have a deep impact on the physics of the ICM (see e.g. Schekochihin et al. 2005; Lazarian 2006) and on the properties of non-thermal components in galaxy clusters (see e.g. Brunetti and Lazarian 2011). The presence of turbulent gas motions in the ICM is suggested by measurements of the Faraday Rotation of the polarization angle of the synchrotron emission from cluster radio galaxies. These studies show that the magnetic field in the ICM is tangled on a broad range of spatial scales (see e.g. Vogt and Enßlin 2005) suggesting the presence of super-Alfvénic motions in the medium. Also independent attempts from X-ray observations of a number of nearby clusters, based on pseudo-pressure maps of cluster cores and on the lack of evidence for resonant scattering effects in the X-ray spectra, provide hints of turbulence in the ICM (see e.g. Churazov et al. 2004). Important constraints on the fraction of the turbulent and thermal energy in the cores of clusters are based on the analysis of line broadening in the emitted X-ray spectra of cool core clusters (Sanders et al. 2010). In the outskirts of clusters, simulations predict that the turbulent pressure support increases relative to the thermal pressure support (Vazza et al. 2010; Paul et al. 2010; Nagai and Lau 2011).

3 Cluster Shocks in X- and γ-Ray Emission

3.1 Cluster Shocks in X-Rays

Regions of high-entropy gas in clusters have been observed by *ROSAT, ASCA, XMM-Newton* and *Chandra* and have been interpreted as the result of shock heating (e.g., Belsole et al.

2005). However, only a handful of shock fronts, exhibiting both a sharp gas density edge and an unambiguous temperature jump, have been found: in the "bullet cluster", 1E 0657-56 (Markevitch et al. 2002), Abell 520 (Markevitch et al. 2005), and Abell 2146 (Russell et al. 2010). Such discoveries are so rare because one has to observe a merger when the shock has not yet moved to the low-brightness outskirts, and is propagating nearly in the plane of the sky, to give us a clear view of the shock discontinuity. In addition, the shocks in A520 and 1E 0657-56 have Mach numbers of $M \sim 2$–3, which provide a big enough gas density jump to enable accurate deprojection. Merger shock fronts have been found in other clusters, e.g., in A3667 (Finoguenov et al. 2010), A754 (Krivonos et al. 2003), A521 (Giacintucci et al. 2008).

Recently, Finoguenov et al. (2010) analysed *XMM-Newton* and *Suzaku* pointings of the cluster Abell 3667, including the NW relic, see Fig. 1. They find a sharp X-ray surface brightness discontinuity at the outer edge of the radio relic, and a significant drop in the hardness of the X-ray emission at the same location. This discontinuity is consistent with a Mach number 2 shock moving at 1200 km/s associated with the merger. Kinetic energy is being dissipated in this shock at a rate of $dE_{\mathrm{KE}}/dt \sim 1.8 \times 10^{45}$ erg/s. The total non-thermal luminosity of the relic is $L_{\mathrm{NT}} \approx 3.8 \times 10^{42}[(3.6\ \mu\mathrm{G}/B)^2 + 1]\ \mathrm{erg\,s^{-1}}$. If this energy is provided by the shock acceleration of electrons, then a fraction $\epsilon \sim 0.2\%$ of the kinetic energy flux through the shock is converted to the non-thermal luminosity. Shock acceleration at the outer edge of the relic and radiative losses as the electrons are advected away from the shock can explain the rapid steepening of the radio spectral index with distance from this edge of the relic. Alternatively, the surface brightness discontinuity and hardening of the X-ray spectrum might be due to Inverse Compton (IC) emission from the relic. In this case, the magnetic field in the relic is about 3 μG. Since the observed X-ray excess from the relic is an upper limit to IC emission, this yields a lower limit on the relic magnetic field of $\geq$3 μG. This is a remarkably strong magnetic field at this large projected distance (2.2 Mpc) from the cluster center, but is consistent with Faraday rotation through the relic observed towards two background radio galaxies.

The bow shock in 1E 0657-56 offers a unique experimental setup to determine how long it takes for post-shock electrons to come into thermal equilibrium with protons in the ICM (Markevitch and Vikhlinin 2007). Currently, X-ray observations only yield the electron temperature but not the ion temperature. However, one can use the measured gas density jump at the shock front and the pre-shock electron temperature to predict the post-shock adiabatic and instant-equilibration electron temperatures, using the adiabatic and the Rankine-Hugoniot jump conditions, respectively, and compare them with the data. Furthermore, if the downstream velocity of the shocked gas flowing away from the shock is known, one can deduce how the flow spreads out the electron temperature along the spatial coordinate in the plane of the sky. The Mach number of the 1E 0657-56 shock is sufficiently high, such that the adiabatic and shock electron temperatures are sufficiently different (for $M \leq 2$, they become close and difficult to distinguish, given the temperature uncertainties). Moreover, the distance traveled by the post-shock gas during the time, is well-resolved by *Chandra*. Markevitch and Vikhlinin (2007) found that the temperatures are consistent with instant heating; equilibration on the collisional timescale is excluded, although with a relatively low 95% confidence. The equilibration timescale should be at least 5 times shorter than the Coulomb time. We will revisit the theory of dissipation in collisionless shocks in Sect. 6.

The low surface brightness of the cluster outskirts is challenging for current X-ray telescopes (e.g. Böhringer and Werner 2010; Ettori and Molendi 2010). Projected temperature

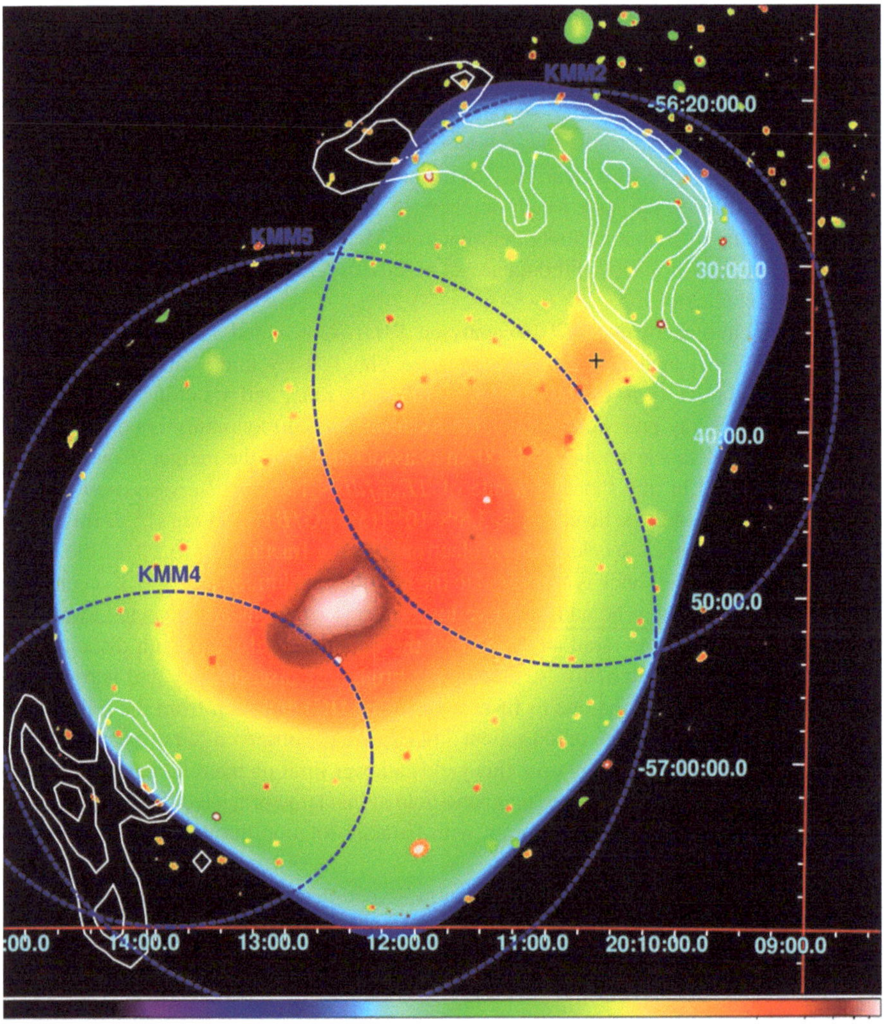

Fig. 1 Wavelet reconstruction of the 0.5–2 keV X-ray mosaic of A3667 from *XMM-Newton*. There is a sharp discontinuity in the X-ray surface brightness to the NW; the *blue ellipses* show the location, name and size of three main galaxy components (KMM5, KMM2, and KMM4) of A3667 from the analysis of Owers et al. (2009). The *black cross* marks the center of KMM2 group. The *white contours* show the wavelet reconstruction of the SUMMS 843 MHz radio image in the region of radio relics. The *contours* are drawn at the 1, 3, 9, and 20 mJy/beam levels. From Finoguenov et al. (2010)

profiles to $\sim 0.5 r_{200}$ of 15 nearby ($z < 0.2$) clusters were derived from *XMM-Newton* observations by Pratt et al. (2007) (see also Vikhlinin et al. 2005).

The low and well-constrained background levels make *Suzaku* the prime instrument for the study of X-ray emission from cluster outskirts. *Suzaku* is in a low orbit within Earth's magnetopause providing significantly lower and stable particle background compared to *Chandra* and *XMM-Newton*. However, the calibration of the instruments can sometimes be difficult. The low surface brightness outskirts of A2204 around r_{200} were studied by Reiprich et al. (2009) with *Suzaku*. Simionescu et al. (2011) determined ICM properties out to the edge of the Perseus Cluster. Contrary to earlier findings, the cluster baryon fraction is consistent with the expected universal value at half of the virial radius. The apparent baryon fraction exceeds the cosmic mean at larger radii, suggesting a clumpy distribution of the gas for radii larger than half of the virial radius. Also with *Suzaku*, George et al. (2009) studied X-ray emission from the outskirts of the cluster of galaxies PKS 0745-191 at $z = 0.1028$. They determined radial profiles of density, temperature, entropy, gas fraction, and mass beyond the virial radius out to $\sim 1.5 r_{200}$. The temperature is found to decrease by roughly 70% from 0.3–1 r_{200}. *Suzaku* observations provide evidence for departure from hydrostatic equilibrium around and even before the virial radius as it is seen in the galaxy cluster Abell 1795 (e.g. Bautz et al. 2009). *Suzaku* observations of PKS 0745-191 revealed one shock front possibly indicating that this cluster is still accreting material, likely along a filament. Evidence for nonthermal pressure support suggests that bulk motions from merger activity could be making a significant contribution to the gas energy in the outskirts of this cluster (George et al. 2009). There is some evidence for the presence of an excess above the dominating thermal emission of the ICM hot gas in both X-rays and EUV (Rephaeli et al. 2008; Durret et al. 2008; Million and Allen 2009). However, a recent search by Ajello et al. (2009) of hard X-ray excess above the thermal emission in a large sample of clusters with *Swift Burst Alert Telescope* does not show significant nonthermal hard X-ray emission. The only exception is Perseus whose high-energy emission Ajello et al. (2009) attributed to the central galaxy NGC 1275. In the next years, as more cluster outskirts are being observed with *Suzaku*, more conclusive results on shocks and nonthermal components are expected.

Compared to X-ray observations, the outer regions of clusters can potentially be better studied by Sunyaev-Zel'dovich (SZ) observations because the SZ effect depends only on the electron density to the first power, while the X-ray emission depends on the electron density squared. Facilities such as the Atacama Cosmology Telescope (ACT), the South Pole Telescope (SPT) and the Planck satellite are searching for the SZ signal of galaxy clusters and have found some interesting first results. For example, it was found that the SZ signal caused by clusters is by a factor of ~ 2 smaller than predicted by models of clusters that suggest that the pressure by the electrons has been overpredicted Lueker et al. (2010). This is presumably caused by a substantial nonthermal pressure at cluster outskirts, most of it is likely to be turbulent pressure (see e.g. Shaw et al. 2010). This picture is also supported by simulations (Vazza et al. 2010; Paul et al. 2010; Nagai and Lau 2011) which we review in Sect. 5.

Given the advancements in the X-ray and SZ observations of the cluster outer regions, as well as the growing evidence of missing thermal energy in the ICM and the possible negative implications for cosmological tests, a more detailed study of the kinetic processes in cluster envelopes is necessary.

3.2 γ-Ray Observations

Recently, Ackermann et al. (2010) reported on the search for GeV emission from clusters of galaxies using the *Large Area Telescope* on the *Fermi Gamma-ray Space Telescope*. Only upper limits on the photon flux were reported in the range 0.2–100 GeV toward a sample of around 30 observed clusters (typical values (1–5) $\times 10^{-9}$ ph cm^{-2} s^{-1}), considering both point-like and spatially resolved models for the high-energy emission. The authors concluded that the volume-averaged relativistic hadron-to-thermal energy density ratio is below 5–10% in several clusters. Also, using *High Energy Stereoscopic System* (*H.E.S.S.*) observations of Coma cluster, Aharonian et al. (2009) placed an upper limit $\sim 10^{-13}$ ph cm^{-2} s^{-1}

at photon energies above 5 TeV for the Coma core of 0.2 degrees in radius constraining the multi-TeV particle population in the cluster. In summary, at this time the γ-ray observations remain inconclusive with respect to the origin of diffuse radio emission in clusters, but this is bound to change soon as the upper limits are expected to decrease.

4 Cluster Shocks in Radio Emission

A number of diffuse, steep-spectrum radio sources without optical identification have been observed in galaxy clusters. The emission from these sources is synchrotron radiation, which indicates the presence of highly relativistic electrons and μG range magnetic fields. Especially radio relics are thought to trace cosmological shocks. Radio relics can be divided into two main groups (Kempner et al. 2004). *Radio gischt* are large elongated, often Mpc-sized, radio sources located in the periphery of merging clusters. They probably trace shock fronts in which particles are accelerated via the diffusive shock acceleration mechanism. Among them are double-relics with the two relics located on both sides of a cluster center (e.g., Bonafede et al. 2009; van Weeren et al. 2009b; Venturi et al. 2007; Bagchi et al. 2006; Röttgering et al. 1997). According to DSA, the integrated radio spectrum should follow a single power-law. *Radio phoenices* are related to radio-loud AGN. Fossil radio plasma from a previous episode of AGN activity is thought to be compressed by a merger shock wave which boosts, both, the magnetic field inside the plasma as well as the momenta of the relativistic particles. As a result the radio plasma brightens in synchrotron emission. In contrast to the radio gischt, the phoenices have a steep curved spectrum[2] indicating an aged population of electrons.

The sizes of relics and the distance to the cluster centre vary significantly. Examples for radio relics with sizes of 1 Mpc or even larger have been observed in Coma (the prototype relic source 1253 + 275 (Giovannini et al. 1991), Abell 2255 (Feretti et al. 1997) and Abell 2256 (Röttgering et al. 1994), which contain both a relic and a halo (as do Abell 225, Abell 521, Abell 754, Abell 1300, Abell 2255 and Abell 2744). The cluster Abell 3667 contains two very luminous, almost symmetric relics with a separation of more than 5 Mpc (Röttgering et al. 1997), as does ZwCl 2341.1 + 0000 (van Weeren et al. 2009c; Giovannini et al. 2010), and Abell 2345 and Abell 1240 (Bonafede et al. 2009). The clusters A115 and A1664 show relics only at one side of the elongated X-ray distribution (Govoni et al. 2001). Recently, another double radio relic was found in the galaxy cluster ZwCl 0008.8 + 5215 (van Weeren et al. 2010) which we will discuss in detail below.

The relic with the best evidence for shock acceleration found to date is located in the northern outskirts of the merging galaxy cluster CIZA J2242.8 + 5301 ($z = 0.1921$), see Fig. 2. The relic is located at a distance of 1.5 Mpc from the cluster center and spans 2 Mpc in length. The relic shows a clear unambiguous spectral index gradient towards the cluster center. The spectral index, measured between 2.3 and 0.61 GHz, steepens from −0.6 to −2.0 across the width of the narrow relic. The gradient is visible over the entire 2 Mpc length of the relic, something that has never been observed before. This is a crucial observation as it constitutes clear evidence for diffusive shock acceleration and spectral ageing of relativistic electrons in an outward moving shock. The relic is strongly polarized at the 50–60% level, indicating an ordered magnetic field, and polarization magnetic field vectors are aligned with the relic (see Fig. 3). In the southern part of the cluster a second fainter broader relic

[2] $F_\nu \propto \nu^\alpha$, with α the spectral index.

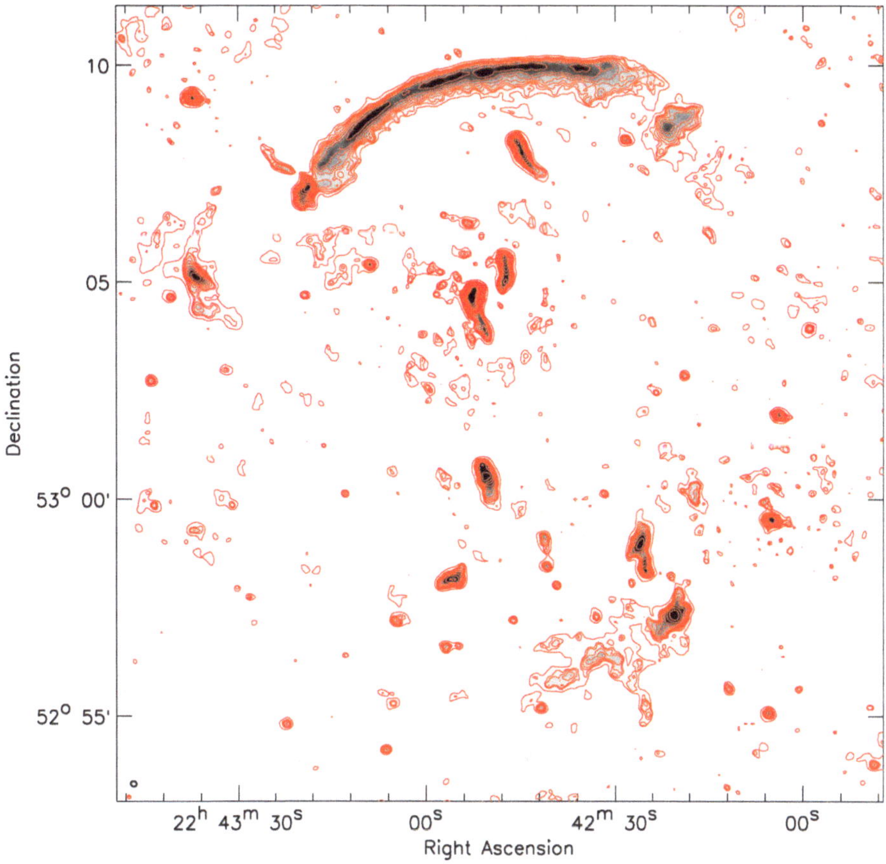

Fig. 2 Radio map (GMRT 325 MHz) of the relic in CIZA J2242.8 + 5301. From van Weeren et al. (2010)

is found and the elongated radio relics are orientated perpendicular to the major axis of the cluster's elongated ICM. This is exactly as expected for a binary cluster merger event in which this second southern relic traces the shock wave that travels in the opposite direction from the first one. Furthermore, a faint halo of diffuse radio emission is seen extending all the way towards the cluster center connecting the two radio relics. This emission extends over 3.1 Mpc, making it by far the largest known diffuse radio source in a cluster to date.

The spectral index at the front of the relic is $\alpha = -0.6 \pm 0.05$. In simple shock acceleration theory, α is related to the Mach number via $\alpha = -(3 + M^2)/(2M^2 - 1)$ (Rosswog and Brüggen 2007), which yields a Mach number of $4.6^{+1.3}_{-0.9}$. Subsequent hydrodynamic simulations of this shock front have indicated slightly lower Mach number of around 3. Using the $L_X - T$ scaling relation for clusters (Markevitch 1998), we estimate the average temperature of the ICM to be ~9 keV. Behind the shock front the temperature is likely to be higher. Using the redshift, downstream velocity, spectral index, and characteristic synchrotron timescale, the width of the relic is given by

$$l_{\text{relic}} \approx 110\ \text{kpc} \times \frac{B^{1/2} B_{\text{CMB}}^{3/2}}{B^2 + B_{\text{CMB}}^2} \tag{1}$$

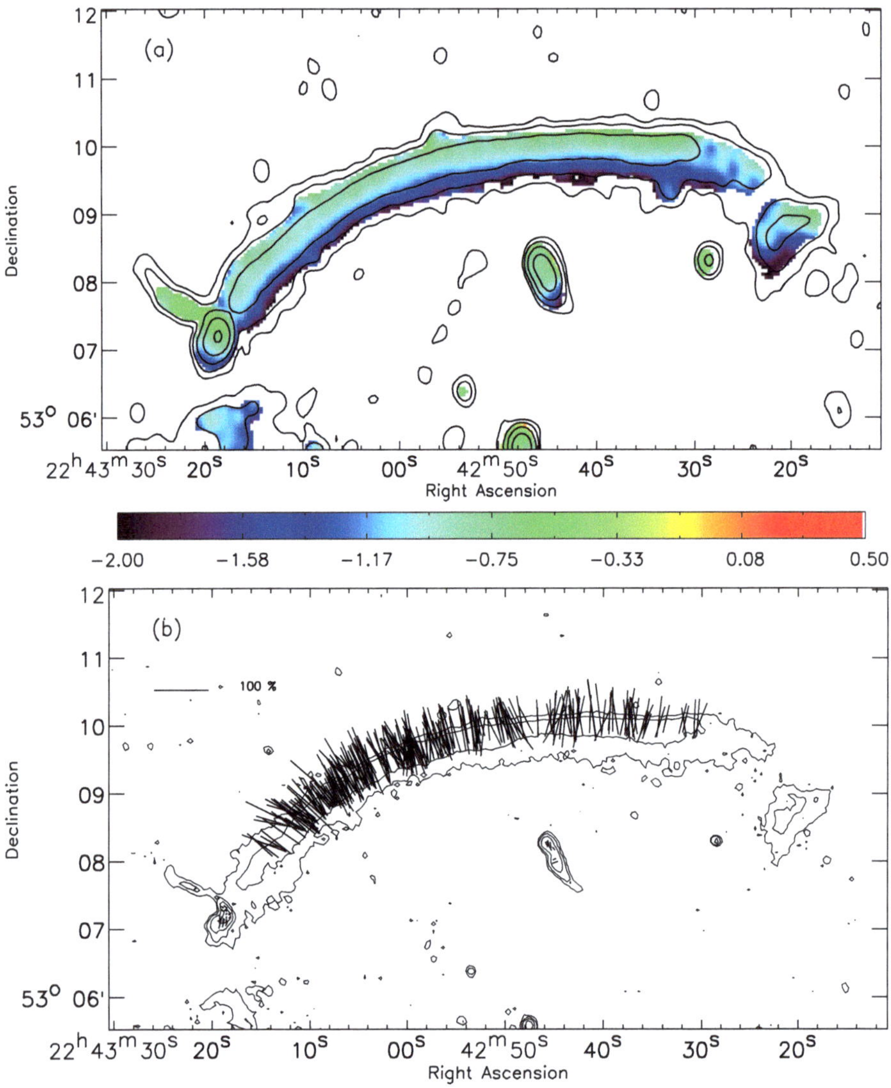

Fig. 3 Radio spectral index and polarization maps of the relic in CIZA J2242.8 + 5301. *Top*: The spectral index was determined using matched observations at 2.3, 1.7, 1.4, 1.2, and 0.61 GHz, fitting a power-law radio spectrum to the flux density measurements. *Bottom*: The polarization electric field vector map was obtained with the VLA at a frequency of 4.9 GHz. The length of the vectors is proportional the polarization fraction, which is the ratio between the total intensity and total polarized intensity. A reference vector for 100% polarisation is drawn in the *top left* corner. From (van Weeren et al. 2010)

for a relic seen edge-on without any projection effects. $B_{\rm CMB}$ is the equivalent field strength of the IC scattering from the Cosmic Microwave Background, which is known, and thus the measurement of $l_{\rm relic}$ from the radio maps, directly constrains the magnetic field. From the 610 MHz image (the image with the best signal to noise ratio and highest angular resolution), the relic has a deconvolved width (FWHM) of 55 kpc. Taking into account projection effects, van Weeren et al. (2010) conclude that the magnetic field strength at the location of the bright

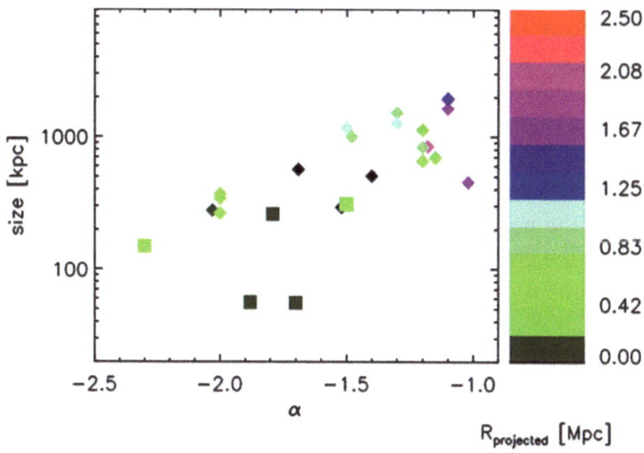

Fig. 4 Spectral index of radio relics versus their size. *Squares* are the proposed radio phoenices. *Diamonds* represent the radio relics tracing merger shocks where particles are being accelerated by shock acceleration. The *color coding* is according to the projected distance from the cluster center. For the relics represented by *black symbols* we could not obtain a reliable projected distance to the cluster center. From van Weeren et al. (2009a)

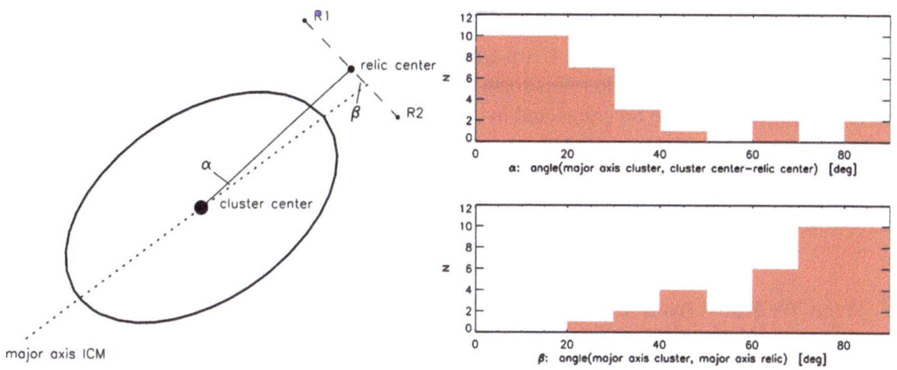

Fig. 5 *Left*: Schematic illustration of the angle between the major axis of the ICM and the line relic center-cluster center (α), and the angle between the relic orientation and major axis of the ICM (β). *Top Right*: Histogram of angles between the major axis of the X-ray emission and the line connecting the cluster center with the center of the relic. *Bottom Right*: Histogram of angles between the major axis of the X-ray emission and the relic's major axis. From van Weeren et al. (2011)

radio relic lies between 5 and 7 μG. As in A3667, this measurement points towards μG-scale fields in merger shocks found at large distances from the cluster centre.

Even though the sample of known radio relics is still small, one can start to find correlations between size, location and spectral index of these unique sources, which can eventually be compared to simulations of relic formation. The spectral index of the radio relics, versus the physical size is shown in Fig. 4. The projected distance from the cluster center is color coded. van Weeren et al. (2010) find that on average the smaller relics have steeper spectra. Such a trend is in line with predictions from shock statistics derived from cosmological simulations (Skillman et al. 2008; Battaglia et al. 2009; Hoeft et al. 2008). They find that larger shock waves occur mainly in lower-density regions and have larger Mach numbers, and consequently shallower spectra. Conversely, smaller shock waves are more likely to be found in cluster centers and have lower Mach numbers and steeper spectra. We note that more spectral index measurements of high quality are needed to confirm the correlation between physical size and spectral index.

A further confirmation of the binary merger origin of most radio relics comes from measuring the angle of orientation of the relic's main axis with respect to the major axis of the X-ray surface brightness map of the host cluster. We have computed the angle α between the major axis of the ICM and the line cluster center-radio relic center. The resulting histogram is shown in Fig. 5. From this histogram we see that relics are preferably found along the major axis of the ICM. This is in line with the simple picture that shock waves propagate outwards along the merger axis. We also calculate the angle β between the major axis of the ICM and the relic's major axis. Figure 18 shows that most relics are oriented perpendicular to the ICM major axis, also in agreement with a shock origin for radio relics.

In the next years, simulations should be able to reproduce these trends, and thus constrain the efficiencies of shock acceleration and the magnetic field evolution in clusters. At the same time, it is predicted that *LOFAR* will find of around 100 new radio relics (Hoeft et al. 2008). This will help address some remaining puzzles that surround radio relics, most importantly:

- Which processes accelerate electrons so efficiently at relatively low Mach number ($M \sim$ 2–4) shocks?
- What produces the magnetic fields inside relics? Both the inferred field strengths as well as the observed polarisation of the radio emission need to be explained.
- Why do some relics have very sharp edges while others appear very fuzzy?
- Under which conditions do relics form? When do we see single and when double relics?
- Some relics appear to be connected to cluster-wide radio halo emission. Is there a physical connection between the two?

In the next section, we will review the current state of cosmological simulations that include a treatment for shock acceleration of cosmic rays.

5 What We Know from Simulations

5.1 Shocks in Cosmological Simulations

The statistics of cosmological shocks in the large-scale structure of the Universe has been derived from simulations using, both, Eulerian hydrodynamic codes (e.g. Miniati et al. 2000, 2001; Ryu et al. 2003; Kang et al. 2007; Skillman et al. 2008; Vazza et al. 2009a, 2009b, 2010; Molnar et al. 2009) and smoothed particle hydrodynamic codes (Pfrommer et al. 2007; Hoeft et al. 2008) as well as a number of semi-analytical (Gabici and Blasi 2003; Keshet et al. 2003) studies.

Ryu et al. (2003) performed uniform grid simulations with limited resolution, in which cluster core regions were not properly resolved, while surrounding outskirts were fairly well reproduced (see Fig. 6). Hence, the shocks identified in these simulations are mostly those in cluster outskirts. A cell is determined to have a shock if it meets the following requirements: (1) $\nabla \cdot \mathbf{v} < 0$, (2) $\nabla T \cdot \nabla s > 0$, (3) $T_2 > T_1$ and (4) $\rho_2 > \rho_1$, where $\mathbf{v}$ is the velocity field, T is the temperature, ρ is the density, and $s = T/\rho^{\gamma-1}$ is, in X-ray astronomy parlance, the entropy. The subscripts 1 and 2 denote up- and downstream quantities, respectively. To find shocks, one loops through rows of cells along each of the coordinate axes and identifies one-dimensional shock structures in each direction. Figure 7 shows the surface area, S, of identified shocks with the preshock gas temperature $T_1 > 10^7$ K, normalized by the entire volume of the simulation box, at the present epoch as a function of shock Mach number. The quantity S provides a measure of shock frequency. The Mach number of shocks expected to be found in cluster outskirts is low with $M \lesssim 3$. The frequency increases to weakest possible shocks with $M \sim 1$.

Vazza et al. (2009b) identifies shocks in cosmological simulations using a method based on velocities (instead of temperatures). They found that the overall differential distribution

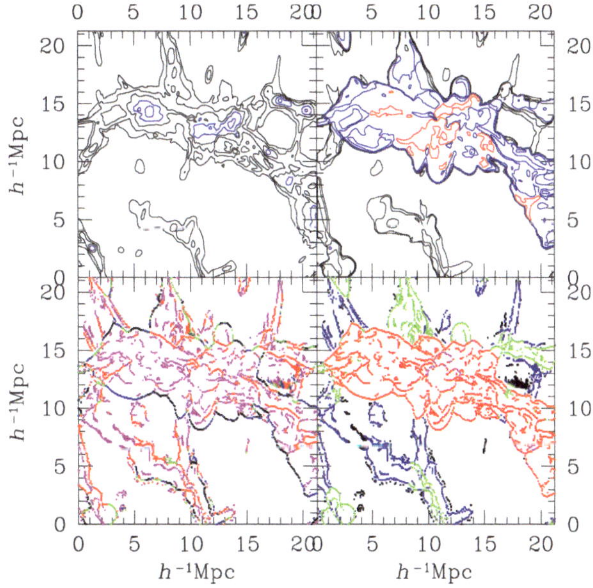

Fig. 6 Two-dimensional slice around a region containing two clusters/groups at $z = 0$. *Top panels* show the distribution of gas density (*left*) and temperature (*right*). *Bottom panels* show the locations of the shocks, which are *color-coded* according to shock Mach number M (*left*) or shock speed v_s (*right*). In the *left panel*, the *colors* are coded as follows: *black* for $M > 100$, *blue* for $30 < M < 100$, *green* for $10 < M < 30$, *red* for $3 < M < 10$, and *magenta* for $M < 3$. In the *right panel*, the *colors* are coded as follows: *black* for $v_s < 15\ \mathrm{km\,s^{-1}}$, *blue* for $15 < v_s < 65\ \mathrm{km\,s^{-1}}$, *green* for $65 < v_s < 250\ \mathrm{km\,s^{-1}}$, *red* for $250 < v_s < 1000\ \mathrm{km\,s^{-1}}$, and *magenta* for $v_s < 1000\ \mathrm{km\,s^{-1}}$

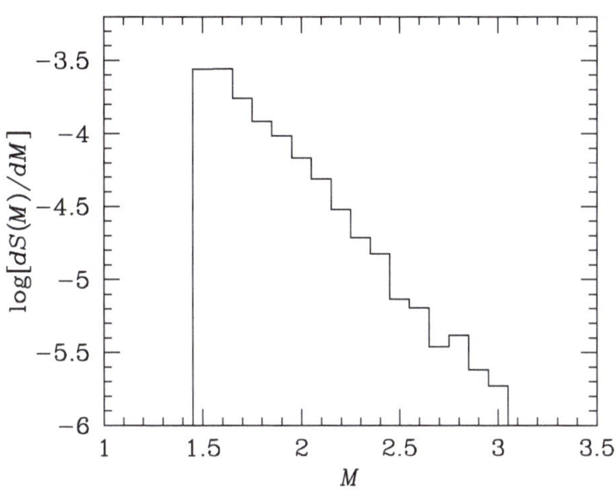

Fig. 7 Frequency of shocks expected to be found at $z = 0$ in cluster outskirts as a function of M. Figure from D. Ryu

of shocks with their Mach number in the cosmic volume is very steep, with $\beta \sim -1.6$ (with $MdN/dM \propto M^{\beta}$), and the bulk of the detected shocks at any Mach number is found in low-density regions. The Mach number distribution of detected shocks becomes increasingly steeper with ambient density: $\beta \sim -3$ to -4 is found in clusters and their outskirts. They find that before the epoch of re-ionization, $z > 6$, roughly 30% of the simulated volume is shocked. Then as soon as re-ionization plays a role, the temperature of the simulated volume increases and the Mach number distribution of shocks at redshift $z \sim 3$–6 undergoes a dramatic change becoming very steep and dominated by weak shocks. The bulk of the energy in their simulations is dissipated in galaxy clusters which contribute about 75% of the total energy dissipation (about 80% if the contribution from cluster outskirts is included), while filaments contribute about 15% of the total energy dissipation.

In line with previous numerical studies, relatively weak shocks are found to dominate the process of energy dissipation in the simulated cosmic volume, although they find a larger ratio between weak and strong shocks with respect to previous studies. The bulk of energy is dissipated at shocks with Mach number $M \sim 2$ and the fraction of strong shocks decreases with increasing density of the cosmic environments.

Skillman et al. (2008) improved the method employed by Ryu et al. (2003) that can produce errors when examining shocks whose direction of motion is not oriented along a coordinate axis. They find that the Mach number evolution can be interpreted as a method to visualize large-scale structure formation. Shocks with $M < 5$ typically trace mergers and complex flows, while $5 < M < 20$ and $M > 20$ generally follow accretion onto filaments and galaxy clusters, respectively.

5.2 Cosmological Simulations with Particle Acceleration

In recent years a number of cosmological simulations have started to include some treatment of cosmic ray physics with particle acceleration at shocks.

Using a shock-finding formalism tuned to smoothed-particle hydrodynamics simulations, Hoeft et al. (2008) have analyzed the MareNostrum Universe simulation which has 2×1024^3 particles in a 500 h^{-1} Mpc box. In addition, they have used the formalism derived in Hoeft and Brüggen (2007) to produce artificial radio maps of massive clusters. Several clusters were found to show radio objects with similar morphologies to observed large-scale radio relics (see Fig. 8), whereas about half of the clusters show only very little radio emission. In agreement with observational findings, the maximum diffuse radio emission of galaxy clusters depends strongly on their X-ray temperature. Moreover, it was found that the so-called accretion shocks cause only very little radio emission.

Using a particle-mesh and Eulerian hydrodynamics code, Ryu et al. (2003) have been studying the gas thermalization and CR acceleration efficiencies of shocks in cosmological grid simulations. In order to quantify the energy dissipation at cosmological shocks, the incident shock kinetic energy flux, $F_{\rm kin} = (1/2)\rho_1 v_s^3$ where ρ_1 is the preshock gas density, and v_s the shock velocity, was calculated for each shock, and the results are shown in Fig. 7.

Then, the kinetic energy flux through shock surfaces, normalized by the entire volume of the simulation box, as a function of shock Mach number was calculated. The resulting kinetic energy flux is shown by the dotted line in Fig. 9. The kinetic energy flux through shock surfaces is larger for weaker shocks; energetically weaker shocks are more important processing more shock energy. The acceleration efficiency for cosmic rays, η, depends on the Mach number and the injection parameter, ϵ_B, which is defined as $\epsilon_B \equiv B_0/B_\perp$, the ratio of the mean magnetic field strength aligned with the shock normal to the amplitude of the post-shock turbulent wave field. In addition, it is assumed that the CR population is isotropized with respect to the local Alfvénic wave turbulence, which would in general drift upstream at the Alfvén speed with respect to the bulk plasma. This reduces the velocity difference between upstream and downstream scattering centers compared to the bulk flow, leading to less efficient shock acceleration. Moreover, the dissipation of Alfvén turbulence heats the inflowing plasma in the precursor, which leads to weakening of the subshock strength. (See Kang and Jones 2007 for further discussions and references.)

Figure 10 shows the gas thermalization and CR acceleration efficiencies, defined as $\delta(M) \equiv F_{\rm th}/F_{\rm kin}$ and $\eta(M) \equiv F_{\rm CR}/F_{\rm kin}$, respectively, where $F_{\rm th}$ and $F_{\rm CR}$ are the gas thermal and CR energy flux, respectively, generated at shocks; they are from numerical simulations of quasi-parallel shocks with speed $v_s = M \cdot 150$ km s^{-1} ($T_1 = 10^6$ K), and based on a shock acceleration model of Kang and Jones (2007) with $\epsilon_B = 0.25$. At strong shocks with

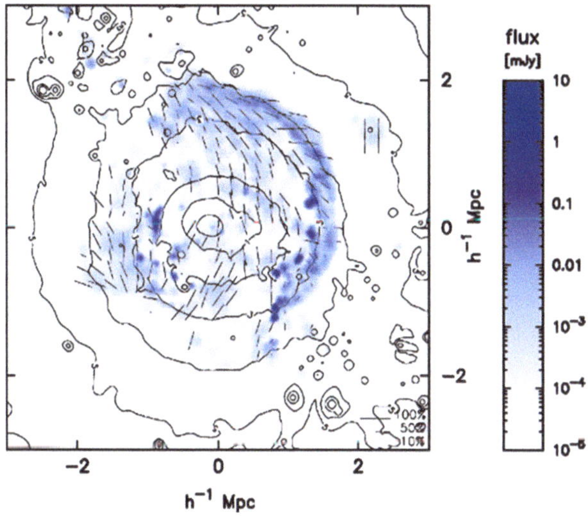

Fig. 8 Radio emission from a galaxy cluster taken from the Mare Nostrum simulation and resimulated at higher resolution. The *vectors* denote the degree and the direction of polarisation of the radio emission. The *contours* indicate the X-ray emission. From M. Hoeft

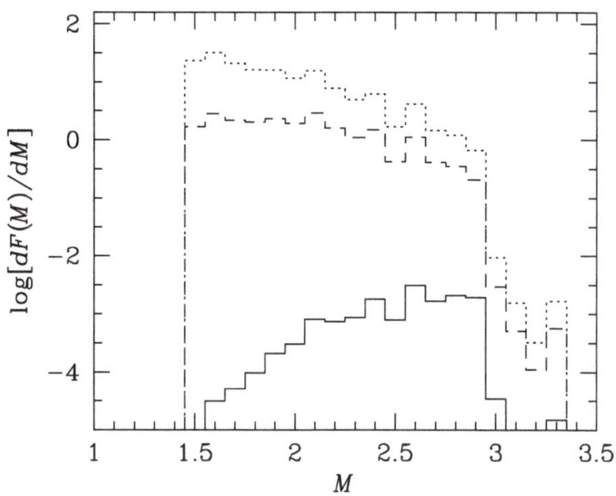

Fig. 9 *Dotted line*: Kinetic energy flux per comoving volume passing through surfaces of shocks in cluster outskirts as a function of M at $z = 0$. *Dashed line*: Gas thermal energy dissipated at the shocks as a function of M. *Solid line*: CR energy accelerated at the shocks as a function of M. All the quantities are in units of 10^{40} ergs $s^{-1}(h^{-1}\,\mathrm{Mpc})^{-3}$

$M \gtrsim 10$, the injection rate is high enough so that the CR acceleration efficiency nearly saturates and becomes almost independent of the parameter ϵ_B. At weak shocks, on the other hand, the level of injection lies in the regime where the CR acceleration efficiency increases with the injection rate; consequently, η depends sensitively on ϵ_B. Note that the value of η presented here is $\sim 1/2$ of that presented in Kang et al. (2007) for $M \lesssim 5$, while it remains about the same for stronger shocks.

By adopting the efficiencies in Fig. 10, the gas thermal and CR energy fluxes dissipated at cosmological shocks as a function of shock Mach number were calculated in the same way the kinetic energy flux through shock surfaces was calculated. The resulting thermal and CR energy fluxes are shown with dashed and solid lines in Fig. 9. For the generation of thermal energy, still weaker shocks are more important; shocks with $M \lesssim 2.5$ contribute the most to the thermal energy. On the other hand, the generation of CR energy is more efficient at shocks with $M \gtrsim 2$. These indicate that in cluster outskirts, while shocks with $M \lesssim 2$ are more abundant, shocks with $M \gtrsim 2$ would be more frequently detected as radio relics. We

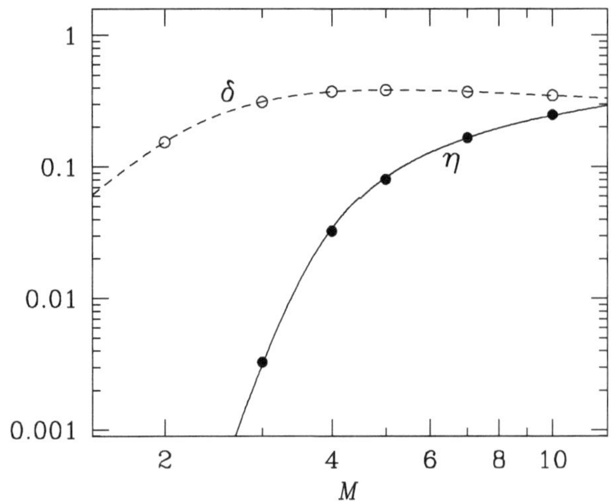

Fig. 10 Gas thermalization efficiency, $\delta(M)$, and CR acceleration efficiency, $\eta(M)$, as a function of Mach number. *Open* and *filled circles* are the values estimated from numerical simulations based on a shock acceleration model and dashed and solid lines are the fits

note that the results in Fig. 9 are consistent with those estimated for cosmological shocks in all the regions of the IGM; while shocks around $M \sim 2$ contribute most to the generation of gas thermal energy, shocks around $M \sim 3$ contribute most to the generation of CR energy (Ryu et al. 2003; Kang et al. 2007).

At weak shocks, DSA is known to be rather inefficient and the CR pressure remains dynamically insignificant, partly because the injection from thermal to nonthermal particles is inefficient (e.g., Kang et al. 2002). Recently, Kang and Ryu (2010) found that at weak shocks expected to form in clusters much less than 10^{-3} of particles are injected into CRs and much less than $\sim 1\%$ of the shock ram pressure is converted into the downstream pressure of CR protons, so the particle acceleration is virtually negligible.

This has been followed up in recent AMR simulations by Vazza (2011, in preparation) that implement prescriptions for CR acceleration at shocks. This work suggests that the shock acceleration efficiency needs to be increased in order to explain the occurrence of radio relics. Kang et al. (2007) have already suggested that pre-existing populations of cosmic rays can boost the acceleration efficiency significantly, and this may be part of the answer. Thus the observations of cluster radio relics can inform us about the microphysics in weak shocks. This will be the subject of the next section.

6 Microphysics of Shocks

6.1 Plasma Physics in CR-Modified Shocks

The heating processes in collisionless shocks are complex phenomena. The *irreversible* transformation of a part of the kinetic energy of the ordered bulk motion of the upstream flow into the energy of the random motions of plasma particles in the downstream flow in collisionless shocks is due to the plasma instabilities in the thin dissipation region where the wave-particle interactions provide momentum and energy redistribution between different components.

The front of a strong collisionless shock wave may consist of a precursor and a viscous velocity discontinuity (subshock) of a local Mach number that is smaller than the total Mach number of the shock wave (Fig. 11). The compression of matter at the subshock can be much lower than the total compression of the medium in the shock wave allowing for a high compression in the precursor. We refer to such shocks as CR-modified. The standard collisional model of a strong single-fluid shock predicts a particle temperature $kT \rightarrow (3/16) \cdot m v_{\rm sh}^2$ for $\gamma_g = 5/3$. However, this is not generally valid in multi-fluid plasma and CR-modified shocks since the total shock compression ratio $R_{\rm t}(v_{\rm sh})$ depends on the shock velocity, $v_{\rm sh}$, in the strong shock limit.

In a simplified steady-state model, a strong CR-modified shock can be parameterized by the total Mach number of the shock $\mathcal{M}_{\rm tot}$ and the Mach number of the subshock $\mathcal{M}_{\rm sub}$. Then the downstream ion temperature $T_{\rm i}^{(2)}$ can be estimated for the CR modified shock of a given velocity $v_{\rm sh}$ if the total compression ratio, $R_{\rm t}$, is known:

$$T_{\rm i}^{(2)} \approx \phi(\mathcal{M}_{\rm sub}) \cdot \frac{\mu\, v_{\rm sh}^2}{\gamma_{\rm g} R_{\rm t}^2(v_{\rm sh})}, \quad \text{where } \phi(\mathcal{M}_{\rm sub}) = \frac{2\gamma_{\rm g}\mathcal{M}_{\rm sub}^2 - (\gamma_{\rm g} - 1)}{(\gamma_{\rm g} - 1)\mathcal{M}_{\rm sub}^2 + 2}. \tag{2}$$

The total compression ratio, $R_{\rm t}$, in (2) depends on the precursor heating rate and it is the main parameter in the steady state CR modified shock to determine the postshock ion temperature $T_{\rm i}^{(2)}$ (Bykov 2005; Bykov et al. 2008a).

Shocks may also generate turbulence through cosmic-ray driven instabilities. In the shock precursor, this turbulence is certain to play a critical role in non-linear models of strong, CR-modified shocks. Different non-linear approaches to the modeling of the large-scale structure of a shock undergoing efficient cosmic ray acceleration (e.g., Bell and Lucek 2001; Amato and Blasi 2006; Vladimirov et al. 2006, 2008; Bell 2004; Zirakashvili et al. 2008; Kang et al. 2009) have been proposed. Many of these have predicted the presence of strong MHD turbulence in the shock precursor. An exact modeling of the shock structure in a turbulent medium, including nonthermal particle injection and acceleration, requires a nonperturbative, self-consistent description of a multi-component and multi-scale system including the strong MHD-turbulence dynamics.

A Monte-Carlo model of nonlinear diffusive shock acceleration accounting for magnetic field amplification through resonant instabilities induced by accelerated particles, and including the effects of dissipation of turbulence upstream of a shock and the subsequent precursor plasma heating was studied by Vladimirov et al. (2008). Feedback effects between the plasma heating due to turbulence dissipation and particle injection are strong, adding to the nonlinear nature of efficient shock acceleration.

Describing the turbulence damping in a parametrized way, Vladimirov et al. (2008) reach two important results: first, for conditions typical of supernova remnant shocks, even a small amount of dissipated turbulence energy ($\sim$10%) is sufficient to significantly heat the precursor plasma, and second, the heating upstream of the shock leads to an increase in the injection of thermal particles at the subshock by a factor of a few. In their results, the response of the non-linear shock structure to the boost in particle injection prevented the efficiency of particle acceleration and magnetic field amplification from increasing. The non-linear models with turbulence generation and dissipation may lead to a scenario in which particle injection boost due to turbulence dissipation results in more efficient acceleration and even stronger amplified magnetic fields than without dissipation. In Fig. 11 the effects are illustrated for a stationary, non-linear Monte-Carlo simulation of a forward shock in a supernova remnant of velocity of 5,000 km/s accelerating particles up to maximal energies of about 10^5 GeV (Vladimirov et al. 2008). The shock velocity is higher than is expected for cluster accretion shocks but the simulation can be used to investigate the qualitative behavior and to obtain the appropriate scalings. The maximal energy depends on the ratio of the shock scale size and particle escape physics that is simply modeled by a free escape boundary in

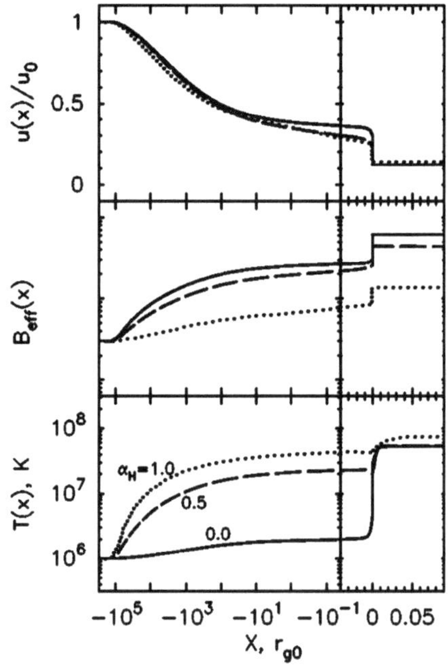

Fig. 11 Results of non-linear simulations in the case of the far upstream temperature $T = 10^6$ K with different values of turbulence dissipation parameter α_H taken from Vladimirov et al. (2008). The *solid*, *dashed* and *dotted lines* correspond, respectively, to $\alpha_H = 0$, 0.5 and 1.0. The plotted quantities are the bulk flow speed $u(x)$, the effective amplified magnetic field $B_{\text{eff}}(x)$ and the thermal gas temperature $T(x)$. The shock is located at $x = 0$. There is a change from the logarithmic to the linear scale at $x = -0.05$. The distances are measured in relativistic proton gyroradius unit r_{g0}

the Monte Carlo simulation. Ultra-high energy CR acceleration above 10^9 GeV by cluster accretion shocks was shown by Norman et al. (1995) and Kang et al. (1996) to be a plausible scenario. In that case the scale size of CR precursor with magnetic field amplification can be about 100 kpc.

Vladimirov et al. (2008) demonstrated that the effect of turbulent dissipation on the thermal plasma is evident in the values of the pre-subshock temperature T_1, the downstream temperature T_2, and the volume-averaged precursor temperature $\langle T(x < 0) >\rangle$ (the averaging takes place over the CR precursor region). The temperatures were calculated from the thermal particle pressure using the ideal gas law. The value of the pre-subshock temperature T_1 depends drastically on the level of the turbulent dissipation α_H, increasing from $\alpha_H = 0$ to $\alpha_H = 0.5$ by a factor of 11 in the case of $T = 10^6$ K. It is less in the case of $T \sim 10^6$ K than for $T \sim 10^4$ K because the efficiency of the CR streaming instability in generating the magnetic turbulence is less for the smaller Mach number. The values of the temperature as high as T_1 are achieved upstream only near the subshock; the volume-averaged upstream temperature, $\langle T(x < 0)\rangle$, is significantly lower. The factor by which the average temperature $\langle T(x < 0)\rangle$ increases in case of $T = 10^6$ K is about 2.3. The multi-fluid processes described above can preheat the accreting gas. An important prediction of the models of strong shocks is the possibility to amplify an initial seed magnetic field by a few orders of magnitude. CR currents and CR pressure gradients upstream of the strong shock can drive magnetic fluctuations on the shock precursor scale length. The precursor scale size is 10^9 times larger than the subshock transition region where strong, small-scale magnetic field fluctuations are directly produced by instabilities of super-Alfvénic bulk plasma flows. These small-scale fluctuations are responsible for bulk dissipation and the adiabatic amplification of the transverse magnetic field in collisionless shocks. From Fig. 11, it is clear that the magnetic field amplification factor depends on the non-linear processes of particle injection through turbulent dissipation in the shock precursor.

Above, we discussed the ion temperature in multi-fluid shocks. However, cluster X-ray spectra depend primarily on the electron temperature, while MHD-type wave dissipation in the shock precursor may preferentially heat ions. However, the complete $e-i$ Coulomb equilibration requires the system age $\tau_{\mathrm{ei}} \gtrsim 10^{10}\ T_6^{3/2}/n$, where the postshock density n is measured in cm^{-3}, the ion temperature T_6 is in 10^6 K, and τ_{ei} is in seconds Mewe (1990). Since the number density around the cluster virial radius is typically above $10^{-5}\ \mathrm{cm}^{-3}$, the temperature equilibration scale length in the accretion shock precursor is below 100 kpc. The electron temperature downstream of the shock depends on collisionless electron heating e.g. Bykov and Uvarov (1999), Rakowski et al. (2008), Bykov et al. (2008b). Non-resonant interactions of the electrons with strong nonlinear fluctuations generated by kinetic instabilities of the ions in the transition region inside the shock front may play the main role in the heating and pre-acceleration of the electrons, as it was shown in the model by Bykov and Uvarov (1999). They calculated the electron energy spectrum in the vicinity of shock waves and showed that the heating and pre-acceleration of the electrons occur on a scale of the order of several hundred ion inertial lengths l_i in the vicinity of a viscous discontinuity. Although the electron distribution function is significantly out of equilibrium near the shock front, its low energy part can be approximated by a Maxwellian distribution.

The effect of the non-adiabatic electron heating efficiency, β_e, on the degree of non-equipartition was studied recently by Wong and Sarazin (2009). They have shown that for a cluster with a mass of $M_{\mathrm{vir}} \sim 1.2 \times 10^{15} M_{\odot}$, electron and ion temperatures differ by less than a percent within the virial radius. The difference is 20% for a non-adiabatic electron heating efficiency of $1/1800$ to 0.5 at $\sim$1.4 of the virial radius. Beyond this radius, the non-equipartition effect depends rather strongly on β_e (Wong and Sarazin 2009), and such a strong dependence at the shock radius can be used to distinguish shock heating models or constrain the shock heating efficiency of electrons.

As opposed to semi-analytic kinetic theory methods (Achterberg et al. 2001; Keshet and Waxman 2005) or Monte Carlo test-particle simulations (Niemiec and Ostrowski 2004; Ellison and Double 2004), PIC simulations can tackle the problem of particle acceleration in shocks without the need for simplifying assumptions about the nature of the magnetic turbulence or the details of wave-particle interactions. Multi-dimensional PIC simulations of relativistic un-magnetized shocks have been presented, e.g. by Spitkovsky (2008) for electron-ion flows. Kato and Takabe (2010) present a two-dimensional particle-in-cell simulation to investigate weakly magnetized perpendicular shocks. The simulated thermal energies of electrons and ions in the downstream region are not in equipartition but their temperature ratio $\beta_e = T_e/T_i \sim 0.3$–$0.4$ is high enough indicating rather efficient non-adiabatic heating of the electrons in the shock transition region.

Using 2.5D PIC simulations, Sironi and Spitkovsky (2010) find that in "subluminal" shocks, where relativistic particles can escape ahead of the shock along the magnetic field lines, ions are efficiently accelerated via a Fermi-like mechanism. The scattering is provided by short-wavelength non-resonant modes produced by Bell's instability (Bell 2004), whose growth is seeded by the current of shock-accelerated ions that propagate ahead of the shock. Upstream electrons enter the shock with lower energy than ions, so they are more strongly tied to the field. As a result, only about 1% of the incoming electrons are Fermi-accelerated at the shock before they are advected downstream, where they populate a steep power-law tail. Thus, efficient electron heating is the universal property of relativistic electron-ion shocks, but significant nonthermal acceleration of electrons is hard to achieve in magnetized flows and requires weakly magnetized shocks, where magnetic fields self-generated via the Weibel instability are stronger than the background field.

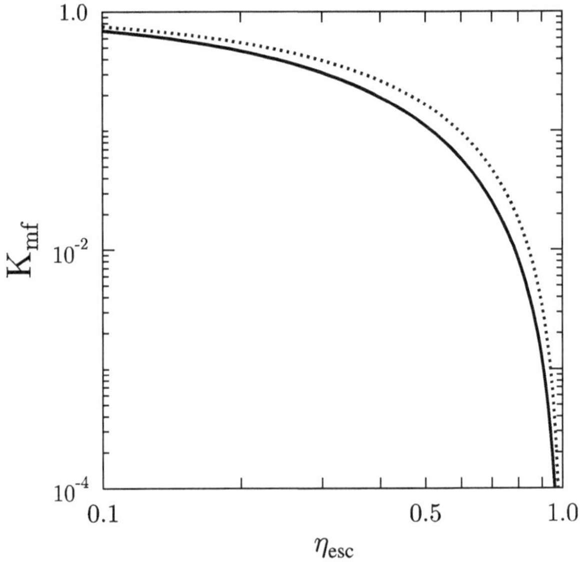

Fig. 12 Normalized post-shock gas entropy, $K_{mf} = K_0\, T/\rho^{(\gamma_g-1)}$, as a function of the CR particle energy escape flux Q_{esc} carried by energetic particles. The dimensionless flux is defined as $\eta_{esc} = Q_{esc}/\Pi_{kin}$, where $\Pi_{kin} = \rho_1 v_{sh}^3/2$. The *upper curve* (*dotted*) corresponds to an effective adiabatic exponent $\gamma_g = 4/3$ (relativistic gas), while the *lower* (*solid*) *curve* corresponds to $\gamma_g = 5/3$. The postshock entropy is normalized to $K_{mf}(\eta_{esc} = 0)$

6.2 Entropy in CR-Modified Accretion Shocks

A distinctive feature of CR-modified shocks is their high gas compression factor $R_t(v_{sh})$ that can be much higher than the single fluid shock limit of $(\gamma_g+1)/(\gamma_g-1)$. Since the postshock gas entropy for a strong multi-fluid shock scales as $R_t(v_{sh})^{-(\gamma_g+1)}$, it can be significantly reduced compared to the single-fluid adiabatic shock of the same velocity (e.g., Bykov et al. 2008a). The effects are due to energetic particle acceleration and magnetic field generation. Energetic particles can penetrate into the shock upstream region. They are coupled with the upstream gas through fluctuating magnetic fields (including the Alfvén waves). Magnetic field dissipation provides gas preheating and entropy production in the shock precursor. Energetic particles deplete the momentum distribution in the gas resulting in a higher gas compression and a reduced temperature. In Fig. 12 the post-shock gas entropy is shown as a function of the fraction η_{esc}. The curves were calculated for a strong shock with no precursor heating i.e. $\alpha_H = 0$.

In galaxy clusters, entropy profiles have been measured by Pratt et al. (2010) at least out to R_{1000} in 31 nearby galaxy clusters from the representative *XMM-Newton* cluster structure survey (REXCESS) and out to R_{500} in thirteen systems. The effects of CRs on the entropy production in shocks illustrated above can be of importance for accretion shocks that are expected to be located at larger radii. It will require deeper exposures of the next generation telescopes to measure the entropy profiles at these radii. The entropy in the inner cluster regions can be also affected by internal shocks if the shocks are efficient accelerators of CRs. The current entropy profile simulations (e.g. Borgani et al. 2008; Borgani and Kravtsov 2009) ought to be extended to include CR physics.

7 Summary

It is only now, with low-frequency radio telescopes, long exposures with high-resolution X-ray satellites and γ-ray telescopes, that we are beginning to learn about the physics of cluster outskirts. In the coming years SZ telescopes are going to deliver further insights into the plasma physics of these special regions in the Universe. The last years have already shown tremendous progress with detections of shocks, estimates of magnetic field strengths and constraints on the particle acceleration efficiency. The main points of this review are listed below:

- Cosmological shocks are collisionless: Coulomb collisions are not sufficient to provide the viscous dissipation of the incoming flow, and collective effects play a major role.
- Collisionless shocks generate energetic, non-thermal particles that can penetrate far into the upstream flow. The particles decelerate the flow and preheat the gas. They can also efficiently generate strong fluctuating magnetic fields in the upstream region. This turbulence, which may result from cosmic-ray driven instabilities, in the shock precursor, is certain to play a critical role in non-linear models of strong, CR-modified shocks.
- A few merger shocks have been identified in X-rays, exhibiting both a sharp gas density edge and an unambiguous temperature jump; in the "bullet cluster", 1E 0657-56 (Markevitch et al. 2002), A520 (Markevitch et al. 2005) and Abell 2146 (Russell et al. 2010).
- On the basis of X-ray observations, microphysical properties of shocks can be derived. Markevitch and Vikhlinin (2007) find that the temperatures across merger shocks are consistent with instant heating; equilibration between electrons and protons on the collisional timescale is excluded. The electron temperature downstream of the shock depends on collisionless electron heating e.g. Bykov and Uvarov (1999), Rakowski et al. (2008), Bykov et al. (2008b). Non-resonant interactions of the electrons with strong nonlinear fluctuations generated by kinetic instabilities of the ions in the transition region inside the shock front may play the main role in the heating and pre-acceleration of the electrons. Future X-ray observations extending to the virial radius or even close to the shock radius should be able to detect signatures of non-equipartition at shocks.
- *Suzaku* observations provide evidence for departures from hydrostatic equilibrium around and even before the virial radius as it is seen in the galaxy cluster Abell 1795 (e.g., Bautz et al. 2009). Evidence for nonthermal pressure support suggests that bulk motions from merging could be making a significant contribution to the gas energy in the outskirts of this cluster (George et al. 2009).
- Recently discovered radio relics provide very strong evidence for shock acceleration at merger shock waves in galaxy clusters. Van Weeren et al. (2010) find that on average smaller radio relics have steeper spectra. Such a trend is in line with predictions from shock statistics derived from cosmological simulations (Skillman et al. 2008; Battaglia et al. 2009; Hoeft et al. 2008). They find that larger shock waves occur mainly in lower-density regions and have larger Mach numbers, and consequently shallower spectra.
- Observations of radio relics in combination with X-ray observations suggest magnetic field strengths in relics of >3 μG and efficiencies of 0.2% of the kinetic energy dissipated in a shock going into relativistic electrons (Finoguenov et al. 2010; van Weeren et al. 2010).
- Radio relics also suggest that the efficiency of shock acceleration at weak shocks is higher than predicted in theories of diffusive shock acceleration.
- The bulk of the energy in their simulations is dissipated in galaxy clusters which contribute about 75% of the total energy dissipation (about 80% the contribution from cluster outskirts is included), while filaments contribute about 15% of the total energy dissipation. In agreement with observational findings, the maximum diffuse radio emission of galaxy clusters depends strongly on their X-ray temperature. While shocks around $M \sim 2$ contribute most to the generation of gas thermal energy, shocks

around $M \sim 3$ contribute most to the generation of CR energy (Ryu et al. 2003; Kang et al. 2007).

– The effects of CRs on entropy production in shocks as illustrated above can be of importance for accretion shocks that are expected to be located at larger radii. It will require deeper exposures of the next generation telescopes to measure the entropy profiles at these radii.

Many open questions remain. So far still little is known about shock acceleration and magnetic field generation in clusters as well as the microstructure of shocks in very dilute plasmas.

Most likely, all these processes are related and their understanding requires input from, both, observations and theory. The next years will see a rapid growth in observational data. On the theoretical side, we expect substantial progress from new techniques that simulate particle acceleration in a magnetohydrodynamical setting. Fully kinetic particle-in-cell simulations provide a powerful tool for exploration of the structure of collisionless shocks from first principles, thus determining self-consistently the interplay between shock-generated waves and accelerated particles (Sironi and Spitkovsky 2010). Results from these simulations can then serve as input for cosmological simulations that include CR-physics.

Acknowledgements We thank an anonymous referee for a very careful read of our manuscript. MB acknowledges support by the DFG Research Unit "Magnetisation of Interstellar and Intergalactic Media: The Prospects of Low-Frequency Radio Observations". DR was supported in part by R&D Program through the National Fusion Research Institute of Korea funded by the Government funds. A.M.B. was supported in part by RBRF grants 09-02-12080, 11-02-00429a, by the RAS Presidium Programm, and the Russian government grant 11.G34.31.0001 to Sankt-Petersburg State Politechnical University. He performed some of the simulations at the Joint Supercomputing Centre (JSCC RAS) and the Supercomputing Centre at Ioffe Institute, St. Petersburg. We thank Aurora Simionescu, Matthias Hoeft and Franco Vazza for helpful comments on the manuscript. Finally, the authors would like the convenors of the workshop in Bern for their initiative and excellent work and the staff at ISSI for their support.

References

A. Achterberg, Y.A. Gallant, J.G. Kirk, A.W. Guthmann, Particle acceleration by ultrarelativistic shocks: theory and simulations. Mon. Not. R. Astron. Soc. **328**, 393–408 (2001). doi:10.1046/j.1365-8711.2001.04851.x

M. Ackermann, M. Ajello, A. Allafort, L. Baldini, J. Ballet, G. Barbiellini, D. Bastieri, K. Bechtol, R. Bellazzini, R.D. Blandford, P. Blasi, E.D. Bloom, E. Bonamente, GeV gamma-ray flux upper limits from clusters of galaxies. Astrophys. J. Lett. **717**, 71–78 (2010). doi:10.1088/2041-8205/717/1/L71

F. Aharonian, A.G. Akhperjanian, G. Anton, U. Barres de Almeida, A.R. Bazer-Bachi, Y. Becherini, B. Behera, K. Bernlöhr, C. Boisson, A. Bochow, V. Borrel, E. Brion, J. Brucker, P. Brun, R. Bühler, T. Bulik, I. Büsching, T. Boutelier, P.M. Chadwick, A. Charbonnier, R.C.G. Chaves, A. Cheesebrough, Constraints on the multi-TeV particle population in the Coma galaxy cluster with HESS observations. Astron. Astrophys. **502**, 437–443 (2009). doi:10.1051/0004-6361/200912086

M. Ajello, P. Rebusco, N. Cappelluti, O. Reimer, H. Böhringer, J. Greiner, N. Gehrels, J. Tueller, A. Moretti, Galaxy clusters in the Swift/Burst alert telescope era: hard X-rays in the intracluster medium. Astrophys. J. **690**, 367–388 (2009). doi:10.1088/0004-637X/690/1/367

E. Amato, P. Blasi, Non-linear particle acceleration at non-relativistic shock waves in the presence of self-generated turbulence. Mon. Not. R. Astron. Soc. **371**, 1251–1258 (2006). doi:10.1111/j.1365-2966.2006.10739.x

W.I. Axford, E. Leer, G. Skadron, The acceleration of cosmic rays by shock waves, in *International Cosmic Ray Conference*. International Cosmic Ray Conference, vol. 11 (1977), p. 132

J. Bagchi, F. Durret, G.B.L. Neto, S. Paul, Giant ringlike radio structures around galaxy cluster Abell 3376. Science **314**, 791–794 (2006). doi:10.1126/science.1131189

N. Battaglia, C. Pfrommer, J.L. Sievers, J.R. Bond, T.A. Enßlin, Exploring the magnetized cosmic web through low-frequency radio emission. Mon. Not. R. Astron. Soc. **393**, 1073–1089 (2009). doi:10.1111/j.1365-2966.2008.14136.x

M.W. Bautz, E.D. Miller, J.S. Sanders, K.A. Arnaud, R.F. Mushotzky, F.S. Porter, K. Hayashida, J.P. Henry, J.P. Hughes, M. Kawaharada, K. Makashima, M. Sato, T. Tamura, Suzaku observations of Abell 1795: Cluster emission to r_{200}. Publ. Astron. Soc. Jpn. **61**, 1117 (2009)

A.R. Bell, The acceleration of cosmic rays in shock fronts. I. Mon. Not. R. Astron. Soc. **182**, 147–156 (1978a)

A.R. Bell, The acceleration of cosmic rays in shock fronts. II. Mon. Not. R. Astron. Soc. **182**, 443–455 (1978b)

A.R. Bell, Turbulent amplification of magnetic field and diffusive shock acceleration of cosmic rays. Mon. Not. R. Astron. Soc. **353**, 550–558 (2004). doi:10.1111/j.1365-2966.2004.08097.x

A.R. Bell, S.G. Lucek, Cosmic ray acceleration to very high energy through the non-linear amplification by cosmic rays of the seed magnetic field. Mon. Not. R. Astron. Soc. **321**, 433–438 (2001). doi:10.1046/j.1365-8711.2001.04063.x

E. Belsole, J. Sauvageot, G.W. Pratt, H. Bourdin, Merging clusters of galaxies observed with XMM-Newton. Adv. Space Res. **36**, 630–635 (2005). doi:10.1016/j.asr.2005.01.029

E.W. Bertschinger, Nonlinear growth of perturbations in an Einstein-De Sitter cosmology. PhD thesis, Princeton University, NJ (1984)

R. Blandford, D. Eichler, Particle acceleration at astrophysical shocks—a theory of cosmic-ray origin. Phys. Rep. **154**, 1 (1987). doi:10.1016/0370-1573(87)90134-7

R.D. Blandford, J.P. Ostriker, Particle acceleration by astrophysical shocks. Astrophys. J. Lett. **221**, 29–32 (1978). doi:10.1086/182658

H. Böhringer, N. Werner, X-ray spectroscopy of galaxy clusters: studying astrophysical processes in the largest celestial laboratories. Astron. Astrophys. Rev. **18**, 127–196 (2010). doi:10.1007/s00159-009-0023-3

A. Bonafede, G. Giovannini, L. Feretti, F. Govoni, M. Murgia, Double relics in Abell 2345 and Abell 1240. Spectral index and polarization analysis. Astron. Astrophys. **494**, 429–442 (2009). doi:10.1051/0004-6361:200810588

S. Borgani, A. Kravtsov, *Cosmological Simulations of Galaxy Clusters* (2009)

S. Borgani, A. Diaferio, K. Dolag, S. Schindler, Thermodynamical properties of the ICM from hydrodynamical simulations. Space Sci. Rev. **134**, 269–293 (2008). doi:10.1007/s11214-008-9317-4

M. Brüggen, S. Heinz, E. Roediger, M. Ruszkowski, A. Simionescu, Shock heating by Fanaroff–Riley type I radio sources in galaxy clusters. Mon. Not. R. Astron. Soc. **380**, 67–70 (2007). doi:10.1111/j.1745-3933.2007.00351.x

G. Brunetti, A. Lazarian, Acceleration of primary and secondary particles in galaxy clusters by compressible MHD turbulence: from radio haloes to gamma-rays. Mon. Not. R. Astron. Soc. **410**, 127–142 (2011). doi:10.1111/j.1365-2966.2010.17457.x

G. Brunetti, P. Blasi, R. Cassano, S. Gabici, High energy emission from galaxy clusters and particle acceleration due to MHD turbulence, in *American Institute of Physics Conference Series*, ed. by D. Bastieri, R. Rando. American Institute of Physics Conference Series, vol. 1112 (2009), pp. 129–137. doi:10.1063/1.3125773

A.M. Bykov, Multi-fluid shocks in clusters of galaxies: entropy, $\sigma_v T$, MT and $L_X T$ scalings. Adv. Space Res. **36**, 738–746 (2005). doi:10.1016/j.asr.2005.01.052

A.M. Bykov, Y.A. Uvarov, Electron kinetics in collisionless shock waves. J. Exp. Theor. Phys. **88**, 465–475 (1999)

A.M. Bykov, K. Dolag, F. Durret, Cosmological shock waves. Space Sci. Rev. **134**, 119–140 (2008a). doi:10.1007/s11214-008-9312-9

A.M. Bykov, S.M. Osipov, D.C. Ellison, Cosmic ray current driven turbulence in shocks with efficient particle acceleration: the oblique, long-wavelength mode instability. Mon. Not. R. Astron. Soc. **410**, 39–52 (2011). doi:10.1111/j.1365-2966.2010.17421.x

A.M. Bykov, F.B.S. Paerels, V. Petrosian, Equilibration processes in the warm-hot intergalactic medium. Space Sci. Rev. **134**, 141–153 (2008b). doi:10.1007/s11214-008-9309-4

A. Cattaneo, S.M. Faber, J. Binney, A. Dekel, J. Kormendy, R. Mushotzky, A. Babul, P.N. Best, M. Brüggen, A.C. Fabian, C.S. Frenk, A. Khalatyan, H. Netzer, A. Mahdavi, J. Silk, M. Steinmetz, L. Wisotzki, The role of black holes in galaxy formation and evolution. Nature **460**, 213–219 (2009). doi:10.1038/nature08135

E. Churazov, W. Forman, C. Jones, R. Sunyaev, H. Böhringer, XMM-Newton observations of the Perseus cluster—II. Evidence for gas motions in the core. Mon. Not. R. Astron. Soc. **347**, 29–35 (2004). doi:10.1111/j.1365-2966.2004.07201.x

K. Dolag, A.M. Bykov, A. Diaferio, Non-thermal processes in cosmological simulations. Space Sci. Rev. **134**, 311–335 (2008). doi:10.1007/s11214-008-9319-2

L.O. Drury, An introduction to the theory of diffusive shock acceleration of energetic particles in tenuous plasmas. Rep. Prog. Phys. **46**, 973–1027 (1983). doi:10.1088/0034-4885/46/8/002

F. Durret, J.S. Kaastra, J. Nevalainen, T. Ohashi, N. Werner, Soft X-ray and extreme ultraviolet excess emission from clusters of galaxies. Space Sci. Rev. **134**, 51–70 (2008). doi:10.1007/s11214-008-9313-8
D.C. Ellison, G.P. Double, Diffusive shock acceleration in unmodified relativistic, oblique shocks. Astropart. Phys. **22**, 323–338 (2004). doi:10.1016/j.astropartphys.2004.08.005
S. Ettori, S. Molendi, *X-Ray Observations of Cluster Outskirts: Current Status and Future Prospects* (2010)
L. Feretti, H. Boehringer, G. Giovannini, D. Neumann, The radio and X-ray properties of Abell 2255 Astron. Astrophys. **317**, 432–440 (1997)
C. Ferrari, F. Govoni, S. Schindler, A.M. Bykov, Y. Rephaeli, Observations of extended radio emission in clusters. Space Sci. Rev. **134**, 93–118 (2008). doi:10.1007/s11214-008-9311-x
A. Finoguenov, C.L. Sarazin, K. Nakazawa, D.R. Wik, T.E. Clarke, XMM-Newton observation of the northwest radio relic region in A3667. Astrophys. J. **715**, 1143–1151 (2010). doi:10.1088/0004-637X/715/2/1143
D.C. Fox, A. Loeb, Do the electrons and ions in X-ray clusters share the same temperature? Astrophys. J. **491**, 459–466 (1997). doi:10.1086/305007
S.R. Furlanetto, A. Loeb, Intergalactic magnetic fields from quasar outflows. Astrophys. J. **556**, 619–634 (2001)
S. Gabici, P. Blasi, Nonthermal radiation from clusters of galaxies: the role of merger shocks in particle acceleration. Astrophys. J. **583**, 695–705 (2003). doi:10.1086/345429
M.R. George, A.C. Fabian, J.S. Sanders, A.J. Young, H.R. Russell, X-ray observations of the galaxy cluster PKS0745-191: to the virial radius, and beyond. Mon. Not. R. Astron. Soc. **395**, 657–666 (2009). doi:10.1111/j.1365-2966.2009.14547.x
S. Giacintucci, T. Venturi, G. Macario, D. Dallacasa, G. Brunetti, M. Markevitch, R. Cassano, S. Bardelli, R. Athreya, Shock acceleration as origin of the radio relic in A 521? Astron. Astrophys. **486**, 347–358 (2008). doi:10.1051/0004-6361:200809459
G. Giovannini, L. Feretti, C. Stanghellini, The Coma cluster radio source 1253 + 275, revisited. Astron. Astrophys. **252**, 528–537 (1991)
G. Giovannini, A. Bonafede, L. Feretti, F. Govoni, M. Murgia, The diffuse radio filament in the merging system ZwCl 2341.1 + 0000. Astron. Astrophys. **511**, 5 (2010). doi:10.1051/0004-6361/200913983
F. Govoni, L. Feretti, G. Giovannini, H. Böhringer, T.H. Reiprich, M. Murgia, Radio and X-ray diffuse emission in six clusters of galaxies. Astron. Astrophys. **376**, 803–819 (2001). doi:10.1051/0004-6361:20011016
M. Hoeft, M. Brüggen, Radio signature of cosmological structure formation shocks. Mon. Not. R. Astron. Soc. **375**, 77–91 (2007). doi:10.1111/j.1365-2966.2006.11111.x
M. Hoeft, M. Brüggen, G. Yepes, S. Gottlöber, A. Schwope, Diffuse radio emission from clusters in the MareNostrum Universe simulation. Mon. Not. R. Astron. Soc. **391**, 1511–1526 (2008). doi:10.1111/j.1365-2966.2008.13955.x
F.C. Jones, D.C. Ellison, The plasma physics of shock acceleration. Space Sci. Rev. **58**, 259–346 (1991). doi:10.1007/BF01206003
H. Kang, T.W. Jones, Self-similar evolution of cosmic-ray-modified quasi-parallel plane shocks. Astropart. Phys. **28**, 232–246 (2007). doi:10.1016/j.astropartphys.2007.05.007
H. Kang, D. Ryu, Diffusive shock acceleration in test-particle regime. Astrophys. J. **721**, 886 (2010)
H. Kang, D. Ryu, T.W. Jones, Cluster accretion shocks as possible acceleration sites for ultra-high-energy protons below the Greisen cutoff. Astrophys. J. **456**, 422 (1996). doi:10.1086/176666
H. Kang, T.W. Jones, U.D.J. Gieseler, Numerical studies of CR injection. Astrophys. J. **579**, 337 (2002)
H. Kang, D. Ryu, T.W. Jones, Self-similar evolution of cosmic-ray modified shocks: the cosmic-ray spectrum. Astrophys. J. **695**, 1273–1288 (2009). doi:10.1088/0004-637X/695/2/1273
H. Kang, D. Ryu, R. Cen, J.P. Ostriker, Cosmological shock waves in the large-scale structure of the Universe: nongravitational effects. Astrophys. J. **669**, 729–740 (2007). doi:10.1086/521717
T.N. Kato, H. Takabe, Nonrelativistic collisionless shocks in weakly magnetized electron-ion plasmas: two-dimensional particle-in-cell simulation of perpendicular shock (2010)
J.C. Kempner, E.L. Blanton, T.E. Clarke, T.A. Enßlin, M. Johnston-Hollitt, L. Rudnick, Conference note: a taxonomy of extended radio sources in clusters of galaxies, in *The Riddle of Cooling Flows in Galaxies and Clusters of Galaxies*, ed. by T. Reiprich, J. Kempner, N. Soker (2004), p. 335
U. Keshet, E. Waxman, Energy spectrum of particles accelerated in relativistic collisionless shocks. Phys. Rev. Lett. **94**(11), 111102 (2005). doi:10.1103/PhysRevLett.94.111102
U. Keshet, E. Waxman, A. Loeb, V. Springel, L. Hernquist, Gamma rays from intergalactic shocks. Astrophys. J. **585**, 128–150 (2003). doi:10.1086/345946
R.A. Krivonos, A.A. Vikhlinin, M.L. Markevitch, M.N. Pavlinsky, A possible shock wave in the intergalactic medium of the cluster of galaxies A754. Astron. Lett. **29**, 425–428 (2003). doi:10.1134/1.1589859

G.F. Krymskii, A regular mechanism for the acceleration of charged particles on the front of a shock wave. Dokl. Akad. Nauk SSSR **234**, 1306–1308 (1977)

R.M. Kulsrud, R. Cen, J.P. Ostriker, D. Ryu, The protogalactic origin for cosmic magnetic fields. Astrophys. J. **480**, 481 (1997)

A. Lazarian, Enhancement and suppression of heat transfer by MHD turbulence. Astrophys. J. Lett. **645**, 25–28 (2006). doi:10.1086/505796

M. Lueker, C.L. Reichardt, K.K. Schaffer, O. Zahn, P.A.R. Ade, K.A. Aird, B.A. Benson, L.E. Bleem, J.E. Carlstrom, C.L. Chang, H. Cho, T.M. Crawford, A.T. Crites, T. de Haan, M.A. Dobbs, E.M. George, N.R. Hall, N.W. Halverson, G.P. Holder, W.L. Holzapfel, J.D. Hrubes, M. Joy, R. Keisler, L. Knox, A.T. Lee, E.M. Leitch, J.J. McMahon, J. Mehl, S.S. Meyer, J.J. Mohr, T.E. Montroy, S. Padin, T. Plagge, C. Pryke, J.E. Ruhl, L. Shaw, E. Shirokoff, H.G. Spieler, B. Stalder, Z. Staniszewski, A.A. Stark, K. Vanderlinde, J.D. Vieira, R. Williamson, Measurements of secondary cosmic microwave background anisotropies with the south pole telescope. Astrophys. J. **719**, 1045–1066 (2010). doi:10.1088/0004-637X/719/2/1045

M.A. Malkov, L. O'C Drury, Nonlinear theory of diffusive acceleration of particles by shock waves. Rep. Prog. Phys. **64**, 429–481 (2001). doi:10.1088/0034-4885/64/4/201

M. Markevitch, The L X-T relation and temperature function for nearby clusters revisited. Astrophys. J. **504**, 27 (1998). doi:10.1086/306080

M. Markevitch, A. Vikhlinin, Shocks and cold fronts in galaxy clusters. Phys. Rep. **443**, 1 (2007)

M. Markevitch, A.H. Gonzalez, L. David, A. Vikhlinin, S. Murray, W. Forman, C. Jones, W. Tucker, A textbook example of a bow shock in the merging galaxy cluster 1E 0657-56. Astrophys. J. Lett. **567**, 27–31 (2002). doi:10.1086/339619

M. Markevitch, F. Govoni, G. Brunetti, D. Jerius, Bow shock and radio halo in the merging cluster A520. Astrophys. J. **627**, 733–738 (2005). doi:10.1086/430695

M.V. Medvedev, L.O. Silva, M. Fiore, R.A. Fonseca, W.B. Mori, Generation of magnetic fields in cosmological shocks. J. Korean Astron. Soc. **37**, 533–541 (2004)

R. Mewe, Ionization of hot plasmas, in *NATO ASIC Proc. 300: Physical Processes in Hot Cosmic Plasmas*, ed. by W. Brinkmann, A.C. Fabian, F. Giovannelli (1990), pp. 39–65

E.T. Million, S.W. Allen, Chandra measurements of non-thermal-like X-ray emission from massive, merging, radio halo clusters. Mon. Not. R. Astron. Soc. **399**, 1307–1327 (2009). doi:10.1111/j.1365-2966.2009.15359.x

F. Miniati, D. Ryu, H. Kang, T.W. Jones, R. Cen, J.P. Ostriker, Properties of cosmic shock waves in large-scale structure formation. Astrophys. J. **542**, 608–621 (2000). doi:10.1086/317027

F. Miniati, D. Ryu, H. Kang, T.W. Jones, Cosmic-ray protons accelerated at cosmological shocks and their impact on groups and clusters of galaxies. Astrophys. J. **559**, 59–69 (2001). doi:10.1086/322375

S.M. Molnar, N. Hearn, Z. Haiman, G. Bryan, A.E. Evrard, G. Lake, Accretion shocks in clusters of galaxies and their SZ signature from cosmological simulations. Astrophys. J. **696**, 1640–1656 (2009). doi:10.1088/0004-637X/696/2/1640

D. Nagai, E. Lau, *Gas Clumping in the Outskirts of Lambda-CDM Clusters* (2011)

J. Niemiec, M. Ostrowski, Cosmic-ray acceleration at relativistic shock waves with a "realistic" magnetic field structure. Astrophys. J. **610**, 851–867 (2004). doi:10.1086/421730

C.A. Norman, D.B. Melrose, A. Achterberg, The origin of cosmic rays above 10 18.5 eV. Astrophys. J. **454**, 60 (1995). doi:10.1086/176465

M.S. Owers, W.J. Couch, P.E.J. Nulsen, Substructure in the cold front cluster Abell 3667. Astrophys. J. **693**, 901 (2009)

S. Paul, L. Iapichino, F. Miniati, J. Bagchi, K. Mannheim, *Evolution of Shocks and Turbulence in Major Cluster Mergers* (2010)

V. Petrosian, A.M. Bykov, Particle acceleration mechanisms. Space Sci. Rev. **134**, 207–227 (2008). doi:10.1007/s11214-008-9315-6

V. Petrosian, A. Bykov, Y. Rephaeli, Nonthermal radiation mechanisms. Space Sci. Rev. **134**, 191–206 (2008). doi:10.1007/s11214-008-9327-2

C. Pfrommer, T.A. Enßlin, V. Springel, M. Jubelgas, K. Dolag, Simulating cosmic rays in clusters of galaxies—I. Effects on the Sunyaev-Zel'dovich effect and the X-ray emission. Mon. Not. R. Astron. Soc. **378**, 385–408 (2007). doi:10.1111/j.1365-2966.2007.11732.x

G.W. Pratt, H. Böhringer, J.H. Croston, M. Arnaud, S. Borgani, A. Finoguenov, R.F. Temple, Temperature profiles of a representative sample of nearby X-ray galaxy clusters. Astron. Astrophys. **461**, 71–80 (2007). doi:10.1051/0004-6361:20065676

G.W. Pratt, M. Arnaud, R. Piffaretti, H. Böhringer, T.J. Ponman, J.H. Croston, G.M. Voit, S. Borgani, R.G. Bower, Gas entropy in a representative sample of nearby X-ray galaxy clusters (REXCESS): relationship to gas mass fraction. Astron. Astrophys. **511**, 85 (2010). doi:10.1051/0004-6361/200913309

C.E. Rakowski, J.M. Laming, P. Ghavamian, The heating of thermal electrons in fast collisionless shocks: the integral role of cosmic rays. Astrophys. J. **684**, 348–357 (2008). doi:10.1086/590245

T.H. Reiprich, D.S. Hudson, Y. Zhang, K. Sato, Y. Ishisaki, A. Hoshino, T. Ohashi, N. Ota, Y. Fujita, Suzaku measurement of Abell 2204's intracluster gas temperature profile out to 1800 kpc. Astron. Astrophys. **501**, 899–905 (2009). doi:10.1051/0004-6361/200810404

Y. Rephaeli, J. Nevalainen, T. Ohashi, A.M. Bykov, Nonthermal phenomena in clusters of galaxies. Space Sci. Rev. **134**, 71–92 (2008). doi:10.1007/s11214-008-9314-7

K. Roettiger, J.O. Burns, J.M. Stone, A cluster merger and the origin of the extended radio emission in Abell 3667. Astrophys. J. **518**, 603–612 (1999). doi:10.1086/307327

S. Rosswog, M. Brüggen, *Introduction to High-Energy Astrophysics* (2007)

H.J.A. Röttgering, M.H. Wieringa, R.W. Hunstead, R.D. Ekers, The extended radio emission in the luminous X-ray cluster A3667. Mon. Not. R. Astron. Soc. **290**, 577–584 (1997)

H.J.A. Röttgering, M.H. Wieringa, R.W. Hunstead, R.D. Ekers, The extended radio emission in the luminous X-ray cluster A3667. Mon. Not. R. Astron. Soc. **290**, 577–584 (1997)

H. Röttgering, I. Snellen, G. Miley, J.P. de Jong, R.J. Hanisch, R. Perley, VLA observations of the rich X-ray cluster Abell 2256. Astrophys. J. **436**, 654–668 (1994). doi:10.1086/174940

H.R. Russell, J.S. Sanders, A.C. Fabian, S.A. Baum, M. Donahue, A.C. Edge, B.R. McNamara, C.P. O'Dea, Chandra observation of two shock fronts in the merging galaxy cluster Abell 2146. Mon. Not. R. Astron. Soc. (2010). doi:10.1111/j.1365-2966.2010.16822.x

D. Ryu, H. Kang, E. Hallman, T.W. Jones, Cosmological shock waves and their role in the large-scale structure of the Universe. Astrophys. J. **593**, 599–610 (2003). doi:10.1086/376723

D. Ryu, H. Kang, J. Cho, S. Das, Turbulence and magnetic fields in the large-scale structure of the Universe. Science **320**, 909 (2008). doi:10.1126/science.1154923

J.S. Sanders, A.C. Fabian, R.K. Smith, J.R. Peterson, A direct limit on the turbulent velocity of the intracluster medium in the core of Abell 1835 from XMM-Newton. Mon. Not. R. Astron. Soc. **402**, 11–15 (2010). doi:10.1111/j.1745-3933.2009.00789.x

A.A. Schekochihin, S.C. Cowley, R.M. Kulsrud, G.W. Hammett, P. Sharma, Plasma instabilities and magnetic field growth in clusters of galaxies. Astrophys. J. **629**, 139–142 (2005). doi:10.1086/431202

S.J. Schwartz, M.F. Thomsen, S.J. Bame, J. Stansberry, Electron heating and the potential jump across fast mode shocks. J. Geophys. Res. **93**, 12923–12931 (1988)

L.D. Shaw, D. Nagai, S. Bhattacharya, E.T. Lau, Impact of cluster physics on the Sunyaev-Zel'dovich power spectrum. Astrophys. J. **725**, 1452–1465 (2010). doi:10.1088/0004-637X/725/2/1452

A. Simionescu, E. Roediger, P.E.J. Nulsen, M. Brüggen, W.R. Forman, H. Böhringer, N. Werner, A. Finoguenov, The large-scale shock in the cluster of galaxies Hydra A. Astron. Astrophys. **495**, 721–732 (2009). doi:10.1051/0004-6361:200811071

A. Simionescu, S.W. Allen, A. Mantz, N. Werner, Y. Takei, R.G. Morris, A.C. Fabian, J.S. Sanders, P.E.J. Nulsen, M.R. George, G.B. Taylor, *Baryons at the Edge of the X-Ray Brightest Galaxy Cluster* (2011)

L. Sironi, A. Spitkovsky, *Particle Acceleration in Relativistic Magnetized Collisionless Electron-Ion Shocks* (2010)

S.W. Skillman, B.W. O'Shea, E.J. Hallman, J.O. Burns, M.L. Norman, Cosmological shocks in adaptive mesh refinement simulations and the acceleration of cosmic rays. Astrophys. J. **689**, 1063–1077 (2008). doi:10.1086/592496

A. Spitkovsky, Particle acceleration in relativistic collisionless shocks: Fermi process at last? Astrophys. J. Lett. **682**, 5–8 (2008). doi:10.1086/590248

R.J. van Weeren, H.J.A. Röttgering, M. Brüggen, A. Cohen, A search for steep spectrum radio relics and halos with the GMRT. Astron. Astrophys. **508**, 75–92 (2009a). doi:10.1051/0004-6361/200912501

R.J. van Weeren, H.J.A. Röttgering, J. Bagchi, S. Raychaudhury, H.T. Intema, F. Miniati, T.A. Enßlin, M. Markevitch, T. Erben, Radio observations of ZwCl 2341.1+0000: a double radio relic cluster. Astron. Astrophys. **506**, 1083–1094 (2009b). doi:10.1051/0004-6361/200912287

R.J. van Weeren, H.J.A. Röttgering, J. Bagchi, S. Raychaudhury, H.T. Intema, F. Miniati, T.A. Enßlin, M. Markevitch, T. Erben, Radio observations of ZwCl 2341.1+0000: a double radio relic cluster. Astron. Astrophys. **506**, 1083–1094 (2009c). doi:10.1051/0004-6361/200912287

R.J. van Weeren, H.J.A. Röttgering, M. Brüggen, M. Hoeft, Particle acceleration on megaparsec scales in a merging galaxy cluster. Science **330**, 347 (2010). doi:10.1126/science.1194293

R.J. van Weeren, H.J.A. Röttgering, M. Brüggen, M. Hoeft (2011)

F. Vazza, G. Brunetti, C. Gheller, Shock waves in Eulerian cosmological simulations: main properties and acceleration of cosmic rays. Mon. Not. R. Astron. Soc. **395**, 1333–1354 (2009a). doi:10.1111/j.1365-2966.2009.14691.x

F. Vazza, G. Brunetti, A. Kritsuk, R. Wagner, C. Gheller, M. Norman, Turbulent motions and shocks waves in galaxy clusters simulated with adaptive mesh refinement. Astron. Astrophys. **504**, 33–43 (2009b). doi:10.1051/0004-6361/200912535

F. Vazza, G. Brunetti, C. Gheller, R. Brunino, Massive and refined: a sample of large galaxy clusters simulated at high resolution. I: thermal gas and properties of shock waves. New Astron. **15**, 695–711 (2010). doi:10.1016/j.newast.2010.05.003
T. Venturi, S. Giacintucci, G. Brunetti, R. Cassano, S. Bardelli, D. Dallacasa, G. Setti, GMRT radio halo survey in galaxy clusters at $z = 0.2$–0.4. I. The REFLEX sub-sample. Astron. Astrophys. **463**, 937–947 (2007). doi:10.1051/0004-6361:20065961
A. Vikhlinin, M. Markevitch, S.S. Murray, C. Jones, W. Forman, L. Van Speybroeck, Chandra temperature profiles for a sample of nearby relaxed galaxy clusters. Astrophys. J. **628**, 655–672 (2005). doi:10.1086/431142
A. Vikhlinin, A.V. Kravtsov, R.A. Burenin, H. Ebeling, W.R. Forman, A. Hornstrup, C. Jones, S.S. Murray, D. Nagai, H. Quintana, A. Voevodkin, Chandra cluster cosmology project. Astrophys. J. **692**, 1060 (2009)
A. Vladimirov, D.C. Ellison, A. Bykov, Nonlinear diffusive shock acceleration with magnetic field amplification. Astrophys. J. **652**, 1246–1258 (2006). doi:10.1086/508154
A.E. Vladimirov, A.M. Bykov, D.C. Ellison, Turbulence dissipation and particle injection in nonlinear diffusive shock acceleration with magnetic field amplification. Astrophys. J. **688**, 1084–1101 (2008). doi:10.1086/592240
C. Vogt, T.A. Enßlin, A Bayesian view on Faraday rotation maps seeing the magnetic power spectra in galaxy clusters. Astron. Astrophys. **434**, 67–76 (2005). doi:10.1051/0004-6361:20041839
K. Wong, C.L. Sarazin, Effects of the non-equipartition of electrons and ions in the outskirts of relaxed galaxy clusters. Astrophys. J. **707**, 1141–1159 (2009). doi:10.1088/0004-637X/707/2/1141
Y.B. Zeldovich, A hypothesis, unifying the structure and the entropy of the Universe. Mon. Not. R. Astron. Soc. **160**, 1 (1972)
V.N. Zirakashvili, V.S. Ptuskin, H.J. Völk, Modeling Bell's nonresonant cosmic-ray instability. Astrophys. J. **678**, 255–261 (2008). doi:10.1086/529579

Space Sci Rev (2012) 166:215–230
DOI 10.1007/s11214-011-9782-z

Magnetic Fields in Galaxies

Rainer Beck

Received: 13 December 2010 / Accepted: 15 April 2011 / Published online: 19 May 2011

Abstract Radio synchrotron emission, its polarization and its Faraday rotation are powerful tools to study the strength and structure of magnetic fields in galaxies. Unpolarized emission traces turbulent fields which are strongest in spiral arms and bars (20–30 μG) and in central starburst regions (50–100 μG). Such fields are dynamically important, e.g. they can drive gas inflows in central regions. Polarized emission traces ordered fields which can be regular or anisotropic random, generated from isotropic random fields by compression or shear. The strongest ordered fields of 10–15 μG strength are generally found in interarm regions and follow the orientation of adjacent gas spiral arms. Ordered fields with spiral patterns exist in grand-design, barred and flocculent galaxies, and in central regions of starburst galaxies. Faraday rotation measures (RM) of the diffuse polarized radio emission from the disks of several spiral galaxies reveal large-scale patterns, which are signatures of regular fields generated by a mean-field dynamo. However, in most spiral galaxies observed so far the field structure is more complicated. Ordered fields in interacting galaxies have asymmetric distributions and are an excellent tracer of past interactions between galaxies or with the intergalactic medium. Ordered magnetic fields are also observed in radio halos around edge-on galaxies, out to large distances from the plane, with X-shaped patterns. Future observations of polarized emission at high frequencies, with the EVLA, the SKA and its precursors, will trace galactic magnetic fields in unprecedented detail. Low-frequency telescopes (e.g. LOFAR and MWA) are ideal to search for diffuse emission and small RMs from weak interstellar and intergalactic fields.

Keywords Synchrotron emission · Radio telescopes · Faraday rotation · Dynamo action · Galaxies: spiral structure · Galaxies: magnetic fields

1 Introduction

Magnetic fields are a major agent in the interstellar medium and also control the density and distribution of cosmic rays. Cosmic rays accelerated in supernova remnants can provide

R. Beck (✉)
Max-Planck-Institut für Radioastronomie, Auf dem Hügel 69, 53121 Bonn, Germany
e-mail: rbeck@mpifr-bonn.mpg.de

the pressure to drive galactic outflows and buoyant loops of magnetic fields via the Parker instability. Outflows from starburst galaxies in the early Universe may have magnetized the intergalactic medium.

The detection of ultrahigh-energy cosmic rays (UHECRs) with the AUGER observatory and the anisotropic distribution of their arrival directions (Abreu et al. 2010) calls for a proper model of particle propagation. As UHECR particles are deflected by large-scale regular fields and scattered by turbulent fields, the structure and the extent of the fields in the disk and halo of the Milky Way need to be known, but the present data do not allow safe conclusions (Noutsos, this issue). The view onto external spiral galaxies can help to model the field structure of the Milky Way.

2 Origin of Magnetic Fields

In spite of our increasing knowledge of cosmic magnetic fields, many important questions, especially their origin and evolution, their strength and structure in intergalactic space, their first occurrence in young galaxies, and their dynamical importance for galaxy evolution remain unanswered.

The origin of the first magnetic fields in the Universe is a mystery (Widrow 2002). *Seed fields* may be "primordial", generated during a phase transition in the early Universe (Caprini et al. 2009), or originate from the time of cosmological structure formation by the Weibel instability (Lazar et al. 2009), or from injection by the first stars or jets generated by the first black holes (Rees 2005), or from the Biermann mechanism in the first supernova remnants (Hanayama et al. 2005). The strength of intergalactic fields is at least 10^{-16} G and their filling factor is larger than 60%, as derived from high-energy γ-ray observations with HESS and FERMI which indicate that the secondary particles are deflected by the intergalactic fields (Dolag et al. 2011).

The most promising mechanism to sustain magnetic fields in the interstellar medium of galaxies is the dynamo (Beck et al. 1996) (see also Brandenburg et al., this issue). A small-scale dynamo in protogalaxies may have amplified seed fields to several μG strength (the energy level of turbulence) within less than 10^8 yr (Schleicher et al. 2010). To explain the generation of large-scale fields in galaxies, the mean-field dynamo has been developed. It is based on turbulence, differential rotation and helical gas flows (*α-effect*), driven by supernova explosions (Gressel et al. 2008). The mean-field dynamo in galaxy disks predicts that within a few 10^9 yr large-scale regular fields are generated from μG turbulent fields (Arshakian et al. 2009), forming spiral patterns (*modes*) with different azimuthal symmetries in the disk and vertical symmetries in the halo (see Sect. 6.1). Global numerical models of galaxies (Gissinger et al. 2009; Hanasz et al. 2009) confirm the basic results of the mean-field approximation.

The mean-field dynamo generates large-scale helicity with a non-zero mean in each hemisphere. As total helicity is a conserved quantity, the dynamo is quenched by the small-scale fields with opposite helicity unless these are removed from the system (Shukurov et al. 2006). Outflows are probably essential for effective mean-field dynamo action.

3 Measuring Magnetic Fields in Galaxies

Magnetic fields need illumination to be detectable. *Polarized emission* at optical, infrared, submillimeter and radio wavelengths holds the clue to measure magnetic fields in galaxies.

Optical linear polarization is a result of extinction by elongated dust grains in the line of sight which are aligned in the interstellar magnetic field (the *Davis-Greenstein effect*). The E-vector runs parallel to the field. However, light can also be polarized by scattering, a process unrelated to magnetic fields and hence a contamination that is difficult to subtract from the diffuse polarized emission from galaxies, e.g. in M 51 (Scarrott et al. 1987). Optical polarization data of about 5500 selected stars in the Milky Way yielded the orientation of the large-scale magnetic field near the sun (Fosalba et al. 2002). Together with measurements of stellar distances, a 3-D analysis of the magnetic field within about 5 kpc from the sun is possible, but more data are needed.

Linearly polarized emission from elongated dust grains at infrared and submillimeter wavelengths is not affected by polarized scattered light. The B-vector is parallel to the magnetic field. The field structure can be mapped in gas clouds of the Milky Way (Tang et al. 2009) and in galaxies, e.g. in the halo of M 82 (Greaves et al. 2000).

Most of what we know about interstellar magnetic fields comes through the detection of radio waves. *Zeeman splitting* of radio spectral lines directly measures the field strength in gas clouds of the Milky Way (Heiles, this issue) and in starburst galaxies (Robishaw et al. 2008). The intensity of *synchrotron emission* is a measure of the number density of cosmic-ray electrons in the relevant energy range and of the strength of the total magnetic field component in the sky plane. The assumption of energy equipartition between these two components allows us to calculate the total magnetic field strength from the synchrotron intensity (see Sect. 4.1).

Linearly polarized synchrotron emission emerges from ordered fields in the sky plane. As polarization "vectors" are ambiguous by 180°, they cannot distinguish *regular (coherent) fields*, defined to have a constant direction within the telescope beam, from *anisotropic fields*, which are generated from turbulent fields by compressing or shearing gas flows and frequently reverse their direction within the telescope beam. Unpolarized synchrotron emission indicates *turbulent (random) fields* which have random directions in 3-D and have been amplified and tangled by turbulent gas flows.

The intrinsic degree of linear polarization of synchrotron emission is about 75%. The observed degree of polarization is smaller due to the contribution of unpolarized thermal emission, which may dominate in star-forming regions, by *Faraday depolarization* along the line of sight and across the beam (Sokoloff et al. 1998), and by geometrical depolarization due to variations of the field orientation within the beam.

The polarization vector is rotated in a magnetized thermal plasma by *Faraday rotation*. If Faraday rotation is small (in galaxies typically at wavelengths shorter than a few centimeters), the B-vector of polarized emission gives the intrinsic field orientation in the sky plane, so that the magnetic pattern can be mapped directly (Beck 2005). The rotation angle is proportional to the square of the wavelength λ^2 and to the *Rotation Measure (RM)*, which is related to the *Faraday depth (FD)*, see below. As the rotation angle is sensitive to the sign of the field direction, only regular fields give rise to Faraday rotation, while anisotropic and random fields do not. Measurements of the Faraday rotation from multi-wavelength observations allow to determine the strength and direction of the regular field component along the line of sight. Dynamo modes of regular fields can be identified from the pattern of polarization angles and of RMs of the diffuse polarized emission of galaxy disks (see Sect. 6.1).

The rotation angle $\Delta\chi$ is proportional to the square of the wavelength λ^2 and the *Faraday depth (FD)*, defined as the line-of-sight integral over the product of the plasma density and the strength of the field component along the line of sight. The *rotation measure (RM)* is defined as $RM = \Delta\chi/\Delta\lambda^2$. If the rotating region is located in front of the emitting region

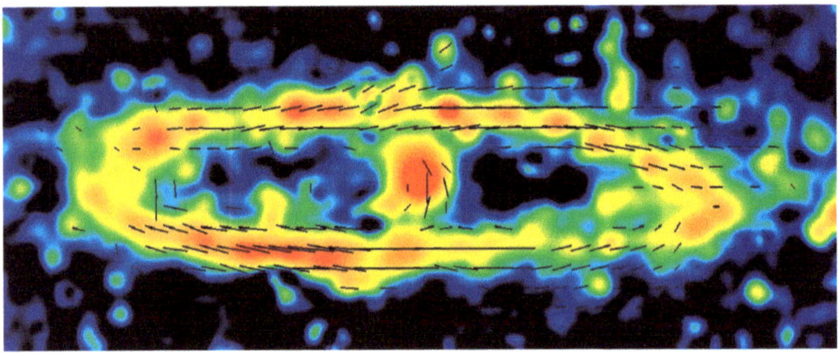

Fig. 1 Total radio intensity (*colors*) and B-vectors (corrected for Faraday rotation) in the Andromeda galaxy (M 31), observed at 6 cm with the Effelsberg telescope (Berkhuijsen et al. 2003)

(*Faraday screen*), $RM = FD$. In case of one region with emission and rotation, $RM \approx FD/2$. Distinct emitting and rotating regions located along the line of sight generate a spectrum of FD components. In such cases, multi-channel spectro-polarimetric radio data are needed that can be Fourier-transformed into Faraday space, called *RM Synthesis* (Brentjens and de Bruyn 2005). If the medium has a relatively simple structure, the 3-D structure of the magnetized interstellar medium can be determined (*Faraday tomography*).

A grid of RM measurements of polarized background sources is another powerful tool to study magnetic field patterns in galaxies (Stepanov et al. 2008). A large number of background sources is required to recognize the field patterns, to separate the Galactic foreground contribution and to account for intrinsic RMs of the background sources.

4 Total Galactic Magnetic Fields

4.1 Total Field Strengths

The typical average *equipartition strength* of the total magnetic field (Beck and Krause 2005) in spiral galaxies is about 10 μG, assuming energy equipartition between cosmic rays and magnetic fields. Radio-faint galaxies like M 31 (Fig. 1) and M 33, our Milky Way's neighbors, have weaker total magnetic fields (about 5 μG), while gas-rich spiral galaxies with high star-formation rates, like M 51 (Fig. 4), M 83 (Fig. 5) and NGC 6946 (Fig. 10), have total field strengths of 20–30 μG in their spiral arms. The strongest total fields of 50–100 μG are found in starburst galaxies, like M 82 (Klein et al. 1988) and the "Antennae" NGC 4038/9 (Chyży and Beck 2004), and in nuclear starburst regions, like in the centers of barred galaxies (Beck et al. 2005).

If energy losses of cosmic-ray electrons are significant, especially in starburst regions or massive spiral arms, the equipartition values are lower limits (Beck and Krause 2005) and are probably underestimated in starburst galaxies by a factor of a few (Thompson et al. 2006). Field strengths of 0.5–18 mG were detected in starburst galaxies by the Zeeman effect in the OH megamaser emission line at 18 cm (Robishaw et al. 2008). These values refer to highly compressed gas clouds and are not typical for the diffuse interstellar medium.

The relative importance of various competing forces in the interstellar medium can be estimated by comparing the corresponding *energy densities*. The mean energy densities of the total (mostly turbulent) magnetic field and the cosmic rays in NGC 6946 (Fig. 2) and M 33 are $\simeq 10^{-11}$ erg cm^{-3} and $\simeq 10^{-12}$ erg cm^{-3}, respectively (Beck 2007; Tabatabaei et

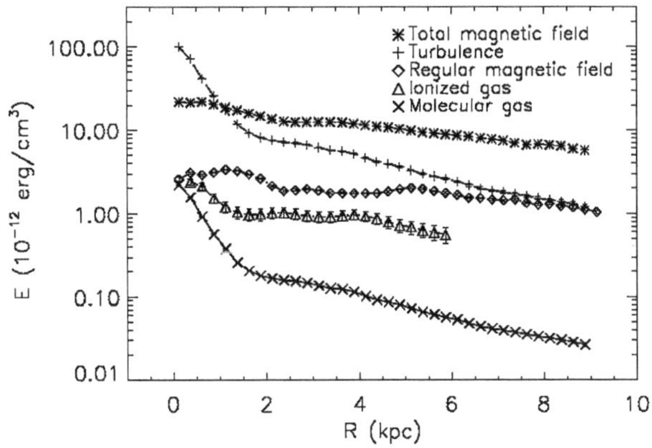

Fig. 2 Radial variation of the energy densities in NGC 6946: total magnetic field E_B $(B_t^2/8\pi)$, regular magnetic field $(B_{reg}^2/8\pi)$, turbulent motion of the neutral gas E_{turb} $(0.5\rho_n v_{turb}^2$, where $v_{turb} \approx$ 7 km/s), thermal energy of the ionized gas E_{th} $(0.5 n_e k T_e)$ and thermal energy of the molecular gas E_n $(0.5\rho_n k T_n)$, determined from observations of synchrotron and thermal radio continuum and the CO and HI line emissions (Beck 2007)

al. 2008), similar to that of the turbulent gas motions across the whole star-forming disk, but about 10 times larger than that of the ionized gas (*low-beta plasma*). Magnetic fields are dynamically important. The total magnetic energy density may even dominate in the outer galaxy where the equipartition field strength is an underestimate due to energy losses of the cosmic-ray electrons. The energy density of the regular magnetic field decreases even more slowly than that of the total field. Although the star-formation activity is low in the outer disk, the magneto-rotational instability (MRI) may serve as the source of turbulence required for dynamo action (Sellwood and Balbus 1999).

4.2 The Radio-Infrared Correlation

The integrated luminosity of the total radio continuum emission at centimeter wavelengths (frequencies of a few GHz), which is mostly of nonthermal synchrotron origin, and the far-infrared (FIR) luminosity of star-forming galaxies are tightly correlated. This correlation is one of the tightest correlations known in astronomy. It extends over five orders of magnitude (Bell 2003) and is valid in starburst galaxies to redshifts of at least 3 (Seymour et al. 2008). Hence the total radio emission can serve as a tracer of magnetic fields and of star formation out to large distances. The correlation requires that total (mostly turbulent) magnetic fields and star formation are connected, so that the field strength exceeds several 100 μG in distant galaxies (Murphy 2009). The tightness needs multiple feedback mechanisms which are not yet understood (Lacki et al. 2010).

The total radio and far-infrared (FIR) or mid-IR (MIR) intensities are also highly correlated within galaxies. The exponent of the correlation in M 51 was found to be different in the central region, spiral arms and interarm regions (Dumas et al. 2011). The radio-infrared correlation can be presented as a correlation between turbulent field strength and star-formation rate (Fig. 3). In contrast, the ordered field is either uncorrelated with the star-formation rate, or anticorrelated in galaxies where the ordered field is strongest in interarm regions with low star formation (Fig. 10). A wavelet cross-correlation analysis for M 33 showed that the radio-FIR correlation holds at all scales down to 1 kpc (Tabatabaei et al. 2007). The correlation in the Large Magellanic Cloud (LMC) breaks down below scales of about 50 pc (Hughes et al. 2006), probably due to the diffusion of cosmic-ray electrons.

Fig. 3 Correlation between the strength of the total equipartition field (dominated by the turbulent field) and star-formation rate per area (determined from the 24 μm infrared intensities) within the galaxy NGC 4254 (Chyży 2008)

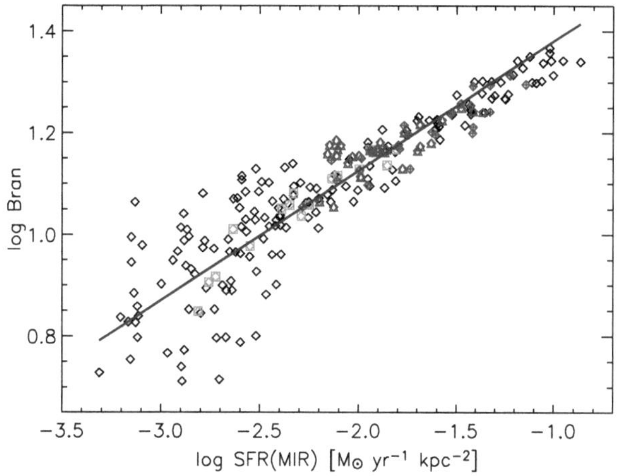

5 Structure of Ordered Galactic Magnetic Fields

5.1 Spiral Galaxies

Ordered (regular and/or anisotropic) field traced by polarized synchrotron emission form spiral patterns in almost every galaxy (Beck 2005), even in ring galaxies (Chyży and Buta 2008), in flocculent galaxies without massive spiral arms (Soida et al. 2002) and in the central regions of galaxies and in circum-nuclear gas rings of barred galaxies (Beck et al. 2005). Ordered fields are generally strongest (10–15 μG) in the regions *between* the optical spiral arms and oriented parallel to the adjacent spiral arms, in some galaxies forming *magnetic arms*, like in IC 342 (Krause 1993) and NGC 6946 (Fig. 10), with exceptionally high degrees of polarization (up to 50%). These are probably generated by a large-scale dynamo (see Sect. 6.1). In galaxies with strong density waves like M 51 (Fig. 4) and M 83 (Fig. 5) enhanced ordered (anisotropic) fields occur at the inner edges of the inner optical arms, in the interarm regions and in the outer optical arms. The observed smooth spiral patterns with significant pitch angles (10°–40°) indicate a general decoupling between magnetic fields and the gas flow, as predicted by mean-field dynamo action. There is no other model to explain the magnetic spiral patterns in many types of galaxies.

The typical degree of radio polarization within the spiral arms is only a few %; hence the field in the spiral arms must be mostly tangled or randomly oriented within the telescope beam, the width of which corresponds to a few 100 pc. Turbulent fields in spiral arms are probably generated by turbulent gas motions related to star formation activity or by a small-scale dynamo.

At wavelengths of around 20 cm, a striking asymmetry of the polarized emission occurs along the major axis of all 12 spiral galaxies observed so far with sufficiently high sensitivity that have inclinations of less than about 60°. The emission is almost completely depolarized by Faraday dispersion always around that side of the major axis which is the kinematically receding one (positive radial velocities) (Braun et al. 2010). In strongly inclined galaxies, both sides of the major axis become Faraday-depolarized at around 20 cm, where most of the polarized emission from the disk is Faraday-depolarized and the emission from the halo field dominates. The asymmetry is still visible at 11 cm wavelength, but disappears at smaller wavelengths. Modeling shows that both, disk and halo field, are of even-symmetry type, as predicted by dynamo models.

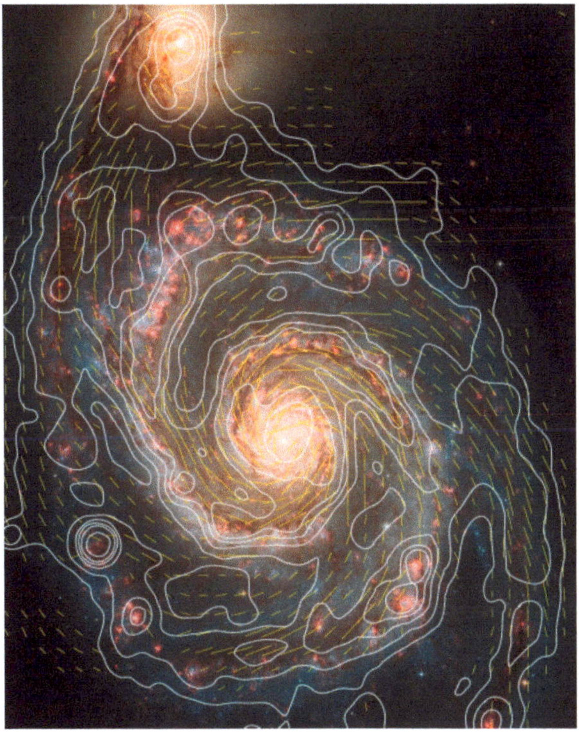

Fig. 4 Total radio emission (*contours*) and *B*-vectors of M 51, combined from observations at 6 cm wavelength with the VLA and Effelsberg telescopes and smoothed to 15" resolution (Fletcher et al. 2011), overlaid onto an optical image from the HST (Copyright: MPIfR Bonn and *Hubble Heritage Team*. Graphics: *Sterne und Weltraum*)

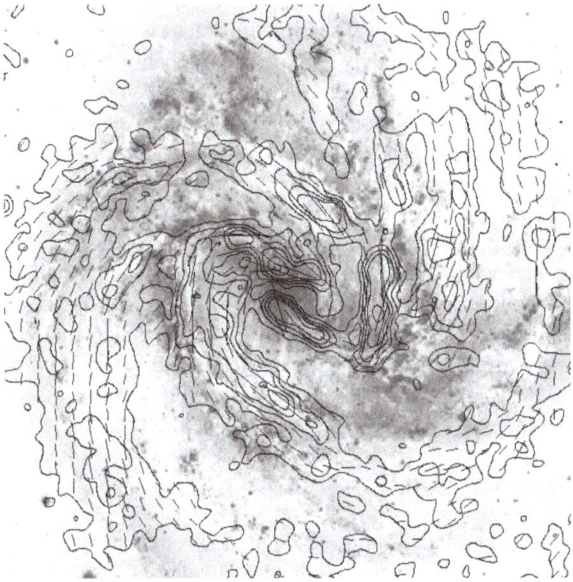

Fig. 5 Polarized radio emission (*contours*) and *B*-vectors of M 83, combined from observations at 6 cm wavelength with the VLA and Effelsberg telescopes and smoothed to 15" resolution (Beck, unpublished), overlaid onto an optical image from Dave Malin (Anglo Australian Observatory)

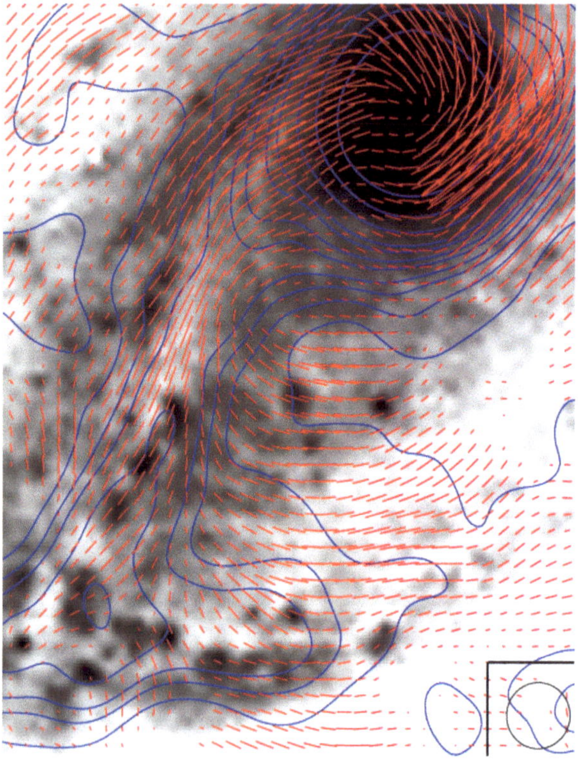

Fig. 6 Total radio emission (*contours*) and B-vectors of the barred galaxy NGC 1097, observed at 6 cm wavelength with the VLA and smoothed to 10" resolution (Beck et al. 2005). The background optical image is from Halton Arp (Copyright: MPIfR Bonn and Cerro Tololo Observatory)

5.2 Barred Galaxies

In galaxies with massive bars the field lines follow the gas flow (Fig. 6). As the gas rotates faster than the bar pattern of a galaxy, a shock occurs in the cold gas that has a small sound speed, while the flow of warm, diffuse gas is only slightly compressed but sheared. The ordered field is also hardly compressed, probably coupled to the diffuse gas and strong enough to affect its flow (Beck et al. 2005). The ordered field is also strong in the upstream region (south of the center in Fig. 6), oriented almost perpendicular to the bar. The polarization pattern in barred galaxies can be used as a tracer of shearing gas flows in the sky plane and complements spectroscopic measurements of radial velocities.

The central regions of barred galaxies are often sites of ongoing intense star formation and strong magnetic fields that can affect the gas flow. NGC 1097 hosts a bright ring with about 1.5 kpc diameter and an active nucleus in its center (Fig. 6). The ordered field in the ring has a spiral pattern and extends towards the nucleus. The orientation of the innermost spiral field agrees with that of the spiral dust filaments visible on optical images. Magnetic stress in the circumnuclear ring due to the strong total magnetic field (about 50 μG) can drive gas inflow (Balbus and Hawley 1998) at a rate of several $M_{\odot}$/yr, which is sufficient to fuel the activity of the nucleus (Beck et al. 2005).

5.3 Flocculent and Irregular Galaxies

Flocculent galaxies have disks but no prominent spiral arms. Nevertheless, spiral magnetic patterns are observed in all flocculent galaxies, indicative that the mean-field dynamo works

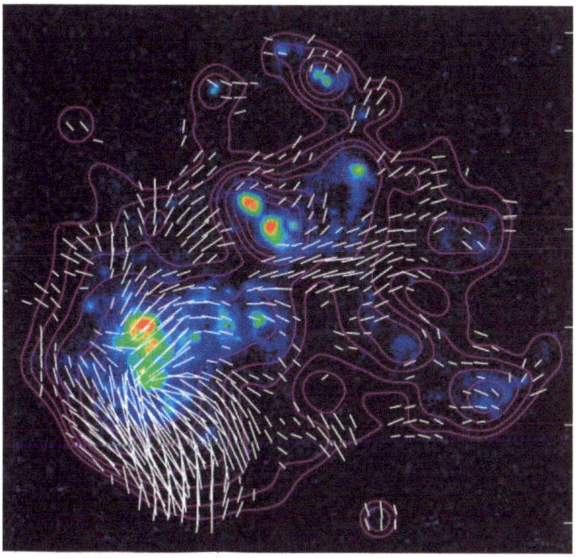

Fig. 7 Total radio intensity (*contours*) and B-vectors of the dwarf irregular galaxy IC 10, observed at 6 cm with the VLA (Chyży 2005). The background $H\alpha$ image is from Dominik Bomans (Bochum University)

independently of density waves. Ordered magnetic fields with strengths similar to those in grand-design spiral galaxies have been detected in the flocculent galaxies M 33, NGC 3521, NGC 5055 and in NGC 4414 (Soida et al. 2002), and also the mean degree of polarization (corrected for the differences in spatial resolution) is similar between grand-design and flocculent galaxies (Knapik et al. 2000).

Radio continuum maps of irregular, slowly rotating galaxies may reveal strong total equipartition magnetic fields, e.g. in the Magellanic-type galaxy NGC 4449, with a partly ordered field of about 7 μG strength and a spiral pattern (Chyży et al. 2000). Faraday rotation shows that this ordered field is mostly regular and the mean-field dynamo is operating. Dwarf irregular galaxies with almost chaotic rotation do not have any regular fields and only spots of faint polarized emission (Fig. 7). The turbulent field strengths are several times smaller than in spiral galaxies (Chyży et al. 2011), while they are of comparable in starburst dwarfs, e.g. in NGC 1569 with 10–15 μG (Kepley et al. 2010), where star formation activity is sufficiently high for the operation of the small-scale dynamo (Beck et al. 1996).

5.4 Interacting Galaxies

Gravitational interaction between galaxies leads to asymmetric gas flows, compression, shear, enhanced turbulence, and outflows. Compression and shear of gas flows can also modify the structure of galactic and intergalactic magnetic fields. In particular, fields can become aligned along the compression front or perpendicular to the velocity gradients. Such gas flows make turbulent fields highly anisotropic.

The classical interacting galaxy pair is NGC 4038/39, the "Antennae" (Chyży and Beck 2004). It shows bright, extended radio emission filling the whole system. In the interaction region between the galaxies, where star formation did not yet start, and at the northeastern edge of the system, the magnetic field is partly ordered, probably the result of compression and shearing motions along the tidal tail, respectively. Particularly strong, almost unpolarized emission comes from a region of violent star formation, hidden in dust, at the southern end of a dense cloud complex extending between the galaxies. The average total magnetic

Fig. 8 Polarized radio intensity (*contours*) and B-vectors of the Virgo galaxy NGC 4535, observed at 6 cm with the Effelsberg telescope (Weżgowiec et al. 2007). The background optical image is from the Digital Sky Survey

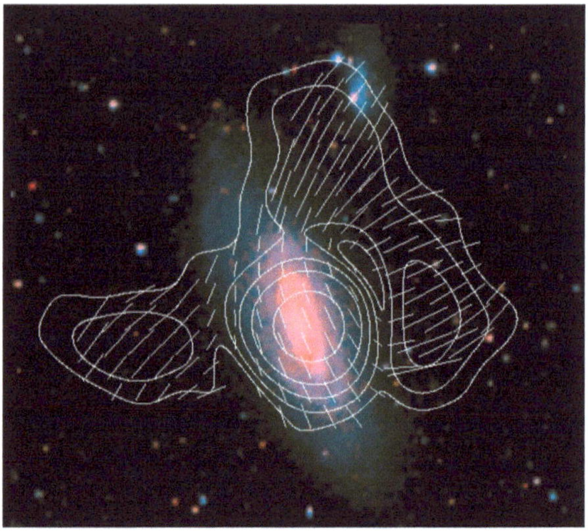

Fig. 9 Polarized radio intensity (*contours*) and B-vectors of the Virgo galaxy NGC 4569, observed at 6 cm with the Effelsberg telescope (Chyży et al. 2006). The background optical image is from the Digital Sky Survey

field is stronger than in normal spirals, but the mean degree of polarization is unusually low, implying that the fields are tangled in the regions of violent star formation.

Interaction with a dense intergalactic medium also imprints unique signatures onto magnetic fields and thus the radio emission. The Virgo cluster is a location of especially strong interaction effects (Figs. 8 and 9), and almost all cluster galaxies observed so far show asymmetries of their polarized emission because the outer magnetic fields were compressed (Vollmer et al. 2007). Ordered fields are an excellent tracer of past interactions between galaxies or with the intergalactic medium.

5.5 Edge-on Galaxies

Nearby galaxies seen "edge-on" generally show a disk-parallel field near the disk plane. As a result, polarized emission can be detected from distant, unresolved galaxies if the inclination is larger than about 20° (Stil et al. 2009). This opens a new method to search for ordered fields in distant galaxies. High-sensitivity observations of edge-on galaxies like NGC 891 (Krause 2009) and NGC 253 (Heesen et al. 2009) revealed vertical field components in the halo forming an X-shaped pattern (see Haverkorn and Heesen, this issue).

5.6 Early-Type Galaxies

Spiral galaxies of type Sa and S0 and elliptical galaxies without an active nucleus have very little star formation and hence do not produce cosmic rays that could emit synchrotron emission. The only deep observation of a Sa galaxy, M 104 with a prominent dust ring, revealed weak, ordered magnetic fields (Krause et al. 2006). Large-scale regular magnetic fields may exist in differentially rotating galaxies without star formation because turbulence can be generated by the magneto-rotational instability (MRI) (Sellwood and Balbus 1999). Their detection may become possible via RM grids of background sources with future radio telescopes (see Sect. 7).

6 Faraday Rotation

6.1 Large-Scale Regular Fields

Spiral fields are generated by compression, by shear in interarm regions, or by dynamo action (Sect. 2). Large-scale patterns of Faraday rotation measures (RM) are signatures of the mean-field dynamo and can be identified from diffuse polarized emission of the galaxy disks (Krause 1990) or from RM data of polarized background sources (Stepanov et al. 2008). If several dynamo modes are superimposed, a Fourier analysis of the RM variation is needed. The resolution of present-day observations is sufficient to identify 2–3 modes.

The disks of about a dozen nearby spiral galaxies reveal large-scale RM patterns. The Andromeda galaxy M 31 (Fig. 1) is the prototype of a dynamo-generated axisymmetric spiral disk field (Fletcher et al. 2004). Other candidates for a dominating axisymmetric disk field (dynamo mode $m = 0$) are the nearby spiral IC 342 (Krause et al. 1989a) and the irregular Large Magellanic Cloud (LMC) (Gaensler et al. 2005). Dominating bisymmetric spiral fields (dynamo mode $m = 1$) are rare, as predicted by dynamo models. Faraday rotation in NGC 6946 (Fig. 11) and in other similar galaxies with magnetic arms can be described by a superposition of two azimuthal dynamo modes ($m = 0$ and $m = 2$) with about equal amplitudes where the quadrisymmetric spiral mode is phase shifted with respect to the density wave (Beck 2007).

However, the spiral pattern of magnetic fields cannot be solely the result of mean-field dynamo action. If the beautiful spiral pattern of M 51 seen in radio polarization (Fig. 4) were only due to a regular field, its line-of sight component should generate a conspicuous large-scale pattern in Faraday rotation, which is not observed. This means that a large amount of the ordered field is anisotropic and probably generated by compression and shear of the non-axisymmetric gas flows in the density-wave potential. The anisotropic field is strongest at the positions of the prominent dust lanes on the inner edge of the inner gas spiral arms, due to compression of turbulent fields in the density-wave shock. Regular fields (dynamo modes

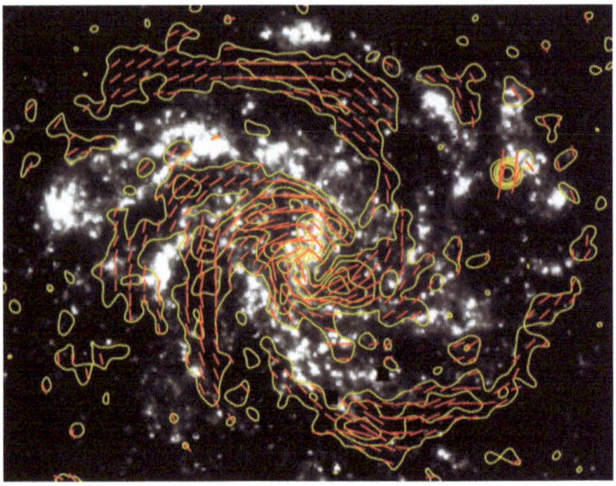

Fig. 10 Polarized radio emission (*contours*) and *B*-vectors of NCC 6946, combined from observations at 6 cm wavelength with the VLA and Effelsberg telescopes and smoothed to 15" resolution (Beck 2007), overlaid onto an Hα image from Anne Ferguson (Copyright: MPIfR Bonn; graphics: *Sterne und Weltraum*)

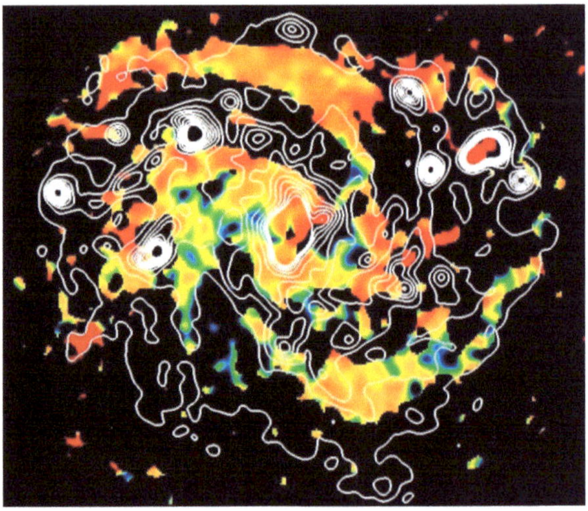

Fig. 11 Total radio intensity at 6 cm wavelength (*contours*) and Faraday rotation measures between 3.6 cm and 6 cm wavelengths (*colors*) in NGC 6946, derived from combined observations with the VLA and Effelsberg telescopes (Beck 2007). The colour scale ranges from -75 rad/m^2 to $+175$ rad/m^2. The average rotation measure of about $+50$ rad/m^2 is caused by the foreground medium in the Milky Way

$m = 0$ and $m = 1$) also exist in the disk of M 51, but are much weaker than the anisotropic field (Fletcher et al. 2011).

In many other observed galaxy disks no clear patterns of Faraday rotation were found. Either several dynamo modes are superimposed and cannot be distinguished with the limited sensitivity and resolution of present-day telescopes, or the timescale for the generation of large-scale modes is longer than the galaxy's lifetime (Arshakian et al. 2009).

By measuring the signs of the RM distribution and the velocity field on both sides of a galaxy's major axis, the inward and outward directions of the radial component of the axisymmetric field can be easily distinguished (Krause and Beck 1998). Dynamo models predict that both signs have the same probability, which is confirmed by observations. The axisymmetric fields of M 31, IC 342, NGC 253 and the axisymmetric field component in NGC 6946 point inwards, while those of NGC 4254, NGC 5775 and the axisymmetric component of the disk field in M 51 point outwards.

While the azimuthal symmetry of the magnetic field is known for many galaxies, the vertical symmetry (even or odd) is much harder to determine. The RM patterns of even and odd modes are similar in mildly inclined galaxies. The field of odd modes reverses its sign above and below the galactic plane. The symmetry type becomes only visible in strongly inclined galaxies, as the RM sign above and below the plane (see Haverkorn and Heesen, this issue).

6.2 Field Reversals

Large-scale field reversals at certain radial distances from a galaxy's center, like those in the Milky Way (Noutsos, this issue), have not been detected in spiral galaxies, although high-resolution RM maps of Faraday rotation are available for many spiral galaxies. In M 81 the dominating BSS field implies two large-scale reversals (Krause et al. 1989b). A satisfying explanation for the reversals in the Milky Way is still lacking. They may be restricted to a thin layer near to the plane and hence are hardly visible in the average RM data of external galaxies along the line of sight. Secondly, the reversals in the Milky Way may be of limited azimuthal extent and are difficult to observe in external galaxies with present-day telescopes. In the barred galaxy NGC 7479, where a jet serves as a bright polarized background and high-resolution observations were possible with high signal-to-noise ratio, several reversals on 1–2 kpc scale were detected in the foreground disk of the galaxy (Laine and Beck 2008). Thirdly, the reversals in the Milky Way may be part of a disturbed field structure, e.g. due to interaction with the Magellanic clouds.

7 Outlook

Future high-resolution, high-sensitivity observations at high frequencies with the *Extended Very Large Array* (EVLA) and the planned *Square Kilometre Array* (SKA) will directly map the detailed field structure and the interaction with the gas. The low-frequency radio telescopes *Low Frequency Array* (LOFAR) and the *Murchison Widefield Array* (MWA) (under construction) are suitable instruments to search for extended synchrotron radiation at the lowest possible levels in outer galaxy disks and halos and the transition to intergalactic space. Low frequencies are also ideal to search for small Faraday rotation measures from weak interstellar and intergalactic fields (Beck 2009).

Total synchrotron emission, signature of total magnetic fields, can be detected with the SKA out to very large redshifts for starburst galaxies, depending on luminosity and magnetic field strength (Fig. 12). However, for fields weaker than 3.25 μG $(1+z)^2$, energy loss of cosmic-ray electrons is dominated by the inverse Compton effect with CMB photons, so that their energy appears mostly in X-rays and not in the radio range. On the other hand, for strong magnetic fields the energy range of the electrons emitting at 1.4 GHz shifts to low energies where ionization and bremsstrahlung losses become dominant. In summary, the mere detection of synchrotron emission at high redshifts will constrain the allowed magnetic field strengths in young galaxies.

If polarized emission from galaxies is too weak to be detected, the method of *RM grids* towards background QSOs can still be applied. Here, the distance limit is given by the polarized flux of the background QSO which can be much higher than that of the intervening galaxy. Significant regular fields of several μG strength were already detected in distant galaxies (Bernet et al. 2008). Mean-field dynamo theory predicts RMs from evolving regular fields with increasing coherence scale at $z \leq 3$ (Arshakian et al. 2009). The RM values are reduced by the redshift dilution factor of $(1+z)^{-2}$.

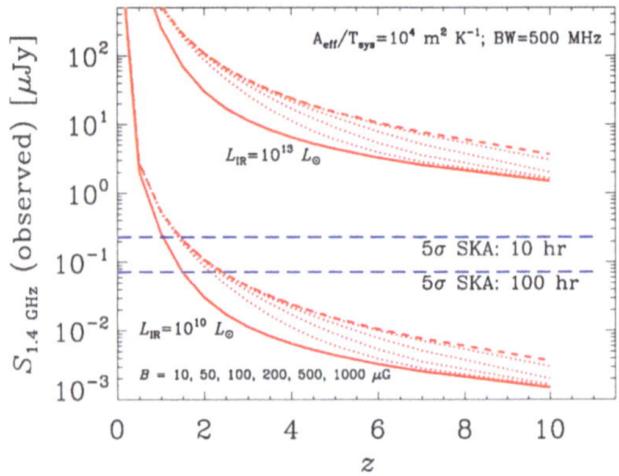

Fig. 12 Total synchrotron emission at 1.4 GHz as a function of redshift z and magnetic field strength B, and the 5σ detection limits for 10 h and 100 h integration time with the SKA (Murphy 2009)

Fig. 13 Simulation of RMs towards background sources (*white points*) in the region of M 31 observable with the SKA within 1 h. Optical emission from M 31 is shown in *red*, diffuse radio continuum intensity in *blue* and diffuse polarized intensity in *green* (from Gaensler, priv. comm.)

Faraday rotation in the direction of QSOs allows to determine the strength and pattern of a regular field in an intervening galaxy (Kronberg et al. 1992). A reliable model for the field structure of galaxies needs RM values from a large number of polarized background sources, hence large sensitivity and/or high survey speed.

The planned all-sky POSSUM survey at 1.4 GHz with the *Australia SKA Pathfinder* (ASKAP) telescope (under construction) with 30 deg^2 field of view (Gaensler et al. 2010) will measure about 100 RM values from polarized extragalactic sources per square degree within 10 h integration time. Similarly long integrations with the EVLA and with MeerKAT will show more sources, but their fields of view are smaller.

The SKA "Magnetism" Key Science Project plans to observe an all-sky survey (at least 10^4 deg^2) around 1 GHz with 1 h integration per field which will be able to detect sources of 0.5–1 μJy flux density and measure at least 1500 RMs per square degree. This will contain at least 2×10^7 RMs from compact polarized extragalactic sources at a mean spacing of $\simeq 90''$ (Gaensler et al. 2004). More than 10 000 RM values are expected in the area of M 31 (Fig. 13) and will allow the detailed reconstruction of the 3-D field structure in this and many other nearby galaxies, while simple patterns of regular fields can be recognized out to distances of about 100 Mpc (Stepanov et al. 2008).

We are entering a golden era of magnetic field observations in galaxies.

Acknowledgements ISSI Bern and Prof. Dr. André Balogh is acknowledged for hospitality and financial support. The author thanks Dr. Marita Krause for careful reading of the manuscript.

References

P. Abreu et al. (The Pierre Auger Collaboration), Astropart. Phys. **34**, 314 (2010)
T.G. Arshakian, R. Beck, M. Krause, D. Sokoloff, Astron. Astrophys. **494**, 21 (2009)
S.A. Balbus, J.F. Hawley, Rev. Mod. Phys. **70**, 1 (1998)
R. Beck, in *Cosmic Magnetic Fields*, ed. by R. Wielebinski, R. Beck (Springer, Berlin, 2005), p. 41
R. Beck, Astron. Astrophys. **470**, 539 (2007)
R. Beck, Rev. Mex. Astron. Astrof. **36**, 1 (2009)
R. Beck, M. Krause, Astron. Nachr. **326**, 414 (2005)
R. Beck, A. Brandenburg, D. Moss, A. Shukurov, D. Sokoloff, Annu. Rev. Astron. Astrophys. **34**, 155 (1996)
R. Beck, A. Fletcher, A. Shukurov et al., Astron. Astrophys. **444**, 739 (2005)
E.F. Bell, Astrophys. J. **586**, 794 (2003)
E.M. Berkhuijsen, R. Beck, P. Hoernes, Astron. Astrophys. **398**, 937 (2003)
M.L. Bernet, F. Miniati, S.J. Lilly, P.P. Kronberg, M. Dessauges-Zavadsky, Nature **454**, 302 (2008)
A. Brandenburg et al., Space Sci. Rev. (2011, this issue)
R. Braun, G. Heald, R. Beck, Astron. Astrophys. **514**, A42 (2010)
M.A. Brentjens, A.G. de Bruyn, Astron. Astrophys. **441**, 1217 (2005)
C. Caprini, R. Durrer, E. Fenu, J. Cosmol. Astropart. Phys. **11**, 001 (2009)
K.T. Chyży, in *From 30 Doradus to Lyman Break Galaxies*, ed. by R. de Grijs, R.M. González Delgado (Springer, Berlin, 2005), p. 12
K.T. Chyży, Astron. Astrophys. **482**, 755 (2008)
K.T. Chyży, R. Beck, Astron. Astrophys. **417**, 541 (2004)
K.T. Chyży, R.J. Buta, Astrophys. J. **677**, L17 (2008)
K.T. Chyży, R. Beck, S. Kohle, U. Klein, M. Urbanik, Astron. Astrophys. **356**, 757 (2000)
K.T. Chyży, M. Soida, D.J. Bomans et al., Astron. Astrophys. **447**, 465 (2006)
K.T. Chyży, M. Weżgowiec, R. Beck, D.J. Bomans, Astron. Astrophys. **529**, A94 (2011)
K. Dolag, M. Kachelriess, S. Ostapchenko, R. Tomas, Astrophys. J. **727**, L4 (2011)
G. Dumas, E. Schinnerer, F.S. Tabatabaei et al., Astron. J. **141**, 41 (2011)
A. Fletcher, E.M. Berkhuijsen, R. Beck, A. Shukurov, Astron. Astrophys. **414**, 53 (2004)
A. Fletcher, R. Beck, A. Shukurov, E.M. Berkhuijsen, C. Horellou, Mon. Not. R. Astron. Soc. **412**, 2396 (2011)
P. Fosalba, A. Lazarian, S. Prunet, J.A. Tauber, Astrophys. J. **546**, 762 (2002)
B.M. Gaensler, R. Beck, L. Feretti, in *Science with the Square Kilometer Array*, ed. by C. Carilli, S. Rawlings. New Astr. Rev., vol. 48 (Elsevier, Amsterdam, 2004), p. 1003
B.M. Gaensler, M. Haverkorn, L. Staveley-Smith et al., Science **307**, 1610 (2005)
B.M. Gaensler, T.L. Landecker, A.R. Taylor, Bull. Am. Astron. Soc. **42**, 470 (2010)
C. Gissinger, S. Fromang, E. Dormy, Mon. Not. R. Astron. Soc. **394**, L84 (2009)
J.S. Greaves, W.S. Holland, T. Jenness, T.G. Hawarden, Nature **404**, 732 (2000)
O. Gressel, D. Elstner, U. Ziegler, G. Rüdiger, Astron. Astrophys. **486**, L35 (2008)
M. Hanasz, D. Wóltański, K. Kowalik, Astrophys. J. **706**, L155 (2009)
H. Hanayama, K. Takahashi, K. Kotake et al., Astrophys. J. **633**, 941 (2005)
M. Haverkorn, V. Heesen, Space Sci. Rev. (2011, this issue). doi:10.1007/s11214-011-9757-0
V. Heesen, M. Krause, R. Beck, R.-J. Dettmar, Astron. Astrophys. **506**, 1123 (2009)
C. Heiles, Space Sci. Rev. (2011, this issue)
A. Hughes, T. Wong, R. Ekers et al., Mon. Not. R. Astron. Soc. **370**, 363 (2006)
A.A. Kepley, S. Mühle, J. Everett et al., Astrophys. J. **712**, 536 (2010)
U. Klein, R. Wielebinski, H. Morsi, Astron. Astrophys. **190**, 41 (1988)
J. Knapik, M. Soida, R.-J. Dettmar, R. Beck, M. Urbanik, Astron. Astrophys. **362**, 910 (2000)
F. Krause, R. Beck, Astron. Astrophys. **335**, 789 (1998)
M. Krause, in *Galactic and Intergalactic Magnetic Fields*, ed. by R. Beck et al. (Kluwer, Dordrecht, 1990), p. 187

M. Krause, in *The Cosmic Dynamo*, ed. by F. Krause et al. (Kluwer, Dordrecht, 1993), p. 303
M. Krause, Rev. Mex. Astron. Astrof. **36**, 25 (2009)
M. Krause, E. Hummel, R. Beck, Astron. Astrophys. **217**, 4 (1989a)
M. Krause, R. Beck, E. Hummel, Astron. Astrophys. **217**, 17 (1989b)
M. Krause, R. Wielebinski, M. Dumke, Astron. Astrophys. **448**, 133 (2006)
P.P. Kronberg, J.J. Perry, E.L.H. Zukowski, Astrophys. J. **387**, 528 (1992)
B.C. Lacki, T.A. Thompson, E. Quataert, Astrophys. J. **717**, 1 (2010)
S. Laine, R. Beck, Astrophys. J. **673**, 128 (2008)
M. Lazar, R. Schlickeiser, R. Wielebinski, S. Poedts, Astrophys. J. **693**, 1133 (2009)
E. Murphy, Astrophys. J. **706**, 482 (2009)
A. Noutsos, Space Sci. Rev. (2011, this issue)
M.J. Rees, in *Cosmic Magnetic Fields*, ed. by R. Wielebinski, R. Beck (Springer, Berlin, 2005), p. 1
T. Robishaw, E. Quataert, C. Heiles, Astrophys. J. **680**, 981 (2008)
S.M. Scarrott, D. Ward-Thompson, R.F. Warren-Smith, Mon. Not. R. Astron. Soc. **224**, 299 (1987)
D.R.G. Schleicher, R. Banerjee, S. Sur et al., Astron. Astrophys. **522**, A115 (2010)
J.A. Sellwood, S.A. Balbus, Astrophys. J. **511**, 660 (1999)
N. Seymour, T. Dwelly, D. Moss et al., Mon. Not. R. Astron. Soc. **386**, 1695 (2008)
A. Shukurov, D. Sokoloff, K. Subramanian, A. Brandenburg, Astron. Astrophys. **448**, L33 (2006)
M. Soida, R. Beck, M. Urbanik, J. Braine, Astron. Astrophys. **394**, 47 (2002)
D.D. Sokoloff, A.A. Bykov, A. Shukurov et al., Mon. Not. R. Astron. Soc. **299**, 189 (1998). Erratum in Mon. Not. R. Astron. Soc. **303**, 207 (1999)
R. Stepanov, T.G. Arshakian, R. Beck, P. Frick, M. Krause, Astron. Astrophys. **480**, 45 (2008)
J.M. Stil, M. Krause, R. Beck, A.R. Taylor, Astrophys. J. **693**, 1392 (2009)
F. Tabatabaei, R. Beck, M. Krause et al., Astron. Astrophys. **466**, 509 (2007)
F. Tabatabaei, M. Krause, A. Fletcher, R. Beck, Astron. Astrophys. **490**, 1005 (2008)
J.-W. Tang, P.T.P. Ho, P.M. Koch et al., Astrophys. J. **700**, 251 (2009)
T.A. Thompson, E. Quataert, E. Waxman, N. Murray, C.L. Martin, Astrophys. J. **645**, 186 (2006)
B. Vollmer, M. Soida, R. Beck et al., Astron. Astrophys. **464**, L37 (2007)
M. Weżgowiec, M. Urbanik, B. Vollmer et al., Astron. Astrophys. **471**, 93 (2007)
L.M. Widrow, Rev. Mod. Phys. **74**, 775 (2002)

Space Sci Rev (2012) 166:231–261
DOI 10.1007/s11214-011-9775-y

Magnetic Fields in Supernova Remnants and Pulsar-Wind Nebulae

Stephen P. Reynolds · B.M. Gaensler · Fabrizio Bocchino

Received: 21 September 2010 / Accepted: 23 March 2011 / Published online: 16 April 2011

Abstract We review the observations of supernova remnants (SNRs) and pulsar-wind nebulae (PWNe) that give information on the strength and orientation of magnetic fields. Radio polarimetry gives the degree of order of magnetic fields, and the orientation of the ordered component. Many young shell supernova remnants show evidence for synchrotron X-ray emission. The spatial analysis of this emission suggests that magnetic fields are amplified by one to two orders of magnitude in strong shocks. Detection of several remnants in TeV gamma rays implies a lower limit on the magnetic-field strength (or a measurement, if the emission process is inverse-Compton upscattering of cosmic microwave background photons). Upper limits to GeV emission similarly provide lower limits on magnetic-field strengths. In the historical shell remnants, lower limits on B range from 25 to 1000 μG. Two remnants show variability of synchrotron X-ray emission with a timescale of years. If this timescale is the electron-acceleration or radiative loss timescale, magnetic fields of order 1 mG are also implied. In pulsar-wind nebulae, equipartition arguments and dynamical modeling can be used to infer magnetic-field strengths anywhere from ~5 μG to 1 mG. Polarized fractions are considerably higher than in SNRs, ranging to 50 or 60% in some cases; magnetic-field geometries often suggest a toroidal structure around the pulsar, but this is not universal. Viewing-angle effects undoubtedly play a role. MHD models of radio emission in shell SNRs show that different orientations of upstream magnetic field, and different assumptions about electron acceleration, predict different radio morphology. In the remnant of

S.P. Reynolds (✉)
Physics Department, North Carolina State University, Raleigh, NC, USA
e-mail: reynolds@ncsu.edu

B.M. Gaensler
Sydney Institute for Astronomy, School of Physics, The University of Sydney, Sydney, NSW 2006, Australia
e-mail: bryan.gaensler@sydney.edu.au

F. Bocchino
INAF-Osservatorio Astronomico di Palermo, Piazza del Parlamento 1, 90134 Palermo, Italy
e-mail: bocchino@astropa.inaf.it

SN 1006, such comparisons imply a magnetic-field orientation connecting the bright limbs, with a substantial density gradient across the remnant.

Keywords Supernova remnants · Pulsar-wind nebulae · Magnetic-field amplification

1 Shell Supernova Remnants: Review

1.1 Introduction

Supernova remnants and pulsar-wind nebulae are prominent Galactic synchrotron sources at radio and, often, X-ray wavelengths. The spatial and spectral analysis of the synchrotron emission can be used to deduce or constrain magnetic-field strengths and orientations. In this review, "supernova remnant" (SNR) will be used for shell (i.e., non-pulsar-driven) remnants; if the emission is due to a pulsar, whether the object is young (like the Crab Nebula) or old (as in Hα bow-shock nebulae) the object will be termed a pulsar-wind nebula (PWN). Thus a SNR may contain a PWN, a combination sometimes called a "composite" SNR.

SNRs are primarily radio objects. In the Milky Way, 274 are listed in Green's (2009) well-known catalogue (http://www.mrao.cam.ac.uk/surveys/snrs/); most are known only by radio emission. SNRs are well-studied only in Local Group galaxies where they are bright and large enough to be well-imaged with radio interferometers. Galactic remnants range in angular size from less than $2'$ to many degrees, and represent a range in ages from about 100 years (G1.9+0.3) to over 10^5 yr, though age estimates are difficult for very old objects.

1.2 SNR Dynamics

SNRs evolve through various phases as their shock waves decelerate in the ambient medium. Very soon after an explosion (either Type Ia or core-collapse) deposits $\sim 10^{51}$ erg into the ISM, a shock wave is driven into the surrounding circumstellar medium (CSM: modified by the progenitor) or unmodified interstellar medium (ISM). Core-collapse SNe eject several solar masses with a range of velocities, typically of order 5,000 km s^{-1}. Type Ia's are thought to represent the thermonuclear disruption of a white dwarf, so eject 1.4 $M_{\odot}$ at about 10,000 km s^{-1}. Not long after the explosion, deceleration of the outer blast wave causes the formation of a reverse shock as inner ejecta are forced to decelerate. This begins the ejecta phase of evolution, characterized by the presence of both the forward shock (blast wave) heating ISM or CSM, and the reverse shock heating ejecta. Cooling timescales are much longer than dynamical timescales, so this phase is essentially adiabatic. Depending on the density structure in the ejecta and surrounding material, the shock radius may evolve as $t^{0.6} - t^{0.9}$ (Chevalier 1982). Eventually the reverse shock moves in to the center of the SNR and, after what may be extensive reverberations, disappears. By this time (several thousand years for typical parameters), several times the ejected mass have been swept up by the blast wave, and the remnant is settling into the Sedov phase, well described by the similarity solution for a point explosion in a uniform medium, with shock radius obeying $R \propto t^{0.4}$. This phase is adiabatic as well. Eventually, cooling times in shocked ISM become comparable to the remnant age, and shocks become radiative: lossy and optically prominent. This typically occurs when shock velocities v_s drop to around 200 km s^{-1}. Denser regions around the remnant periphery may become radiative before other parts; even young SNRs (like Kepler, SN 1604) can show radiative shocks in some regions. Once the bulk of the blast wave is radiative, the deceleration is more marked. The remnant interior will remain hot for some time, producing what is known as a "pressure-driven snowplow." Eventually, the remnant becomes

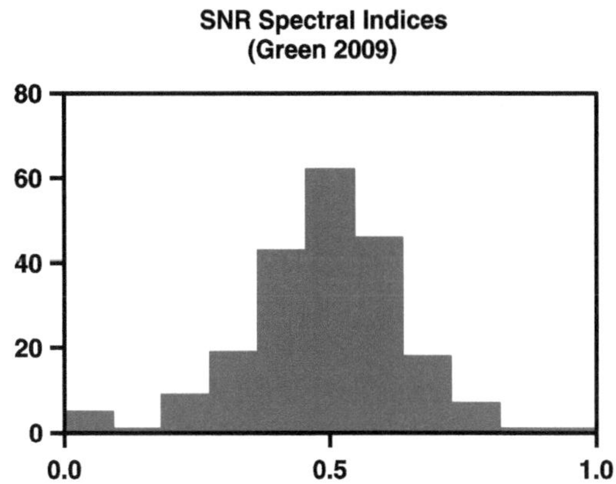

Fig. 1 Histogram of shell SNRs with fairly well-measured radio spectral indices, from Green (2009). PWNe are excluded

sufficiently confused as to lose its identity, and the remaining kinetic energy is dissipated as sound waves in the ISM. Various blast-wave solutions are described in Ostriker and McKee (1988); particular applications to SNRs are made in Truelove and McKee (1999).

1.3 Radio inferences

The prevalence of radio emission from adiabatic-phase SNRs, with low shock compression ratios, indicates that particle acceleration must be ongoing; simple compression of ambient Galactic magnetic field and cosmic-ray electrons would produce both insufficiently bright emission and the wrong spectrum (Reynolds 2008). Since 1977, the traditional explanation for this electron acceleration has been diffusive shock acceleration (DSA; see Blandford and Eichler 1987 for a comprehensive review). Electrons emitting synchrotron radiation at radio wavelengths have energies of order 1–10 GeV ($E = 14.7\sqrt{\nu(\mathrm{GHz})/B(\mu\mathrm{G})}$ GeV). The observed radio power-law spectra $S_\nu \propto \nu^{-\alpha}$, with $\alpha \sim 0.5$ (see Fig. 1), imply from synchrotron theory a power-law energy distribution of electrons KE^{-s} electrons cm^{-3} erg^{-1} with $s = 2\alpha + 1 \sim 2$. Radio observations do not allow the separate deduction of energy in electrons and in magnetic field; the synchrotron emissivity of this power-law distribution of electrons is proportional to $KB^{(s+1)/2}$, or roughly to the product of energy density in electrons and in magnetic field. If particle and magnetic-field energies were in equipartition (the minimum-energy state producing an observed spectrum), the magnetic-field strength could be deduced. For a spherical remnant of radius R pc at distance D kpc with $\alpha = 0.5$,

$$B = 20\big((1+k)S_9 D_{\mathrm{kpc}}^2/(\phi R_{\mathrm{pc}}^3)\big)^{2/7}\ \mu\mathrm{G}, \tag{1}$$

where S_9 is the flux density at 1 GHz in Jy, ϕ is the volume filling factor, and k is the ratio of energy density in ions to that in electrons (e.g., Pacholczyk 1970). But there is no obvious reason that equipartition should hold: SNRs are typically very inefficient synchrotron radiators, with a very small fraction of the 10^{51} erg required to produce the observed emission. Thermal energies are very much larger than either particle or field energy. Thus equipartition magnetic-field strengths are really just a measure of (approximately) remnant surface brightness.

The question of magnetic-field amplification in strong SNR shocks, to be discussed at length below, is intimately related to the question of shock modification by cosmic-ray pressure (efficient shock acceleration, or nonlinear DSA). Thus evidence for nonlinear DSA is related to estimation of magnetic-field strengths. One prediction of this theory is the gradual

Fig. 2 Schematic velocity profile of a shock wave with upstream velocity u_1 and downstream u_2. *Dotted line*: test-particle shock (velocity discontinuity). *Solid line*: shock modified by cosmic rays

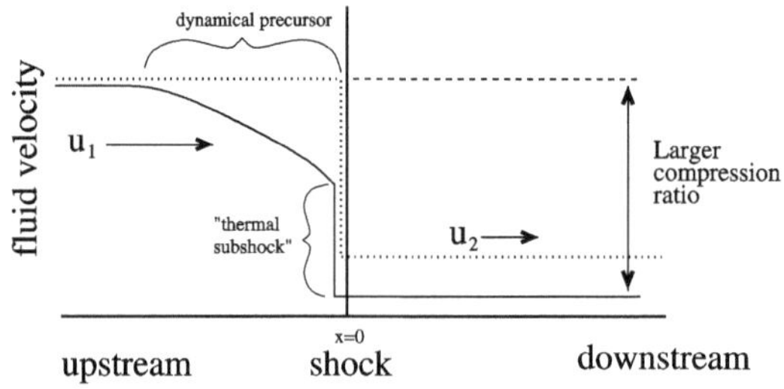

deceleration of incoming flow (in the shock frame) by forward-diffusing cosmic rays, so that the compression ratio r varies continuously from a possibly large value (far upstream velocity/downstream velocity, perhaps 10 or more) to about 3, at which point most calculations show the development of a "thermal subshock" on a length scale of thermal proton gyroradius, much smaller than other length scales. (See Fig. 2.) In test-particle (inefficient) DSA, well-known results give the spectrum of accelerated particles as a power-law with energy index s depending only on the shock compression ratio, $s = (r+2)/(r-1)$. Since strong adiabatic shocks are expected to have $r = 4$ (for adiabatic index $\gamma = 5/3$), we predict $s = 2$ which implies $\alpha = 0.5$, close to the mean of the distribution of over 200 SNR values from Green's catalogue. However, the dispersion in that distribution is substantial (see Fig. 1); explaining values of α of 0.6–0.7, typical for young SNRs, or as low as 0.3, sometimes seen in older SNRs, is not straightforward. Low Mach-number shocks will have lower compressions: in fact, for Mach number $\mathcal{M}$, $r^{-1} = (\gamma-1)/(\gamma+1) + 2/((\gamma+1)\mathcal{M}^2)$. For $\mathcal{M} < 10$, this can significantly increase the steepness of the predicted spectrum; but this is demonstrably not the case in the young remnants with the steeper spectra. No variation in shock Mach number can explain $\alpha < 0.5$; traditional explanations involve confusing thermal emission or other instrumental effects, but remain unconvincing.

A consistent explanation for steeper radio spectra in young remnants is available, however. Particles of a particular energy will diffuse (on average) a certain distance ahead of an efficient shock, where they will see a particular effective compression ratio, which fixes the slope of the distribution function near that energy. If, as expected, more energetic particles diffuse further, they will see larger compression ratios and the spectrum will become flatter. Thus one predicts spectra flattening (hardening) to higher energy, an effect first pointed out by Eichler (1979). This effect was quantified in a Monte Carlo simulation of electron acceleration, and compared with radio observations of young SNRs, which did seem to show the effect (Ellison and Reynolds 1991; Reynolds and Ellison 1992; see Fig. 3). The prediction for the emitted synchrotron spectrum depends on the magnetic-field strength; stronger fields mean lower-energy particles producing radiation in a particular radio-astronomical bandpass. Reynolds & Ellison found that explaining the relatively steep spectra of Tycho's and Kepler's SNRs required magnetic fields of up to 1 mG and above, though with substantial errors.

The degree of order in the magnetic field in SNRs can be directly inferred from the degree of polarization of synchrotron radiation (subject to observational issues such as Faraday depolarization and resolution effects [beam depolarization]). The degree of linear polarization of power-law synchrotron radiation from electrons in a uniform magnetic field is given by $P = (s+1)/(s+7/3)$, so that $s = 2 \Rightarrow P = 69\%$. Such high values are never observed in SNRs; a few older SNRs such as DA 530 show $P \sim 50\%$ in some regions (Landecker et al. 1999, Fig. 4), but historical shells show much lower values, 10–15% (see references in Reynolds and Gilmore 1993). Very roughly, the observed polarized

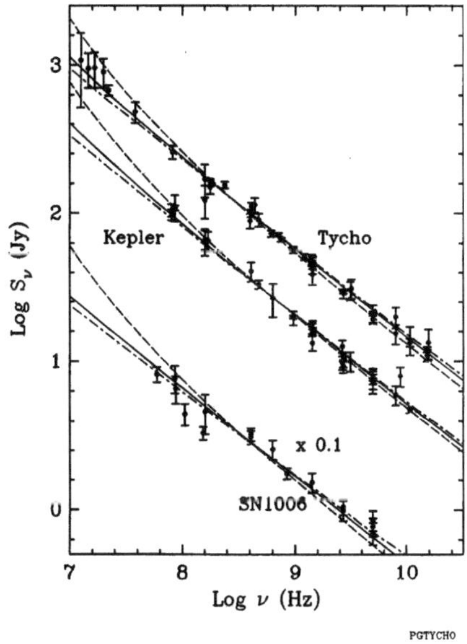

Fig. 3 Observed total flux observations of Tycho's and Kepler's SNRs and SN 1006, with model spectra superposed. All three spectra derive from the same electron spectrum but with varied magnetic field. For Tycho and Kepler, the *dashed line* corresponds to $B = 10$ mG, the *solid line* to 1 mG, and the *dot-dashed line* to 0.1 mG; for SN 1006, the *dashed, solid,* and *dot-dashed lines* correspond to $B = 10$ mG, 0.1 mG, and 1 μG, respectively. The *dotted line* shows the spectrum that would be produced if electrons had the same energy distribution as protons, showing how differently the nonlinear shock treats electrons and protons (Reynolds and Ellison 1992)

fraction over the maximum possible gives an estimate of the fractional energy in the ordered component of the magnetic field (Burn 1966); thus in Tycho, for instance (see Fig. 5, Reynoso et al. 1997), less than 15% of the magnetic energy is in an ordered component. However, the direction of that component is radial, a peculiar property shared by the other historical shells SN 1006, Cas A, Kepler, and G11.2-0.3. The nature of this radial component is unclear; while radial motions such as predicted to occur at the Rayleigh-Taylor unstable contact discontinuity between shocked ejecta and shocked ISM might enhance radial field, this effect should occur somewhat interior, while radial fields are observed immediately at remnant edges (e.g., Tycho: Dickel et al. 1991). Various MHD simulations have been performed in an attempt to understand this effect (Jun and Norman 1996; Jun and Jones 1999). In older remnants such as DA 530, magnetic fields are generally confused or tangential. Tangential fields can readily be explained by simple compression in radiative shocks with large r, and this is the conventional explanation.

1.4 X-Ray and Gamma-Ray Inferences

The discovery of X-ray synchrotron emission from SNRs dates from the early 1980s (Toor 1980; Becker et al. 1980; Reynolds and Chevalier 1981) but observational and theoretical confusion prevented its wide acceptance until the unmistakable observations with *ASCA* of the remnant of SN 1006, demonstrating conclusively that central emission was thermal while the bright limbs showed lineless spectra well described by power laws (Koyama et al. 1995). (See Reynolds (2008) for a detailed review of high-energy nonthermal emission from SNRs.) At this time, four Galactic SNRs show X-ray spectra dominated by synchrotron

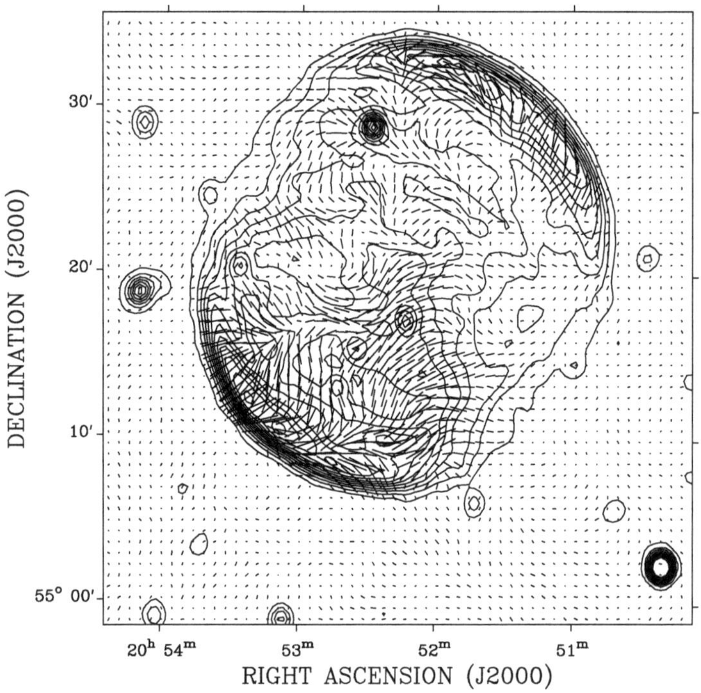

Fig. 4 Radio contours and polarized E-vectors at 1420 MHz for DA 530 (Landecker et al. 1999). Direction of the sky-plane component of the magnetic field is at right angles to the directions of the E-vectors, i.e., predominantly tangential in the bright limbs. Foreground Faraday rotation has been corrected for, though it is small

emission: G1.9+0.3, SN 1006, G347.3-0.5 (RX J1713.7-3946), and G266.2-1.2 ("Vela Jr." or RX J0852-4622). Of these, the first two are young, with symmetric X-ray morphologies, and show thermal X-ray emission from fainter regions. SN 1006 is widely accepted to be the remnant of a Type Ia supernova (evidence includes the presence of Balmer-dominated optical emission from the forward shock, and the high Galactic latitude of 15°); the evidence is not as firm for G1.9+0.3, but it is also suspected to be a Type Ia remnant (Reynolds et al. 2008). The other two have much larger angular sizes and considerably more irregular morphologies (Slane et al. 1999, 2001); they also show no trace of thermal X-rays, putting severe upper limits on the presence of any thermal gas (Ellison et al. 2010). Both also contain X-ray point sources, suggesting a core-collapse origin. In addition to these four objects, all other historical shells (Tycho = SN 1572, Kepler = SN 1604, Cas A (unseen SN around 1680), G11.2-0.3 = SN 386, and RCW 86 = SN 185?) show synchrotron emission in regions, typically (but not always) "thin rims" at the remnant periphery. All, in addition to SN 1006, also show hard continua in the integrated X-ray flux seen by the (non-imaging) RXTE satellite (Allen et al. 1999). Detections above 8 keV have also been reported by INTEGRAL (Cas A, Renaud et al. 2006; SN 1006, Kalemci et al. 2006; G347.3-0.5, Krivonos et al. 2007) and by non-imaging instruments BeppoSAX (Cas A to 80 keV; Favata et al. 1997) and *Suzaku* HXD (Cas A [to 40 keV], Maeda et al. 2009; G347.3-0.5 [to 40 keV], Tanaka et al. 2008; and Tycho [to 30 keV], Tamagawa et al. 2009). Arguments that the radiation is synchrotron essentially turn on the exclusion of all other possible processes (see arguments

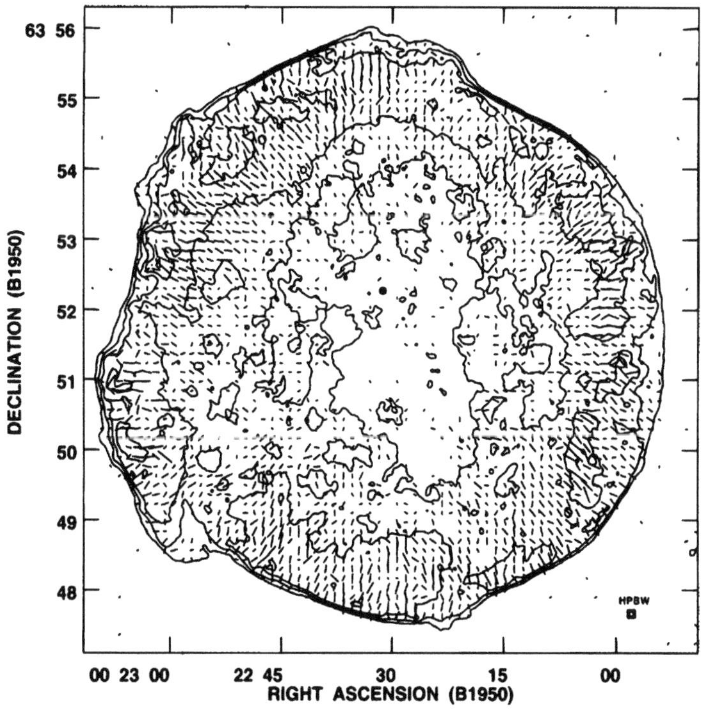

Fig. 5 Polarization vectors in Tycho's SNR at 1375 MHz (Reynoso et al. 1997). These have been corrected for foreground Faraday rotation and rotated by 90° to show the direction of the magnetic field

summarized in Reynolds 2008). Synchrotron photons at 4 keV energy ($\nu \sim 10^{18}$ Hz) imply electron energies of order $100(B/10\ \mu\text{G})^{1/2}$ TeV.

Power-law fits to the X-ray spectra are much steeper than radio spectra, indicating that the X-rays come from the cutting-off tail of the electron distribution. Three limitations might be imagined for the maximum energy to which electrons can be accelerated due to DSA: finite age or size of the remnant, lack of MHD scattering waves above some $\lambda(\text{max})$, or radiative losses. The first two mechanisms would restrict ion acceleration as well, but radiative losses would affect electrons only. The cutoffs are likely to be exponentials or modified exponentials, so that the electron distribution is roughly given by $N(E) = KE^{-s}e^{-E/E_{\text{max}}}$. In a uniform magnetic field, such a distribution would give rise to a spectrum $S_\nu \propto \nu^{-\alpha}e^{-\sqrt{\nu/\nu_c}}$ approximately, dropping off rather slowly. (A more careful calculation in the case of synchrotron losses predicts a spectrum with the same exponential factor $e^{-\sqrt{\nu/\nu_c}}$ (Zirakashvili and Aharonian 2007).) Here $\nu_c \propto E_{\text{max}}^2 B$ is the "rolloff" frequency corresponding to the peak frequency emitted by electrons with energy E_{max}.

In the cutoff part of the spectrum, we observe the competition between acceleration and loss rates. In standard DSA, the time for a particle to reach a relativistic energy E (from $E_i \ll E$) is given by $\tau(E) \sim \kappa/u_s^2$ where $\kappa \equiv \lambda_{\text{mfp}}c/3$ is the diffusion coefficient. It is often assumed, but not demanded by the data, that κ is "Bohm-like," that is, that the mean free path λ_{mfp} is proportional to the particle gyroradius $r_g \equiv E/eB$ (for extreme-relativistic energies). If we write $\lambda_{\text{mfp}} = \eta r_g$, with η the "gyrofactor," depending on the amplitude of scattering turbulence, then $\eta = 1$ is Bohm diffusion and $\eta = \text{const.} > 1$ is "Bohm-like". In commonly used "quasi-linear theory," $\eta(E) \equiv (\delta B/B)^{-2}$ where δB is the

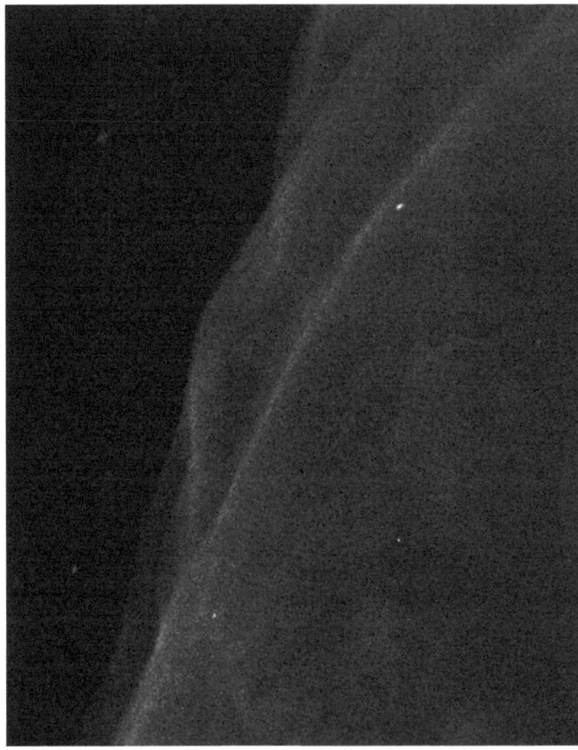

Fig. 6 Raw *Chandra* counts image of the NE region of the limb of SN 1006, showing "thin rims" of synchrotron radiation (Long et al. 2003). The brightness contrast of the very sharp rims is about a factor of 2. The inner rim has a clear Hα counterpart (Winkler et al. 2003)

amplitude of MHD turbulence resonant with particles of energy E. (So $\eta =$ const. means a particular spectrum of turbulence, equal energy per unit logarithmic bandwidth.) Bohm diffusion is often taken as a limiting case, $\delta B \sim B$, but it has been argued that $\delta B \gg B$ is the possible outcome of some cosmic-ray driven instabilities (Bell and Lucek 2001; Bell 2004). Relevant for using X-ray spectra as a diagnostic of magnetic fields is that high fields give short mean free paths, small diffusion coefficients, rapid acceleration, and higher maximum particle energies.

Two classes of argument for high magnetic fields in young SNRs come from the last decade of observations with high-resolution, high throughput X-ray observatories: *Chandra*, *XMM-Newton,* and *Suzaku.* A morphological argument is based on the commonly seen phenomenon of "thin rims," in which synchrotron X-ray emission occurs at remnant peripheries in very narrow tangential features coincident with the shock location as inferred, e.g., from Hα observations (see Fig. 6). Bamba et al. (2003) and Vink and Laming (2003) argued that the small radial extent resulted from synchrotron losses on electrons, to infer strong amplification of B:

$$B \sim 200(u_s/1000\ \mathrm{km\,s^{-1}})^{2/3}(w/0.01\ \mathrm{pc})^{-2/3}\ \mathrm{G} \tag{2}$$

(Parizot et al. 2006), where w is the filament width. Values of B from 60 to 230 μG have been inferred for Tycho, Kepler, SN 1006, Cas A, and G347.3-0.5 (Parizot et al. 2006). These field strengths apply only to regions where X-ray synchrotron emission is seen, of course. An alternative explanation for the disappearance of synchrotron emission such a short distance behind the shock is that the magnetic field is primarily turbulent, and decays on a short lengthscale (Pohl et al. 2005). Tests attempting to discriminate between these

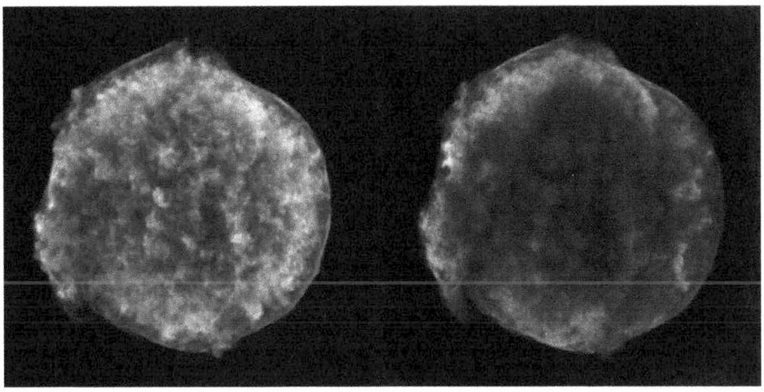

Fig. 7 *Left*: *Chandra* image of Tycho's SNR (CXC). *Right*: VLA image at 1420 MHz (Reynoso et al. 1997). Note the thin rims in the X-ray image (e.g., NE quadrant), with some thin radio rims as well

two possibilities somewhat favor the radiative-loss explanation, but are not yet conclusive (Cassam-Chenaï et al. 2007). One particular difficulty is the presence of thin rims in radio images as well (see Fig. 7); since radio-emitting electrons have enormously longer loss timescales, disappearance of the magnetic field is the only alternative. It is possible that both processes operate; theoretical models for strong magnetic-field amplification in modified shocks produce turbulent magnetic field which might well have some decay mechanism to compete with electron energy losses.

Another argument for (locally) strong magnetic fields is the discovery of variations in brightness on timescales of a few years of small features in X-ray synchrotron emission in Cas A (Patnaude and Fesen 2007) and G347.3-0.5 (Uchiyama 2007). Again, factors of 2 drop in brightness on a timescale of years require $B \sim 1$ mG. Brightness increases on similar timescales are also observed; demanding acceleration times of years gives similar estimates for B. However, very strong levels of magnetic turbulence also naturally predict these kinds of fluctuations (Bykov et al. 2008, 2009) without necessarily requiring quite as high values of $B_{\rm rms}$. However, the remnant of SN 1006 shows no such small features or brightness variations (Katsuda et al. 2010), perhaps as a result of a Type Ia origin with more-or-less uniform surrounding material. (The absence of short-term variations does not imply absence of magnetic-field amplification; very high B values could be generated in a quasi-steady state behind the shock, without producing short-term brightness fluctuations.)

New arguments relative to magnetic-field strengths in SNRs have resulted from the new generation of GeV-TeV observational capabilities. The *Fermi* Gamma-Ray Space Telescope has been mapping the sky between about 0.2 and 300 GeV since its launch in 2008, with gradually improving statistics on any given region of sky. One important early result is the detection of Cas A between 0.5 and 50 GeV (Abdo et al. 2010a, see Fig. 8). The emission may be either leptonic, that is, electron bremsstrahlung plus inverse-Compton upscattering of local photon fields, or hadronic, from the decay into photons of π^0 mesons produced in inelastic collisions between cosmic-ray protons and local thermal gas. The bremsstrahlung and hadronic contributions require knowledge of the gas density; the IC contribution requires knowledge of appropriate photon fields. The leptonic model shown in Fig. 8 assumes a mean density of 26 cm^{-1}, which requires a mean magnetic field (averaged over the entire emitting region of Cas A) of about 0.12 mG. If the actual emission is hadronic, the IC contribution must be lower (fewer relativistic electrons), demanding a larger magnetic field to produce the observed synchrotron fluxes at radio energies. (This argument was originally made based on early upper limits above 100 MeV from COS-B and SAS-2; Cowsik and Sarkar (1980) derived a lower limit of about 1 mG.)

Similarly, any TeV detection or upper limit places lower limits on the mean magnetic field in regions containing relativistic electrons. So far, four SNRs have been imaged with the HESS air-Čerenkov telescope array in Namibia. These include three of the synchrotron-

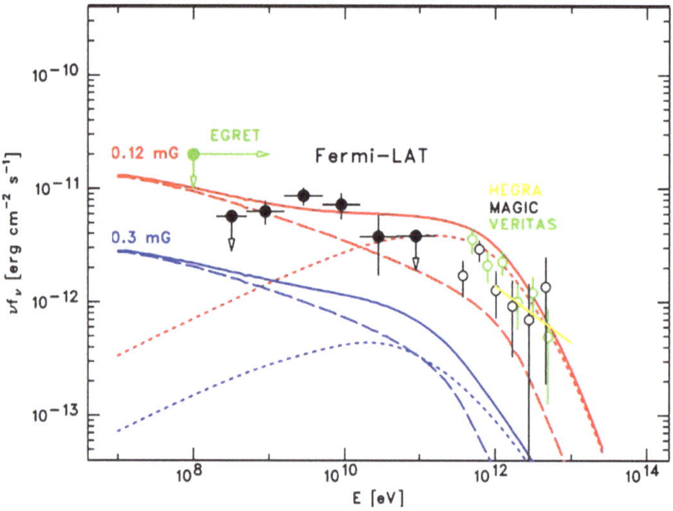

Fig. 8 Spectrum of GeV emission from Cas A as seen by *Fermi* (Abdo et al. 2010a). The *solid curves* are leptonic models (*dots*, IC; *dashes*, bremsstrahlung) with magnetic-field values shown. If the emission is actually hadronic, the magnetic-field strengths must be larger to suppress the IC component

dominated remnants: SN 1006, G347.3-0.5, and Vela Jr. (The fourth, G1.9+0.3, is too small for HESS imaging and too faint for a constraining upper limit.) In addition, the (probably) historical shell SNR RCW 86 has been detected (Aharonian et al. 2009). In all cases, the TeV morphology tracks the X-ray morphology surprisingly closely. However, detailed attempts to model the emission as either due to hadronic or to leptonic processes run into significant problems. For hadronic models, required target gas densities seem to exceed measurements or upper limits on local thermal gas, while simple leptonic models do not do a good job reproducing the spectral shape and may also require implausibly small filling factors of magnetic field. However, lower limits to $\langle B \rangle$ do not depend much on the detailed modeling. In SN 1006 (Acero et al. 2010), $B \gtrsim 30$ μG is required; for G347.3-0.5, a leptonic model (Lazendić et al. 2004) demands a filling factor of magnetic field of only 1% to avoid over-predicting synchrotron emission. Within that 1% of the volume, the field is a modest 15 μG. (These estimates are relatively insensitive to model details; even a model that does not use DSA at all as a particle acceleration mechanism, but rather stochastic acceleration in the downstream region, fits the most recent GeV and TeV data for G347.3-0.5 with a leptonic model with $B \sim 12$ μG (Fan et al. 2010).) Hadronic models typically invoke much higher fields (Zirakashvili and Aharonian (2010) find $B = 127$ μG in G347.3-0.5).

HESS has also detected several other sources coincident with shell SNRs, such as CTB37B (HESS J1713-381; Aharonian et al. 2008) and HESS J1731-347 (Acero et al. 2009). These may be further members of the nonthermal-X-ray class, older SNRs interacting with molecular clouds, or PWNe. They await further observational clarification.

The question of magnetic-field amplification is intimately connected with that of efficient particle acceleration through the proposals of Bell and Lucek (2001) and Bell (2004) that cosmic-ray driven instabilities can greatly increase the magnetic-field strength. Evidence for high magnetic fields is then taken as indirect evidence for an energetically significant component of cosmic rays (which must of necessity be ions) accelerated in the shock. This argument can be reversed: if evidence is found for efficient ion acceleration, then magnetic fields are likely to be amplified.

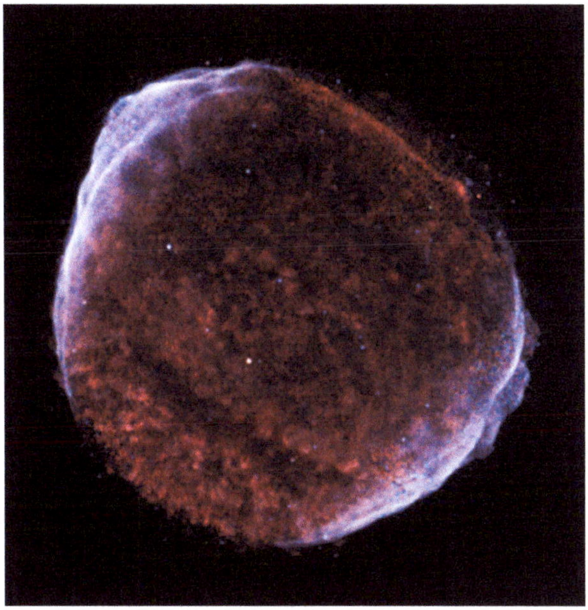

Fig. 9 SN 1006 with *Chandra* (NASA/CXC). *Red*, 0.50–0.91 keV; *cyan*, 0.91–1.34 keV; *blue*, 1.34–3.00 keV. Blue-white rims are line-free synchrotron emission; red interior is dominated by oxygen ejecta

1.5 Obliquity Dependence

The nature of magnetic-field amplification at strong shocks is a subject of intense theoretical activity. Observations can provide powerful constraints on some aspects of such amplification. In particular, two SNRs, SN 1006 and G1.9+0.3, show very simple bilaterally symmetric morphology which may contain clues to the dependence of particle acceleration and magnetic-field amplification processes on the shock obliquity angle $\theta_{\rm Bn}$ between the mean upstream magnetic field direction and the shock velocity. Section 3 discusses in detail a model for the radio morphology of SN 1006, as representative of this bilaterally symmetric class of SNRs (BSNRs).

The most natural explanation for the bilateral symmetry evident in Figs. 9 and 11 is the variation in $\theta_{\rm Bn}$ one would encounter at different locations as a spherical shock encounters a roughly uniform magnetic field. If the field lies in the plane of the sky, the range in obliquities would be 0 to $\pi/2$; if there is a line-of-sight component, the range drops until it is zero for **B** along the line of sight. Later in this chapter we describe a detailed calculation of the radio synchrotron morphologies resulting from different assumptions about the nature of the upstream magnetic field and of particle acceleration. Here we focus on the X-ray spectral variations each remnant displays.

A simple model adequately characterizing most synchrotron X-ray emission from shell SNRs is the "srcut" model in the XSPEC software package, which is the spectrum emitted by a power-law electron distribution with an exponential cutoff. The frequency corresponding to the cutoff energy $E_{\max}$ is $\nu_{\rm roll}$, the "rolloff" frequency (so called because the synchrotron spectrum from this distribution actually steepens rather gradually). $\nu_{\rm roll}$ serves as a good indicator of the maximum energy reached by electrons at different locations, although depending on the mechanism limiting that maximum energy, $\nu_{\rm roll}$ has different dependencies on physical parameters.

If the acceleration rate of particles varies with shock obliquity (independently of any obliquity-variations of the magnetic field), we can parameterize such effects in terms of

Fig. 10 Rolloff frequencies measured in 10″ deep (radially; much larger in the azimuthal direction) rectangular regions around the outer edge of SN 1006 (*red points*: an interior rim) (Katsuda et al., in preparation). The exposure time is much longer in the NE region, so boxes are smaller

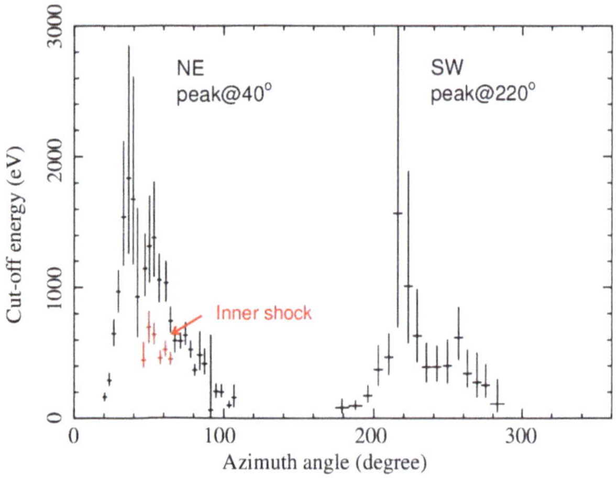

a quantity R_J: $R_J(\theta_{Bn}, \eta, r) \equiv \tau(\theta_{Bn})/\tau(\theta_{Bn} = 0)$. (This means that the value of η is that for $\theta_{Bn} = 0$; any obliquity-dependence is subsumed into R_J.) We scale to typical values for young SNRs: $u_{8.5} \equiv u_{sh}/3000\ \mathrm{km\,s^{-1}}$; $t_3 \equiv t/1000$ yr; $B_{10} \equiv B/10\ \mu$G; and $\lambda_{17} \equiv \lambda_{max}/10^{17}$ cm. Then the values of ν_{roll} for each limiting mechanism obey (Reynolds 2008)

$$h\nu_{roll}(\mathrm{age}) \sim 0.4\, u_{8.5}^4\, t_3^2\, B_{10}^3\, (\eta R_J)^{-2}\ \mathrm{keV} \tag{3}$$

$$h\nu_{roll}(\mathrm{esc}) \sim 2\, B_{10}^3\, \lambda_{17}^2\ \mathrm{keV} \tag{4}$$

$$h\nu_{roll}(\mathrm{loss}) \sim 2\, u_{8.5}^2\, (\eta R_J)^{-1}\ \mathrm{keV}. \tag{5}$$

The effective limiting mechanism is the one producing the lowest value of E_{max}; this may vary around the remnant periphery. Note that if radiative losses limit electron acceleration, the resulting ν_{roll} is independent of the magnetic-field strength.

Figure 10 shows the variation in ν_{roll} around the periphery of SN 1006 (Katsuda et al., in preparation); similar data for G1.9+0.3 (Fig. 11) at much lower resolution (Reynolds et al. 2009) are shown in Fig. 12. Variations by large factors occur. Equations (3)–(5) above show that achieving variation by over an order of magnitude is not easy, especially for loss-limited acceleration. Even in the absence of any cosmic-ray-driven magnetic-field amplification, one would expect a variation in B by about the shock compression ratio (4 or more) simply due to compression of the tangential component of B due to flux-freezing. While there are no comprehensive predictions, calculations of magnetic-field amplification normally assume parallel shocks ($\theta_{Bn} \sim 0$), with the process being less effective at higher obliquities. The strong B-dependence for age or escape-limited acceleration could easily explain the observations—but with the serious side effect of predicting that the resulting maximum energies ($E_{max} \sim 10$–100 TeV) would apply to ions as well as electrons, putting at least these two SNRs out of the running for producing cosmic rays up to the "knee", the slight steepening in the integrated spectrum of Galactic cosmic rays at Earth. Loss-limited maximum energies do depend on the shock speed, but the required variation by over 30 in ν_{roll} would demand an implausible factor of 5 variation in the current shock velocity. Anisotropic diffusion (Jokipii 1987) could produce more rapid acceleration at quasi-perpendicular shocks than quasi-parallel ($R_J(\theta_{Bn}) < R_J(0)$), easily explaining the required amount of variation (Fig. 13). However, most workers believe that this effect may not occur in highly turbulent media. In addition, electron injection into the DSA process may be more difficult at

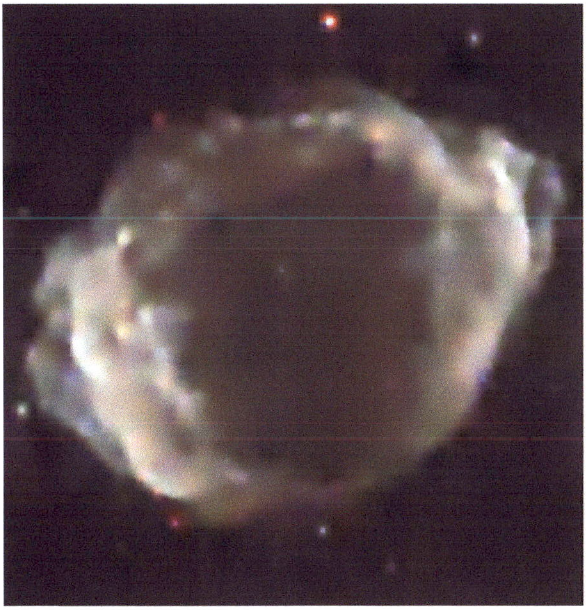

Fig. 11 G1.9+0.3, the Galaxy's youngest SNR (*Chandra*; Borkowski et al. 2010). *Red*, 1–3 keV; *green*, 3–5 keV; *blue*, 5–8 keV. Except for faint thermal emission in the center and north, the emission is dominantly synchrotron

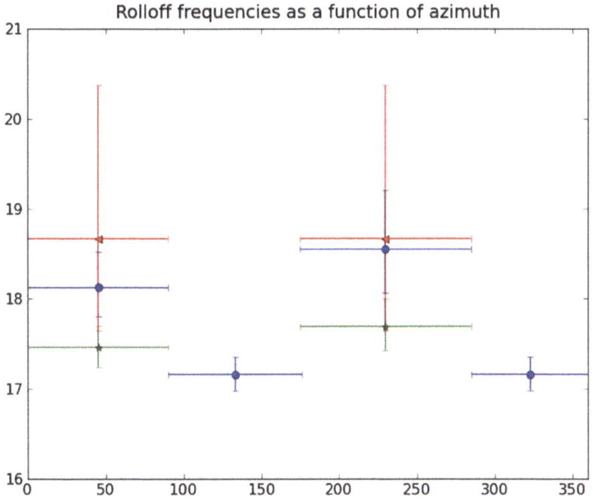

Fig. 12 Variations of $\nu_{\rm roll}$ with radius and azimuth, for G1.9+0.3. *Red*: "Ears" extending beyond the main shell to E and W. *Blue*: main shell. *Green*: Just interior to bright regions of main shell. Note the similar trend at much poorer resolution to that seen in Fig. 10

quasi-perpendicular shocks. As we show later in this chapter, the radio morphology may be more accurately reproduced with a quasi-parallel model, leaving the explanations of Figs. 10 and 12 as an open question.

1.6 Summary

We summarize here the inferences on magnetic-field strength and geometry described above.

1. From radio observations, equipartition values of magnetic field strength are in the $\sim$10 μG range, but there is little physical motivation to assume equipartition.

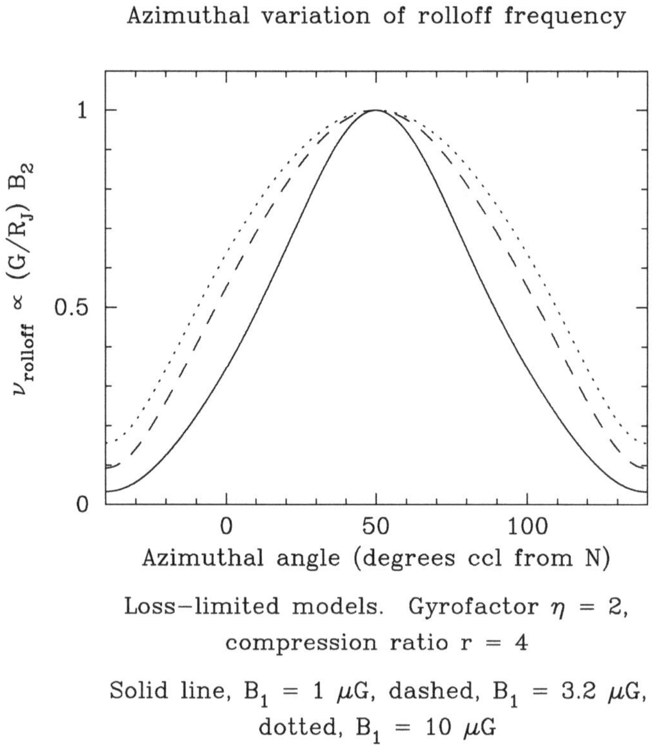

Fig. 13 Variation of ν_{roll} with azimuth (Reynolds 1998), according to the anisotropic diffusion model of Jokipii (1987). The required amount of variation of ν_{roll} can be produced, but only if the shock is perpendicular at the bright limbs. B_1 and B_2 are the upstream and downstream magnetic-field strengths, and G is the ratio of an electron's energy-loss time in a field B_1 to that in the actual obliquity-dependent combination of B_1 and B_2 as it scatters back and forth across the shock; see Reynolds (1998)

2. Radio polarization studies show that in young SNRs, the magnetic field is largely disordered, with a small radial preponderance. In older, larger SNRs, the field is often disordered but sometimes tangential.
3. Curvature (spectral hardening to higher frequency) is observed in the radio spectra of Tycho and Kepler. A nonlinear shock acceleration model can explain this with magnetic field strengths of 0.1–1 mG (average over the emitting regions).
4. Thin rims of X-ray synchrotron emission in a few young remnants require $B \sim$ 50–200 μG in the rims, if they are due to synchrotron losses on down-stream-convecting electrons. However, thin radio rims are sometimes seen as well; they require that the magnetic field disappear somehow, presumably because it is a wave field which damps.
5. Brightening and fading of small X-ray synchrotron features in G347.3-0.5 and Cas A require $B \sim 1$ mG, if they represent acceleration and loss times for electrons. Fields smaller by a factor of several are possible if the fluctuations are due to strong magnetic turbulence.
6. Large azimuthal variations in the rolloff frequency in SN 1006 and G1.9+0.3 are difficult to explain for a conventional picture of loss-limited acceleration in parallel shocks.
7. For Cas A, the detection at GeV energies with *Fermi* requires $B \gtrsim 0.1$ mG to avoid overproducing the GeV emission with electron bremsstrahlung.

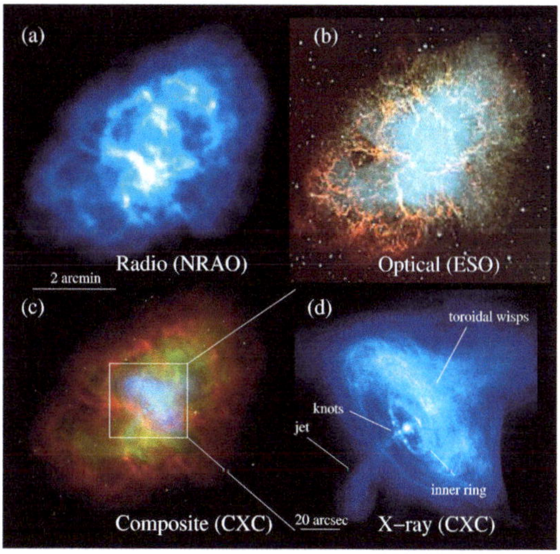

Fig. 14 Images of the Crab Nebula

8. TeV emission seen in four shell SNRs is not well explained by either leptonic or hadronic processes. However, if it is hadronic, the magnetic fields implied are of order 100 μG, while leptonic models require much lower fields.

2 Magnetic Fields in Pulsar-Wind Nebulae

2.1 Introduction

All young pulsars are slowing down in their rotation. The corresponding rate of change of rotational kinetic energy is enormous, typically in the range $\dot{E} = 10^{32}$–10^{39} erg s^{-1}. In most cases only a negligible amount of energy is contained in the electromagnetic radiation corresponding to the neutron star's observed pulsations. Rather, the bulk of the rotational power is usually deposited into a relativistic magnetized particle wind that flows outward from the pulsar.

External pressure will cause this wind to abruptly decelerate at a termination shock. Beyond the termination shock, the pulsar wind thermalizes in pitch angle and radiates synchrotron emission, resulting in a pulsar wind nebula (PWN). The best known PWN is the Crab Nebula (Fig. 14), powered by the central young pulsar B0531+21. The Crab Nebula radiates synchrotron emission across the electromagnetic spectrum. In X-rays, its extent is small, reflecting the relatively short synchrotron lifetimes of electrons emitting at these energies. The optical synchrotron emission is larger, reflecting the longer lifetimes of the corresponding electron population. And finally, the radio emission shows the full extent of the source, with a radiative lifetime exceeding the age of the source ($\sim$ 1000 years).

2.2 Observational Properties

PWNe and SNRs have a variety of common properties: they both result from a supernova explosion, and they both are characterized by their polarized radio synchrotron emission. This has led to considerable confusion in the literature. For example, the Crab Nebula is often described as a "supernova remnant", despite the shell or blast-wave that would normally correspond to a SNR remaining as yet unidentified in this source (Frail et al. 1995;

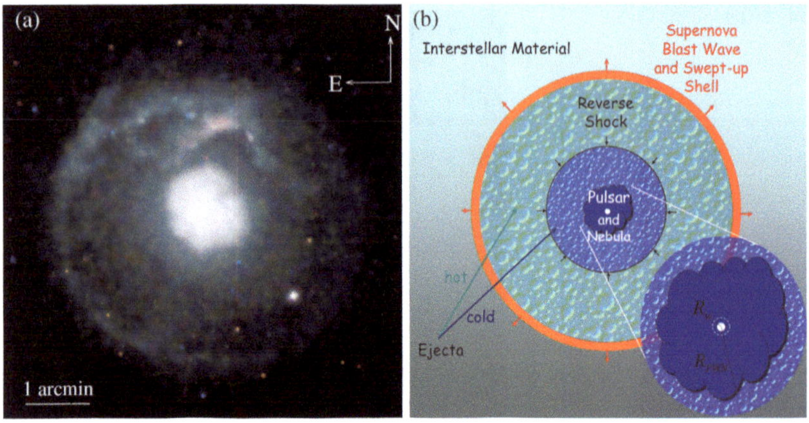

Fig. 15 *Left*: Deep *Chandra* image of the composite SNR G21.5-0.9 (Matheson and Safi-Harb 2005). The PWN is the bright central condensation. *Right*: Schematic structure of a composite SNR containing a PWN

Seward et al. 2006; Tziamtzis et al. 2009). Some PWNe have for historical reasons been cataloged as "filled-center supernova remnants" (Green 2009), while others, discovered more recently, have been excluded from such lists. As we describe below, a young SNR may contain a PWN, a combination often called a composite remnant (e.g., G21.5-0.9; Fig. 15).

The key differences between SNRs and PWNe are as follows:

- **Energy Source:** SNRs result from an essentially instantaneous deposition of energy, in the form of a blast wave driven into the ISM by a supernova explosion. In contrast, PWNe have a continuous power source, the bulk relativistic flow of electron/positron pairs from an energetic neutron star.
- **Radio Morphology:** As a direct result of their differing sources of energy, SNRs and PWNe have distinct radio morphologies. SNRs are usually limb-brightened shells of synchrotron emission, while PWNe are typically amorphous or filled-center synchrotron nebulae, brightest at the pulsar's position.
- **Radio Spectral Index:** SNRs usually have relatively steep radio spectral indices, $\alpha \approx$ 0.3–0.8, as shown in Fig. 1. In contrast, PWNe have spectral indices in the range, $\alpha \approx$ 0–0.3, which is too flat to be explained by simple models of diffusive shock acceleration (e.g., Atoyan 1999; Fleishman and Bietenholz 2007; Tanaka and Takahara 2010).
- **Angular Extent:** SNRs are long-lived objects with a wide range of sizes, with angular extents ranging from $\sim 1'$ to $>5^\circ$. PWNe are usually relatively small, with sizes in the range $10''$ to $30'$, although a few older PWNe may be significantly larger (e.g., Vela X; Fig. 16).
- **Fractional Polarization:** At radio frequencies near 1 GHz, SNRs typically have modest amounts of linear polarization, at the level of 5%–10%. PWNe usually have very well-organized magnetic fields, with correspondingly higher polarization fractions, in the range 30%–50%.

2.3 PWN Evolution

Theoretical expectations predict three main phases in the evolution of a PWN (Reynolds and Chevalier 1984; Chevalier 1998, 2005), governed in turn by:

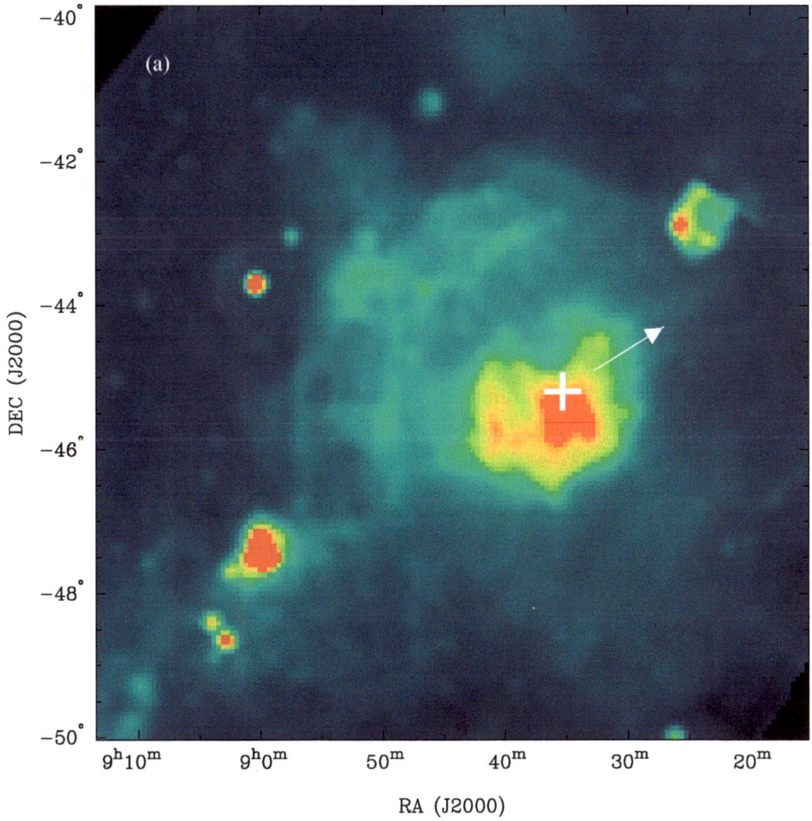

Fig. 16 2.4 GHz Parkes image of the Vela supernova remnant (Duncan et al. 1996). The *white cross* indicates the pulsar, and the *arrow* its proper motion (Dodson et al. 2003); the fact that the pulsar is neither at nor moving away from the PWN's center indicates that the reverse-shock interaction has take place

1. the expansion of the PWN;
2. interaction of the PWN with the surrounding SNR;
3. the motion of the pulsar powering the PWN.

Each of these phases is discussed in more detail below.

2.3.1 Phase 1: Expansion into Unshocked Ejecta

At early stages in a PWN's evolution, the pulsar's spin-down luminosity remains relatively constant, so that there is a steady injection of energy into the PWN. As a consequence, the PWN expands supersonically into the surrounding low-pressure environment, with the radius, R, of the PWN evolving with time, t, as $R \propto t^{6/5}$ (Chevalier 1977). Thus the PWN drives a shock into the inner edge of the expanding SN ejecta (see Fig. 15).

Many pulsars are born with high space velocities, up to and beyond 1000 km s^{-1}. However, because the sound speed inside the PWN is much higher than this, the PWN stays relatively centered on the pulsar, and has a quasi-spherical appearance.

Inside the PWN, there are two broad zones. Close to the pulsar is the unshocked wind zone, in which particles and Poynting flux flow freely outward from the central star. The pressure in this wind drops with increasing distance from the pulsar, until we reach a termination shock at which the wind is confined by external pressure. At this shock the wind particles are accelerated up to even higher Lorentz factors, and have their pitch angles scattered so that they can emit synchrotron radiation. Observations of this emission allow us to infer the composition of the wind at and immediately downstream of the shock. Specifically, the parameter σ corresponds to the ratio of electromagnetic energy to particle energy at the wind shock, and is usually much less than one (Kennel and Coroniti 1984; Gaensler et al. 2002): the shocked wind is particle-dominated ($\sigma \ll 1$), in contrast to the unshocked wind which is expected to be radiation-dominated ($\sigma \gg 1$).

Beyond the termination shock is the second main zone, the emitting region of the PWN, in which the bulk flow continues to decelerate to match the external boundary conditions, and where synchrotron emission is produced.

The morphology of a PWN in this early phase of evolution has been spectacularly revealed by X-ray images of the Crab Nebula taken by the *Chandra X-ray Observatory* (Weisskopf et al. 2000; Hester et al. 2002). These observations show that the Crab Nebula is dominated by a bright X-ray torus. The inner part of this torus is bounded by a bright X-ray ring, thought to correspond to the wind termination shock. To the south of the pulsar, an X-ray jet runs along the axis of the torus, apparently originating very close to the central pulsar before curving slightly at large distances from the center. A faint counter-jet can also be seen to the north.

The torus is thought to result from the Poynting flux concentrated in equatorial regions of the system by the wound-up magnetic field. However, a key point, first noted by Lyubarsky (2002), is that the termination shock radius should decrease with increasing angle from the equator. The axial jet seen for the Crab Nebula is therefore not a jet originating from the pulsar itself, but is part of the post-shock wind (e.g., Bogovalov and Khangoulyan 2002).

Relativistic magnetohydrodynamic simulations show that at mid-latitudes, the external pressure can reverse the post-shock particle flow. The wind flows back along the surface of the termination shock toward the poles (Komissarov and Lyubarsky 2003; Del Zanna et al. 2004; Del Zanna et al. 2006), at which points it is collimated into jets by hoop stress (the "toothpaste effect; Camus et al. 2009). Simulations thus can do a reasonable job of reproducing the morphological properties of the Crab Nebula, as a result of a latitude-dependent termination-shock radius with a flow reversal.

Recently published optical polarimetry of the Crab Nebula using the *Hubble Space Telescope* has revealed spectacular details of this PWN's large-scale magnetic field (Hester 2008). A great deal of fibrous structure seen in polarization corresponds to the local field direction, while the torus of the nebula appears to be confined by a poloidal component of the field.

2.3.2 Phase 2: Interaction with Reverse Shock

A PWN moves into a new phase of evolution when its outer boundary collides with the reverse shock from the SNR in which it is embedded. This typically occurs after ~ 7000 years for the typical case of a 10^{51} erg supernova ejecting 10 solar masses into an ambient ISM of density 1 atom cm^{-3} (Reynolds and Chevalier 1984). The collision with the reverse shock initially compresses the PWN, and the system may reverberate for a couple of cycles before establishing a new equilibrium (Blondin et al. 2001; van der Swaluw et al. 2001; Gelfand et al. 2009)

The compression of the PWN raises the internal magnetic field, resulting in substantial synchrotron burn-off at high energies. The X-ray extent of a PWN in this phase can therefore be rather small. Furthermore, if the pulsar has a high space velocity or if the surrounding ISM is inhomogeneous, then the reverse shock does not collide with all parts of the PWN at the same time. As a result of this asymmetric collision, the pulsar can end up substantially offset from the center of its PWN, and the PWN's radio morphology can take on a chaotic or filamentary appearance (Blondin et al. 2001; van der Swaluw et al. 2004) (see Fig. 16).

2.3.3 Phase 3: Supersonic Motion

Since a pulsar moves ballistically while the surrounding SNR decelerates in its expansion, any pulsar with a significant space velocity will eventually begin to move toward the SNR's rim. For a SNR in the Sedov phase, the sound-speed in the remnant's interior drops with increasing distance from the center. Specifically, van der Swaluw et al. (1998) has shown that for a Sedov SNR, the pulsar's motion will become supersonic when it has moved 68% of the way to the SNR's rim, which occurs at 50% of the total time taken for the pulsar to escape the SNR completely. Once the pulsar becomes supersonic, a new, bow-shock, PWN is formed, which no longer expands and which is externally confined by the ram pressure resulting from the pulsar's motion. Eventually the pulsar escapes the SNR completely, and now drives a bow shock through the ambient ISM (Bucciantini 2002; Chatterjee and Cordes 2002; Gaensler et al. 2004; Bucciantini et al. 2005) (see Fig. 17).

2.3.4 Measuring PWN Magnetic Fields

We can infer the strength of the magnetic field inside PWNe through various techniques: by assuming equipartition between particles and magnetic fields averaged over an assumed synchrotron emitting volume (e.g., Hester et al. 1995), by measuring the radius of the termination shock from a high-resolution X-ray image and combining this with a physical model for the flow to estimate the field value at this location (e.g., Gaensler et al. 2002), or through observations of inverse Compton emission in GeV or TeV gamma-rays (e.g., MSH 15-52: Abdo et al. 2010b; Aharonian et al. 2005). De Jager and Djannati-Ataï (2008) review GeV/TeV emission from PWNe. These approaches show that PWNe can have a wide range of nebular magnetic fields, from $\sim$5 μG to $>$1 mG. The magnetic field strength generally depends strongly on $\dot{E}$, t and σ, as well as on the pulsar braking index.

A variety of magnetic field orientations have been inferred in PWNe from the angles of their radio polarization. Some PWNe show a broadly toroidal magnetic field, some show a radial magnetic field, and in others there is a complex or tangled appearance. Kothes et al. (2006) have suggested that all PWNe have predominantly toroidal magnetic fields, but that these different polarization patterns result from differences in viewing angle. Toroidal polarization vectors, such as seen for the PWN G106.6+2.9, result when the observer looks along the pulsar's spin axis (Kothes et al. 2006). Radial fields, such as seen in the PWN G21.5–0.9 (Reich 2002) correspond to a viewing angle perpendicular to the spin axis. And more complex magnetic field morphologies, as for the PWN 3C 58 (Wilson and Weiler 1976), occur for oblique viewing angles. The complex polarization properties of systems like 3C 58 may also occur as a result of kink instabilities, which can produce sheared loops of magnetic field throughout the nebula (Begelman 1998; Slane et al. 2004).

There are only limited observations of magnetic fields in pulsar bow shocks. The theoretical expectation is that the system forms a toroidal field near the apex of the bow shock, but that the magnetic field is collimated in the trailing magnetotail. Radio polarimetry matching this prediction has recently been observed by Ng et al. (2010) for the bow-shock PWN powered by PSR J1509–5850. However, other magnetic field geometries have also been

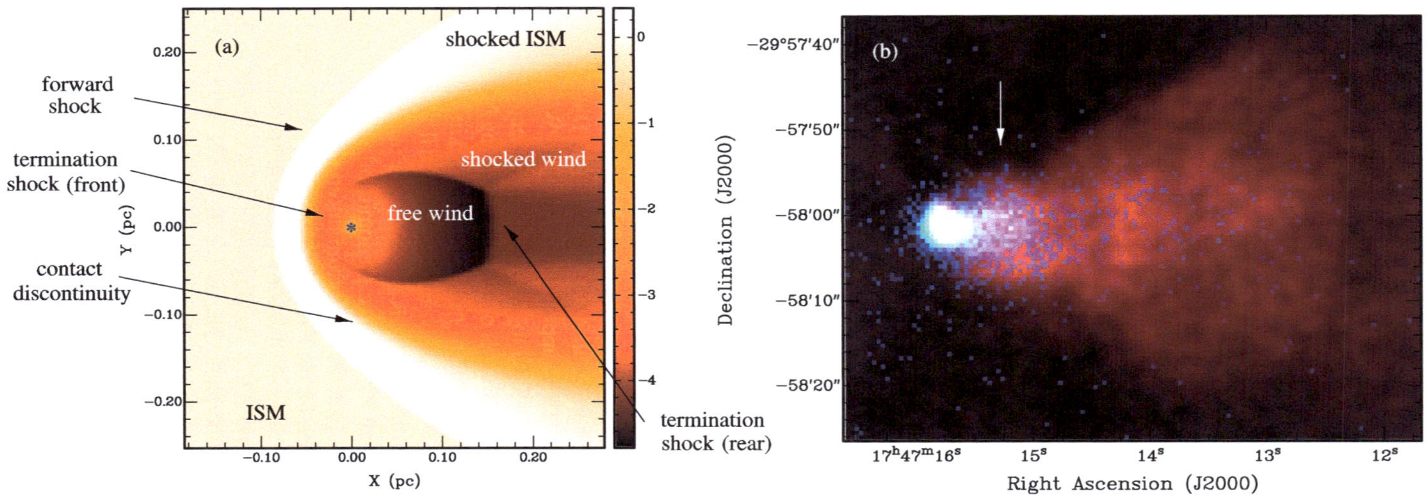

Fig. 17 *Left*: Hydrodynamic simulation of a pulsar bow shock. The pulsar (*asterisk*) is moving from right to left with Mach number 60. *Right*: *Chandra* image (*blue*) and VLA image (*red*) of G359.23-0.82 (the "Mouse"), the bow shock associated with PSR J1747-2958 (Gaensler et al. 2004). The *white arrow* indicates a structure which may be the termination shock

seen (e.g., Yusef-Zadeh and Gaensler 2005), suggesting that the magnetic field structure of bow-shock PWNe may depend on whether the pulsar's spin axis aligns or mis-aligns with its velocity vector (e.g., Vigelius et al. 2007).

2.4 Outstanding Issues and Future Work

There are many aspects of PWNe which are still active areas of investigation. Two particular areas of focus are as follows.

First, several PWNe have now been observed to show substantial time variability in their morphology and intensity (e.g., Pavlov et al. 2001; Hester et al. 2002; DeLaney et al. 2006; De Luca et al. 2007). These observations have provided a rich suite of information, but we now need relativistic magnetohydrodynamic simulations to fully understand the flow conditions and instabilities that produce these effects (Komissarov and Lyubarsky 2004; Del Zanna et al. 2004; Bucciantini et al. 2005; Bucciantini 2010).

Second, while we have a multitude of observations of pulsar winds after they have been shocked, we have very limited data on the composition and transportation mechanism in the unshocked wind zone. Targeted observations and modeling of unique systems such as PSR J0737–3039A/B (the "double pulsar") may provide vital insights in this regard (McLaughlin et al. 2004; Lyutikov and Thompson 2005).

3 Constraints on the Local Interstellar Magnetic Field from Radio Emission of SN 1006

3.1 Introduction

The radio morphology of supernova remnants (SNRs) may be very informative of the conditions of the magnetized environments in which the blast-wave expansion occurs and, in particular, on the acceleration processes at the shock front which give rise to the energetic electrons ultimately responsible of the synchrotron emission in the radio and (possibly) X-ray band. In this context, the bilaterally symmetric or barrel-shaped BSNRs (Kesteven and Caswell 1987; Gaensler 1998) are considered ideal laboratories, because their morphology is definitely the result of the lack of small scale inhomogeneities which may confuse the interpretations. A point-like explosion in a uniform magnetized medium with constant strength and direction of **B** should give rise to a symmetric BSNR whose bright limbs are located where the magnetic field is parallel or perpendicular to the shock speed, if the injection efficiency is greatest where the shock is quasi-parallel or quasi-perpendicular/isotropic, respectively, and if **B** lies in the plane of the sky, whereas different configurations occur at different aspect angles (Fulbright and Reynolds 1990).

However, in real life, BSNRs are often asymmetric, and therefore Orlando et al. (2007) (hereafter Paper I) have recently generalized the study of Fulbright and Reynolds (1990) to the cases in which the explosions occur in a large scale gradient of density or magnetic field, showing that this model is able to reproduce most of the asymmetries observed in real BSNRs. In particular, the radio morphology loses one axis of symmetry, and the limbs are not equally bright (if the gradient runs across the limbs) or they are not opposite and they converge on the side in which the density or the magnetic field is increasing (if the gradient runs between and parallel to the limbs).

It is clear that morphology of BSNRs is tightly coupled to the magnetized environments in which the shock expands, and it is of particular interest to note here the preference of

BSNR symmetry axes to be oriented parallel to the Galactic plane, as reported by Gaensler (1998). It seems therefore possible to study their morphology to derive the geometry of the surrounding magnetic field, thus shedding more light on the microphysics of the particle acceleration processes at the shock front.

The remnant of SN 1006 seems to be the object in which this kind of study may be most fruitful. The uniform environment and the bright limbs visible in most of the electromagnetic spectrum make it a real case study in the field of particle acceleration mechanisms in strong shocks. Indeed, Rothenflug et al. (2004), using a simple and powerful geometrical argument applied to the *XMM-Newton* X-ray image of SN 1006, based on the ratio between the central and the rim luminosity, showed that if the remnant is cylindrically symmetric, the bright limbs are likely to be polar caps (instead of an equatorial belt) and that, therefore, the magnetic field is oriented perpendicular to the bright limbs, in the NE-SW direction. This argument seems to break the dichotomy between the two competing scenarios of the dependence of the electron injection efficiency or electron acceleration rate on $\theta_{\rm Bn}$, preferring a situation in which the injection is most efficient or acceleration most rapid when the field is along the shock speed (quasi-parallel scenario) over the situation in which the field is perpendicular (quasi-perpendicular scenario, see Fulbright and Reynolds 1990)[1].

On one hand, these findings were somehow expected since Völk et al. (2003) already pointed out that in SN 1006 the injection should be maximum at parallel shocks. However, in the light of the uncertainties related to the details of the acceleration processes, several authors still considered the quasi-perpendicular scenario a viable option: Fulbright and Reynolds (1990) argued against the quasi-parallel scenario pointing out that quasi-parallel models often give rise to unobserved morphologies in the radio band; Yamazaki et al. (2004) still considered both models to explain the observed width of selected filaments of SN 1006 observed by Bamba et al. (2003), whereas Amano and Hoshino (2007) developed a quasi-perpendicular model which agrees very well with the same data. Moreover, the same simple geometrical argument of Rothenflug et al. (2004), if applied to the radio image, would be in agreement with the equatorial belt (cf. $R_{\pi/3} = 0.7$ in Sect. 3.2 of Rothenflug et al. 2004). This discrepancy has never been explained in the literature and remain one of the most intriguing open issues in the comparison between models and observations.

Recently, Petruk et al. (2009) have further investigated this issue, showing that, in the framework of a simple model of SN 1006 in terms of a point-like explosion occurring in a uniform density and uniform magnetic field medium, there is no way to reconcile the quasi-parallel scenario and the SN 1006 morphology as observed in the radio band. This contradiction between the radio morphology (suggesting quasi-perpendicular injection efficiency scenario) and X-ray morphology (suggesting quasi-parallel scenario) has prompted us to investigate the effects of non-uniformity of the environment on the observed properties of this remnant, capitalizing on the work of Orlando et al. (2007) on asymmetric BSNRs.

In Sect. 3.2, we will briefly describe the MHD model we have used to reproduce the remnant of SN 1006, which include a small gradient of the magnetic field. In Sect. 3.3, we introduce the methodology for the comparison between modeled and observed radio images of SN 1006, while in Sect. 3.4 we will discuss the results of the comparison, showing that,

[1] Since X-ray synchrotron brightness depends on both the efficiency of electron injection into the acceleration process, and on the rapidity of acceleration to high energies, studies of the X-ray morphology involve a combination of both possible effects, while radio studies are insensitive to acceleration-rate issues, because acceleration to the GeV energies required for radio emission is always extremely rapid compared to evolutionary timescales. Thus variations of radio morphology with obliquity point to electron injection physics alone.

Table 1 MHD models of SN 1006 used in this work

Name	$\nabla|\mathbf{B}|$[a]
UNIFORM_B	0
GRAD1	1.4
GRAD2	1.5
GRAD3	2.0
GRAD4	4.7

[a]The relative variation of the modulus of the magnetic field over a scale of 10 pc

not only can the model invoking a gradient of $|\mathbf{B}|$ reconcile, for the first time, the radio and X-ray morphology of the remnant, but it also provides stringent constraints on the overall geometry of the field in the vicinity of the remnant.

3.2 The Model

Since there is accumulated evidence that the density around SN 1006 is fairly constant, with the exception of the NW sector where an encounter with a dense cloud is occurring (Acero et al. 2007; Miceli et al. 2009), we argue that a model with a $|\mathbf{B}|$ gradient is more appropriate to describe the asymmetries of the limbs of SN 1006. The remnant was modeled as a point-like explosion of 1.4 solar masses of ejecta having a kinetic energy of $E = 1.3 \times 10^{51}$ erg occurring in a uniform density medium with $n_0 = 0.05$ cm^{-3}. The ISMF has a value of 30 μG in the environment of the explosion site and it is assumed to have a gradient along the Y axis. Such a high value (for the location of SN1006 in the Galaxy) has been chosen in order to take into account as a first approximation the effects of magnetic-field amplification, to have a post-shock $|\mathbf{B}|$ of the order of ~100 μG in our simulation, in agreement with observations of some SN1006 filaments. Moreover, the magnetic field has a gradient along the Y axis. The direction of the magnetic field is in the XY plane. $\nabla|\mathbf{B}|$ values used in our simulations are reported in Table 1. The gradient in the initial conditions has been modeled with a dipole located along the Y axis, exactly as in Orlando et al. (2007), to which the reader is referred for further details on the model and numerical code used for the simulations. The simulations were stopped at $t = 1002$ yr, checking that the shock velocity and remnant radius are compatible with the observed values (4600 km s^{-1} and 9.6 pc, respectively).

3.3 Comparison Between Models and Observations

3.3.1 The Angles Defining the Viewing Geometry

Several contributions in the literature show that the observed morphology of a SNR emitting synchrotron radiation is strongly affected by the angle between the dominant direction of **B** and the line of sight, $\phi_{\mathbf{B}}$. Moreover, in our circumstance, since we also have a gradient of the magnetic field, we are forced to consider the dependence of the observed morphology on the angle between the direction of the gradient and the plane of the sky, $\phi_{\nabla|\mathbf{B}|}$. The definitions of $\phi_{\mathbf{B}}$ and $\phi_{\nabla|\mathbf{B}|}$ are sketched in Fig. 18. For the purposes of this work, we have chosen to synthesize our maps in the following way: first, we apply a rotation of $\phi_{\nabla|\mathbf{B}|}$ degrees around an axis passing through the center of the remnant and parallel to the limbs, where positive angles mean that regions of increasing **B** are closer to us. Next, we apply a rotation of $\phi_{\mathbf{B}}$ degrees around an axis passing through the remnant center and lying in the plane of the sky. Others rotation schemes give similar results. The adopted values of $\phi_{\mathbf{B}}$ are from 0° to 90° in steps of 2° and the ones of $\phi_{\nabla|\mathbf{B}|}$ are from 0° to 90° in steps of 15°. Therefore, for

Fig. 18 *Left panel*: Definition of $\phi_{\mathbf{B}}$. *Right panel*: Definition of $\phi_{\nabla|\mathbf{B}|}$

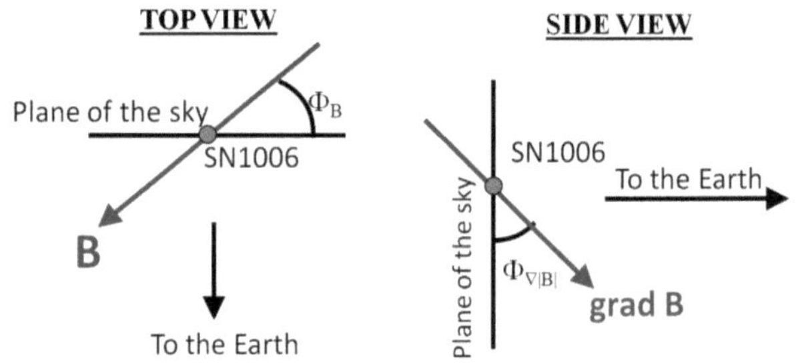

each model, we have generated 315 maps encompassing all the combinations of the relevant angles.

3.3.2 The Morphological Parameters

The exploration of the parameter space is a challenging task, because it involves the morphological comparison between many model maps and the real SN 1006 images. We devised a simple procedure which involves the comparison of the value of 4 morphological parameters calculated both from the synthesized radio emission maps and from the observed radio image of SN 1006. The parameters are the ratio between the maximum and the minimum surface brightness (S_b) around the rim (A), the ratio between the maximum around the rim and the center of the remnant (C), the ratio of S_b of the two bright limbs (R_{max}), and the angular separation between the limbs (θ_D). The parameters A, R_{max} and θ_D were also used in Orlando et al. (2007), and we refer to that paper for further discussion. The C parameter is introduced here to measure the luminosity contrast between the brightest rim and the center. Note that the C parameter does not correspond to the $R_{\pi/3}$ parameter used by Rothenflug et al. (2004) to exclude the equatorial belt scenario for SN 1006, because C is measured in a small circular region in the radio map (see below). The parameters A, C and R_{max} are measured using the radio map of Petruk et al. (2009), by averaging the S_b value in circular regions of 45″ radius (Fig. 19). The values we obtained for the observed image of SN 1006 are $A = 2.7 \pm 0.2$, $C = 0.36 \pm 0.03$, $R_{max} = 1.2 \pm 0.1$, and $\theta_D = 135^\circ \pm 10^\circ$. For the measurement of the parameters in the synthesized model radio maps, we used an automated procedure to find the maximum of the two limbs, the minimum between rims along the rim and the central position. Then we used an average in a circular region whose radius is the same percentage of the SNR radius as used in the real radio image (5%). Since the model images have been synthesized using the same number of pixels per radius of the remnant, this procedure ensures that the model and observed values of the parameters are comparable.

3.4 Results

3.4.1 Quasi-Parallel Scenario

In Fig. 20, we show the $A - C$, $R_{max} - C$ and $\theta_D - C$ scatter plots computed in the synthesized radio map of our SN 1006 model including a weak gradient of the magnetic field (**B** varies by a factor of 1.4 over 10 pc), and considering a quasi-parallel scenario for the obliquity dependence of the electron injection efficiency. We also overplotted the values of the parameters derived using the real radio map of the remnant of Petruk et al. (2009). Though the agreement is not exactly perfect in all the plots, we can note that we can define a very limited region of the parameter space $\phi_{\mathbf{B}}$–$\phi_{\nabla|\mathbf{B}|}$ which is compatible with the

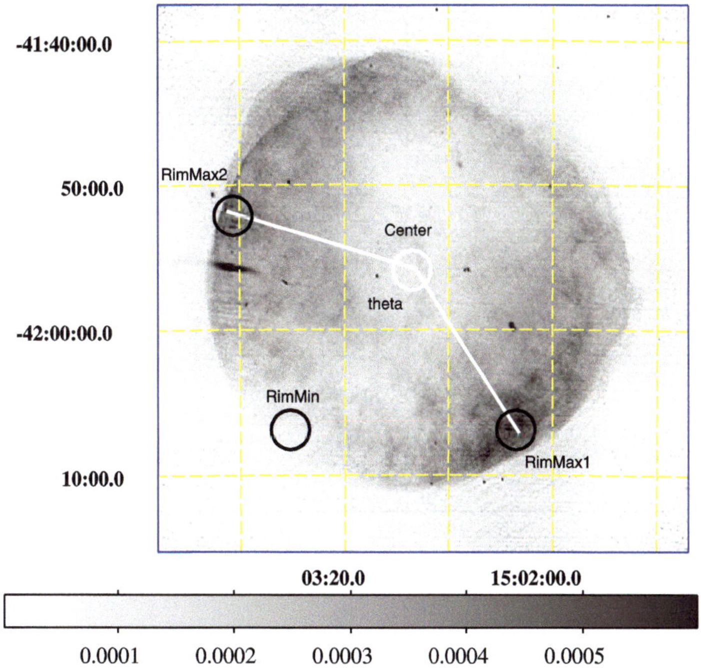

Fig. 19 Radio map of SN 1006 at 1.4 GHz, with the four regions in which the surface brightness of the remnant has been averaged for the purpose of computation of the four morphological parameters: $A = RIM_{max}/RIM_{min}$, $C = center/RIM_{max}$, $R_{max} = RIM_{max1}/RIM_{max2}$ and θ_D, the aperture angle between the limb maxima. Adapted from Petruk et al. (2009)

observed values of A, C, R_{max} and θ_D. This means that the observed radio morphology of SN 1006 is overall compatible with a quasi-parallel scenario for this remnant, if we include a weak gradient of **B**. A nonzero gradient is an essential ingredient to reconcile the radio morphology with the quasi-parallel scenario, and this is shown in Fig. 21, in which we show the A–C scatter plot for various $\phi_{\mathbf{B}}$ angles, in the uniform **B** case and quasi-parallel, quasi-perpendicular and isotropic scenarios[2]. In this case, this plot suggests that quasi-parallel models do not reproduce the observed parameters at all, unlike the quasi-perpendicular and isotropic scenarios, as already pointed out by Petruk et al. (2009).

Let us now come back to Fig. 20. A remarkable result is that the comparison between the quasi-parallel model and the observation strongly excludes a situation in which the polar caps are in the plane of the sky ($\phi_{\mathbf{B}} = 90°$) or along the line of sight ($\phi_{\mathbf{B}} = 0°$). The latter geometry would cause a centrally brightened morphology instead of two limbs, as already discussed by Orlando et al. (2007). The best-fit values of the aspect angle we derive from Fig. 20 and a conservative estimate of their uncertainties are $\phi_{\mathbf{B}} = 38° \pm 4°$ and $\phi_{\nabla|\mathbf{B}|} = 15° \pm 15°$.

The synthesized radio map of the best fit model is shown in Fig. 22, along with the observed radio map. Both maps have been smoothed to hide the small scale features and to focus the comparison on large spatial scales. In fact, we don't expect a pixel to pixel match between model and observations, given the simple assumptions behind the model. Indeed,

[2]In the uniform **B** case, $R_{max} = 1$ and $\theta_D = 180°$ always.

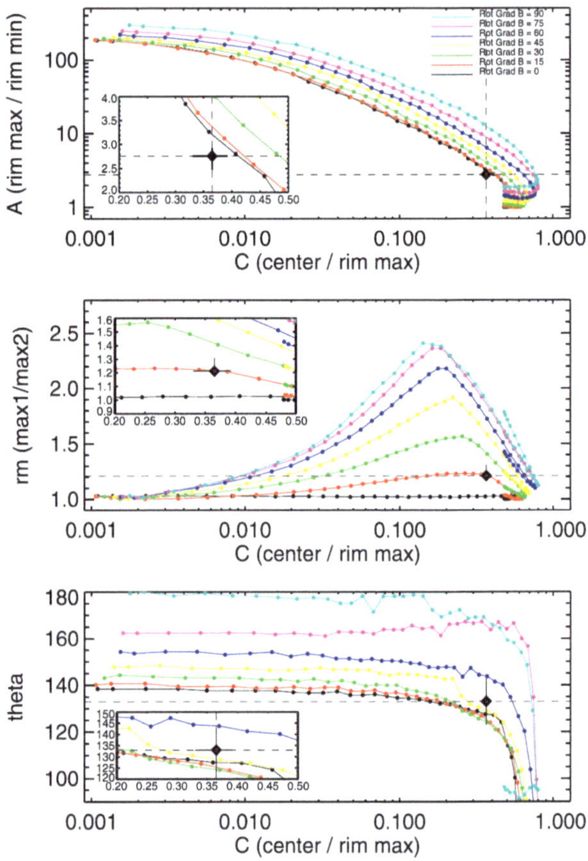

Fig. 20 *Top panel*: A–C scatter plot for our models, assuming the quasi-parallel scenario and an age of 1000 yr. Different colors correspond to different values of $\phi_{\nabla|\mathbf{B}|}$, where each dot corresponds to a given value of $\phi_{\mathbf{B}}$ (from 0° to 90° in steps of 2°. We overplotted the (A, C) values measured for SN 1006 (the *inset* shows a zoom near this point). *Middle panel*: Same as *top panel* but for the R_{max}–C pair. *Lower panel*: Same as *top panel* but for the θ_D–C pair

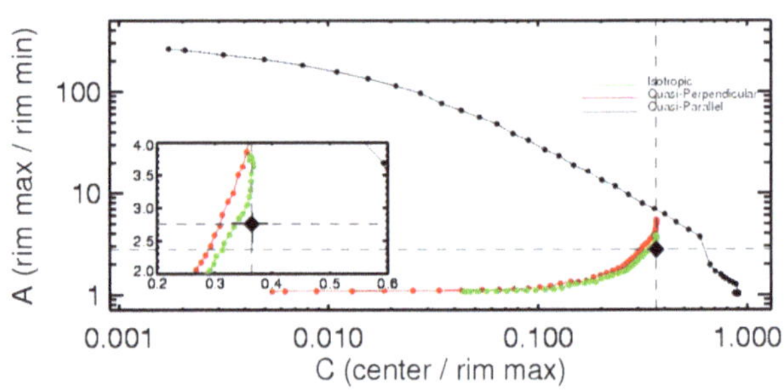

Fig. 21 Same as Fig. 20, top panel, but for the model with uniform magnetic field

the large structures of the observed radio emission are very well recovered by the best-fit model and the two images look similar.

The comparison with other gradients, namely the GRAD2, GRAD3 and GRAD4 models, shows that it is not possible to find a satisfactory fit for all the parameters. In particular, for higher values of the gradient, the angle θ_D is underestimated, so GRAD1 is the only model which gives us an overall good fit.

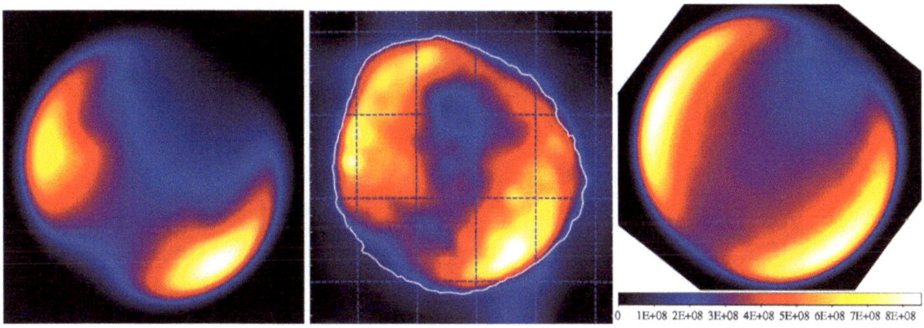

Fig. 22 *Left panel*: Synthesized radio image at 1 GHz for our best-fit model (GRAD1 in Table 1, and quasi–parallel injection efficiency scenario, $\phi_{\mathbf{B}} = 38^\circ \pm 4^\circ$ and $\phi_{\nabla|\mathbf{B}|} = 15^\circ \pm 15^\circ$), smoothed with a sigma of $2'$. *Central panel*: radio map of SN 1006 of Fig. 19, but smoothed with the same sigma used for the model image. *Right panel*: same as the *left panel*, but for the best-fit quasi-perpendicular scenario ($\phi_{\mathbf{B}} \sim 38^\circ$)

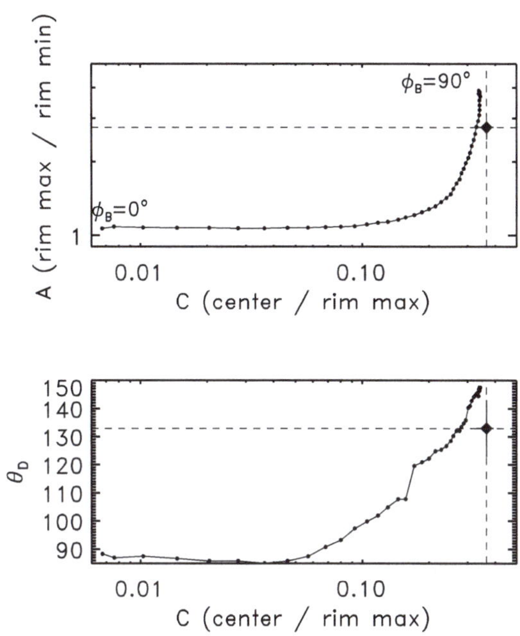

Fig. 23 *Top panel:* Same as Fig. 20 (top panel) but for quasi-perpendicular scenario. There is no $\phi_{\nabla|\mathbf{B}|}$ angle in this case, because $\mathbf{B}$ is always aligned with $\nabla|\mathbf{B}|$. *Lower panel*: Scatter plot of θ_D–C parameters. We overplotted the values observed in SN 1006

3.4.2 Quasi-Perpendicular Scenario

We also produced model images in the quasi-perpendicular injection scenario, and we repeated the analysis described in the previous paragraphs. In this case, the direction of the $\nabla|\mathbf{B}|$ is aligned with the direction of $\mathbf{B}$, so the angle $\phi_{\nabla|\mathbf{B}|}$ is the same as $\phi_{\mathbf{B}}$, and we did not consider it any further. Another consequence of this is that R_{max} is always 1, so this parameter cannot give any diagnostic. The results are shown in Fig. 23, and it seems that a good fit can be found for $\phi_{\mathbf{B}} \sim 70^\circ$ (a value in agreement with the azimuthal profile analysis of Petruk et al. 2009), even if the model points in the θ_D–C diagram seems to be more distant from the observed points than any scatter plot in Fig. 20, thus indicating a better fit in the quasi-parallel case than the quasi-perpendicular case.

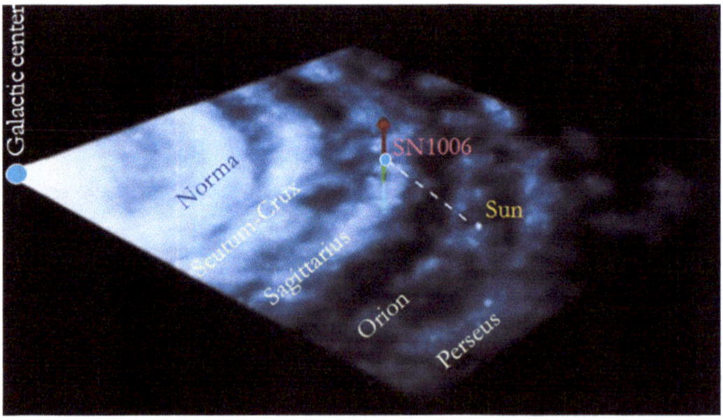

Fig. 24 Artist's impression of the magnetic field at the location of SN 1006 in our Galaxy. The *red* and *green arrows* mark the directions of the field and its gradient as derived by the best-fit model for the synchrotron radio emission of SN 1006, assuming a quasi-parallel scenario, derived in this work. The figure is to scale

3.5 Discussion

The proposed method of comparison between models of synchrotron radio emission and the real observational data of bilateral SNRs is based on the calculation of 4 morphological parameters, and we have seen that we get a good fit for a quasi-parallel injection efficiency scenario if we include a gradient of **B** and a worse fit in case of quasi-perpendicular scenario with the same gradient. This means that an agreement is found between the overall observed morphology of SN 1006 and our model, which corresponds to well defined values of the viewing geometry angles ($\phi_{\mathbf{B}} = 38° \pm 4°$ and $\phi_{\nabla|\mathbf{B}|} = 15° \pm 15°$). We know that the angle between the SN 1006 axis of symmetry and the Galactic plane is roughly 90°, and that its distance is 2.2 kpc and its Galactic latitude is 14.6°. Combining all this information, we can plot the direction of the magnetic field and its gradient in a 3D representation of the Galactic disk. This is shown in Fig. 24.

Remarkably, the direction of $\nabla|\mathbf{B}|$ points down toward the plane and the direction of **B** is aligned with the direction of the spiral arm near the remnant. This is indeed a very reasonable configuration for the magnetic field at this position in the Galaxy, because models of Galactic **B** indicates that at this location above the plane the field still retains spiral arms features, even if having smaller amplitudes. Therefore, it is tempting to conclude that we are sampling the large scale field of the Galaxy. However, we note that the best-fit value of $\nabla|\mathbf{B}|$ (i.e. a variation of 1.4 times over 10 pc) seems to be high to be associated with the large scale **B** of the Galaxy. We have used the model of Sun et al. (2008) to compute the expected variation of the azimuthal field in the Galaxy at the location of SN 1006. In particular, we have used their *ASS+RING* model which seems to be favored by rotation measures of pulsars in the plane. At a distance of ~550 pc from the plane for the SN 1006 remnant, the large-scale field is expected to vary by a factor of 1.4 on scales of more than 100 pc, much longer than the scale-length required by our model fit. This opens up the possibility that we are actually sampling the random magnetic field component. Minter and Spangler (1996) reports length scales for this component of the order of few pc, which is compatible with the variations we derive.

3.6 Summary and Conclusion

The synchrotron radio emission of the archetypical bilateral supernova remnant SN 1006 is compared against an MHD and emission model for this remnant, including a gradient of the external magnetic field, particle acceleration and its obliquity dependence, namely quasi-perpendicular, quasi-parallel, and isotropic scenarios for the injection efficiency. In order to explore the parameter space, which is very large due to the necessity of including the viewing geometry in the model-data comparison, we developed a simplified procedure based on the computation of four morphological parameters. We have found a very good fit with a model assuming quasi-parallel injection efficiency and an aspect angle $\phi_{\mathbf{B}} = 38^\circ \pm 4^\circ$ between the direction of **B** the line of sight, and $\phi_{\nabla|\mathbf{B}|} = 15^\circ \pm 15^\circ$ between the plane of the sky and the direction of the gradient of the magnetic field, and a variation of **B** of about 1.4 times over a scale of 10 pc. A worse fit is obtained with quasi-perpendicular scenario. The overall morphology of the observed radio emission at 1.4 GHz is correctly recovered by our best-fit model. The projected direction of **B** and $\nabla|\mathbf{B}|$ in the Galaxy are along the spiral arm and toward the plane respectively, which is in very good agreement with the expected direction of the large scale Galactic **B**. However, the implied gradient is too high to be associated with the large-scale Galactic **B** and more typical of the random magnetic field components.

The application of our method to selected samples of bilateral supernova remnants may yield independent estimates of the geometry of the Galactic field at several locations, which can be useful to understand the field topology in our Galaxy.

Acknowledgements SPR acknowledges support from NASA and NSF for supernova-remnant research. FB thanks his collaborators S. Orlando, O. Petruk and M. Miceli for their work on the modeling of the radio morphology of SN 1006. BMG acknowledges the support of a Federation Fellowship from the Australian Research Council.

References

A.A. Abdo et al., Astrophys. J. **710**, L92 (2010a)
A.A. Abdo et al., Astrophys. J. **714**, 927 (2010b)
F. Acero, J. Ballet, A. Decourchelle, Astron. Astrophys. **475**, 883 (2007)
F. Acero et al. (HESS collaboration), 31st ICRC (2009). arXiv:0907.0642
F. Acero et al. (HESS collaboration), Astron. Astrophys. **516**, A62 (2010)
F. Aharonian et al. (HESS collaboration), Astron. Astrophys. **435**, L17 (2005)
F. Aharonian et al. (HESS collaboration), Astron. Astrophys. **486**, 829 (2008)
F. Aharonian et al. (HESS collaboration), Astrophys. J. **692**, 1500 (2009)
G.A. Allen, E.V. Gotthelf, R. Petre, Proc. 26th ICRC **3**, 480 (1999)
T. Amano, M. Hoshino, Astrophys. J. **661**, 190 (2007)
A.M. Atoyan, Astron. Astrophys. **346**, L49 (1999)
A. Bamba et al., Astrophys. J. **589**, 827 (2003)
R.H. Becker et al., Astrophys. J. **240**, L33 (1980)
M.C. Begelman, Astrophys. J. **493**, 291 (1998)
A.R. Bell, Mon. Not. R. Astron. Soc. **353**, 550 (2004)
A.R. Bell, S.G. Lucek, Mon. Not. R. Astron. Soc. **321**, 433 (2001)
R.D. Blandford, D. Eichler, Phys. Rep. **154**, 1 (1987)
J.M. Blondin, R.A. Chevalier, D.M. Frierson, Astrophys. J. **563**, 806 (2001)
S.V. Bogovalov, D.V. Khangoulyan, Astron. Lett. **28**, 373 (2002)
K.J. Borkowski et al., Astrophys. J. **724**, L161 (2010)
N. Bucciantini, Astron. Astrophys. **387**, 1066 (2002)
N. Bucciantini (2010). arXiv:1005.4781
N. Bucciantini, E. Amato, L. Del Zanna, Astron. Astrophys. **434**, 189 (2005)
D.J. Burn, Mon. Not. R. Astron. Soc. **133**, 67 (1966)

A.M. Bykov, Y.A. Uvarov, D.C. Ellison, Astrophys. J. **689**, L133 (2008)
A.M. Bykov et al., Mon. Not. R. Astron. Soc. **399**, 1119 (2009)
N.F. Camus, S.S. Komissarov, N. Bucciantini, P.A. Hughes, Mon. Not. R. Astron. Soc. **400**, 1241 (2009)
G. Cassam-Chenaï et al., Astrophys. J. **665**, 315 (2007)
S. Chatterjee, J.M. Cordes, Astrophys. J. **575**, 407 (2002)
R.A. Chevalier, in *Supernovae*, ed. by D.N. Schramm (Reidel, Dordrecht, 1977), pp. 53–61
R.A. Chevalier, Astrophys. J. **258**, 790 (1982)
R.A. Chevalier, Mem. Soc. Astron. Ital. **69**, 977 (1998)
R.A. Chevalier, Astrophys. J. **619**, 839 (2005)
R. Cowsik, S. Sarkar, Mon. Not. R. Astron. Soc. **191**, 855 (1980)
O.C. De Jager, A. Djannati-Ataï (2008), arXiv:0803.0116v1
A. De Luca, R.P. Mignani, P.A. Caraveo, G.F. Bignami, Astrophys. J. **667**, L77 (2007)
L. Del Zanna, E. Amato, N. Bucciantini, Astron. Astrophys. **421**, 1063 (2004)
L. Del Zanna, D. Volpi, E. Amato, N. Bucciantini, Astron. Astrophys. **453**, 621 (2006)
T. DeLaney, B.M. Gaensler, J. Arons, M.J. Pivovaroff, Astrophys. J. **640**, 929 (2006)
J.R. Dickel, W.J.M. van Breugel, R.G. Strom, Astron. J. **101**, 2151 (1991)
R. Dodson et al., Astrophys. J. **596**, 1137 (2003)
A.R. Duncan et al., Mon. Not. R. Astron. Soc. **280**, 252 (1996)
D. Eichler, Astrophys. J. **229**, 419 (1979)
D.C. Ellison, S.P. Reynolds, Astrophys. J. **382**, 242 (1991)
D.C. Ellison et al., Astrophys. J. **712**, 287 (2010)
Z. Fan, S. Liu, C.L. Fryer, Mon. Not. R. Astron. Soc. **406**, 1337 (2010)
F. Favata et al., Astron. Astrophys. **324**, L49 (1997)
G.D. Fleishman, M.F. Bietenholz, Mon. Not. R. Astron. Soc. **376**, 625 (2007)
D.A. Frail, N.E. Kassim, T.J. Cornwell, W.M. Goss, Astrophys. J. **454**, L129 (1995)
M.S. Fulbright, S.P. Reynolds, Astrophys. J. **357**, 591 (1990)
B.M. Gaensler, Astrophys. J. **493**, 781 (1998)
B.M. Gaensler et al., Astrophys. J. **569**, 878 (2002)
B.M. Gaensler et al., Astrophys. J. **616**, 383 (2004)
J.D. Gelfand, P.O. Slane, W. Zhang, Astrophys. J. **703**, 2051 (2009)
D.A. Green, Bull. Astron. Soc. India **37**, 45 (2009)
J.J. Hester, Annu. Rev. Astron. Astrophys. **46**, 127 (2008)
J.J. Hester et al., Astrophys. J. **448**, 240 (1995)
J.J. Hester et al., Astrophys. J. **577**, L49 (2002)
J.R. Jokipii, Astrophys. J. **313**, 842 (1987)
B.-I. Jun, T.W. Jones, Astrophys. J. **511**, 774 (1999)
B.-I. Jun, M.L. Norman, Astrophys. J. **472**, 245 (1996)
E. Kalemci et al., Astrophys. J. **644**, 274 (2006)
S. Katsuda et al., Astrophys. J. **723**, 383 (2010)
C.F. Kennel, F.V. Coroniti, Astrophys. J. **283**, 694 (1984)
M.J. Kesteven, J.L. Caswell, Astron. Astrophys. **183**, 118 (1987)
S.S. Komissarov, Y.E. Lyubarsky, Mon. Not. R. Astron. Soc. **344**, L93 (2003)
S.S. Komissarov, Y.E. Lyubarsky, Mon. Not. R. Astron. Soc. **349**, 779 (2004)
R. Kothes, W. Reich, B. Uyanıker, Astrophys. J. **638**, 225 (2006)
K. Koyama et al., Nature **378**, 255 (1995)
R. Krivonos et al., Astron. Astrophys. **475**, 775 (2007)
T.L. Landecker et al., Astrophys. J. **527**, 866 (1999)
J.S. Lazendić et al., Astrophys. J. **602**, 271 (2004)
K.S. Long et al., Astrophys. J. **586**, 1162 (2003)
Y. Lyubarsky, Mon. Not. R. Astron. Soc. **329**, L34 (2002)
M. Lyutikov, C. Thompson, Astrophys. J. **634**, 1223 (2005)
Y. Maeda et al., Publ. Astron. Soc. Jpn. **61**, 1217 (2009)
H. Matheson, S. Safi-Harb, Adv. Space Res. **35**, 1099 (2005)
M.A. McLaughlin et al., Astrophys. J. **616**, L131 (2004)
M. Miceli et al., Astron. Astrophys. **501**, 239 (2009)
A.H. Minter, S.R. Spangler, Astrophys. J. **458**, 194 (1996)
C.-Y. Ng, B.M. Gaensler, S. Chatterjee, S. Johnston, Astrophys. J. **712**, 596 (2010)
S. Orlando et al., Astron. Astrophys. **470**, 927 (2007)
J.P. Ostriker, C.F. McKee, Rev. Mod. Phys. **60**, 1 (1988)
A.G. Pacholczyk, *Radio Astrophysics* (Freeman, San Francisco, 1970)
E. Parizot et al., Astron. Astrophys. **453**, 387 (2006)

D.J. Patnaude, R.A. Fesen, Astron. J. **133**, 147 (2007)
G.G. Pavlov, O.Y. Kargaltsev, D. Sanwal, G.P. Garmire, Astrophys. J. **554**, L189 (2001)
O. Petruk et al., Mon. Not. R. Astron. Soc. **393**, 1034 (2009)
M. Pohl, H. Yan, A. Lazarian, Astrophys. J. **626**, L101 (2005)
M. Renaud et al., Astrophys. J. **647**, L41 (2006)
W. Reich, in *Proceedings of the 270. Heraeus Seminar on Neutron Stars, Pulsars and Supernova Remnants*, ed. by H.L.W. Becker, J. Trümper (MPE, Garching, 2002), pp. 1–12
S.P. Reynolds, Astrophys. J. **493**, 375 (1998)
S.P. Reynolds, Annu. Rev. Astron. Astrophys. **46**, 89 (2008)
S.P. Reynolds, R.A. Chevalier, Astrophys. J. **245**, 912 (1981)
S.P. Reynolds, R.A. Chevalier, Astrophys. J. **278**, 630 (1984)
S.P. Reynolds, D.C. Ellison, Astrophys. J. **399**, L75 (1992)
S.P. Reynolds, D.A. Gilmore, Astron. J. **106**, 272 (1993)
S.P. Reynolds et al., Astrophys. J. **680**, L41 (2008)
S.P. Reynolds et al., Astrophys. J. **695**, L149 (2009)
E. Reynoso et al., Astrophys. J. **491**, 816 (1997)
R. Rothenflug et al., Astron. Astrophys. **425**, 121 (2004)
F.D. Seward, P. Gorenstein, R.K. Smith, Astrophys. J. **636**, 873 (2006)
P.O. Slane et al., Astrophys. J. **525**, 357 (1999)
P.O. Slane et al., Astrophys. J. **548**, 814 (2001)
P. Slane, D.J. Helfand, E. van der Swaluw, S.S. Murray, Astrophys. J. **616**, 403 (2004)
X.H. Sun, W. Reich, A. Waelkens, T.A. Enßlin, Astron. Astrophys. **477**, 573 (2008)
T. Tamagawa et al., Publ. Astron. Soc. Jpn. **61**, S167 (2009)
S.J. Tanaka, F. Takahara, Astrophys. J. **715**, 1248 (2010)
T. Tanaka et al., Astrophys. J. **685**, 988 (2008)
A. Toor, Astron. Astrophys. **85**, 184 (1980)
J.K. Truelove, C.F. McKee, Astrophys. J. Suppl. Ser. **120**, 299 (1999)
A. Tziamtzis, M. Schirmer, P. Lundqvist, J. Sollerman, Astron. Astrophys. **497**, 167 (2009)
Y. Uchiyama, Nature **449**, 576 (2007)
E. van der Swaluw, A. Achterberg, Y.A. Gallant, Mem. Soc. Astron. Ital. **69**, 1017 (1998)
E. van der Swaluw, A. Achterberg, Y.A. Gallant, G. Tóth, Astron. Astrophys. **380**, 309 (2001)
E. van der Swaluw, T.P. Downes, R. Keegan, Astron. Astrophys. **420**, 937 (2004)
M. Vigelius et al., Mon. Not. R. Astron. Soc. **374**, 793 (2007)
J. Vink, J.M. Laming, Astrophys. J. **584**, 758 (2003)
H.J. Völk, E.G. Berezhko, L.T. Ksenofontov, Astron. Astrophys. **409**, 563 (2003)
M.C. Weisskopf et al., Astrophys. J. **536**, L81 (2000)
A.S. Wilson, K.W. Weiler, Astron. Astrophys. **53**, 89 (1976)
P.F. Winkler et al., Astrophys. J. **585**, 324 (2003)
R. Yamazaki et al., Astron. Astrophys. **416**, 595 (2004)
F. Yusef-Zadeh, B.M. Gaensler, Adv. Space Res. **35**, 1129 (2005)
V.N. Zirakashvili, F.A. Aharonian, Astron. Astrophys. **465**, 695 (2007)
V.N. Zirakashvili, F.A. Aharonian, Astrophys. J. **708**, 965 (2010)

Space Sci Rev (2012) 166:263–280
DOI 10.1007/s11214-011-9796-6

The Role of Magnetic Fields in the Interstellar Medium of the Milky Way

Evidence from the Diffuse Polarized Radio Emission

T.L. Landecker

Received: 5 November 2010 / Accepted: 7 June 2011 / Published online: 2 July 2011

Abstract Synchrotron radiation is generated throughout the Milky Way. It fills the sky, and carries with it the imprint of the magnetic field at the point of origin and along the propagation path. Observations of the diffuse polarized radio emission should be able to provide information on Galactic magnetic fields with detail matching the angular resolution of the telescope. I review what has been learned from existing data, but the full potential cannot be realized from current observations because they do not adequately sample the frequency structure of the polarized emission, or they lack information on large-scale structure. I discuss three surveys, each overcoming one of these limitations, and show how use of complementary data on other ISM tracers can help elucidate the role of magnetic fields in interstellar processes. The focus of this review is on the small-scale field, on sizes comparable with the various forms of interaction of stars with their surroundings. The future is bright for this field of research as new telescopes are being built, designed for the survey mode of observation, equipped for wideband, multichannel polarization observations.

Keywords Galactic radio emission · Polarization · Magnetic fields · Faraday rotation · Rotation measure synthesis · Interstellar medium

1 Introduction

Magnetic fields pervade the interstellar medium (ISM) of the Milky Way. They are a significant reservoir of energy and play crucial roles in many interstellar processes because they are locked into the ISM plasma and add significantly to its pressure. Magnetic fields control the star-formation process, are essential to particle acceleration, and play a central role in the return of matter from stars to the ISM. Nevertheless, we still know very little about the details of these and other interactions of magnetic field with the ISM. It is usual to make a division between the large-scale field, on scales of the spiral arms,

T.L. Landecker (✉)
Dominion Radio Astrophysical Observatory, National Research Council Canada, Penticton, BC, Canada, V2A 6K3
e-mail: tom.landecker@nrc.gc.ca

and the small-scale (quasi-random) field, on scales of stellar and other activity within the arms. At the Solar radius the large-scale field in the spiral arms is $\sim$2 μG and the small-scale random component is $\sim$3 μG (Sun et al. 2008). That the random field is larger than the regular field immediately implies strong interaction of interstellar processes with the field. Here I will discuss mostly the small-scale field and will concentrate on the information that observations yield on the role of magnetic fields in interstellar processes. General reviews of the magnetic field in the Milky Way and other galaxies abound (Heiles 1995; Beck 2001, 2009) but there are few reviews of the subjects dealt with here. I will emphasize both what is known and what is unknown, indicating fruitful directions for research.

Of the tracers of the Galactic magnetic field, synchrotron radio emission appears to be one of the most promising: it is generated everywhere in the Galaxy and there is Faraday rotation along virtually every path through the Galaxy as the emission propagates through the magnetized plasma. The polarization state is quite easy to measure at wavelengths from 1 cm to 1 m, and magnetic-field information should be available from such data with the full angular resolution of the telescope and potentially from the entire Galactic volume. Other tools for probing the field are more limited. Measurements of starlight polarization can give information on fields as a function of distance to a limit of a few kpc (Fosalba et al. 2002), and infrared observations promise a longer reach (Clemens et al. 2011). The Faraday rotation of signals from pulsars and extragalactic sources has given much information on large-scale fields (Han et al. 1997; Brown and Taylor 2001; Brown et al. 2003, 2007; Han 2009; Brown 2011; Van Eck et al. 2011; Noutsos 2011, this issue), but the spatial sampling is very sparse with present-day telescopes. Pulsars have associated distance information, but the Faraday rotation of extragalactic sources indiscriminately sums the entire line of sight through the Galaxy. Zeeman splitting of emission lines probes molecular clouds very well. Zeeman observations of H I may ultimately prove to be the best tool for measuring magnetic fields in the general ISM, but to date they have been successful in only a few hundred directions (Heiles and Robishaw 2009).

However, observations of the diffuse Galactic polarized emission have not lived up to their promise. On the technical side, we have not sampled the sky adequately, in either the spatial or frequency domains (and we have only recently come to understand this). On the scientific side, we have needed complementary observations of other components of the ISM that have not been available. Despite these limitations, much has been achieved, and I review what has been learned. Three datasets break free from these limitations to some extent, and I present significant results from them. We are on the cusp of major advances in this field, and I comment on the progress that will come from surveys underway and planned with the next generation of radio telescopes.

2 The Data

The first detections of polarized emission from the Galaxy (Westerhout et al. 1962; Wielebinski et al. 1962) were followed by investigations of the polarized sky at various wavelengths. By 1976 the number of polarization observations of extensive areas had reached 18—listed by Spoelstra (1984). The most thorough were the surveys carried out with the Dwingeloo telescope at five frequencies from 408 to 1411 MHz. Observed between 1961 and 1966, and published by Brouw and Spoelstra (1976), these surveys have careful absolute calibration and correction for instrumental effects, and they appear to cover most of the northern sky. However, at 1411 MHz the 1726 pointings sample only $\sim$3% of the sky accessible to the telescope. At the lowest frequency, 408 MHz, with a broader beam, this fraction rises to $\sim$20%.

Table 1 Surveys of the Galactic polarized emission

Survey	Ref.	ν [MHz]	θ [']	A-S	S-A	Large Scale	RM Synth.	Coverage [Sr]
WSRT-150	(1)	139–153	2	y		n	y	1×10^{-2}
WENSS	(2)	325	8	y	y	y?	n	0.3
WSRT-350	(3)	341–375	5			n	n	1.6×10^{-3}
WSRT-RMS	(4)	324–387	4			n	y	2.6×10^{-2}
Dwingeloo	(5)	408	138		y	y	n	7.0
		465	120		y	y	n	7.0
		610	90		y	y	n	7.0
		800	60		y	y	n	7.0
		1411	36		y	y	n	7.0
EMLS	(6)	1400	9.4		y	n	n	2.3
DRAO 26m	(7)	1410	36		y	y	n	9.1
Villa-Elisa	(8)	1420	36		y	y	n	5.2
CGPS	(9)	1420	1	y	y	y	n	0.37
SGPS	(10)	1336–1432	2	y		n	n	9.6×10^{-2}
Eff-11	(11, 12)	2695	5			n	n	0.21
Pks-11	(13)	2417	10		y	some	n	0.71
PGMS	(14)	2300	9		y	n	y	8.7×10^{-2}
Urumqi	(15, 16)	4800/4963	10		y	y?	n	0.32
WMAP-23	(17)	23000	120			y	n	12.6

References: (1) Bernardi et al. (2009), (2) Schnitzeler et al. (2007), (3) Haverkorn et al. (2003), (4) Schnitzeler et al. (2009), (5) Brouw and Spoelstra (1976), (6) Reich et al. (2004), (7) Wolleben et al. (2006), (8) Testori et al. (2008), (9) Landecker et al. (2010), (10) Haverkorn et al. (2006), (11) Junkes et al. (1987), (12) Duncan et al. (1999), (13) Duncan et al. (1997), (14) Carretti et al. (2010), (15) Sun et al. (2007), (16) Gao et al. (2010), (17) Hinshaw et al. (2009)

Notes: (a) ν, frequency in MHz. (b) θ, beamwidth in arcminutes. (c) A-S denotes aperture synthesis, S-A denotes single antenna. (d) y or n under the heading "Large Scale" indicates that large-scale structure is or is not represented. y? indicates that large-scale structure has been incorporated incompletely and/or derived from observations at another frequency—see discussion in Sect. 3.2. (e) y under RM Synth. indicates that the data are adequate for application of RM Synthesis. (f) Coverage gives the sky area observed in steradians

In the 1980s polarization observations using large single-antenna radio telescopes began again, initiated with the Effelsberg 100-m Telescope, starting with the work of Junkes et al. (1987), this time with more thorough or even complete sampling, but usually concentrating on the Galactic plane. The most recent of this series, the Effelsberg Medium Latitude Survey (EMLS—Reich et al. 2004) extends coverage to $-20° < b < 20°$ at 1400 MHz. These have been joined by surveys using aperture-synthesis telescopes, achieving higher angular resolution. Table 1 lists surveys, and observations of lesser scope, that give valuable information on the polarization of the Galactic synchrotron emission. Note that the CGPS survey incorporates data from the EMLS and the DRAO 26 m survey with aperture-synthesis data: see Sect. 5.2 for more detail. The surveys of Table 1 represent a major observational effort, and we have learned much from them (reviewed in Sect. 4). Nevertheless, they have major shortcomings. In the next section I discuss what really needs to be measured in order to make advances in the astrophysics of the magneto-ionic medium (MIM).

3 Technical Requirements

The polarized sky does not resemble the total-intensity sky. The reason, understood from the beginning, is that Faraday rotation occurs along the propagation path between the source of the emission and the telescope. When polarization angle, θ [rad], is measured as a function of wavelength, λ [m],

$$\Delta\theta = \zeta\lambda^2 = 0.81\lambda^2 \int n_e B_{\parallel}\, \mathrm{d}l \tag{1}$$

where ζ [rad m^{-2}] is the Faraday depth (Burn 1966), n_e [cm^{-3}] is the electron density, $B_{\parallel}$ [μG] is the line-of-sight component of the magnetic field, and the integral extends along the entire line of sight. Under simple circumstances, a Faraday screen distinct from and in front of a background synchrotron emitter, $\zeta = \mathrm{RM}$, the rotation measure.

If Faraday rotation is dominant in shaping the polarized sky, then we will not understand the ISM by simply measuring θ at one wavelength: *the physcially significant quantity is Faraday depth (or rotation measure), the product of magnetic field strength and electron density*. Measuring this quantity immediately implies making polarization measurements at a number of frequencies. In Sect. 3.1 I discuss the sampling in the frequency domain that is required to probe polarization structure.

The Milky Way is all around us, and synchrotron emission is generated throughout. Galactic radio emission therefore has structure on all scales from large to small. If the large-scale structure is missing or under-represented in polarization measurements the polarization angle cannot be established; this is a major defect in many polarization datasets and limits their value for astrophysical interpretation. I discuss spatial sampling in Sect. 3.2.

3.1 Adequate Sampling in Frequency

Synchrotron radiation is broadband by nature. To determine the spectrum of total-intensity emission from the Milky Way or from a typical extragalactic source requires only a handful of measurements spanning the microwave spectrum: synchrotron emission can be characterized by a small number of parameters (as can mixed thermal and synchrotron emission). On the other hand, the Faraday rotation effects along the propagation path are usually not so simple. For example, Reich et al. (2004) and Gao et al. (2010) present data on a large area near the anticentre at 1.4 GHz and 4.8 GHz respectively, with very similar angular resolution ($\sim 10'$). The total-intensity images are unmistakably similar, but there is little resemblance in polarization structure at the two frequencies. Towards the inner Galaxy matters are worse. The probable cause is the interaction of a number of Faraday-rotation phenomena, such as beam depolarization and depth depolarization—differential Faraday rotation—(Burn 1966; Sokoloff et al. 1998). At the lower frequencies the ISM can become Faraday thick, and telescopes operating at widely separated frequencies may sample different volumes.

The question then arises: what sampling in frequency *is* required to characterize the extended polarized emission? We do not know the full answer yet, and probably will not until the Galactic plane has been extensively re-observed with adequate tools. Nevertheless, it is quite clear that we need polarimetric data measured in many adjacent channels across wide bands. The need for data of this kind, and methods for their analysis were laid out by Burn (1966), but technical limitations made acquisition and analysis of such data impossible until recently. Such methods are now easily implemented as Rotation Measure Synthesis (Brentjens and de Bruyn 2005; Heald 2009).

In (1) $\zeta = \mathrm{RM}$ under very simple circumstances, where a source of polarized emission lies behind the Faraday rotating screen, and this is an adequate assumption when these conditions are approximately true (for example when studying a compact extragalactic source seen through the Galactic ISM). In the Galaxy, emission and Faraday rotation are thoroughly intermixed, and both occur virtually everywhere; under these conditions θ does not have a linear dependence on λ^2. RM Synthesis may still be able to recover the polarization structure along the line of sight (not completely, but to a useful extent). Choosing a value for ζ, the observed polarization vector in each channel is rotated by $\zeta\lambda^2$ and the rotated vectors are coherently added to obtain an image of polarized intensity at that value of Faraday depth.

We do not know exactly how much resolution in Faraday depth is required, but we infer from existing data (Schnitzeler et al. 2009) that a value of $\sim$10 rad m^{-2} is useful for regions off the Galactic plane. This implies observing down to frequencies of order 300 MHz: the resolution in Faraday depth depends on the longest wavelength observed—Brentjens and de Bruyn (2005), Schnitzeler et al. (2009). Such observations have their own complications, including ionospheric Faraday rotation and heavy RFI. Furthermore, turbulence in the MIM will produce dispersion in Faraday depth, leading to depolarization (Burn 1966) that is especially significant at long wavelengths since the effect depends on λ^4. Observations near 300 MHz may have limited value at low Galactic latitudes. The GMIMS survey (Sect. 5.4) will extend down to 300 MHz, and analysis of these data will answer some of the questions asked here.

Nor do we know how thoroughly we need to sample λ^2 space. For example, it is conceivable that sampling a moderate number of narrow channels across a wide total bandwidth is adequate. The rotation measure spread function (Brentjens and de Bruyn 2005) will have bad sidelobes, but these can be removed (partly) by deconvolution, a CLEAN process (Heald 2009). This process interpolates in λ^2 space, filling in the missing frequency channels from information at other frequencies. Modelling may give useful guidance, but, at this stage, without experience in analyzing actual data, it is impossible to decide how risky this is. Given the complexity that we find in the ISM, it seems advisable to strive for the fullest frequency coverage possible: measured data is always better than reconstructed data. Existing analyses of MIM structures where emission and Faraday rotation are mixed are confined to those situations that can be treated analytically (Burn 1966; Sokoloff et al. 1998) and these are simple situations.

The absence of information at *short* wavelengths from RM Synthesis data imposes an upper limit to the largest Faraday-depth structure that can be detected. A Faraday slab, in which emission and rotation both occur, produces smooth, extended features in RM space. Such a feature will be "differentiated" and break up into delta functions where sharp transitions in RM occur at the front and back of the slab. This is analogous to the problem in interferometric imaging where the absence of data from short baselines hides large physical structure (Brentjens and de Bruyn 2005; Schnitzeler et al. 2009). A further limitation of RM Synthesis is its inability to recover the intrinsic polarization angle (the angle at $\lambda = 0$) because it is impossible to make measurements at negative values of λ^2. Frick et al. (2010, 2011) discuss the use of symmetry arguments and wavelet-based RM Synthesis to help overcome this problem.

The product of RM Synthesis is a data cube of complex numbers, a set of images of polarized intensity and polarization angle at different values of Faraday depth. The sensitivity of each image is determined by the entire bandwidth of the observations, which is usually very large. As a consequence, the new generation of instruments is producing polarization images whose sensitivity far exceeds anything that existed hitherto. With this great sensitivity and with the ability to analyze the MIM in a new way, we are on the cusp of a new era in the study of the extended polarized emission from the Milky Way.

3.2 Sampling Structure on All Scales

Our understanding of other constituents of the ISM has been transformed by the attainment of angular resolution of order one arcminute in recent extensive surveys along the Galactic plane, the Canadian Galactic Plane Survey—the CGPS—(Taylor et al. 2003), and its companions, the Southern Galactic Plane Survey—the SGPS—(McClure-Griffiths et al. 2001), and the VLA Galactic Plane Survey—VGPS—(Stil et al. 2006). We can expect the same to be true of the magnetic field.

How much angular resolution is adequate for studying the polarized ISM? The glib answer is that the highest resolution is never enough, that there are always problems just beyond the reach of the telescope. As a guideline, one arcminute seems a reasonable target because this is the stellar separation in the nearest spiral arms. If we can understand arcminute data we will be ready to take on the next challenge.

Such resolution at decimetre wavelengths demands use of aperture-synthesis (A-S) techniques. Since A-S telescopes are not sensitive to the largest structures, it is immediately apparent that single-antenna (S-A) data will have to be incorporated. A problem appears at once, because adequate S-A polarization data are not available. To date, only one polarization survey has met this challenge, the CGPS Polarization Survey (Landecker et al. 2010)—see Sect. 5.2. High quality wideband spectropolarimetric data obtained with S-A telescopes will be vital if progress is to be made in this field (see Sect. 5.4).

Why do many of the surveys made with S-A telescopes not include large-scale structure (Table 1) when such telescopes are inherently able to detect structure on all scales? When making polarization observations with S-A telescopes there is no point in the sky where the polarized signal is zero: there is always a contribution from instrumental polarization and from ground emission (not itself polarized, but converted to apparently polarized emission by the antenna sidelobes). Measuring ground emission and instrumental polarization on an absolute scale is difficult. This problem is avoided in the EMLS (Reich et al. 2004) (and most other S-A polarization surveys) by observing the sky one small patch at a time. Over that small patch ground emission will be approximately constant and instrumental polarization will be stable over the short time of the observation. Subtracting the mean value of Q and U from the observations then removes these effects, but it also filters off large structure. The extent of this filtering depends on the size of the patches that make up the observations. In the case of the EMLS this is typically $10^\circ \times 10^\circ$, and polarized structures of $\sim 5^\circ$ and larger are seriously under-represented. This feature of the EMLS is fully discussed by Reich et al. (2004)—see Fig. 3 of that paper—and emphasized again by Reich (2006). The extent to which large structure is filtered off when scanning with a S-A telescope depends critically on the extent of the scans that make up the data: the very best result is obtained by observing the entire sky (see, for example, Carretti 2011).

The ISM is turbulent, and electron density, field direction and field strength have spatial variability. As a result Faraday rotation can break large, smooth total-intensity structures into smaller polarization features. If this effect is sufficiently strong there will be virtually no large structure in polarization images. Is it then necessary to add S-A data to A-S data? Available evidence suggests that it is. Large-scale structure is definitely present in the CGPS polarization survey at 1420 MHz (see Fig. 8 of Landecker et al. 2010). From S-A radio telescope data we see that significant polarized emission remains in a 2° beam at 408 MHz (Brouw and Spoelstra 1976) and in a 1.5° beam at 240 MHz (Wilkinson 1973). However, it is hard to generalize about this question.

Despite everything that I have said here, datasets that miss broad structure are still valuable for the *identification* of interesting polarization structures—e.g. Rudnick and Brown (2009).

4 What Have We Learned?

The observations listed in Table 1 may have their limitations, but we have learned much from their data. In this section I discuss some of the conclusions reached from observations of the diffuse polarized emission from the Milky Way.

4.1 Sensitivity of Polarization Observations to Ionized Gas

A typical radio telescope operating at decimetre wavelengths is more sensitive to the Faraday rotation in a volume of ISM plasma than to its bremsstrahlung. Consider an ionized region of extent 10 pc with electron density 0.5 cm^{-3} in a magnetic field with a line-of-sight component of 2 μG (typical ISM conditions). At 1.4 GHz this region produces a Faraday rotation of 20°, which is easy to detect. The same region produces an emission measure of 2.5 cm^{-6} pc, a challenge for the most sensitive radio telescope to detect in total intensity. This is the origin of the frequent comment that the polarized sky does not resemble the total-intensity sky. However, the sensitivity of Hα surveys is superior. The Wisconsin Hα Mapper—WHAM—(Haffner et al. 2003) can reach levels of emission measure ten times lower than most radio telescopes, and the polarized sky often bears a stronger resemblance, either correlation or anti-correlation, to the Hα sky (limited, of course, by optical extinction). A detailed study of these resemblances covering a large area has not yet been made.

4.2 The Distance to Polarization Features

As in any astronomical endeavor, determining distance is important in assessing the physical properties of a polarization feature. Physical size is significant, for example, in establishing scale sizes of turbulence. The path length through a Faraday rotating structure must be known if a magnetic field estimate is to be derived from a Faraday depth measurement (and, of course, independent data on electron density are also needed).

H I absorption is a powerful technique for determining (or limiting) distance to continuum emitters. This can be applied to polarized emission, but not without difficulty. On the positive side, H I emission is unpolarized, so confusion from small emission regions in the off-source area is not a problem (this is the limiting factor in total-intensity absorption work). On the negative side, signal-to-noise ratios are very low. Dickey (1997) has applied this technique to determine the distance to the extended Galactic polarized emission in a small target area at $\ell = 329.5°$, $b = 5°$: the derived distance was 2 kpc. The only further application of this difficult technique has been its use by Kothes et al. (2004) to determine H I absorption distances to supernova remnants of low surface brightness.

Depolarization by H II regions can be used as an aid in distance determination. If the outline of the H II region corresponds with an area of reduced polarized intensity, the association is usually firm, and the polarization feature definitely lies behind the H II region (whose distance is often independently known). There are several caveats. First, some H II regions have quite regular fields, and will not necessarily depolarize (see Sect. 4.4). Second, the depolarization due to an H II region will not necessarily be total: the ISM on the near side of the H II region can be a synchrotron emitter, and itself a polarized emitter. In the CGPS survey (Landecker et al. 2010, Sect. 5.2) there are many instances in which Local-arm H II regions depolarize more distant Perseus-arm features. The foreground emission is relatively smooth, and the emission coming from larger distances shows structure on smaller angular scales; an unambiguous distance limit can then be deduced.

Two effects combine to determine a "polarization horizon": polarized signals originating beyond the polarization horizon cannot be received (Uyanıker et al. 2003). First, the ISM is Faraday thick and signals from further away are depolarized. Second, the telescope beam embraces an ever larger physical volume as the beam expands with distance: signals from more distant volumes suffer progressively increasing beam depolarization. The distance to the polarization horizon is beamwidth dependent, frequency dependent, and direction dependent. In the CGPS survey, with a $1'$ beam at 1420 MHz, the polarization horizon is at about 2 kpc between $\ell = 66^\circ$ and $\ell = 110^\circ$, so the polarization features detected arise within the Local arm. Beyond $\ell = 110^\circ$ the horizon moves into the Perseus arm, and towards the anticentre the telescope is seeing to the limit of the Galaxy (Landecker et al. 2010). Most of the imaging of the Galactic plane done to date has given us a very local view of the MIM. RM Synthesis can overcome some of the effects of Faraday thickness, but only high angular resolution can conquer beam depolarization.

RM Synthesis can offer information on distance, revealing the relative placement of RM features along the line of sight. It cannot give actual distances unless combined with other information and, possibly, combined with modelling.

4.3 Modelling and Statistical Approaches to Characterizing the Magneto-Ionic Medium

Modelling the total-intensity and polarized sky is a powerful technique for integrating many lines of astrophysical data.

On the basis of the 408 MHz all-sky map of Haslam et al. (1982), Beuermann et al. (1985) modelled the synchrotron emission from the Milky Way as a thin disk of scale height 150 pc within a thick disk whose scale height is 1.5 kpc (adapted to a distance to the Galactic centre of 8.5 kpc). The synchrotron scale height is produced by the combined effects of the scale heights of the magnetic field and of cosmic-ray electrons, and so the scale height of the magnetic field cannot be derived directly from the Beuermann et al. (1985) model. However, the scale height of the magneto-ionic disk may yield the desired information if coupled with measurements of the scale height of thermal electrons, recently estimated as 1.8 kpc (Gaensler et al. 2008). Measurements of the scale height of the RM disk of 1.4 kpc (Simard-Normandin and Kronberg 1980) are supported by recent measurements by Rae and Brown (2011), who find a value of 1.2 kpc in the outer Galaxy. Investigation of the roles of spiral structure and the Galactic warp is needed before the latter value can be translated to a scale height for the magnetic field. This is a developing subject.

The thorough work of Sun et al. (2008) is the first modelling effort to take radio polarization data into account, both RMs of extragalactic sources and data on diffuse emission. This is the most valuable contribution to understanding the magnetic field in the Galactic disk of recent years (see Sun and Reich (2010) for an improved estimate of the values of the halo field). The properties of the MIM, for which the evidence comes from radio polarization data, are crucial in the modelling. From this work, the large-scale field in the local vicinity has strength ~ 2 μG, and its configuration is well determined. The small-scale, random field strength is ~ 3 μG. The total field is lower than the equipartition value of ~ 6 μG (Heiles 1995; Beck 2009), but the estimates are consistent within the errors. Strong depolarization evident between 23 GHz (Hinshaw et al. 2009) and 1.4 GHz (Wolleben et al. 2006) in a strip $-30^\circ < b < 30^\circ$ can be explained with a small filling factor of thermal electrons and a coupling of the random field with electron density (although an exact coupling factor could not be established).

The features visible in polarization images have often been described as "chaotic", "disordered" or "patchy". This structure is generally understood as the product of Faraday rotation in a turbulent magneto-ionic medium, and statistical analysis of its structure is a useful approach to uncovering the physics of the turbulence.

Observations of pulsar scintillation suggest that electron density variations on very small (sub-pc) scales exhibit Kolmogorov turbulence (Stinebring et al. 2000). On larger scales, Haverkorn et al. (2008) investigated structure functions of RMs of extragalactic sources. RM fluctuations are present in interarm regions up to scales of 100 pc, but in the arms there are no fluctuations beyond a few pc, much smaller than expected (values confirmed by estimates based on depolarization of extragalactic sources). These two very different values for the outer scales of the turbulence imply two different driving forces. Within the arms the dominant source of energy injection into turbulence is star-related activity (star formation, H II regions, stellar winds). Larger-scale effects appear to dominate in the interarm.

Sun and Reich (2009) have generated model radio images of total and polarized emission and RM at arsecond resolution at various Galactic latitudes, based on their earlier modelling (Sun et al. 2008). These can provide the basis for comparison with observations via structure function or other techniques, and were created for use in simulations of the Square Kilometre Array.

"Canals" are long, thin features in polarization images where the polarized intensity drops to zero. Typically one beamwidth wide and many times as long, canals have apparently random locations and shapes. They were recognized first in A-S polarization images, but are also found in S-A data. The high-pass filtering of structure by A-S telescopes increases their number, but they nevertheless represent something real in the MIM. They have been variously interpreted as the boundaries between regions of different RM (Haverkorn et al. 2000, 2004), locations where differential Faraday rotation happens to lead to total depolarization (Shukurov and Berkhuijsen 2003), or the locations of shock fronts (Haverkorn and Heitsch 2004; Fletcher and Shukurov 2006). Haverkorn and Heitsch (2004) used magnetohydrodynamic simulations of the turbulent ISM to show that shocks can generate RM gradients sufficiently steep to produce the observed effects.

While the small-scale field is quasi-random, Brown and Taylor (2001), examining RMs of extragalactic sources seen through the disk, found evidence of preferential alignment with the large-scale field. Their result applies to the outer Galaxy; such studies should be extended to other directions.

4.4 H II Regions

The effects of H II regions on polarization images have been discussed to some extent in Sect. 4.2. Intuitively one would expect that H II regions are depolarizers. Within them electron density is high and the ionized gas is generally turbulent, driven by winds and heating from the ionizing stars; field strengths should be high. The combination should result in strong Faraday rotation that changes quickly on small physical scales, leading to beam depolarization. Many observations confirm this view (Gray et al. 1999; Gaensler et al. 2001; Landecker et al. 2010), and it is further supported by the data of Stil and Taylor (2007) and Stil et al. (2011) who reported that some H II regions depolarize even compact extragalactic sources that are seen through them.

On the other hand there is recent evidence from the Urumqi survey at 4.8 GHz (Sun et al. 2007; Gao et al. 2010) that at least some H II regions act as Faraday screens but not as depolarizers. These authors report regions with electron densities a few electrons per cm^3 and fields up to $\sim 15\ \mu G$, considerably stronger than the usual values of a few μG. It is not yet clear what mechanism can lead to these strong, well-organized fields. However, these regions seem to be closer to the regular Warm Ionized Medium than they are to classical H II regions, normally thought of as dense volumes ionized by one or more powerful stars.

If H II regions totally depolarize emission arising beyond them, then we can study synchrotron emissivity over local paths of known length on the near side (Uyanıker et al. 2003).

This is a new method that complements earlier results from absorption by H II regions at low frequencies (Roger et al. 1999; Nord et al. 2006) or from modelling (Beuermann et al. 1985; Sun et al. 2008).

Given the high sensitivity of radio telescopes to Faraday rotation, polarization data can be used to probe low-density envelopes of H II regions. Gray et al. (1999) examined the low-density ionized zone around W4. The W4 H II region is a strong depolarizer; at the edge of the depolarization zone the field strength is ~20 μG, and apparently scales as the square-root of electron density (but the data for this work do not include information on large structure, and the conclusions need to be revisited). These conditions rotate background emission but do not totally depolarize it.

Polarization observations can trace disk-halo interactions from H II complexes. The walls of the superbubble driven by the stars that excite W4 are detectable as depolarization features. West et al. (2007) estimated a magnetic field strength of 3 to 5 μG in the bubble walls based on an optical estimate of electron density (~0.4 cm^{-3}) and the observed degree of depolarization (~50%). Again, the data used in this work do not include large-scale structure. Duncan et al. (1997) observed depolarization "plumes" emanating from H II regions at $\ell = 288°$ and $\ell = 292°$ at latitudes just below $b = 0°$, which they interpreted as vertical flows of ionized gas.

4.5 Photodissocation Regions

Wolleben and Reich (2004) observed Faraday rotation associated with the Taurus molecular cloud at 1400 and 1660 MHz (using the standard techniques of the EMLS). Information on large structure was available at 1400 MHz, and reasonable assumptions were used to derive such information at 1660 MHz. The data were fitted well with a model in which background emission is Faraday rotated in a thin "screen" of partially ionized material, a photodissociation region (PDR) on the surface of the molecular material, and adds vectorially to the foreground. Optical data place limits on electron density, making it possible to estimate field strength. The deduced magnetic field in the PDR is strong, ~20 μG, and it is well ordered. Local synchrotron emissivity deduced from the modeled foreground emission is quite high in relation to other estimates. The field estimate within this PDR has possibly been confirmed by H I Zeeman observations (Heiles and Robishaw 2009). Apparently similar PDR phenomena have been detected by Gao et al. (2010) on the edges of the W5 H II region.

4.6 Planetary Nebulae

Two planetary nebulae (PNe) have been identified as Faraday rotating objects in the CGPS survey (Table 1). The amounts of ionized material in these objects are small, of the order of 1 $M_\odot$, yet, surprisingly, they generate unmistakable polarization signatures. Ransom et al. (2008) discuss Sh-2-216 ($\ell = 158.34°$, $b = 0.2°$)—see Fig. 1. Ransom et al. (2010) present data for DeHt 5 ($\ell = 111.1°$, $b = 11.6°$). Both objects appear as depolarization features: background polarized emission is Faraday rotated on passage through the ionized material and adds destructively to foreground emission.

Optical observations of Sh-2-216 yield good values of distance (allowing path length through the Faraday screen to be calculated) and electron density. The magnetic field can then be tightly constrained. The field in the shell is ~5 μG. This is probably interstellar field, slightly compressed, not stellar field (Ransom et al. 2008).

The progenitor of DeHt 5 has a high proper motion through the ISM. The PN itself is a depolarization feature. Ransom et al. (2010) also identify a series of patches of reduced

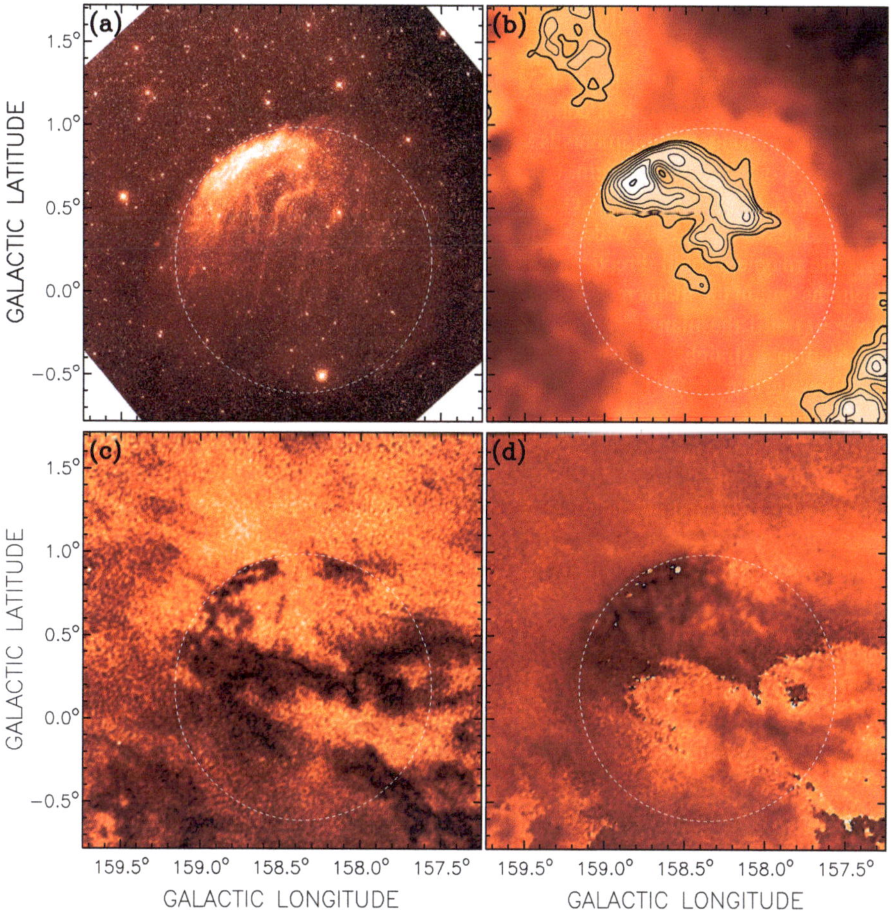

Fig. 1 The planetary nebula Sh 2-216 seen in optical emission (*top left*), total intensity at 1420 MHz (*top right*), polarized intensity (*bottom left*), and polarization angle (*bottom right*). The *dotted circle* outlines the optical emission. The central star has proper motion towards the bright optical rim in the north-east. An arc of depolarization corresponds to that rim. A superimposed feature confuses the polarized signal over the southern half of the object

polarized intensity *behind* this PN as relics of the winds from the progenitor during its AGB phase, prior to the formation of the PN. We are seeing a "magnetic wake" left behind as the star moves through the ISM. The wake consists of moving ionized material which drags field lines with it. The change in the line-of-sight component of the field generates the observed polarization signature.

There are several general conclusions. First, PNe could be used as probes of the Galactic magnetic field, as there is often useful optical data on distance, size, and electron density. High resolution aperture-synthesis telescopes (the VLA today, and especially the SKA in the future) could make similar observations of many PNe, reaching right across the Galaxy. Second, observations like these show that the (plentiful) low-mass stars contribute significant quantities of ionized material to the ISM.

4.7 Correlation of Polarization and H I Features

The new images of the Galactic ISM emerging from the CGPS, SGPS, and VGPS show that filamentary structure is a common feature of H I, in both cold and warm phases. The abundance of long, thin filaments is very suggestive that magnetic fields have played a role in forming or maintaining them. Only one observation bears on this question, the remarkable data of McClure-Griffiths et al. (2006). Long, thin filaments of cold gas, up to 17 pc long and less than 0.2 pc wide, were observed in self-absorption. The strands are aligned with the magnetic field direction deduced from polarization of starlight. The implied field strengths within the filaments, estimated using the Chandrasekhar and Fermi (1953) method, are ≥ 30 μG if the filaments are magnetically dominated (a field of this strength can hold the filaments straight against the effects of turbulence). Note that Heiles and Robishaw (2009) argue that the CF method overestimates the field. The features discussed by McClure-Griffiths et al. (2006) comprise cold gas. It is plausible that cold, dense filaments carry quite strong fields, since field is correlated with density (Sun et al. 2008), but that leaves open the question of the widespread filamentary features seen in emission from the (presumably) warm H I. Densities are lower and fields are probably lower too. No systematic search has been made for correlations between field and either cold or warm H I filaments; both starlight polarization data and radio polarization data will be useful.

5 Three Surveys that Overcome (Some of) the Limitations of Existing Data

In this section I describe three surveys that overcome some of the limitations that beset many of the efforts in Table 1.

5.1 Rotation Measure Synthesis with the Westerbork Telescope

Schnitzeler et al. (2007, 2009) have pioneered the use of RM Synthesis in the study of the extended Galactic emission. Using the Westerbork SRT with frequency coverage of 324 to 387 MHz yielded remarkably large coverage of the λ^2 domain, $0.6\ \mathrm{m}^2 < \lambda^2 < 0.9\ \mathrm{m}^2$, and excellent resolution in Faraday depth of 12 rad m^{-2}. They observed an area $7° \times 7°$ centred at $\ell \approx 181°$, $b \approx 20°$.

However, the absence of information at *short* wavelengths from these data imposed an upper limit to the largest Faraday depth that can be detected of ~5 rad m^{-2}, smaller than the resolution in Faraday depth. This study therefore focused on a region with simple structure, the Galactic anticentre at high latitudes. Here the line of sight probes the Local arm and, possibly, the lower reaches of the Galactic halo. The regular component of the Galactic field is essentially perpendicular to the line of sight, producing relatively little Faraday rotation. These observations, then, have mostly probed the random field component. On the majority of the lines of sight RM spectra show only one peak, or two peaks with one dominant. The RM structure is indeed simple (at the resolution in Faraday space of these observations—increasing resolution will at some point encounter finer scales in the turbulence of the ISM).

Again exploiting the simplicity of the anticentre Schnitzeler et al. (2009) used WHAM Hα data (Haffner et al. 2003) to provide values of emission measure. Translating those to dispersion measure led to values of the line-of-sight component of the magnetic field. The angular resolution is limited by the 1° beam of the WHAM data, but coverage of the field is complete since there is little optical extinction. The deduced field values ($0 < \langle B_{\parallel} \rangle < 0.75$ μG) are larger than the anticipated value of the regular field (nearly zero in this direction), but substantially smaller than the 3 μG random component estimated by Sun et al. (2008), possibly because of the 1° effective resolution.

Extragalactic sources seen in this region show RM values comparable to the Faraday depths of the extended emission, indicating that the emitting and Faraday rotating regions are entirely within the polarization horizon (Sect. 4.2). This is not the case in more complex directions (towards the inner Galaxy), indicating again the simplicity of these lines of sight in the anticentre. However, this comparison is possible over only half the area. In the other half the number of polarized background sources detected was insufficient for any definite conclusion. It is instructive that measurements of the extended Galactic emission can give almost complete sampling of Faraday depth while simultaneous observations of extragalactic sources with the same telescope can give only sparse data.

5.2 The CGPS Polarization Survey

The Canadian Galactic Plane Survey (Taylor et al. 2003, the CGPS) set out to observe the major constituents of the ISM with arcminute resolution.[1] As part of the CGPS, the polarized emission at 1420 MHz along the northern Galactic plane has been imaged using the DRAO Synthesis Telescope (Landecker et al. 2010). Observations at 1420 MHz in Stokes parameters I, Q, and U cover $\ell = 66°$ to $\ell = 175°$ over a range $-3° < b < 5°$, with an extension from $\ell = 101°$ to $\ell = 116°$ up to $b = 17.5°$. Angular resolution is $\sim 1'$. The survey brings together data from three telescopes. The DRAO Synthesis Telescope (Landecker et al. 2000) provided arcminute resolution, and the images were accurately corrected for instrumental polarization across the field of view (Reid et al. 2008). Structures larger in size than about $20'$, undersampled or missing in the A-S data, were incorporated from the Effelsberg Medium-Latitude Survey—the EMLS—(Reich et al. 2004) and structures larger than $\sim 5°$, missing from the EMLS data because of the observing and processing techniques used (Sect. 3.2), were derived from the survey made with the DRAO 26-m Telescope (Wolleben et al. 2006). Evaluation of the relative scales of three sets of observations with quite distinct telescopes has an accuracy of about $\pm 10\%$, set by the accuracy of knowledge of the S-A beamshapes (Landecker et al. 2010). This is the first extensive survey of the Galactic polarized emission to incorporate S-A with A-S data, and, covering 1060 square degrees with 1.5×10^7 independent data points, is the largest polarization survey ever.

5.3 Results from the CGPS Polarization Survey

Some of the polarized features found in the CGPS polarization survey are discussed in Landecker et al. (2010); in this section I discuss only one. Figure 2 shows a polarized intensity (PI) image from the CGPS together with an H I image at $v_{lsr} = -20\ \mathrm{km\,s^{-1}}$. There is a very clear association of the polarized arc that starts at $\ell = 162°$, $b = 2.4°$, and curves across the top of the PI image, with an H I feature. Total extent of these features is $162° \leq \ell \leq 170°$, 8° in longitude.

The association with H I immediately provides a kinematic distance, approximately 1.9 kpc, which places this object in the Perseus arm and implies a linear extent of about 350 pc. The interpretation (Kothes et al. in preparation) sees this as a large stellar-wind bubble (SWB), the product of a cluster of massive stars. In the SWB model of Weaver et al.

[1] The Canadian Galactic Plane Survey database is accessible to the worldwide astronomy community at http://www3.cadc-ccda.hia-iha.nrc-cnrc.gc.ca/cgps/.

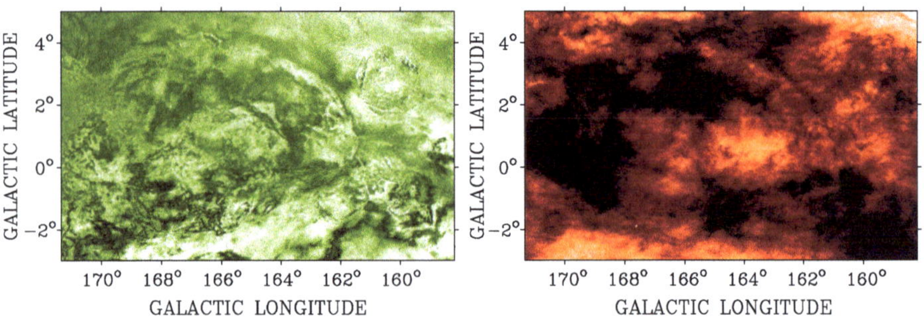

Fig. 2 CGPS data, 1420 MHz, angular resolution 1′. Polarized intensity (*left*) from 0 (*black*) through *green* to 500 mK (*white*), and H I at $v_{lsr} = -20\ \mathrm{km\,s^{-1}}$ from 20 (*white*) through *red* to 80 K (*black*)

(1977) there is a central hot-wind region where the temperature, T, is 10^6 K and the electron density, n_e, is 0.01 cm^{-3}, surrounded by a shocked-wind region where $T = 10^5$ K and n_e is higher (in this paragraph all numbers should be read as approximate). A very simple Faraday rotation model with a toroidal magnetic field of strength 30 μG in the shocked wind zone and $n_e = 0.2\ \mathrm{cm}^{-3}$ can explain the PI and polarization angle manifestations of this object. Synchrotron emission from a smooth background is Faraday rotated in the shocked-wind zone, and vector addition with polarized foreground emission produces increases and decreases in PI that define the structure seen in the figure.

Once detected, other manifestations of the SWB became evident: X-rays from hot gas in the interior: evidence of the expansion of the H I shell from absorption spectra of background compact sources seen through it: a clear association with the supernova remnant G166.0+4.3. The SWB has an age of 1–2 × 10^7 years, but all surrounding objects are younger, $\sim 1 \times 10^6$ years, and there is evidence that the large SWB has triggered star formation in its environs.

This result has several implications. First, the strong toroidal magnetic field evident in this SWB is probably generated by winding up a radial field arising from stellar winds. The field may have been enhanced by magnetized material carried out from massive stars by the winds. Compression by the expansion may have amplified the field further. Second, this is an example of a large and significant object that remained hidden until detected by its Faraday rotation, and a powerful demonstration of the sensitivity of polarization observations as detectors of ionized gas. Third, modelling of the region would be meaningless, and so interpretation would be very difficult, if data on broad structure were not incorporated in the images (the object can be seen in the EMLS data (Reich et al. 2004) but interpretation could not proceed from those data alone). Fifth, arcminute resolution was instrumental in the recognition of the SWB, and helped refine model parameters. Sixth, most significantly, observations of this object would be very difficult to understand without complementary data on other ISM tracers.

5.4 The Global Magneto-Ionic Medium Survey

The Global Magneto-Ionic Medium Survey—GMIMS—(Wolleben et al. 2009) sets out to map the polarized emission from the entire sky, North and South, using large single antennas over the entire frequency range 300 to 1800 MHz. It brings together three technologies, (a) wideband feeds for reflector antennas, (b) FPGA-based spectropolarimeters with several thousand frequency channels, and (c) RM Synthesis. The survey task has been divided

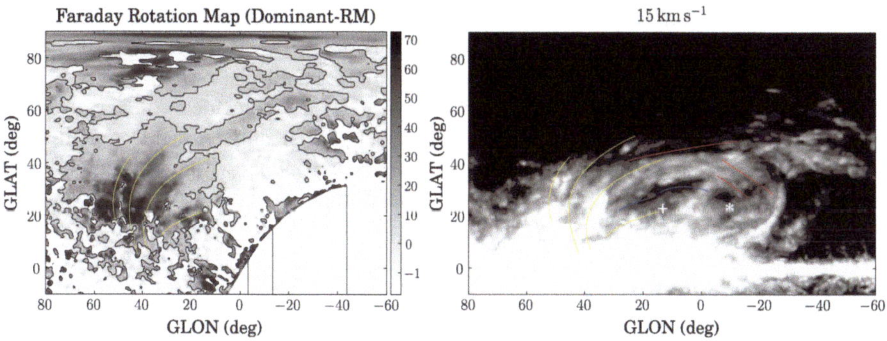

Fig. 3 The *left image* shows Faraday depth in a 90° × 140° region with angular resolution 1°. For each pixel the position in rad m^{-2} of the strongest peak in the RM-Synthesis spectrum is shown, with black indicating the most positive values. Superimposed *yellow curves* trace four "filaments" of strong positive RM. The *right image* of the same area shows H I at $v_{lsr} = 15$ km s^{-1} from the LAB Survey. The filaments from the *left panel* are reproduced (*yellow*), and *red lines* trace filaments detected at negative values of RM

Table 2 Surveys of the Galactic polarized emission underway but not published

Survey	Ref.	ν [MHz]	θ [']	A-S	S-A	Large Scale	RM Synth.
GALFACTS	(1)	1225–1525	3		y	y	y
GMIMS-HBN	(2)	1277–1762	40–30		y	y	y
GMIMS-HBS	(3)	1300–1800	10		y	y	y
GMIMS-LBS	(3)	300–470 & 660–870	60–20		y	y	y
SPASS	(4)	2300	8.9		y	y	y
CBASS	(5)	5000	44		y	y	

References: (1) Taylor and Salter (2011), (2) Wolleben et al. (2010a), (3) Wolleben et al. (2009), (4) Carretti (2011), (5) King et al. (2010)

Notes: (a) Headings and abbreviations are as in Table 1. (b) GMIMS surveys are labeled HBN (High-Band North), HBS (High-Band South), and LBS (Low-Band South). Mid-Band surveys, covering ~800 to ~1300 MHz, have not been started yet

between the Northern and Southern hemispheres and the frequency band has been divided into three sub-bands, dictated by receiver technology (see Table 2). The survey technique of Carretti (2011) will be used wherever possible, and intensity scales will be absolutely calibrated.

5.5 Results from the Global Magneto-Ionic Medium Survey

GMIMS High-Band North using the DRAO Telescope (Wolleben et al. 2010a) represents the first use of RM Synthesis with a single-antenna telescope. The value of such data, even at an angular resolution of ~1°, is revealed by the data shown in Fig. 3 (reproduced from Wolleben et al. 2010b). Once again, complementary data on other ISM constituents have been an indispensable aid to interpretation.

The left panel shows a map of Faraday depth derived from RM Synthesis of the data cube. At each pixel of the image, the Faraday depth of the strongest polarized emission is plotted.

Filaments are identifiable in this image (traced by lines superimposed on the grayscale). These convincingly coincide with one side of an H I bubble, 100×200 pc in size, seen in the LAB Survey (Kalberla et al. 2005). The boundary of the H I bubble at $\ell \approx 40°$ coincides with emission with positive RM, while the boundary near $\ell \approx -25°$ coincides with negative RM. This suggests a magnetic field wrapped around the bubble. The implied field strength in the shell is between 20 and 34 μG. The H I bubble may be expanding within the dense wall of the local bubble, the strong field arising from further compression of an already compressed plasma.

6 Concluding Discussion

The observations reviewed have provided opportunities to measure line-of-sight magnetic fields in a variety of ISM situations. The field values, sometimes as high as 30 μG, are well above the local value of ~2 μG, presumably as a result of compression. As evidence like this accumulates it will inform theory-based models of the ISM. An example is provided by the work of de Avillez and Breitschwerdt (2005). Using MHD code, they have run a massive computation tracking the evolution of the magnetized ISM, driven by star formation, stellar winds, supernovae, etc. After 3.5×10^8 years the total field averages to 4.4 μG, it is concentrated in the walls of bubbles and superbubbles whose expansion is controlled by it, but field strength remains uncorrelated with density. The filling factor of hot ($\geq 10^{5.5}$ K) gas is ~20%. We are on the verge of having polarization data useful for comparison with all-encompassing models like this.

The magnetic field should not be considered alone, but as a constituent of the ISM. Polarization data will always be easier to understand when seen in the context of observations of other constituents of that complex system.

7 The Future

Advances in understanding the interplay of magnetic fields and the ISM will come from observations with high angular resolution, full coverage of all spatial scales, and excellent resolution in Faraday depth. These requirements will be met by combining A-S and S-A data.

Table 2 lists polarization surveys being carried out with existing telescopes, and surveys approved and funded. All employ single-antenna telescopes. The GMIMS surveys will provide excellent frequency coverage and, consequently, very good resolution in Faraday depth, but only with the angular resolution available with S-A telescopes. The only survey whose angular resolution approaches the desired one arcminute is GALFACTS.

Beyond that lies the prospect of the new aperture-synthesis telescopes, LOFAR (already operating), ASKAP, and MeerKAT. All are designed as survey telescopes. All have high angular resolution, and all will produce wideband multi-channel data amenable to RM synthesis. The POSSUM survey (Gaensler et al. 2010), using the ASKAP telescope, will provide spectropolarimetric data at least from 1130 to 1430 MHz. All surveys with these A-S telescopes will require combination with single-antenna data (with the possible exception of LOFAR at the low end of its frequency range). Interpretation of the wealth of new data will benefit from the approach promoted here, one that views the magnetic field as a constituent of the ISM, and interprets polarization observations in conjunction with data on other ISM tracers. I have made many suggestions in Sect. 4 of topics where further investigation could be profitable, and I fully expect that the new instruments will lead to rapid progress along such paths in understanding the role that magnetic fields play in the Galactic ecosystem.

References

R. Beck, Space Sci. Rev. **99**, 243 (2001)
R. Beck, Astrophys. Space Sci. Trans. **5**, 43 (2009)
G. Bernardi, A.G. de Bruyn, M.A. Brentjens et al., Astron. Astrophys. **500**, 965 (2009)
K. Beuermann, G. Kanbach, E.M. Berkhuijsen, Astron. Astrophys. **153**, 17 (1985)
M.A. Brentjens, A.G. de Bruyn, Astron. Astrophys. **441**, 1217 (2005)
W.N. Brouw, T.A.T. Spoelstra, Astron. Astrophys. Suppl. **26**, 129 (1976)
J.C. Brown, in *ASP Conference Series*, vol. 438 (2011) p. 216
J.C. Brown, A.R. Taylor, Astrophys. J. Lett. **563**, 31 (2001)
J.C. Brown, A.R. Taylor, B.J. Jackel, Astrophys. J. Suppl. **145**, 213 (2003)
J.C. Brown, M. Haverkorn, B.M. Gaensler et al., Astrophys. J. **663**, 258 (2007)
B.J. Burn, Mon. Not. R. Astron. Soc. Lett. **133**, 67 (1966)
E. Carretti, in *ASP Conference Series*, vol. 438 (2011), p. 277
E. Carretti, M. Haverkorn, D. McConnell et al., Mon. Not. R. Astron. Soc. Lett. **405**, 1670 (2010)
S. Chandrasekhar, E. Fermi, Astrophys. J. **118**, 116 (1953)
D.P. Clemens, A. Pinnick, M. Pavel et al., in *Bulletin of the American Astronomical Society*, vol. 43 (2011), 241-15
M.A. de Avillez, D. Breitschwerdt, Astron. Astrophys. **436**, 585 (2005)
J.M. Dickey, Astrophys. J. **488**, 258 (1997)
A.R. Duncan, R.F. Haynes, K.L. Jones, R.T. Stewart, Mon. Not. R. Astron. Soc. Lett. **291**, 279 (1997)
A.R. Duncan, P. Reich, W. Reich, E. Fürst, Astron. Astrophys. **350**, 447 (1999)
A. Fletcher, A. Shukurov, Mon. Not. R. Astron. Soc. Lett. **371**, 25 (2006)
P. Fosalba, A. Lazarian, S. Prunet, J.A. Tauber, Astrophys. J. **564**, 762 (2002)
P. Frick, D. Sokoloff, R. Stepanov, R. Beck, Mon. Not. R. Astron. Soc. Lett. **401**, 24 (2010)
P. Frick, D. Sokoloff, R. Stepanov, R. Beck, Mon. Not. Roy. Astron. Soc., 557 (2011)
B.M. Gaensler, J.M. Dickey, N.M. McClure-Griffiths et al., Astrophys. J. **549**, 959 (2001)
B.M. Gaensler, G.J. Madsen, S. Chatterjee, S.A. Mao, Publ. Astron. Soc. Aust. **25**, 184 (2008)
B.M. Gaensler, T.L. Landecker, A.R. Taylor, POSSUM Collaboration, in *Bulletin of the American Astronomical Society*, vol. 41 (2010), p. 515
X.Y. Gao, W. Reich, J.L. Han et al., Astron. Astrophys. **515**, 64 (2010)
A.D. Gray, T.L. Landecker, P.E. Dewdney et al., Astrophys. J. **514**, 221 (1999)
L.M. Haffner, R.J. Reynolds, S.L. Tufte et al., Astrophys. J. Suppl. **149**, 405 (2003)
J.L. Han, in *IAU Symposium*, vol. 259 (2009), p. 455
J.L. Han, R.N. Manchester, E.M. Berkhuijsen et al., Astron. Astrophys. **322**, 98 (1997)
C.G.T. Haslam, C.J. Salter, H. Stoffel, W.E. Wilson, Astron. Astrophys. Suppl. **47**, 1 (1982)
M. Haverkorn, F. Heitsch, Astron. Astrophys. **421**, 1011 (2004)
M. Haverkorn, P. Katgert, A.G. de Bruyn, Astron. Astrophys. **356**, 13 (2000)
M. Haverkorn, P. Katgert, A.G. de Bruyn, Astron. Astrophys. **403**, 1031 (2003)
M. Haverkorn, P. Katgert, A.G. de Bruyn, Astron. Astrophys. **427**, 549 (2004)
M. Haverkorn, B.M. Gaensler, N.M. McClure-Griffiths et al., Astrophys. J. Suppl. **167**, 230 (2006)
M. Haverkorn, J.C. Brown, B.M. Gaensler, N.M. McClure-Griffiths, Astrophys. J. **680**, 362 (2008)
G. Heald, in *IAU Symposium*, vol. 259 (2009), p. 591
C. Heiles in *ASP Conference Series*, vol. 80 (1995), p. 507
C. Heiles, T. Robishaw, in *IAU Symposium*, vol. 259 (2009), p. 579
G. Hinshaw, J.L. Weiland, R.S. Hill et al., Astrophys. J. Suppl. **180**, 225 (2009)
N. Junkes, E. Fürst, W. Reich, Astron. Astrophys. Suppl. **69**, 451 (1987)
P.M.W. Kalberla, W.B. Burton, D. Hartmann et al., Astron. Astrophys. **440**, 775 (2005)
O.G. King, C. Copley, R. Davies et al., in *Society of Photo-Optical Instrumentation Engineers (SPIE) Conference Series*, vol. 7741 (2010)
R. Kothes, T.L. Landecker, M. Wolleben, Astrophys. J. **607**, 855 (2004)
T.L. Landecker, P.E. Dewdney, T.A. Burgess et al., Astron. Astrophys. Suppl. **145**, 509 (2000)
T.L. Landecker, W. Reich, R.I. Reid et al., Astron. Astrophys. **520**, 80 (2010)
N.M. McClure-Griffiths, A.J. Green, J.M. Dickey et al., Astrophys. J. **551**, 394 (2001)
N.M. McClure-Griffiths, J.M. Dickey, B.M. Gaensler et al., Astrophys. J. **652**, 1339 (2006)
M.E. Nord, P.A. Henning, R.J. Rand et al., Astron. J. **132**, 242 (2006)
A. Noutsos, Space Sci. Rev. (2011, this issue)
K.M. Rae, J.C. Brown, in *ASP Conference Series*, vol. 438 (2011), p. 229
R.R. Ransom, B. Uyanıker, R. Kothes, T.L. Landecker, Astrophys. J. **684**, 1009 (2008)
R.R. Ransom, R. Kothes, M. Wolleben, T.L. Landecker, Astrophys. J. **724**, 946 (2010)
W. Reich, in *Cosmic Polarization*, ed. by R. Fabbri (2006), p. 91

W. Reich, E. Fürst, P. Reich et al., in *The Magnetized Interstellar Medium, Antalya, Turkey*, ed. by B. Uyanıker, W. Reich, R. Wielebinski (2004), p. 45
R.I. Reid, A.D. Gray, T.L. Landecker, A.G. Willis, Radio Sci. **43**, 2008 (2008)
R.S. Roger, C.H. Costain, T.L. Landecker, C.M. Swerdlyk, Astron. Astrophys. Suppl. **137**, 7 (1999)
L. Rudnick, S. Brown, Astron. J. **137**, 145 (2009)
D.H.F.M. Schnitzeler, P. Katgert, A.G. de Bruyn, Astron. Astrophys. **471**, 21 (2007)
D.H.F.M. Schnitzeler, P. Katgert, A.G. de Bruyn, Astron. Astrophys. **494**, 611 (2009)
D.H.F.M. Schnitzeler, P. Katgert, M. Haverkorn, A.G. de Bruyn, Astron. Astrophys. **461**, 963 (2007)
A. Shukurov, E.M. Berkhuijsen, Mon. Not. R. Astron. Soc. Lett. **342**, 496 (2003)
M. Simard-Normandin, P.P. Kronberg, Astrophys. J. **242**, 74 (1980)
D.D. Sokoloff, A.A. Bykov, A. Shukurov, E.M. Berkhuijsen, R. Beck, A.D. Poezd, Mon. Not. R. Astron. Soc. Lett. **299**, 189 (1998)
T.A.T. Spoelstra, Astron. Astrophys. **135**, 238 (1984)
J.M. Stil, A.R. Taylor, Astrophys. J. Lett. **663**, 21 (2007)
J.M. Stil, A.R. Taylor, C. Sunstrum, Astrophys. J. **726**, 4 (2011)
J.M. Stil, A.R. Taylor, J.M. Dickey et al., Astron. J. **132**, 1158 (2006)
D.R. Stinebring, T.V. Smirnova, T.H. Hankins et al., Astrophys. J. **539**, 300 (2000)
X.-H. Sun, W. Reich, Astron. Astrophys. **507**, 1087 (2009)
X.-H. Sun, W. Reich, Res. Astron. Astrophys. **10**, 1287 (2010)
X.H. Sun, J.L. Han, W. Reich et al., Astron. Astrophys. **469**, 1003 (2007)
X.H. Sun, W. Reich, A. Waelkens, T.A. Enßlin, Astron. Astrophys. **477**, 573 (2008)
A.R. Taylor, S.J. Gibson, M. Peracaula et al., Astron. J. **125**, 3145 (2003)
A.R. Taylor, C.J. Salter in *ASP Conference Series*, vol. 438 (2011), p. 402
J.C. Testori, P. Reich, W. Reich, Astron. Astrophys. **484**, 733 (2008)
B. Uyanıker, T.L. Landecker, A.D. Gray, R. Kothes, Astrophys. J. **585**, 785 (2003)
C.L. Van Eck, J.C. Brown, J.M. Stil et al., Astrophys. J. **728**, 97 (2011)
R. Weaver, R. McCray, J. Castor et al., Astrophys. J. **218**, 377 (1977)
J.L. West, J. English, M. Normandeau, T.L. Landecker, Astrophys. J. **656**, 914 (2007)
G. Westerhout, C.L. Seeger, W.N. Brouw, J. Tinbergen, Bull. Astron. Inst. Neth. **16**, 187 (1962)
R. Wielebinski, J.R. Shakeshaft, I.I.K. Pauliny-Toth, The Observatory **82**, 158 (1962)
A. Wilkinson, Mon. Not. R. Astron. Soc. Lett. **163**, 147 (1973)
M. Wolleben, W. Reich, Astron. Astrophys. **427**, 537 (2004)
M. Wolleben, T.L. Landecker, W. Reich, R. Wielebinski, Astron. Astrophys. **448**, 411 (2006)
M. Wolleben, T.L. Landecker, E. Carretti et al., in *IAU Symposium* vol. 259 (2009), p. 89
M. Wolleben, T.L. Landecker, G.J. Hovey et al., Astron. J. **139**, 1681 (2010a)
M. Wolleben, A. Fletcher, T.L. Landecker et al., Astrophys. J. Lett. **724**, 48 (2010b)

Space Sci Rev (2012) 166:281–291
DOI 10.1007/s11214-011-9781-0

Galaxies in a Box: A Simulated View of the Interstellar Medium

Frederick A. Gent

Received: 13 December 2010 / Accepted: 19 April 2011 / Published online: 17 May 2011

Abstract We review progress in the development of physically realistic three dimensional simulated models of the galaxy. We consider the scales from star forming molecular clouds to the full spiral disc. Models are computed using hydrodynamic (HD) or magnetohydrodynamic (MHD) equations and may include cosmic ray or tracer particles. The dynamical scales covered, ranging from the full galaxy structure, through the turbulent scales of supernova (SN) explosions, down to cloud collapse and star formation, make it impossible with current computing tools and resources to resolve all of these in one model. We therefore consider a hierarchy of models and how they can be related to enhance our understanding of the complete galaxy.

Keywords ISM · Simulations · Magnetohydrodynamic · MHD · Hydrodynamic · Supernova · Cosmic rays · Magnetic field · Galaxy

1 Introduction

Generally modeling any astrophysical phenomena numerically has proven challenging, even with some considerable simplifications. However with the improvements in computing from the late nineties onwards reasonable 3-dimensional approximations of galactic features have been developed.

The aim of this chapter is to briefly review progress and to summarize some key findings to date. To impose some finite limit on the scope of the review we restrict our attention to three dimensional models, which include realistic parameters on galactic scales. Please attribute any omissions as a reflection of the ignorance of this author and not of the value of that work.

We present a roughly chronological review of models and their methods, objectives and results. This approach evolves out of the earlier 2D study by Rosen et al. (1993), which

F.A. Gent (✉)
School of Mathematics and Statistics, Newcastle University, Herschel Building, Newcastle upon Tyne, NE1 7RU, UK
e-mail: f.a.gent@ncl.ac.uk

investigated the hydrodynamical turbulent structure of the ISM using stellar heating and star formation.

Tomisaka (1998) attempted to model multiple supernovae (SNe) in 3D using a super-bubble approximation of in a stratified ISM with magnetic field. We summarize in Table 1 some key features of the models, which followed subsequently.

2 Review of Galaxy Simulations in 3D

In the following sections we summarize the work of several groups, who have published in some cases several papers over the last decade. For brevity we have cited the more recent publications, whilst including material from earlier papers. Please refer to the cited papers for the full list of papers cited.

2.1 ηMHD[(a)] Stratified Model

Korpi et al. (1999) constructed a comprehensive 3D model of a section of the galactic disc using parameters typical of the solar neighborhood. The simulation included non-ideal MHD applied to a Cartesian grid of $0.5 \times 0.5 \times 2$ kpc^3. The resolution was about 8^3 parsecs^3 (pc).

A vertical gravitational field (Kuijken and Gilmore 1989) and a shearing box to replicate differential rotation were included. Thus a vertically stratified density profile for the interstellar medium (ISM) was applied. Heating was applied via Supernovae (SNe) Types I and II distributed randomly in time and space. Thermal conductivity, magnetic diffusivity and kinematic viscosity were used in the form of hyper diffusion as well as viscosity applied to handle shocks.

The model replicated a multi-phase medium with the warm and hot phases in dynamic pressure equilibrium. The cold phase was not well reproduced through lack of a thermally unstable cooling function and low resolution.

The resultant ISM structure was highly differentiated in Z. Volume filling factors for temperature varied with height. Correlations for mass and velocity varied with temperature. The outward flow of the galactic fountain appeared to be described within the model, but the size of the model was insufficient to recover the global fountain behavior.

In this study the warm phase was characterized by strong turbulent flow, but in the hot phase flows were on the scale of the remnant size enclosing the hot regions. This could be a feature of the numerical limits of the model.

With this model the group also studied the evolution of super-bubbles (SBs) of SNe clusters and how this was affected by the strength of the magnetic field and turbulence. They found that the frequency of SB break outs from the disc to the halo was significantly increased in a turbulent regime compared to previous experiments in a homogeneous medium.

The role of magnetic field strength in confining the bubbles was reduced. This could result from the increased random element of the field due to turbulence. Thus our expectation of the mixing between the disc and the halo would be higher.

2.2 iMHD[(b)] Tall Stratified Model

Building on their (2000) model, de Avillez and Breitschwerdt (2007) applied 3-D HD equations to a Cartesian grid 1 kpc^2 by 8 kpc. They used adaptive mesh refinement (AMR) with a resolution ranging between 1.25^3 and 10^3 pc^3. Also included was a vertical gravitational field and Types Ib+Ic and Type II SNe. 60% of the Type II SNe were clustered in OB associations, exploding within a short time interval and close proximity to each other. Thermal

Table 1 Recent comprehensive 3D numerical models of the ISM

	ηMHD [a] stratified	iMHD[b] tall stratified	ηMHD/CR[c] global disc	iMHD[d] unstratified	HD[e] star formation	iMHD[f] MRI	HD[g] tall stratified	ηMHD [h] stratified
Domain, kpc	$0.5 \times 0.5 \times 2$	$1 \times 1 \times 5$–20	$50 \times 50 \times 10$	0.2^3	1.28^3	0.2^3	$1 \times 1 \times 10$	$0.8 \times 0.8 \times 4$
Res, pc	8	1.25–10	10–100	0.78–1.56	10–20	0.78–1.56	1.95	2–8
Heating	SN II+SN I	SN II+SN I	CR	SN	SN	–	SN II+SN I	SN II+SN I
Temp, K	10^2–10^8	2×10^2–10^6	2×10^2–10^6	10^2–10^8	10^2–10^8	$10^{1.5}$–10^4	2×10^2–10^7	1–10^7
Phases	warm(W)+hot(H)	cold(C)+W+H	C+W+H	W+H	C+W+H	C+W	C+W+H	C+W+H
MHD	non-ideal	hydro & ideal	non-ideal	ideal	hydro	ideal	hydro	non-ideal
Diffusion	hyper $\eta\ \nu\ \chi$	–	η	–	–	χ	–	$\eta\ \nu\ \chi$
Gravity	stellar+halo	stellar+halo	stellar+halo	–	self gravity	linear z	stellar+halo	yes
Rotation	differential	–	rigid & differential	–	–	differential	–	differential
Cosmic rays	–	–	yes	–	–	–	–	–
Star forming	–	–	–	–	yes	–	–	–
Duration	100 Myr	400 Myr	4.8 Gyr	–	350 Myr	2 Gyr	80 Myr	1 Gyr

A summary of three dimensional simulated models of galaxy features by: Korpi et al. (1999) (**a**), de Avillez and Breitschwerdt (2007) (**b**) from 2000, Hanasz et al. (2009) (**c**) from 2004, Balsara et al. (2004) (**d**), Slyz et al. (2005) (**e**), Piontek and Ostriker (2007) (**f**) from 2005, Joung et al. (2009) (**g**), Gressel et al. (2008) (**h**) conductivity and kinematic viscosity were neglected, but a thermally stable radiative cooling function was included.

The model evolved from hydrostatic equilibrium over a period of about 65 Myr to a dynamically steady state. Distinct cold ($T \leq 10^3$ K), warm ($10^3 < T \leq 10^4$ K), warm ionized ($10^4 < T \leq 10^5$ K) and hot ($T > 10^5$ K) phases were co-existing with varying scale heights. The cold phase was restricted to thin irregular strips around the mid-plane. Hot gas dominated above 1.5 kpc and warm gas in the disc below 500 pc.

The model was extended vertically to ± 10 kpc between 2001 and 2007. They identified bubbles and super-bubbles typically up to 120 pc across and 200 pc high breaking out of the disc periodically. The cold sheet like structures that evolved, and which resembled the fragments of SN remnants, were in fact not well correlated with the bubble remnants, but in fact appeared to be a more general effect of the turbulent compressions and convection currents.

For a further extension of the model to MHD an initial uniform 5.8 μG magnetic field in the azimuthal direction was included. The horizontal magnetized disc only briefly delayed the disc-halo cycle observed in the HD simulations and a hot phase volume filling factor of between 17 to 21% resulted both with HD and MHD.

Supersonic and super-Alfvénic flows led to strong MHD shocks. A highly turbulent field evolved with a field strength ranging between 0.1 and 10 μG, which was not strongly correlated to the density fluctuations. This is illustrated in Fig. 1. Magnetic pressure dominated the coldest gas, ram pressure dominant above 200 K and thermal pressure above $10^{5.5}$ K. In cold regions the relation $B \propto \rho^\alpha$ did not hold, with α in the range -0.006 to 0.085.

2.3 HD$^{(e)}$ Star Formation Model

In their 2002 2D model Slyz et al. (2005) investigated the characteristics of the spiral arms through isothermal HD. They studied 23^2 kpc^2 with resolution of 115^2 pc^2, including kinematic viscosity and applying a centrifugal and gravitational potential.

They varied the temperature, via the sound speed, the density profiles and rates of galactic rotation. Given the typical sound speed throughout the galaxy is far exceeded by the angular velocity of the gas relative to the spiral arms, it might not have been expected to have much effect on the structure. They found that the non-axisymmetric pattern of the spirals diminished with increasing sound speed. At levels above 25 to 30 km s^{-1} they faded away. As sound speeds reduced the mass in the spiral arms increased.

They also found that the inner galaxy was more sensitive to changes in the fraction of galactic mass in the disc. As the disc fraction increased, velocities at the centre increased massively, whilst the structure in the outer regions was fairly insensitive to the disc fraction.

Subsequently Slyz et al. (2005) conducted a more localized analysis of star formation and SN feedback with a 3D hydrodynamical model. They used a 1.28^3 kpc^3 Cartesian grid together with a particle mesh to track the stellar mass. The model included a thermally unstable cooling function and self-gravity. Stratification and any external gravity were not included.

At the start a random Gaussian velocity perturbation was applied to the ISM of uniform density and temperature. They found that the regime saturated to a multi-phase medium, with velocity dispersion generally settling to levels consistent with the temperature of the medium: 15 km s^{-1}, 30 and 75 for cold (<2000 K), warm (< 10^5 K) and hot (<4×10^6 K) respectively. Velocities in the very hot (>4×10^6 K) were far more dispersed, with up to 500 km s^{-1} observed. These violent flows stripped away elements of the cold dense star forming regions as they interacted.

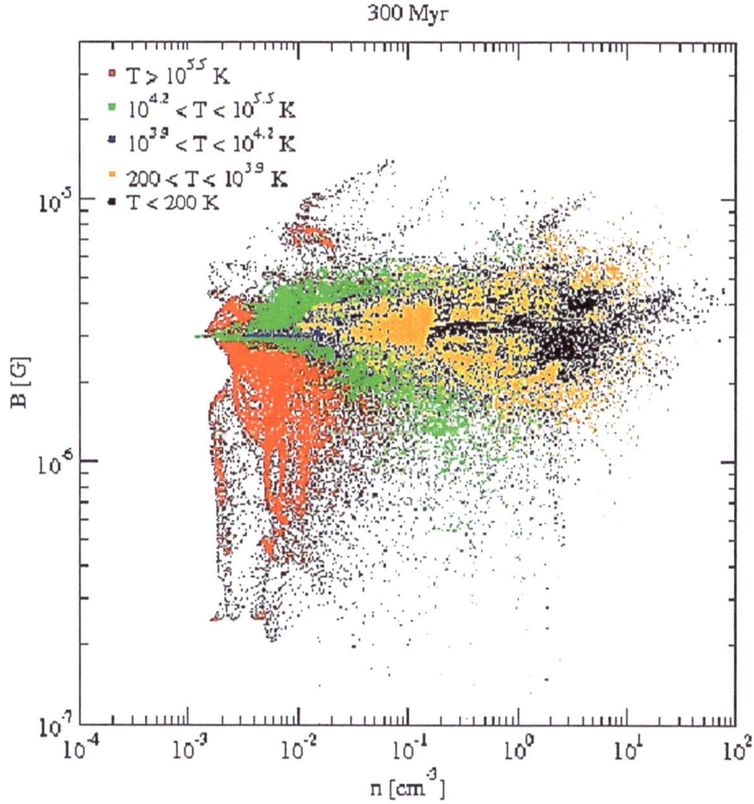

Fig. 1 de Avillez and Breitschwerdt (2007) Scatter plot from study (2005) of B versus ρ for $T \leq 200$ (*black*), $200 < T \leq 10^{3.9}$ (*orange*), $10^{3.9} < T \leq 10^{4.2}$ (*blue*), $10^{4.2} < T \leq 10^{5.5}$ K (*green*), and $T > 10^{5.5}$ K (*red*) regimes at 300 Myr of disc evolution. The points in the plot are sampled at intervals of four points in each direction. Note that during the evolution of the system the field strength broadened its distribution spanning two orders of magnitude after 300 Myr

The distribution of mass was also a feature of the ISM temperature. The bulk of the volume was filled with hot diffuse medium of below 10^{-3} atoms cm^3. The bulk of the mass was restricted to dense cold filamentary, sheet-like or worm-like structures accounting for a very small portion of the volume. In the absence of SN shocks the probability density functions (PDF) for the ISM were well described by the log normal. However with SN feedback a much higher incidence of increased density regions were observed, suggesting that star formation rates might be enhanced in a turbulent ISM.

2.4 iMHD$^{(d)}$ Unstratified Model

Balsara et al. (2004) applied 3D ideal MHD to a (200 pc)^3 region with SN randomly distributed. They included cooling (Raymond et al. 1976) and a constant diffuse heating term. Initial equilibrium was achieved by balancing heating and cooling for some given isothermal uniform density. The cooling considered could reproduce hot and warm phases, but not the cold phase.

They explored the results when varying the initial temperature and density and applying SN rates 8, 12 and 40 × the rate in the solar neighborhood (1 every 1.26 Myr equivalent to 2 per century for the Galaxy). They investigated resolution at about 1.5^3 and 0.75^3 pc^3. Stratification and differential rotation were not considered nor thermal conductivity and kinematic viscosity.

They found evidence of a fast dynamo sensitive to the SN rate. Over a 40 Myr simulation the field was amplified by two orders of magnitude. The growth rate was stronger with higher resolution, due to the additional amplification available at smaller scales.

Taking a turbulent scale from an SN remnant to be 30 pc they compared energy growth at scales above and below this. They found that eventually the growth rates were independent of scale.

Growth rates however were sensitive to SN rates, increasing up to a threshold where the magnetic energy was quenched. The growth of the field depended on the generation of helicity from the SN shocks interacting with the warm ISM. This was suppressed as the ISM became saturated by hot remnants. This effect was not present with a rate of 12 $\times$ but was for 40 $\times$ the galactic rate, so the critical bound for quenching lies within this range. There was also some increased amplification if the SN rate was more intermittent and field growth was most vigorous where the ISM had lower temperature and higher density.

Continuing the work from Balsara et al. (2004), Mac Low et al. (2005) studied the pressure distributions in the turbulent ISM with slightly differing results to those of Slyz et al. (2005). They applied ideal MHD equations in a $(200\ \text{pc})^3$ domain, with 3.12, 1.56 and 0.83 parsec resolution, including a radiative cooling function and uniform uv-heating, but excluding gravity.

They randomly injected SN remnants, at 1 and 4 times the solar neighborhood rate and analysed the evolution of the pressure distribution. They found the distribution of the pressure anomalies was independent of the orientation of the magnetic field due to the super-Alfvénic flows. Typically the hot gas formed discrete clumps, which were enclosed by warm or cold gas. The distribution of the colder dense clusters, tended not to be closely aligned directly to remnant structures. These formed on scales of dozens of parsecs and appeared to result from the general effects of large scale turbulence.

Even with SN driving they found the PDFs of the pressure were well described by the log normal rather than a power law, consistent with random supersonic turbulence. The large spread in the log pressure distribution was in contrast with a model of the ISM in thermal equilibrium. They did not find a global multiphase medium, rather a continuous distribution of temperatures and densities. Formation of dense isolated cloud filaments did not require self gravity, but were the transient results of turbulent flows. Even with 4 times the SN rate hot zones remained unconnected and warm gas remained the largest volume filling factor.

2.5 ηMHD/CR$^{(c)}$ Global Disc Model

From 2002 Hanasz et al. (2009) investigated dynamo action with rigid rotation on a magnetized galactic disc. A uniform azimuthal field was applied to a rotating Cartesian domain $600 \times 1800 \times 600$ pc with resolution $5 \times 20 \times 2.5$ pc. They solved isothermal resistive MHD equations (neglecting kinetic viscosity).

In 2004 they included cosmic rays (CR) randomly injected at spherical locations representing SN remnants. Differential rotation was now included in a $0.5 \times 1 \times 1.2$ kpc^3 Cartesian grid, with a resolution of 10^3 pc^3. The gravitational potential was as described by Ferrière (2001) and the cosmic ray diffusion tensor that of Ryu et al. (2003).

They observed an exponential growth in the magnetic field over a period of about 1000 Myr. Growth in the vertical component dominated, indicating that the CR buoyancy was important to the dynamics. Growth in the azimuthal direction reached 1400 $\times$ the seed field amplitude and 400 $\times$ in the radial direction.

Growth in the ideal regime onset rapidly after 250 Myr, due to magnetorotational instability (MRI). They argued that MRI amplification is suppressed when comparing the resistive MHD to ideal MHD due to magnetic reconnection in the thin current sheets generated by compressions in the field.

They observed that the fastest growing dynamo model resulted from sporadic switching on and off of the SN activity. Increased SN activity enhanced vertical field lines, increasing the loss of CR to the halo. Reduced activity supported their horizontal realignment thus enabling the amplitude of the field to grow. So the irregularity of the SN distribution can add to the dynamo.

Hanasz et al. (2009) investigated the source of the galactic magnetic seed field using their CR-driven dynamo model. Their domain was $50^2 \times 10$ kpc^3 with resolution 100^3 pc^3. They started with an isothermal, vertically and radially stratified density, including a gravity potential and rotational shear, but excluding the central bulge.

Into an unmagnetized ISM they randomly exploded SN remnants. Some 10% of these included a randomly oriented weak dipolar magnetic field. The CR energy was equivalent to 10% of the typical kinetic energy from an SN remnant. Over a 4.8 Gyr simulation they produced a magnetic field, which became efficiently ordered within 2 Gyr, aligned along the spirals and including field reversals. The vertical field evolved an x-shaped configuration. In their movies, vertical outflows could be seen, which transported the counter polarity or helicity out of the system (Fig. 2, small inserts). This seemed to be crucial for dynamo action.

The number and orientation of the spiral configurations was sensitive to the CR diffusion coefficients. They could not be derived from first principles without smaller scales, so the assumed value for diffusivity was by necessity large and unrealistic. Snap shots of the configuration are shown in Fig. 2.

2.6 iMHD$^{(f)}$ MRI Model

The turbulence in the galaxy is not adequately accounted for by SN activity alone so in 2005 Piontek and Ostriker (2007) modeled MRI-driven turbulence in the warm & cold phases of the ISM. They applied ideal MHD to a $(200\ \mathrm{pc})^3$ domain with resolution of 0.78 and 1.56 pc and realistic heating and cooling. They modelled differential rotation at the rate for the solar neighborhood. The growth of turbulence was studied for the ISM with initial uniform density and pressure and with randomly distributed cold spherical clouds amongst the warm ISM. The MRI instability rapidly dominated and the saturated state was independent of the initial state. In their model the thermal instability in the cooling function gives rise to a two-phase structure. They examined to what extent MRI driven turbulence might perturb this into a single phase environment.

They found that the gas separated into two phases with peaks at 100 K and 8000 K, with an increasing spread of gas lying outside the equilibrium as turbulence saturated. The relative proportions of gas in the two phases was dependent on the mean density, with the proportion of mass in the cold phase increasing substantially as the mean density increased.

Pressures were in reasonable equilibrium even in two phases without turbulence, but in the saturated state the pressure dispersion increased five fold. However the two-phase structure was still evident.

Magnetic field growth from 0.26 μG to about 2.5 μG was not sensitive to density nor phase. Away from regions of SN activity MRI could certainly be a factor in the turbulent structure. Extrapolating their results for mean densities typical in the inner galaxy, $R <$ 10 kpc, they would predict turbulent velocity scales of about 4 $\mathrm{km\,s^{-1}}$, which could be a significant component.

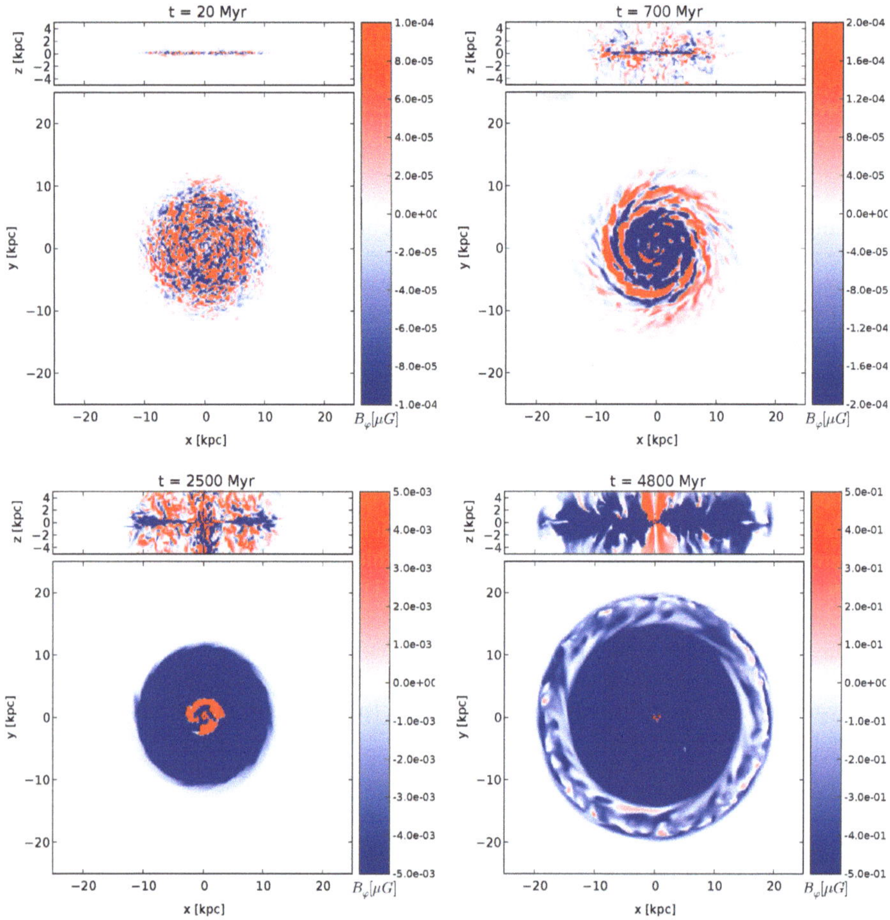

Fig. 2 Distribution of toroidal magnetic field from Hanasz et al. (2009). Unmagnetized regions are *white*, while positive and negative toroidal fields are *shaded red* and *blue*, respectively. Note that the color scale in magnetic field maps is saturated to enhance weaker magnetic field structures in the disc peripherals. The maximum magnetic field strengths are 5.9×10^{-4}, 4.4×10^{-3}, 1.5 and 29 μG at $t = 0.02, 0.7, 2.5$ and 4.8 Gyr respectively

Piontek and Ostriker (2007) included a vertical gravity profile, which now induced a vertical stratification in both density and temperature. The turbulent mixing of the warm ISM had a significantly higher scale height than the cold, which now was more clustered than in the earlier model. This combined with MRI to increase the magnetic pressure and they argued that the MRI could be responsible for suppressing star formation in the less dense outer region of the mid plane, even given the presence of cold clouds.

2.7 HD$^{(g)}$ Tall Stratified Model

From 2006 Joung et al. (2009) conducted an HD study in a vertically stratified ISM, including stellar and halo gravity. They also applied realistic diffuse heating and cooling plus SN turbulence. The vertical profile of the heating was realistically motivated. They neglected thermal conduction and kinematic viscosity, self-gravity and differential rotation.

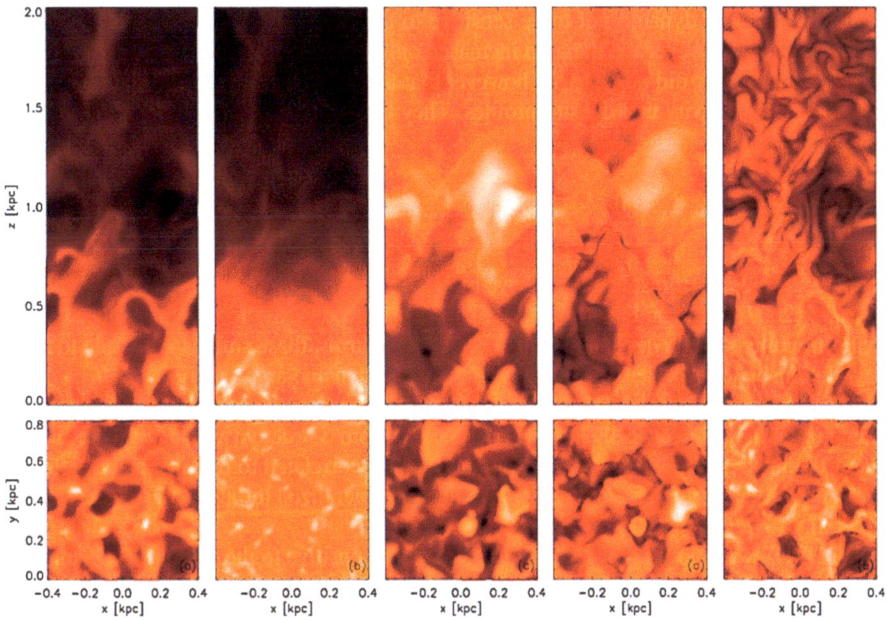

Fig. 3 Vertical slices from above the mid-plane (*upper panels*) and horizontal mid-plane slices (*lower panels*) after $t = 112$ Myr from Gressel et al. (2008). From *left* to *right* [in logarithmic scales]: number density $[-4.82, 1.10]$ cm^{-3}, column density $[17.34, 21.55]$ cm^{-2}, temperature $[2.06, 7.20]$ K, velocity dispersion $[-0.15, 2.77]$ km s^{-1} & magnetic field strength $[-4.16, 0.34]$ μG

They found the scale height of the galactic fountain to be several kpc. The volume filling factor near the mid-plane of the hot gas was inversely related to the magnitude of the diffuse heating, which they tested over a range up to 4 times that of Wolfire et al. (1995). This was due to a shift in the thermal equilibrium position, which excluded the cold phase and increased the volume component of the warm gas.

The vertical density profile was out of agreement with observations which they attributed to lack of pressure, in the absence of magnetic and cosmic ray pressure.

In this study and also from 2009 they found that neither the density nor the kinetic energy wavelengths follow the Kolmogorov power law, but instead are far more widely dispersed.

2.8 ηMHD$^{(h)}$ Stratified Model

Gressel et al. (2008) conducted a non-ideal 3D MHD simulation using the NIRVANA code. Their domain was $0.8^2 \times 4$ kpc^3 and included realistic cooling, thermal conductivity, kinematic viscosity and rotational shear. A snapshot from their simulation is shown in Fig. 3 to illustrate the resolution and structure obtained through such simulations.

They tested a variety of SN rates ($1/4$ to $1 \times \sigma_0$) and rotational frequency (1 to $8 \times \Omega_0$), where σ_0 and Ω_0 denote the observed rates in the solar neighborhood, and measured the resulting dynamo parameters.

A result common to each run, was that the velocity dispersion increased with height up to about 1 kpc, even though the peak of SN activity was much lower. They found that turbulent pumping was directed inward (towards the mid-plane), but balanced by an outward wind of comparable magnitude.

They found the dynamo not to be sensitive to the SN rates considered, 1, 2 & 4 × the observed rates. However the dynamo increased only with galaxy rotation rates 2 & 4 × the observed rate. They did not rule out however that these conclusions could also be sensitive to the assumed gravity and density profiles. They found no dynamo in the models without rotational shear.

3 Summary

At the highest level of resolution we have had the study of Balsara et al. (2004), Piontek and Ostriker (2007) and Mac Low et al. (2005) in 200^3 pc^3 domains. Too small to include large scale structure, such as density stratification or spiral arms, these studies were able to relate the relationship of turbulence from SN or MRI, with dynamo amplification.

They were able conclude that MRI could be a significant source of turbulence in the galaxy and in particular in the outer galaxy away from SN activity. The presence of cold dense filaments, they demonstrated, were not closely correlated to the dynamics of the SN remnants. Instead they appeared to be a product of generalized turbulence and gravitational instability.

Finally they found the dynamo amplification was sensitive to the SN rate up to an upper bound of between 12 & 40 × the solar neighborhood rate. Amplification of the magnetic field was also increased by greater fluctuation in the SN rate. Independent of SN rate the dynamo is quenched once the ISM is saturated by hot gas.

Losing resolution, but gaining structure Korpi et al. (1999), de Avillez and Breitschwerdt (2007), Slyz et al. (2005), Joung et al. (2009) and Gressel et al. (2008) constructed models incorporating the density stratification of the ISM near the plane of the galactic disc.

Unlike the smaller models, the ISM does not become saturated by SN activity. The stratified structure permits the disc to expand and relax as the SN rate fluctuates in time and spatial distribution. This relieves the quenching effect of turbulent saturation on the magnetic dynamo.

However results for the dynamo have been mixed. Korpi et al. (1999) and Gressel et al. (2008) did not find a dynamo with solar neighborhood parameters, and Gressel et al. (2008) only found the dynamo with differential rotation 2 and 4 × the galactic rate. Both these models used non-ideal MHD, where magnetic reconnection may dampen the dynamo through diffusion in the critical cold dynamic filaments. It may conversely enhance the dynamo by facilitating the exchange of the field. The numerical values for resistivity are of course much higher than the realistic values observed in the ISM. de Avillez and Breitschwerdt (2007) did not investigate the dynamo but included a 5 μG magnetic field using ideal MHD within their more recent work.

With a vertical range of 20 kpc de Avillez and Breitschwerdt (2007) were able to include elements of the galactic fountain and without boundary mass losses, could sustain simulations over 400 Myr. They found that an excess of 200 Myr was required to completely saturate and establish an equilibrium state for the disc-halo cycle. They identified the scale height of the disc-halo interface at about ± 1.5 kpc. The additional height was required to allow the hot gas in the halo to cool and rain back to the disc.

Despite outflow boundary conditions, and hence mass losses over the duration of the simulation, Gressel et al. (2008) were able to sustain their simulations over 1 Gyr. Once the turbulent state had evolved they found the boundary flows were minimal, so much of the dynamical cycle could be contained within a domain height of ± 2 kpc.

On the grand scale Hanasz et al. (2009) were able to produce a familiar model of the spiral structure of the galaxy. Restricted to isothermal models, they found the structure of the spirals sensitive to the CR diffusion. In addition the 2D study of Slyz et al. (2005) found the structure was also dependent on the temperature of the ISM, as parameterized by sound speed.

These models have informed our understanding of how the ISM behaves on different scales across a range of parameters. It is also clear that the interaction of temperature, density, magnetic field and cosmic rays combined are significantly altered in the absence of any one of these with unexpected consequences.

Global models of galaxies are indispensable, with sufficiently high resolution to include the small-scale physics. To resolve SNe remnants requires a resolution under 5^3 pc^3. For the random field dynamo we may require scales $< 2^3$ pc^3. The gravitational field (and self-gravitation of the gas) needs to be included, the latter requiring scales below 1 pc. Density waves are expected to interact with the magnetic fields and need to be considered. Arm and inter-arm modeling would require domains extending a few kpc in the galactic plane. To include even the lowest reach of the halo, would require a vertical range of at least 4 kpc. The challenge is to successfully incorporate all these elements over a number of physically meaningful scales.

References

D.S. Balsara, J. Kim, M.M. Mac Low, G.J. Matthews, Astrophys. J. **617**, 339–349 (2004)
M.A. de Avillez, D. Breitschwerdt, Astrophys. J. **665**, 35–38 (2007)
K.M. Ferrière, Astrophys. J. **73**, 1031–1066 (2001)
O. Gressel, D. Elstner, U. Ziegler, G. Rüdiger, Astron. Astrophys. **486**, 35–38 (2008)
M. Hanasz, D. Wóltański, K. Kowalik, Astrophys. J. **706**, 155–159 (2009)
M.K.R. Joung, M.M. Mac Low, G.L. Bryan, Astrophys. J. **704**, 137–149 (2009)
M.J. Korpi, A. Brandenburg, A. Shukurov, I. Tuominen, Å. Nordlund, Astrophys. J. **514**, 99–102 (1999)
K. Kuijken, G. Gilmore, Mon. Not. R. Astron. Soc. **239**, 605–649 (1989)
M.M. Mac Low, D.L. Balsara, J. Kim, M.A. de Avillez, Astrophys. J. **626**, 864–876 (2005)
R.A. Piontek, E. Ostriker, Astrophys. J. **663**, 183–203 (2007)
J.C. Raymond, D.P. Cox, B.W. Smith, Astrophys. J. **204**, 290 (1976)
A. Rosen, J.N. Bregman, M.L. Norman, Astrophys. J. **413**, 137–149 (1993)
D. Ryu, J. Kim, S.S. Hong, T.W. Jones, Astrophys. J. **589**, 338–346 (2003)
A.D. Slyz, J.E.G. Devriendt, G. Bryan, J. Silk, Mon. Not. R. Astron. Soc. **356**, 737–752 (2005)
K. Tomisaka, Mon. Not. R. Astron. Soc. **298**, 797–810 (1998)
M.G. Wolfire, C.F. McKee, D. Hollenbach, A.G.G.M. Tielens, E.L.O. Bakes, Astrophys. J. **443**, 152 (1995)

Space Sci Rev (2012) 166:293–305
DOI 10.1007/s11214-012-9866-4

Magnetic Fields in the Multiphase Interstellar Medium

Carl Heiles · Marijke Haverkorn

Received: 13 December 2010 / Accepted: 26 December 2011 / Published online: 31 January 2012

Abstract We discuss the interstellar magnetic field and its measurement in the different phases of the interstellar medium. The measurement techniques include Faraday rotation, Zeeman splitting, linear polarization from aligned dust grains, and the Chandrasekhar-Fermi method. The phases include the classical ones: molecular clouds, the Cold and Warm Neutral Media, the Warm and Hot Ionized Media. We also include a less well-known phase, the Warm Partially Ionized Medium, which can be surprisingly prominent in Faraday rotation.

Keywords Interstellar medium · Interstellar magnetic field · Faraday rotation · Zeeman splitting · Chandrasekhar-Fermi

1 Introduction: ISM Phases and Measuring Their Magnetic Fields

The Interstellar Medium (ISM) consists broadly of molecular clouds, in which self-gravity is a decisive force; and the diffuse ISM, where gravity is negligible. Here we consider both, but we concentrate on the diffuse ISM. Traditionally, it is considered to be composed of phases, and our discussion will follow this tradition. It is important to note, however, that modern numerical simulations show that the interaction between turbulence and thermal instability can blur the distinction between phases, even to the point where distinct phases become replaced by a continuum of gas properties; we briefly summarize this development in Sect. 1.4.

C. Heiles (✉)
Astronomy Department, University of California at Berkeley, B-20 Hearst Field Annex 3411, Berkeley, CA 94720-3411, USA
e-mail: heiles@astro.berkeley.edu

M. Haverkorn
ASTRON, PO Box 2, 7990 AA Dwingeloo, The Netherlands

M. Haverkorn
Leiden Observatory, Leiden University, PO Box 9513, 2300 RA Leiden, The Netherlands

Considered as phases, we consider that there are five physically distinct phases. They are defined by their observables and, also, explained well by theory. Four of these phases are well established in the literature, and we denote these as "classical" phases; they are either almost fully ionized or almost neutral. A fifth, which is partially ionized, is not generally recognized by most astronomers. We list the five phases' physical properties in Table 1, which is an update and expansion of Table 1 of Heiles (2001). We now describe these phases.

1.1 Five Diffuse ISM Phases

In the diffuse ISM, phases are classically categorized as largely ionized or neutral. The neutral phases come in two flavors, the Cold and Warm Neutral Media (the CNM and WNM), which have typical temperatures ~ 50 and 5000 K. Being largely neutral, magnetic fields in these phases are measurable only with Zeeman splitting of spectral lines and linear polarization of aligned dust grains; there are not enough electrons to produce measurable Faraday Rotation.

The ionized phases also come in two flavors, again classified by temperature: the Warm and Hot Ionized Media (the WIM and the HIM), which have typical temperatures ~ 8000 and $\gtrsim 10^6$ K. Being largely ionized, Faraday rotation is easily measured for the WIM and often the HIM. For Zeeman splitting, the only significant spectral lines are the WIM recombination lines, which are very weak. (However, if one could observe simultaneously a large number of WIM recombination lines, Zeeman splitting could become useful.) For dust grains, the 8000 K temperature and significant electron density probably weaken and effectively remove magnetic alignment of dust grains.

We call the above four diffuse ISM phases "classical" because their existence is very well established by observations in the radio (21-cm line; the LAB survey, Kalberla et al. 2005), optical (Hα line; the WHAM survey, Haffner et al. 2003), the X-ray (the ROSAT survey, Snowden et al. 1997); and optical/UV absorption lines. Also, their observables and properties are well-explained by theory, with well-defined heating and cooling mechanisms (McKee and Ostriker 1977; Wolfire et al. 2003).

There lurks a fifth phase, the Warm Partially Ionized Medium (WPIM). The WPIM is not widely recognized, mainly because its presence is hard to establish observationally, but it might be much more common than generally realized. Heiles (2001) reviewed the evidence for and physical properties of this phase, which he called the McKee/Ostriker WIM (MOWIM). It is well represented by the Local Interstellar Cloud (LIC), whose properties are very well specified in a wonderful series of papers by Redfield and Linsky (e.g. 2004, 2008). It presence is also inferred from optical and UV absorption line spectra of stars, particularly in the excellent series of papers by Fitzpatrick and Spitzer (e.g. 1997). It is neither mainly neutral nor ionized; typically, its ionization fraction $x_e = \frac{n_e}{n_{\rm Hn}} \sim \frac{1}{2}$ (the subscript Hn means H-nuclei, i.e. both neutral and ionized H). Being partially ionized means, also, that it is warm, but probably not quite as warm as the WIM; the average temperature for the four locales discussed by Heiles is $T \sim 5000$ K. This gas doesn't show up in Hα emission because its volume density n_e and column density N_e are too small. However, N_e is large enough so that, with a typical interstellar magnetic field of $B \sim 6$ μG, it produces measurable and recognizable Faraday rotation. This is a new realization, introduced in the present paper for the first time (to our knowledge). In principle, magnetic fields in this fifth phase could be revealed by Zeeman splitting and linear polarization of aligned dust grains, but the typical column densities are probably too small to make these techniques useful.

Table 1 lists physical properties for these ISM phases, including column densities for a typical individual cloud ($N_{\rm cl,Hn}$) and through the Galaxy for a vertical ($b = 90°$) sightline extending to infinity.

Table 1 Properties of phases

Property	CNM	WNM	WIM	WPIM[a]	HIM	HIM Erid[b]
$\frac{P_{th}}{k}$ (cm^{-3} K)	4000	4000	4000	2000	10000	50000
T (K)	50	6000	8000	7000	1.5×10^6	1.5×10^6
n_{Hn} (cm^{-3})	80	0.7	0.25	0.2	0.0034	0.015
$\frac{n_e}{n_{HII}}$	2×10^{-4}	1×10^{-3}	1	$\frac{1}{2}$	1	1
$N_{typ,Hn,20}$	0.5	1	0.08	0.06	0.01	0.06
$N_{typ,e,20}$	1×10^{-4}	1×10^{-3}	0.08	0.03	0.01	0.06
$N_{\perp,Hn,20}$	1.5	1.5	1.0	?	0.1	–
$N_{\perp,e,20}$	3×10^{-4}	1.5×10^{-3}	1.0	?	0.1	–

The subscript Hn means H-nuclei (i.e., both neutral and ionized H).

The subscript $\perp$ means line-of-sight at Galactic latitude 90°.

The subscript typ means a typical individual structure.

[a]From Redfield and Linsky (2004, 2008).

[b]For the Orion/Eridanus superbubble (Guo et al. 1995).

1.2 Approximate Pressure Equality Between the Gas Phases

Approximate pressure equality among these phases is a fundamental basis for their understanding. With pressure equality, the phases don't tend to push each other around and one can think in terms of quasi-static morphologies. Nevertheless, it is a common misconception to simplify this and speak of strict pressure equality, always and everywhere, among all phases. This is wrong, because there are sources of huge dynamical energy for the diffuse ISM—HII regions, stellar winds, supernovae, clusters of supernovae—and the latter three generate pockets of HIM with severe overpressure that produce strong shocks. Not only do these shocks push phases around, but they compress the multiphase gas, magnetic fields, and cosmic rays and, also, convert gas from one phase to another. These processes lie at the heart of the evolution of the ISM and, concomitantly, the large-scale evolution of individual galaxies. So it is the departure from the common misconception of strict pressure equality that governs the evolution of the galaxies and their ISM—and makes life interesting.

When we discuss the ISM it is appropriate and highly useful to parameterize its properties in terms of the thermal pressure, because this is more nearly constant from one phase to another than is the density or temperature. In discussions, especially theorists use the thermal pressure as a proxy for total pressure, which is what should, in fact, be reasonably constant. This is a bit of a swindle because at least in the CNM (and probably other phases too), the thermal pressure is typically smaller than the magnetic and turbulent pressures—not to mention cosmic-ray pressure (see Sect. 1.3 for a discussion of these components). Nevertheless, we proceed by parameterizing physical properties in terms of thermal pressure, and adopt a typical value $P_{th} = 4000$ K (which is a bit too high; see Jenkins and Tripp 2001 and Wolfire et al. 2003). But we need to keep in mind that this parameterization neglects energy density differences that can be an order of magnitude among components.

1.3 Approximate Equipartition Between ISM Components

The crudely comparable thermal pressures of the various gas components extends—together with order-of-magnitude differences—to pressures or energy densities of the (thermal and

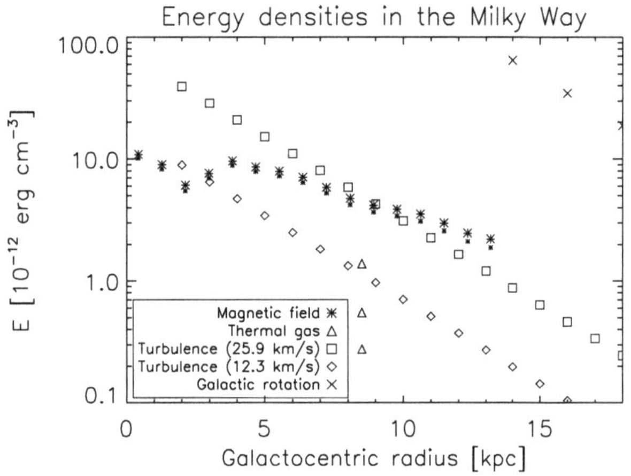

Fig. 1 Energy densities for the various components of the Galactic interstellar medium as a function of Galactocentric radius: the magnetic field according to the revised (*large asterisks*) and the classical equipartition formula (*small asterisks*); gas with a velocity dispersion of 25.9 km s^{-1} (*boxes*) and of 12.3 km s^{-1} (*diamonds*); thermal gas at the solar radius for the WPIM, CNM/WNM/WIM and HIM components (*triangles*); and Galactic rotation (*crosses*)

turbulent) gas pressure versus the energy density in magnetic fields and cosmic rays. On small scales these crude balances break down, but on large scales the energy densities of the three components have the relationships shown in Fig. 1 (see also Beck, *Magnetic Fields in Galaxies*, doi:10.1007/s11214-011-9782z, 2012).

The magnetic energy density $B^2/8\pi$ is derived from the magnetic field strength in the Milky Way as a function of radius. Beck (2001) computed these magnetic field strengths assuming equipartition with the cosmic ray energy density. We also include this calculation according to the equipartition formula revised by Beck and Krause (2005), who replace the uncertain ratio K of the total energies of cosmic ray protons and electrons integrated over a frequency interval with the number density ratio of cosmic ray protons and electrons integrated over an energy interval.

For turbulent gas energy densities ρv^2, we use the estimates of McClure-Griffiths and Dickey (2007). They find three neutral hydrogen velocity components in the fourth Galactic quadrant of $\Delta v = 6.3$ km s^{-1}, $\Delta v = 12.3$ km s^{-1}, and $\Delta v = 25.9$ km s^{-1}. They conclude that the two narrowest components follow each other very well, suggesting that the two thermal phases share the same bulk motions and velocity gradient variations. The broad component increases for Galactocentric radius $R < 4$ kpc, which may be a true increase in cloud-to-cloud motions or streaming motions. Here, we neglect the increase in the inner Galaxy and only include the two broadest components, as presented in Fig. 1. The thermal gas energy density is shown for the WPIM, CNM/WNM/WIM and HIM components with parameters as in Table 1.

The amount of energy stored in Galactic rotation is obviously much higher than the other components. However, the coupling to the energy densities of the other components shown in Fig. 1 is low, except perhaps through spiral density wave shocks or through the Ω-effect as part of the mean field dynamo.

Summarizing, around the Solar radius, the energy densities of the turbulent gas, of cosmic rays, and of magnetic field are crudely comparable—with order-of-magnitude differences—in the Milky Way. Therefore, all components influence each other and knowledge of each component is necessary to fully understand the Galactic ISM. At larger radii, the magnetic field energy density may dominate over the turbulent gas energy density, but the uncertainties involved are too large to make a definite conclusion.

1.4 Are There Phases in the ISM?

We have stolen the title to this subsection from Vazquez-Semadeni (2009), who provides an excellent contemporary critical review of the concept of phases in the turbulent ISM. Our ability to study this question is directly attributable to the power of modern computers used in 3d numerical simulations of interstellar gas dynamics including thermal effects and sometimes magnetic fields. His discussion is based on work by many different groups, with representative examples being Wada and Norman (2007), Tasker and Bryan (2008), Audit and Hennebelle (2010), Tilley and Pudritz (2007), and Mac Low (2004) and collaborators. There are others; this is not a complete list!

A quick summary of his discussion is that, with no turbulence, the mechanism of thermal instability produces well-defined phases. The resulting Probability Density Function (PDF) of volume density shows well-defined peaks, which correspond to the different phases. However, turbulence is in fact induced by supernovae, spiral density waves, and the like. In the extreme, in which turbulence and the accompanying mixing and dynamical processes dominate thermal instability, the PDFs become smooth, and discrete phases become undefined.

Nevertheless, modern discussion of phases like this tend concentrate on volume density and temperature. Specifically, they avoid dealing with the ionization fraction. This is understandable, because ionization is determined not only the standard microphysical and collisional processes that are part and parcel of modern simulations, but also by the UV field (including photons above the 13.6 ev ionization potential of HI) and cosmic rays.

As we will see below, the ionization fraction is crucial for the observations, especially those of Faraday rotation. It is our belief, based on a phenomenological understanding of these processes, that the PDF of ionization fraction should be bimodal—either very highly or very lightly ionized—with a low probability for moderate ionization fractions (i.e., 0.5 ± 0.49 or so) in the WPIM range. Does high turbulence change the PDF of ionization fraction to be smooth, or does the PDF remain bimodal?

For observational interpretation of Faraday rotation, it is crucial to include ionization in future numerical simulations. Regarding density and temperature, the five phases defined in Table 1 are similar to the four traditional ones. What distinguishes the WPIM is its intermediate ionization fraction. Can turbulence change the bimodal PDF of ionization fraction to a smoother PDF, as it does with gas density? Regarding ionization, the jury is still out, and this is a fertile field for future theoretical and numerical work.

2 Faraday Rotation of the Five Phases

2.1 General Considerations Regarding the Five Phases

Consider the typical RM of these phases:

$$RM = 0.81 n_e B_{||} L \text{ rad m}^{-2} \tag{1}$$

where the units on the right-hand side are the usual cm^{-3}, μG, and pc. Rewriting for units of $L_{20} = \frac{L}{10^{20}}$ cm gives

$$RM = 26 n_e B_{||} L_{20} = 26 N_{e,20} B_{||} \tag{2}$$

where $N_{e,20}$ is the electron column density in units of 10^{20} cm^{-2}.

Typical observed values for RM are ~ 10 rad m^{-2}, more and less. For the typical interstellar field strength $B \sim 6$ μG, we have $RM \sim 100 N_{e,20}$. Row 6 of Table 1 gives $N_{\text{typ},e,20}$, the electron column density in units of 10^{20} cm^{-2} for a typical individual interstellar structure for the various phases. The CNM and WNM have no chance of producing RMs of the observed range.

All of the other three phases can be significant contributors. It's not just the WIM, as is commonly thought. The HIM in superbubbles can also be a major player: the Orion/Eridanus superbubble can dominate WIM seen weakly in Hα. Surprisingly, cloud-for-cloud, the WPIM can be a major contributor.

For a typical high-latitude sight line extending to infinity, Row 8 of Table 1 gives the total electron column from the four classical phases $N_{\perp,e,20}$; the WIM value is a somewhat uncertain upper limit because it assumes that electrons are contributed exclusively by the WIM, and none by the WPIM. The WIM probably dominates, but again, the HIM can make a nontrivial contribution.

2.2 Let's Look at Some Data!

The top and bottom panels of Fig. 2 exhibit the two most massive ISM gas components. The top panel shows the CNM/WNM from the 21-cm line from the LAB survey (Kalberla et al. 2005) and the bottom shows WIM from the Hα line from the WHAM survey (Haffner et al. 2003), and also the four major radio-continuum loops. The CNM/WNM image shows only three specific velocity ranges, while the WIM image shows all velocities. The middle panel shows Faraday Rotation Measures, and we will discuss it in some detail below. What's missing in this collage is the HIM, which is revealed by X-ray emission; the most spectacular features are the two nearby HIM-filled superbubbles: the North Polar Spur (whose periphery follows Radio Loop 1) and the Orion/Eridanus superbubble (which lies inside the roughly spherical WIM bubble on the lower right of the bottom panel).

The CNM/WNM image on top covers velocity intervals ranging over tens of km s^{-1} and emphasizes the role of supersonic gas dynamics in the ISM. The WIM image reveals lots of isolated blobs, which are HII regions, as well as gas that is ionized by distant stars, such as the Orion/Eridanus structure in the lower right. This Hα image doesn't do justice to the Hα sky, because it shows only the brighter features. In fact, fascinating structure exists over a huge range of Hα brightnesses. Dimmer features reveal more delicate filamentary structures that probably trace the peripheries of old supernova or superbubble bubbles.

Let's compare the RMs in the middle panel (from Taylor et al. 2009) with the massive ISM phases, the CNM/WNM on top and the WIM on bottom. There are *many* correspondences! They are best seen by blinking the images, but that is not possible for this written, static document. We outline some of the more obvious morphological associations. However, even these are hard to see in Fig. 2 because of the small scale and our attempt to show 3 HI velocity ranges with color—but they are quite evident when blinking individual velocity gray-scale HI images with the RM image.

1. The magenta RM filament near the top of radio loop 3 and the red HI filament.
2. The yellow RM filaments near the bottom, especially left, of radio loop 2 and the red HI.
3. The RM pattern changes abruptly (from red/yellow to light blue) across the upper boundary of radio loop 2.
4. The RM pattern changes abruptly (from yellow to light blue) across the weak green HI filament that is located just below the top of radio loop 2.
5. Small-scale similarities of the blue HI and RM patterns near 2 o'clock on the southern-sky circle [near (GLON, GLAT) = $(-100, 0)$].
6. The Rho-Oph HII region, located near [near (GLON, GLAT) = $(0, 20)$], is very evident on the WHAM and RM images, and dimly visible in green on the HI image.

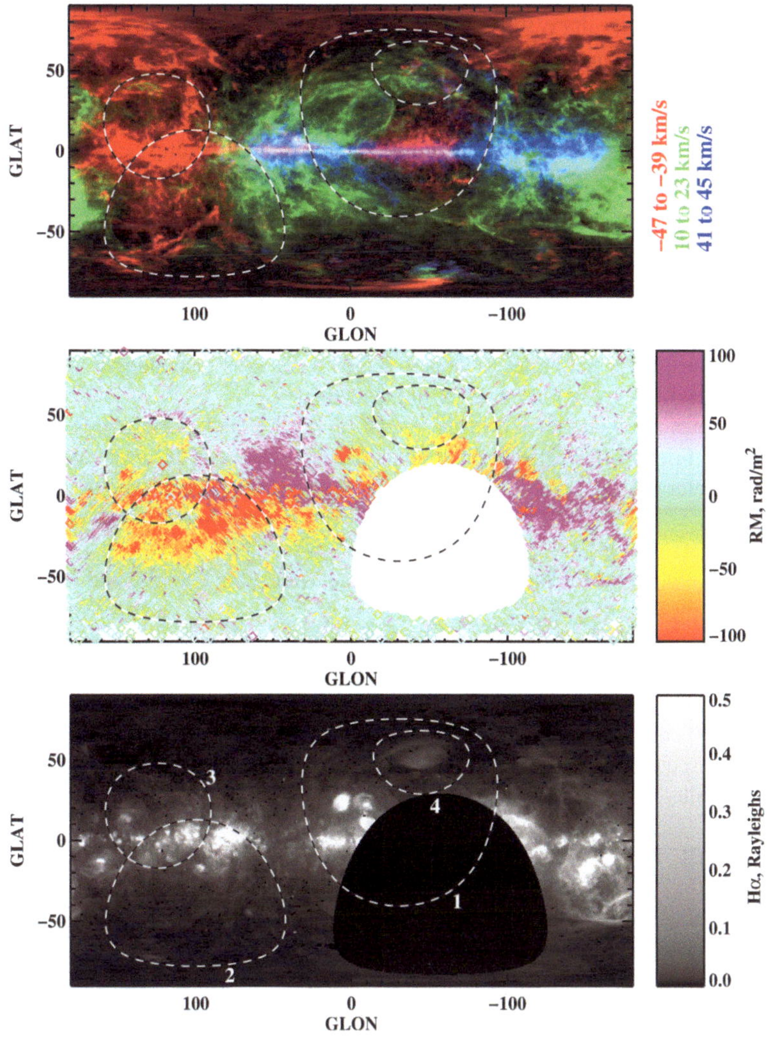

Fig. 2 Images of the LAB 21-cm line at the indicated V_{LSR}s, Faraday rotations, and WHAM Hα line (top to bottom). *Dashed lines* show the four classical radio loops

7. The magenta RM filament in the Orion/Eridanus superbubble, which runs horizontally near (GLON, GLAT) = (−10, −50), is clearly visible in both HI and WHAM images.

Generally, these associations indicate the powerful influence of individual interstellar structures on the observed RMs. Such conclusions have been drawn before (e.g. Wolleben 2007). Interpretations of the RMs as indicating global-scale Galactic field configurations, such as Han (2009) and Taylor et al. (2009), must be regarded with caution.

Given Table 1, the association of RMs and HI filaments is strange: HI is usually too neutral to produce significant RMs. Some of these associations are a distinct change in RM across an HI filament; here, the HI may be a shock-produced edge of an HIM region. Others probably indicate the presence of partially ionized HI or very weak Hα emission associated with the HI structures—in other words, the fifth diffuse ISM phase, the WPIM. RMs may be the best way to identify the WPIM! For structures 1 and 7 above, we estimate field strengths of 10 to 20 μG based on considerations of morphology, pressure, probable electron density, and line-of-sight lengths.

Clearly, the new dense coverage of RMs is a powerful probe of not only the interstellar magnetic field, but also the ISM and its relationship to ISM structures.

3 Zeeman Splitting and Linear Polarization

Two tools exist for measuring magnetic fields in neutral media: linear polarization from absorbing or emitting grains, which samples $B_\perp$, and Zeeman splitting, which samples $B_{||}$. It is more probable for field lines to lie across the line of sight than along it, so $B_\perp$ tends to be stronger than $B_{||}$—1.5 times stronger in the median.

These topics are well covered in the recent review by Heiles and Crutcher (2005), so here we very briefly summarize the highlights of that review and outline some more recent results. We refer the reader there for more extensive discussions and relevant references.

4 Zeeman Splitting in the Neutral Components, Molecular Clouds, and Masers

Crutcher et al. (2010b) applied a Bayesian analysis to 138 Zeeman-splitting measurements in both diffuse HI and dense molecular clouds. In keeping with earlier studies, they find the magnetic field to be independent of n_{Hn} for $n_{\mathrm{Hn}} \lesssim 300$ cm^{-3}, and $B \propto n_{\mathrm{Hn}}^x$, with $x \sim 0.65$, for $n_{\mathrm{Hn}} \gtrsim 300$ cm^{-3}. Figure 3 shows their data, together with the fitted model as the solid line. Their discussion includes aspects of magnetic field in molecular clouds as they relate to cloud stability and star formation.

4.1 Diffuse HI

For diffuse HI, Zeeman splitting has been measured for hundreds of positions in emission, and a dozen or so in absorption. The median value in the CNM is about 6 μG, which makes magnetic pressure dominate thermal gas pressure by a factor ~ 3; magnetic and turbulent energy densities are comparable. Measured field strengths are higher in regions which show shock morphology, such as shell walls. The fields are strong enough to limit the post-shock volume density (i.e., the post-shock magnetic pressure dominates the thermal pressure).

4.2 Molecular Clouds: Pre-star-Formation

In molecular clouds, Zeeman splitting of OH usually traces volume densities of a few thousand cm^{-3}. When HC05 wrote their review, the current set of results indicated quite close equipartition among gravitational, magnetic, and turbulent energy (Basu 2000). This is, in turn, consistent with the theoretical picture of star formation being an evolution of quasi-static equilibrium states in virial equilibrium, with the important energy densities being gravitational, magnetic, and turbulent; gradually, the field slips through the cloud's gas via ambipolar diffusion until the magnetic pressure becomes too small to resist gravity, at which point the cloud begins its inevitable transition to a protostar and then a star. More recent Zeeman-splitting results with the CN molecule, which traces higher densities than OH, are consistent with this picture; typical Zeeman-derived field strengths are ~ 0.6 mG for $n_{H_2} \sim 10^6$ cm^{-3} (Falgarone et al. 2008).

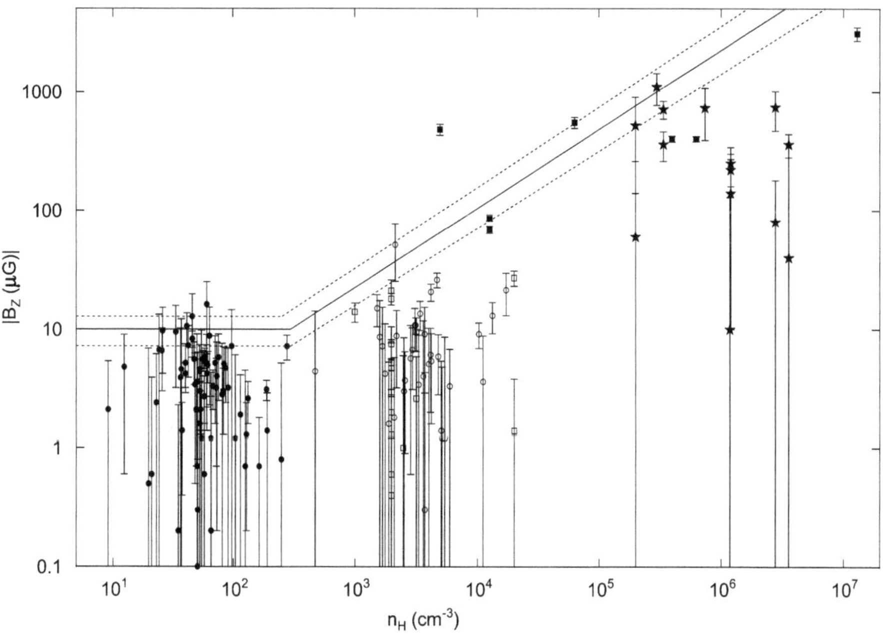

Fig. 3 Crutcher et al.'s (2010b) set of diffuse cloud and molecular cloud Zeeman-splitting measurements of the line-of-sight component $|B_{||}|$ and their 1σ uncertainties, plotted against $n_{\rm Hn}$ diffuse HI and molecular clouds

However, this conventional, convenient, and theoretically satisfying picture of a gradual evolution of virial equilibrium states is now being questioned as a result of observations with enough angular resolution to distinguish between magnetic fields in cloud cores and envelopes. Crutcher et al. (2009) studied four objects and found the mass-to-magnetic flux ratios in the cores to be smaller than those in the envelopes, which is contrary to this picture; consequently, those authors favor a supersonic-turbulence formation mechanism for these four objects. In essence, the emergence of turbulence in this dense-cloud regime is an extension of its role at the lower densities discussed above in Sect. 1.4; numerical simulations at the higher densities include those mentioned above and, in addition, work by others who concentrate exclusive on the larger densities [including, for example, Padoan and Nordlund (2011) and Tilley and Pudritz (2007)].

The observational interpretation and conclusions of Crutcher et al. (2010b) are sharply criticized by Mouschovias and Tassis (2009), and a rather severe controversy has erupted; the situation is not yet resolved (Crutcher et al. 2010a). Perhaps some clouds form and evolve via turbulence and some via the quasi-equilibrium models, as implied by the discussion of Crutcher et al. (2010b). As they say, *"que será, será"*.

4.3 Post-star-Formation: HII Regions and Masers

The Brogan/Crutcher/Troland group (e.g. Brogan and Troland 2001) has measured and mapped Zeeman splitting of HI and OH in absorption against the strong radio continuum for about ten HII regions. This gas resides just outside the ionized gas in a thin, highly compressed shell located between the ionization front and the shock front, known as the PhotoDissociation Region (PDR). Densities reach up to $\sim 10^6$ cm^{-3} and field strengths are tens to hundreds of μG; the density buildup in the PDR is limited by the strong magnetic field. Remarkably, the strong fields in some PDRs are not balanced by the dynamical pressure of the HII region gas, but rather by the radiation pressure of the massive stars that produce the HII region (Pellegrini et al. 2007)!

OH masers are compressed regions within the PDRs. Being denser (a few $\times 10^7$ cm^{-3}), they exhibit much higher field strengths. A reasonable fit is $B \propto n^{1/2}$. An unexpected result is that the field directions in OH masers tend to be the same as directions in the vicinity of the maser (Fish et al. 2003). This result is being used, together with Faraday rotation, to trace the pattern of the global Galactic magnetic field—which is circular or spiral shaped, with a few reversals with Galactocentric radius (the number is uncertain).

Zeeman splitting is seen even in very intense H_2O masers, which are associated with massive star formation. The field has to be strong to be observable because H_2O has no electron angular momentum, so the Landé g factor is ~ 1000 times smaller than for OH. Magnetic fields in H_2O masers tend to be higher than in OH masers, tens of mG with volume densities $n \gtrsim 10^8$ to 10^{10} cm^{-3} (Fiebig and Gusten 1989; Sarma et al. 2002).

5 Linear Polarization in the Neutral Components and Molecular Clouds

5.1 Morphological Patterns

Linear polarization is produced by spinning grains aligned by the interstellar magnetic field. Background starlight is absorbed; the orientation of the polarization is usually parallel to $B_\perp$. Warm grains emit IR radiation, which is usually polarized perpendicular to $B_\perp$. In dense regions where stars have formed, the polarization orientation changes with position in such a way as to suggest the "hourglass" configuration that is expected from the field following the inflowing gas. A particularly nice recent result is the SMA image of polarized emission from dust in the massive star forming region W51 e2/e8 (Tang et al. 2009).

In diffuse gas, measurable starlight polarization occurs even at very low extinctions. Results for ~ 10000 stars show that the nearby global field lies in the Galactic plane and roughly follows the spiral arms (Heiles 1996). Fosalba et al. (2002) present a comprehensive and detailed statistical analysis of these data, deriving the ratio of uniform to random field strengths (~ 0.8) and the angular power spectrum, among other things.

On smaller scales, the field morphology is clearly related to the diffuse gas morphology, particularly in regions where shocks have pushed the gas around. A map of starlight polarization over the whole sky reveals amazing patterns that coincide with the gas morphology (Fig. 4); blinking the two images provides visual correspondences with amazing detail. The most famous example or this correspondence is Radio Loop 1, which is a nearby superbubble blown by clustered supernovae in the Rho-Oph region. Figure 4 clearly shows that HI filaments, $B_\perp$ from starlight polarization, and Radio Loop 1 are all related.

5.2 Linear Polarization and the Chandrasekhar-Fermi Method

Using the Chandrasekhar-Fermi (CF) relation, linear polarization can be used to find not just the orientation of $B_\perp$, but also its strength. For example, if the field energy density totally dominates the gas dynamical energy density, then the field lines approach a force-free configuration in which there is no small-scale fluctuation of the field orientation so the dispersion of field orientation is small. In contrast, if the magnetic energy density is very weak, the magnetic lines follow the random turbulent gas motion and $B_\perp$ exhibits large

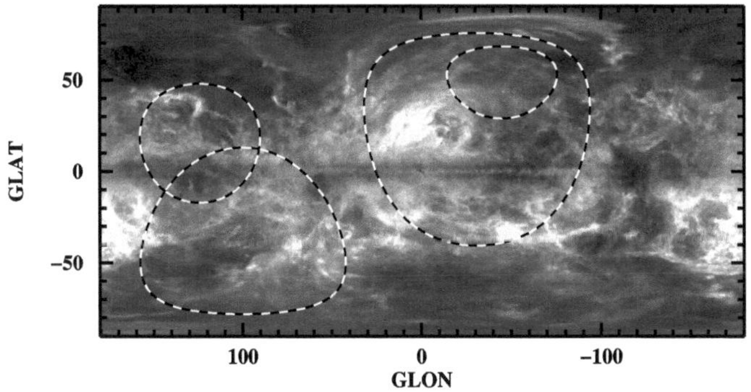

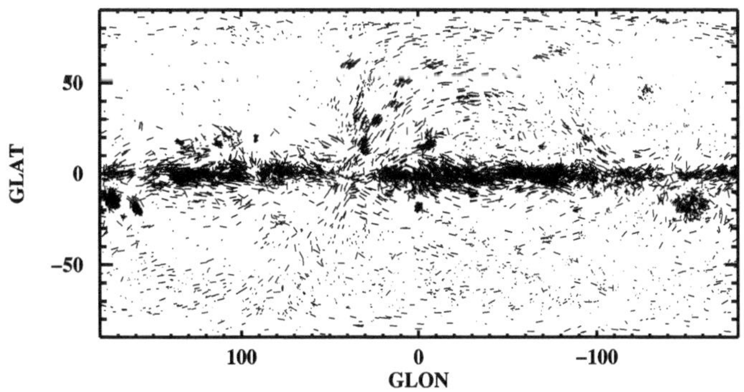

Fig. 4 Images of the LAB 21-cm line at $V_{\mathrm{lsr}} = 4\ \mathrm{km\,s^{-1}}$ and stellar polarization (top to bottom). *Dashed lines on the top panel* show the four classical radio loops

fluctuations. Thus, comparing the dispersion in field orientation to the turbulent gas velocity provides the field strength.

These estimates must be done carefully. One sees many turbulence elements along the line of sight, which tends to reduce the observed dispersion and increase the derived strength. Recently, this problem has been approached by using the structure function of the polarization angle difference (Hildebrand et al. 2009; Houde et al. 2009); this approach provides not only the field strength, but also good values for the uniform-to-random field strength ratio and the scale length of the field fluctuations. Another approach is the combining of numerical simulations of magnetic turbulence and the data. An excellent example is Novak et al. (2009), who applied the technique to their far-IR polarization maps of several molecular clouds.

6 Summary and Perspective

Our title, introductory sections, and ensuing discussion follow the traditional idea of ISM phases in very rough pressure and thermal equilibrium. In particular, Fig. 2 and its discussion emphasize the role of the fifth phase, the WPIM of Table 1, in producing and perhaps dominating the ISM's Faraday Rotation. However, the order-of-magnitude departures from pressure equilibrium and energy density equality (Sects. 1.2 and 1.3) reveal that the ISM is a dynamically interesting medium in which turbulence and large-scale energy input from sources such as supernovae play hugely important roles. Strong turbulence blurs the distinction among phases, and in the extreme invalidates the very notion of phases (Sect. 1.4).

Magnetic field strengths in the neutral atomic components (CNM and WNM) are measured by Zeeman splitting and are typically about 6 μG (Fig. 3); magnetic and turbulent energy densities are comparable, and they dominate thermal pressure. As densities rise, we enter the molecular regime, and as expected from flux freezing, field strengths also rise. This is seen not only in molecular clouds and protostellar regions, but also in compressed regions associated with and/or produced by star formation (Sect. 4.3). The jury is still out regarding the balance between gravitational and magnetic forces in most star formation, and whether turbulence or gradual cloud evolution produces the density enhancements required for gravity to take over and produce collapse (Sect. 4.2). Linear polarization of dust emission and absorption traces out the plane-of-sky orientation of the field, which is important in understanding its influence in dynamical processes.

Acknowledgements It is a pleasure to thank Tim Robishaw for his pleasant and capable collaboration on matters magnetic; and Rainer Beck for excellent commentary and suggestion of several important references. We greatly appreciate the support of the International Space Science Institute in attending the meeting. This work was supported in part by NSF grant AST 027321-003. The Wisconsin H-Alpha Mapper is funded by the National Science Foundation. The research leading to these results has received funding from the European Union's Seventh Framework Programme (FP7/2007-2013) under grant agreement number 239490.

References

E. Audit, P. Hennebelle, Astron. Astrophys. **511**, 76 (2010)
S. Basu, Astrophys. J. **540L**, 103 (2000)
R. Beck, Space Sci. Rev. **99**, 243 (2001)
R. Beck, M. Krause, Astron. Nachr. **326**, 414 (2005)
C.L. Brogan, T.H. Troland, Astrophys. J. **560**, 821 (2001)
R.M. Crutcher, N. Hakobian, T.H. Troland, Astrophys. J. **692**, 844 (2009)
R.M. Crutcher, N. Hakobian, T.H. Troland, Mon. Not. R. Astron. Soc. **402L**, 64 (2010a)
R.M. Crutcher, B. Wandelt, C. Heiles, E. Falgarone, T.H. Troland, Astrophys. J. (2010b, in press)
E. Falgarone, T.H. Troland, R.M. Crutcher, G. Paubert, Astron. Astrophys. **487**, 247 (2008)
D. Fiebig, R. Gusten, Astron. Astrophys. **214**, 333 (1989)
V.L. Fish, M.J. Reid, A.L. Argon, K.M. Menten, Astrophys. J. **596**, 328 (2003)
E.L. Fitzpatrick, L. Spitzer Jr., Astrophys. J. **475**, 623 (1997)
P. Fosalba, A. Lazarian, S. Prunet, J.A. Tauber, Astrophys. J. **564**, 762 (2002)
Z. Guo, D.N. Burrows, W.T. Sanders, S.L. Snowden, B.E. Penprase, Astrophys. J. **453**, 256 (1995)
L.M. Haffner, R.J. Reynolds, S.L. Tufte, G.J. Madsen, K.P. Jaehnig, J.W. Percival, Astrophys. J. Suppl. Ser. **149**, 405 (2003)
J.L. Han, in *Cosmic Magnetic Fields: From Planets to Stars and Galaxies*, ed. by K.G. Strassmeir, A.G. Kosovichev, J.E. Beckman. IAU Symp., vol. 259 (Kluwer, Dordrecht, 2009), p. 455
C. Heiles, Astrophys. J. **462**, 316 (1996)
C. Heiles, in *Tetons 4: Galactic Structure, Stars and the Interstellar Medium*, ed. by C.E. Woodward, M.D. Bicay, J.M. Shull. ASP Conference Series, vol. 231 (Astronomical Society of the Pacific, San Francisco, 2001), p. 294. ISBN: 1-58381-064-1
C. Heiles, R. Crutcher, in *Cosmic Magnetic Fields*, ed. by R. Wielebinski, R. Beck. Lecture Notes in Physics, vol. 664 (Springer, Berlin, 2005), p. 137
R.H. Hildebrand, L. Kirby, J.L. Dotson, M. Houde, J.E. Vaillancourt, Astrophys. J. **696**, 567 (2009)
M. Houde, J.E. Vaillancourt, R.H. Hildebrand, S. Chitsazzadeh, L. Kirby, Astrophys. J. **706**, 1504 (2009)
E.B. Jenkins, T.M. Tripp, Astrophys. J. Suppl. Ser. **137**, 297 (2001)
P.M.W. Kalberla, W.B. Burton, D. Hartmann, E.M. Arnal, E. Bajaja, R. Morras, W.G.L. Pöppel, Astron. Astrophys. **440**, 775 (2005)
M.-M. Mac Low, Astrophys. Space Sci. **289**, 323 (2004)
N.M. McClure-Griffiths, J.M. Dickey, Astrophys. J. **671**, 427 (2007)
C.F. McKee, J.P. Ostriker, Astrophys. J. **218**, 148 (1977)
T.Ch. Mouschovias, K. Tassis, Mon. Not. R. Astron. Soc. **400**, L15 (2009)

G. Novak, J.L. Dotson, H. Li, Astrophys. J. **695**, 1362 (2009)
E.P. Padoan, A. Nordlund, Astrophys. J. **730**, 40 (2011)
E.W. Pellegrini, J.A. Baldwin, C.L. Bogan, M.M. Hanson, N.P. Abel, G.J. Ferland, H.B. Nemala, G. Shaw, T.H. Troland, Astrophys. J. **658**, 1119 (2007)
S.R. Redfield, J.L. Linsky, Astrophys. J. **613**, 1004 (2004)
S.R. Redfield, J.L. Linsky, Astrophys. J. **673**, 283 (2008)
A.P. Sarma, T.H. Troland, R.M. Crutcher, D.A. Roberts, Astrophys. J. **580**, 928 (2002)
S.L. Snowden, R. Egger, M.J. Freyberg, D. McCammon, P.P. Plucinsky, W.T. Sanders, J.H.M.M. Schmitt, J. Trümper, W. Voges, Astrophys. J. **485**, 125 (1997)
Y. Tang, P.T.P. Ho, P.M. Koch, J.M. Girart, S. Lai, R. Rao, Astrophys. J. **700**, 251 (2009)
E.J. Tasker, G.L. Bryan, in *Formation and Evolution of Galaxy Disks*. ASP Conference Series, vol. 396 (2008), p. 105
A.R. Taylor, J.M. Stil, C. Sunstrum, Astrophys. J. **702**, 1230 (2009)
E.D.A. Tilley, R.E. Pudritz, Mon. Not. R. Astron. Soc. **382**, 73 (2007)
E.E. Vazquez-Semadeni, The role of disk-halo interaction in galaxy evolution: outflow vs. infall? (2009). arXiv: 0902.0820
K. Wada, C.A. Norman, Astrophys. J. **660**, 276 (2007)
M.G. Wolfire, C.F. McKee, D. Hollenbach, A.G.G.M. Tielens, Astrophys. J. **587**, 278 (2003)
M. Wolleben, Astrophys. J. **664**, 349 (2007)

Space Sci Rev (2012) 166:307–324
DOI 10.1007/s11214-011-9860-2

The Magnetic Field of the Milky Way from Faraday Rotation of Pulsars and Extragalactic Sources

Aristeidis Noutsos

Received: 14 December 2010 / Accepted: 18 November 2011 / Published online: 10 December 2011

Abstract Faraday rotation towards polarised pulsars and extragalactic sources is the best observable for determining the configuration of the magnetic field of the Galaxy in its plane and also at high latitudes. The Galactic magnetic field plays an important role in numerous astrophysical processes, including star formation and propagation of ultrahigh-energy cosmic rays; it is also an important component in measurements of the cosmological microwave background. This review article provides a brief overview of the latest advancements in the field, from an observer's point of view. The most recent results based on pulsar rotation measures are discussed, which show that we have begun to confidently resolve the main features of the Galactic magnetic field on kiloparsec scales, both in the Solar neighbourhood and at larger distances. As we are currently in great anticipation of polarisation observations with new, state-of-the-art telescopes and hardware, a brief overview of how much this field of research will benefit from the upcoming pulsar surveys is also given.

Keywords Milky Way: magnetic fields · Pulsars: observations · Polarisation

1 Introduction

The properties of the Galactic magnetic field (GMF) have been in the centre of scientific attention since the birth of radio astronomy, in the 1930s. The GMF has been understood to play a role in numerous astrophysical processes: on kpc-scales, it is believed that the GMF contributes to the stability of the interstellar medium (hereafter ISM; Boulares and Cox 1990), and it has even been claimed that it plays a role in the dynamical behaviour of the Galactic disc (Battaner and Florido 2007); also, at smaller scales, the compressed magnetic fields in molecular clouds play an important role in star formation (Heiles and Crutcher 2005)—it is believed that these magnetic fields preserve information related to the large-scale field (Li et al. 2006); last but not least, large-scale fields in the Galaxy, and especially the component of the GMF that is dominant in the Galactic halo, generate pathways for the

A. Noutsos (✉)
Max-Planck-Institut für Radioastronomie, Auf dem Hügel 69, 53121 Bonn, Germany
e-mail: anoutsos@mpifr-bonn.mpg.de

highest-energy cosmic rays from extragalactic sources (Tinyakov and Tkachev 2002; Prouza and Šmída 2003); therefore, understanding the halo fields is crucial in tracing the sources of the most energetic particles in the Universe. Recently, Page et al. (2007) reviewed and modelled the contribution of the Galactic polarised radio foreground to the CMB measurements of WMAP. The polarisation of the Galactic radio emission arises from synchrotron-emitting electrons streaming along the GMF lines but also from the thermal emission of dust particles aligned with the GMF. It is therefore becoming clear that the GMF is an important agent not only for local astrophysical processes, but also for extragalactic astronomy and even cosmology.

2 Tracers of the GMF

2.1 Starlight Polarisation

Starlight can become polarised by scattering or absorption through the ISM, when interstellar dust grains are preferentially aligned along local magnetic field lines. Starlight polarisation can provide the sky-projected component of the magnetic field averaged over the integrated size of the scattering screen between the star and the observer; therefore the measured magnetic fields via this method are weighted by the amount of intervening dust—which is usually unknown. Unfortunately, it is very difficult to detect magnetic fields beyond 2–3 kpc, with this method. Despite this limitation, maps of the magnetic field from starlight polarisation towards several thousands of stars have been produced, which show that the GMF is generally oriented parallel to the Galactic plane (Heiles 2000; Fosalba et al. 2002).

2.2 Polarised Thermal Dust Emission

Magnetically aligned grains can also produce polarised thermal emission in dense regions (e.g. Giant Molecular Clouds or GMCs). The polarised emission is detectable in the mm, sub-mm and far-infrared bands, each band probing a different part of the distribution of dust-particle sizes. As with starlight polarisation, this method gives the sky-projected component of the magnetic field permeating those dense dust regions: typically fields coherent over $\sim$ 1–10 pc can be detected. Currently, the only regions of the ISM that are dense and bright enough to be detectable via this method are molecular clouds. Therefore, the obvious link to observations of the large-scale GMF with this method is the central molecular ring of the Galaxy. However, it has been recently shown with sub-mm observations of GMCs that the field direction surrounding the clouds during their formation tends to be preserved through their volume (Li et al. 2006); and hence such observations could help us bring the direction of the large-scale field to light in certain regions.

2.3 Zeeman Splitting

Zeeman splitting of spectral lines of astrophysical origin generally requires relatively strong magnetic fields ($\sim$ mG), whereas the typical strength of the large-scale GMF is $\sim$ μG. Therefore, Zeeman splitting can be measurable in molecular clouds and masers but is undetectable over larger scales. Hence, this method is very useful for determining the line-of-sight (LOS) component of localised magnetic fields but can directly tell us very little about the large-scale configuration of the GMF. However, observations of Zeeman splitting in molecular clouds and OH masers, spread across the Galactic plane, have resulted in trends in the overall direction of the detected fields, as a function of distance and longitude, that resemble the structure of the large-scale field deduced from other methods (e.g. pulsar rotation measures, hereafter RMs—see below; Han and Zhang 2007). In addition, high-resolution Very Long Baseline Interferometric (VLBI) measurements of Zeeman splitting of OH masers in massive star-forming regions indicate that the field orientation before and after collapse is preserved (Fish and Reid 2006). This implies that the field is decoupled during the collapse of the cloud, and therefore a mechanism like *ambipolar diffusion* is perhaps in operation (Hezareh et al. 2010). In conclusion, it seems that—as in the case of polarised starlight and thermal emission—the magnetic field direction of the large-scale fields is preserved across regions that are as dense as e.g. hydroxyl masers ($\rho \sim 10^7$ cm^{-3}; Reid et al. 1987; Fish et al. 2003).

2.4 Diffuse Synchrotron Emission

The large-scale magnetic field can also be estimated from polarisation observations of the synchrotron emission from the Galactic population of cosmic-ray (relativistic) electrons (see review by Reich 2006). This method provides the sky-projected, transverse component of the field, and like all such methods it can only tell us the orientation but not the direction of the magnetic field. In addition, measurements of the Galactic synchrotron flux, assuming energy equipartition between the cosmic rays and the total magnetic field, can provide an estimate of the ordered field's strength: i.e. the strength of the regular field, possibly arising from a large-scale dynamo, plus any compressed or sheared fields that are generated from the turbulent ISM. The typical estimates of the large-scale field strength from this method tend to be higher than from other methods: a likely reason is that the GMF also consists of the above-mentioned small-scale, turbulent component—of equal or even higher strength—that contributes to the observed polarised emission. Another caveat to this method is the Faraday depolarisation (i.e. the reduction of polarised flux) that occurs as the observer measures the sum of all polarised emission from different emitting regions along the LOS. A remedy to this deleterious effect is to observe at high frequencies (e.g. at 6 cm; note that accessing the Galactic centre requires $\lambda < 1$ cm), as the Faraday effect scales with λ^2. Also, high-frequency observations allow us to resolve the local regions of synchrotron emission, which distort the large-scale picture of the GMF: such observations have revealed a plethora of foreground sources, like e.g. the North Polar Spur (Wolleben et al. 2006).

The development of high frequency resolution backends combined with low-frequency frontends, like LOFAR and the WSRT, have recently made possible to resolve multiple, synchrotron-emitting Faraday screens towards different LOS (Schnitzeler et al. 2009). A crucial, missing component that has been technically unattainable until recently is the application of Faraday-tomography techniques, such as *Rotation Measure Synthesis* and *wavelet tomography*. RM Synthesis can be applied to polarisation data obtained at high frequency resolution. (A description of the method and its application to astrophysical context can be found in Brentjens and de Bruyn 2005.) In this method, the Fourier transform of the observed sum of polarised intensity across the observing band, $P(\lambda^2)$, provides the Faraday dispersion function, $F(\phi) \propto \int_{-\infty}^{+\infty} P(\lambda) \exp(-2i\phi\lambda^2) d\lambda^2$—where ϕ is equivalent to RM—which is the polarised brightness profile of all sources along the LOS, in RM space. Hence, a complex region of polarised emission and Faraday rotation is revealed as a number of separate components at different RMs. A recent improvement on the RM Synthesis method comes from the introduction of wavelets, which allow one to decompose $F(\phi)$—and thus $P(\lambda^2)$—into different RM structures, for different spatial scales; this is beneficial when dealing with complicated Faraday structures along the LOS, involving several regions of polarised emission, but it also helps filter out the 'noisy' small-scale structure (Frick et al. 2010, 2011).

A main advantage of studying the Faraday rotation of the diffuse, polarised emission is that it is observable almost everywhere in the Galactic foreground (although polarisation voids and canals have also been identified; de Bruyn et al. 2006; Haverkorn et al. 2000). Coupled with modern, wavelet-based, Faraday-tomography techniques such observations are potentially a powerful tracer of the GMF on both small and large scales.

The reader is referred to a more detailed review on magnetic fields in the diffuse ISM by Landecker (2011).

2.5 Faraday Rotation of Pulsars and Extragalactic Sources

Faraday rotation of pulsars and of extragalactic point sources is the best available method for mapping the large-scale GMF in the plane but also at high latitudes. Faraday rotation of the plane of the linearly polarised emission occurs when polarised radio waves traverse the ionised component of the magnetised ISM: the warm, ionised ISM (WIM; $T_{\rm WIM} \sim 10{,}000$ K) and, to a lesser degree, the hot, ionised ISM (HIM; $T_{\rm HIM} \sim 10^6$ K) constitute the principal free-electron populations causing Faraday rotation in the Milky Way; the ultra-relativistic population of cosmic rays has a negligible cross-section for this effect. The magnitude of the Faraday rotation is proportional to the square of the emission wavelength and the RM; the latter is equal to the line integral of the magnetic field, B, along the LOS to the polarised source, weighted by the electron density: i.e. $\mathrm{RM} = 0.812 \int_{\rm PSR}^{\oplus} (n_{\rm e}/\mathrm{cm}^{-3})(\boldsymbol{B}/\mu\mathrm{G}) \cdot (\boldsymbol{ds}/\mathrm{pc})$, expressed in rad m^{-2}. Hence, long waves are rotated the most and high-frequency waves remain practically unaffected. The amount of Faraday rotation measured across the observing band, $\lambda_1 - \lambda_2$, is thus given by $\Delta\mathrm{PA} = \mathrm{RM} \cdot \Delta(\lambda^2)$.

The majority of pulsars are significantly linearly polarised, which allows us to easily determine RMs along numerous LOS throughout the Galactic volume: at present, there are roughly 750 pulsar RMs available in the literature (Manchester et al. 2005). Around 85% of the known pulsars are found within a kpc of the Galactic plane and, as expected from the higher stellar density, there is an appreciable concentration of pulsars in the spiral arms. Those pulsars are beneficial for the study of the disc field of the large-scale GMF. The significantly fewer pulsars found at high latitudes are, still, very useful in studies of the halo-field component: e.g. as was done in investigations of the large-scale field's symmetry by Han et al. (1997), using the RMs of high-latitude pulsars ($|b| > 8°$). Thus, the local pulsar population will allow us to completely map the vertical distribution of the GMF in our neighbourhood; furthermore, the high-latitude pulsars accessible to LOFAR will be studied together with remote pulsars in the Galactic disc, from higher-frequency surveys, to investigate whether such a distribution holds elsewhere in the Galaxy.

Moreover, pulsars are compact radio sources and are thus devoid of internal Faraday rotation—which is typically the case for the extended external galaxies—and this allows us to solely sample the magnetic field of the ISM between the source and the observer. An exception is pulsars in supernova shells, whose RMs exhibit variations over monthly timescales due to the magnetised filamentary structure of the shells (Rankin et al. 1988; Moffett and Hankins 1999); but those variations are typically of the order of a few rad m^{-2} and hence hardly affect the overall picture of the magnetic field in that direction. Indeed, there are LOS which pass through numerous, magnetised, local regions, e.g. $\mathrm{H_{II}}$ regions, and those should be avoided in studies of the large-scale GMF with pulsars (Mitra et al. 2003; Nota and Katgert 2010).

Last but not least, the beamed, pulsed emission of pulsars means that the amount of dispersive ISM along the LOS (i.e. the 'warm', 8,000 K electrons) is easily measurable by the delay of the pulses across the observing band: $\Delta t = (4.5 \times 10^6\ \mathrm{ms})\ \mathrm{DM} \cdot \Delta(\lambda^2)$,

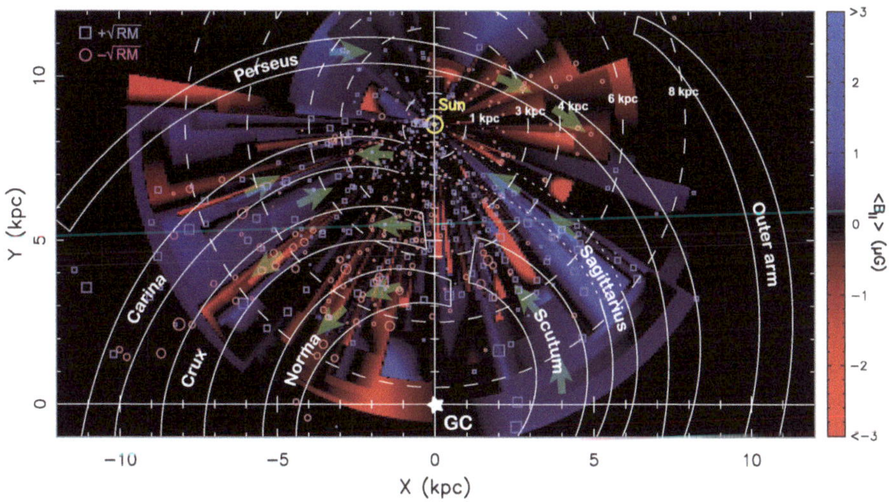

Fig. 1 (Produced by the author) Plan-view map of the large-scale Galactic magnetic field in the plane, within 8 kpc from the Sun (*yellow circle*), from 622 pulsar rotation measures. Blue-graduated regions denote magnetic fields directed towards the Sun (*yellow circle*) and *red*-graduated regions contain fields directed away from the Sun. The field strengths in this map have been interpolated in both the radial and azimuthal directions, in order to fill the empty regions between pulsars. Field strengths $|\langle B_{\parallel}\rangle| > 3$ μG are represented with the maximum colour saturation on the *colour scale*

where DM is the dispersion measure, equivalent to the average number of electrons in a unit column between the pulsar and the observer, i.e. $\mathrm{DM} = \int_{\mathrm{PSR}}^{\oplus} n_e ds$ (usually given in pc cm^{-3}). This information can be used to estimate pulsar distances by using a model of the free-electron distribution of the ISM (e.g. NE2001—Cordes and Lazio 2002); moreover, when combined with the RM, it gives us an estimate of the LOS component of the magnetic field, averaged along the LOS between the pulsar and the observer, weighted by the free-electron density: i.e. $\langle B_{\parallel}\rangle = 1.232$ (RM/DM) μG. Unlike RMs of extragalactic sources, which provide the magnetic field averaged over practically infinite length, pulsar RMs are unique in that respect, as they reveal the average magnetic field only to the finite pulsar distance; using a large number of pulsars, this allows one to map the field as a function of distance (e.g. see Fig. 1).

3 Analysis Techniques

The first attempts to map the large-scale GMF used a few tens of pulsar RMs, most of them from pulsars within 2 kpc of the Sun, to make maps of the strength and direction of the magnetic field as a function of longitude (Manchester 1972, 1974). Later analyses included hundreds of pulsar RMs, as well as extragalactic RMs, which made it possible to track more closely the variation of the field: by plotting the pulsar RM as a function of DM, sub-kpc scale variations in the magnetic field became evident (Lyne and Smith 1989). The scatter in the RM data revealed the small-scale structure of the magnetic field. Moreover, the change in the sign of the RM between local pulsars and those further out suggested that the magnetic field reverses direction. Those early studies of the GMF simply used the RM/DM ratios towards different pulsars, to calculate the LOS magnetic field averaged over the entire distance between the pulsar and the observer. Later work took advantage of denser pulsar regions (mainly the first Galactic quadrant, Q1) and provided estimates of the field variations along different LOS by using the RM–DM gradients between pulsar pairs with small angular separations (i.e. a few degrees): the magnetic field averaged between those pulsar pairs, at distances r_1 and r_2, was then given by $\langle B_{\|}\rangle = [\mathrm{RM}(r_2) - \mathrm{RM}(r_1)]/[\mathrm{DM}(r_2) - \mathrm{DM}(r_1)]$.

In the recent years, the advancement of backend technology together with the development of broadband receivers has motivated pulsar-polarisation censuses (e.g. Noutsos et al. 2008; Han et al. 2009). As a result, it has become possible to confront various theoretical models of the large-scale GMF with multi-parametric fits to the rapidly increasing number of pulsar RMs. The different field geometries typically tested are primarily motivated by the corresponding galactic-dynamo modes: e.g. axisymmetric (ASS; Brown et al. 2007), bisymmetric (BSS; Prouza and Šmída 2003) and quadrisymmetric (QSS; Stepanov et al. 2008) spiral-field geometries are predicted from the different dynamo modes, $m = 1, 2, 3$, respectively.

The most advanced methodology of reconstructing the GMF from RMs uses Fourier deconvolution of the spatial distribution of the RMs, by means of a wavelet function that filters out the unwanted small scales of the field: the method is known as wavelet transform (Grossman and Morlet 1984; Farge 1992; Holschneider 1995) and was recently introduced in GMF mapping by Frick et al. (2001) and Stepanov et al. (2002). For the time being, this method can be applied to non-uniformly distributed RM data scattered throughout a volume but requires that the pulsar separation is at most ~ 0.5 kpc (Stepanov et al. 2002). This condition is easily satisfied within about 3 kpc from the Sun, towards the inner Galaxy, but fails elsewhere. For the purposes of large-scale field modelling, one can apply the wavelet transform to a set of measured RMs but also to the RMs predicted by a model: performing both allows one to then fit the model transform to that of the data and, because of the filtered-out small-scale noise, vastly improve the quality of the fits (Frick et al. 2001).

4 The Turbulent ISM

4.1 WIM and Pulsar Distances

A significant obstacle in modelling the large-scale GMF is the unknown distribution of the free-electron density that impacts on the pulsar distances. Early estimates of pulsar distances based on the measured DM assumed a constant electron density (e.g. Thomson and Nelson 1980; Han and Qiao 1994). Later, Taylor and Cordes (1993) made the first effort to model the WIM in their TC93 model; but that model systematically overestimated pulsar distances—due to the underestimated average value of n_e. Nowadays, the most commonly used density model is the NE2001 model of Cordes and Lazio (2002). Although a clear improvement over the TC93, NE2001 is still typically 20% in error for DM-derived distances in the Galactic disc: precision VLBI parallax measurements of pulsars has shown that the distance to individual pulsars from DM can be wrong by as much as a factor of a few (e.g. Deller et al. 2009); as was shown by Gaensler et al. (2008), the problem is accentuated at high latitudes, where much less information about the ISM is available. Despite follow-up work for an improved model of the ISM (e.g. BMM06; Berkhuijsen et al. 2006; Berkhuijsen and Müller 2008) there is still no reliable model that can accurately reproduce all pulsar distances. The limitations of current ISM modelling were recently highlighted by Van Eck et al. (2011) in their attempt to model the large-scale disc field using pulsar and extragalactic RMs: amongst other reasons, they stated that an improvement to their model is unlikely unless a more detailed model of the Galactic free electrons becomes available.

A key ingredient that is missing from all current ISM models is the information on the small-scale structure. The turbulent, small-scale ISM has a significant impact on both DM and RM measurements. This can be easily seen in maps of the reconstructed GMF from pulsar RM and DM gradients, where small-scale fluctuations are largely responsible for the observed incoherent picture (Fig. 1): the magnitude of the turbulent ISM fields is typically estimated to be from roughly of the same order of magnitude up to twice that of the ordered fields, leading to strong RM fluctuations towards most LOS (Noutsos et al. 2008); in addition, a simple differentiation of RM and DM data along a given LOS amplifies the small-scale noise, as was shown by Ruzmaikin (cf. Ruzmaikin et al. 1988, p. 34); finally, depending on the nature of the ISM turbulence, n_e and B fluctuations can be positively or negatively correlated, which results in over- or underestimation of the magnetic-field strengths from pulsar RMs, as the latter calculation assumes that the above quantities are independent (Beck et al. 2003). It is clear, then, that a model of the small scales needs to be included in future modelling.

The next free-electron density model (NE20—) is already being planned and will be the amalgamation of all available observational data: i.e. dispersion, scattering and rotation measures, Hα maps, synchrotron diffuse emission, astrometric and timing parallaxes, etc. (see e.g. Jaffe et al. 2010). Many of these data have become available from recent all-sky surveys: e.g. the Parkes multi-beam and High Time Resolution Universe (HTRU) northern and southern surveys (Manchester et al. 2001; Keith et al. 2010; Barr 2011). A much better coverage of the sky with pulsar RMs and scattering measures is also expected soon from the LOFAR and Effelsberg all-sky surveys (van Leeuwen and Stappers 2010). All of the above data will be incorporated in multi-parametric fits that will provide more stringent boundary conditions than before, in order to reach a self-consistent solution for the ionised matter and magnetic field distribution simultaneously.

4.2 RM Scatter in Measurements of the Regular Field

Many investigators have noted the large RM scatter in plots of the spatial RM distribution of pulsars and extragalactic sources (e.g. Mitra et al. 2003; Han et al. 2006). As was noted by Noutsos et al. (2008) and later by Nota and Katgert (2010) and Mao et al. (2010), the observed scatter is neither due to measurement errors nor due to polarisation ambiguities in RM measurements. The main contributing factor is the small-scale ($\sim$ 1–100 pc) magnetic-field fluctuations, which manifest themselves in plots of RM versus longitude or latitude, or distance along the LOS to pulsars and extragalactic sources (Mitra et al. 2003; Han et al. 2006; Noutsos et al. 2008; Nota and Katgert 2010).

In an attempt to describe the effects of the turbulent ISM on polarised emission, Gaensler et al. (2001) modelled a region of reduced linearly polarised intensity ('void'; centred on $l = 332\overset{\circ}{.}4$, $b = 1\overset{\circ}{.}4$) on maps of Q4 from the Southern Galactic Plane Survey (SGPS; McClure-Griffiths et al. 2005; Haverkorn et al. 2006). Given the degree of depolarisation seen in the observations of the above test region, it was possible to estimate the RM fluctuations towards the region ($\sigma_{\rm RM} \sim$ 35–50 rad m^{-2}) and combine it with an *upper* limit on the emission measure ($\mathrm{EM} = \int n_e^2 ds < 7000$ pc cm^{-6}), in order to place a *lower* limit on the magnitude of the random fields. The estimated strength of those fields was $B_r \gtrsim 1.3$ μG.

A different approach to determining the influence of the small scales on RM was followed by Mitra et al. (2003), who measured the variation in DM and RM of pulsars behind the H$_{\rm II}$ region S205 (Sharpless 1959). Given the region's extent ($\approx$ 30 pc) and distance ($\approx 900 \pm 300$ pc), the measurements could be used to estimate the magnitude of the magnetic field in S205. The resulting value for the magnetic field was 5.7 μG, which is consistent with alternate estimates of the strength of the small-scale fields (Rand and Kulkarni 1989; Ohno and Shibata 1993; Beck et al. 1996; but see Mao et al. 2010, who derived $B_{\mathrm{r}} \sim 1\ \mu\mathrm{G}$ for the halo, from equipartition).

5 The Large-Scale GMF

5.1 Field Structure

The large-scale, regular GMF is coherent over kpc-scales and permeates the entire Galactic volume. It is so defined as to separate it from the small-scale, turbulent fields ($\lesssim 100$ pc), where a steepening of the magnetic-energy spectrum is observed due to the dissipation of magnetic energy from larger to smaller scales (Han 2004; e.g. Bowers and Li 2007). The large-scale field is associated with the entire Galaxy as whole, thus implying that it was either already present at the Galaxy's formation—and has been later amplified by dynamo action (e.g. Beck et al. 1996; Shukurov 2005; Hanasz et al. 2009)—or was formed in the early stages of Galactic evolution (e.g. Rees 1987, 2006) and has been stretched across the plane by the Galactic rotation, throughout the Galaxy's history.

After the pioneering observations of the Galactic synchrotron emission at 408 MHz by Haslam et al. (1981, 1982) and the subsequent modelling of the emission as the sum of a thin- and a thick-disc component by Beuermann et al. (1985), it has become common practice to treat the large-scale field of the Galaxy in a similar fashion: i.e. as having a thin-disc component of a few hundred pc scale-height, embedded in a thick-disc component of ≈ 2 kpc scale-height—similar to that observed in edge-on galaxies (Krause 2009) and also to the scale-height of the free-electron distribution of the ISM (see e.g., Kachelrießet al. 2007). The value for the latter has been recently revised to 1.8 kpc, based on pulsars with reliable distance estimates (Gaensler et al. 2008), from its earlier value of ~ 1 kpc in the NE2001 free-electron density model of Cordes and Lazio (2002).

Various geometries for the large-scale structure of the GMF have been fitted to pulsar and extragalactic-source RM data (Brown et al. 2007; Noutsos et al. 2008; Sun et al. 2008; Men et al. 2008). However, it is becoming clear that the most likely geometry of the disc field is that of a logarithmic spiral (e.g. Nota and Katgert 2010), similar to those observed in external galaxies (e.g. M51—Fletcher et al. 2011; NGC6946—Beck 2007). Nevertheless, it is interesting to note that, although much effort has gone into modelling the spiral field—with various pitch angles and arm widths tested against the data—the latest research shows that the overall profile of the magnetic field as a function of distance (probed along different lines-of-sight) is insensitive to the details of the spiral field (Nota and Katgert 2010). The principal reasons for our inability to distinguish different geometries, especially in the case of pulsars, are: (a) the sparse sample of pulsar RMs available in the Galactic volume, which leads to noise amplification in maps of the magnetic field from RM–DM gradients; and (b) the absence of adequate modelling of the small-scale component of the magnetic field (and electron density) in the models; this small-scale component can be as much as a factor of a few stronger than the large-scale component, introducing huge uncertainties in the models. A direct consequence of these shortcomings, and a limiting factor in itself, is the uncertainty in pulsar distances, which effectively adds confusion to the reconstructed maps (e.g. see Sect. 5 of Van Eck et al. 2011).

On the bright side, it is worth noting the recent simulations of the regular and turbulent Galactic magnetic field and the free-electron distribution, based on a grid of extragalactic RMs observable with the SKA (Sun and Reich 2009). The simulations predict that arcsecond-resolution RM maps will be possible; these maps show fluctuations of the RM similar to those seen in GMF maps from pulsar RMs. This is a promising result that we need not only expect from the SKA: LOFAR will soon embark on an all-sky polarisation survey at 150 MHz, which is certain to produce an RM grid from Galactic and extragalactic sources that can be used to test the above models.

Beyond the disc field, the dominant component is the halo field. Due to the lack of available polarised sources in its volume, there is little known information about the size of the halo field. In principle, the halo field's extent is observationally limited by the amount of detectable Faraday Rotation through its volume. Alternatively, one can indirectly estimate the scale-height of the halo field by using the vertical distribution of the observed synchrotron emissivity and assume equipartition between the cosmic rays and the magnetic field (Cox 2005): such methods have led to scale-heights of 5–6 kpc for the halo field. Furthermore, the recent surveys of the polarised synchrotron emission, e.g. the Parkes Galactic Meridian Survey (PGMS; Carretti et al. 2009a), are partly aimed at probing the Galactic magnetism of the halo; a detailed discussion of the results from such efforts is presented herein, in Haverkorn and Heesen (2011).

In addition to the planar field in the disc, there are recent measurements of the strength of the vertical component of the magnetic field at the solar radius, towards the north and south Galactic poles, from RMs of polarised extragalactic sources (Mao et al. 2010). These measurements show an incoherent field at the Sun's position, with the field being consistent with zero towards the North Galactic Pole and significantly different from zero ($\sim 0.3\ \mu$G) towards the South Galactic Pole. If these values for the GMF reflect the large-scale field and are not heavily affected by the fields in the local bubble and/or magnetised clouds in the local ISM, then these new results are inconsistent with pure-dipole or quadrupole geometries at the solar radius (see e.g. Han 2007; Sun et al. 2008). It should be stressed, however, that these observations cannot probe the field geometry elsewhere in the Galaxy, as different field configurations may exist in the inner and outer Galaxy (as was suggested, for example, by Van Eck et al. 2011; see Sect. 5.3).

5.2 Field Magnitude

The original polarisation observations of only a few tens of pulsars, as early as the 1970s, revealed a large-scale azimuthal magnetic-field ($l \sim 90°$) within a kpc of the Sun, with a magnitude of the order of 2 μG (Manchester 1974). Follow-up attempts to model the typical magnitude of the random fields consistently resulted in estimates of a factor of a few or higher than that of the regular field (Thomson and Nelson 1980; Lyne and Smith 1989).

The contemporary consensus about the field strength at the solar circle is that direct measurement of Faraday Rotation of polarised pulsars, having a range of longitudes and latitudes, yields an average strength of the large-scale field of $\sim 2\ \mu$G, with a clockwise direction, as seen from the North Galactic Pole (Han 2001; Mitra et al. 2003). Moreover, from starlight polarisation, Heiles (1996) found that this local field is directed towards $l = 83° \pm 4°$. Further out, in Q4, towards the Carina, Crux and Norma arms, Nota and Katgert (2010) used pulsars and extragalactic sources with reliable RM measurements (excluding those behind regions of high n_e fluctuations) and showed that the large-scale field maintains a strength of a few μG independently of galactocentric distance; most previous studies of the GMF either assumed or found a radial dependence of the field strength, with the field magnitude increasing towards the Galactic centre (Han et al. 2006; Brown et al. 2007).

At the same time, the various estimates of the magnitude of the large-scale, regular magnetic field from Faraday rotation of pulsars and extragalactic sources are systematically lower than those from energy-equipartition arguments based on the surface brightness of the polarised Galactic diffuse synchrotron emission: the former typically result in strengths of $\approx$ 1–2 μG, whereas the equipartition strength is roughly twice that, i.e. $\approx$ 4 μG (Frick et al. 2001; Mitra et al. 2003; Nota and Katgert 2010). Given that energy-equipartition from the polarised intensity of the Galactic diffuse emission yields the ordered (regular + anisotropic) fields, the higher magnetic field value perhaps implies that a significant fraction of the local, Galactic ordered fields are anisotropic and turbulent. However, as was shown by Beck et al. (2003), the fluctuations of the magnetic field in a turbulent medium are correlated with those of the electron density. Since the measured values of $\langle B_{\|}\rangle$ are based on the weighted average of the magnetic field by the electron density, which requires that δn_e and δB vary independently, there is an additional systematic error on the estimates of the regular field both from RMs and from equipartition: if these quantities are anti-correlated, the regular component of the field from RMs is underestimated, whereas positive correlation leads to overestimation of the ordered field from the polarised synchrotron emission.

5.3 Field Reversals

Several investigators have reported a number of large-scale reversals in the magnetic field direction between the optical arms of the Galaxy (Han et al. 2006; Brown et al. 2007; Noutsos et al. 2008; Nota and Katgert 2010; Van Eck et al. 2011). Although there is a certain degree of confusion as to which of these reversals are true features of the large-scale GMF and not local fluctuations of the significantly stronger turbulent field, the 'smoke' is beginning to clear: a number of recent publications have confirmed earlier results (Thomson and Nelson 1980; Han and Qiao 1994; Rand and Lyne 1994) indicating a reversal in Q1, between the Orion and the Carina–Sagittarius arm, within 1–2 kpc of the Sun (Weisberg et al. 2004; Han et al. 2006; Noutsos et al. 2008); the reversal changes the clockwise (CW) field of the Orion arm to counter-clockwise (CCW).

For the rest of the Galaxy, the number of reversals has been significantly revised over the recent years. Earlier publications suggested that the Galaxy possesses a field that reverses in every arm–interarm region (Han et al. 2006). More recently, a number of publications claimed a reversal between the Carina–Sagittarius and the Crux arms in Q4 (Brown et al. 2007; Noutsos et al. 2008). However, it has been known that measurements through the Carina region in Q4 show anomalous deviations that disrupt the smoothness of the RM distributions with longitude (e.g. Han et al. 2006; Noutsos et al. 2008; Haverkorn et al. 2008). Nota and Katgert (2010) showed that removal of RMs from pulsars in the direction of the Carina arm alleviates the unusually sharp RM fluctuations and, furthermore, it eliminates the requirement for a reversal between the Carina and Crux arms. It was also stressed that measurements in that direction should be regarded with caution: it is very likely that the Carina–Sagittarius arm possesses an number of H_{II} regions that corrupt the RM measurements—Mitra et al. (2003) showed that an H_{II} region with a 2° angular size and 30-pc linear size can produce a $\Delta RM \sim 250\ \mathrm{rad\,m^{-2}}$ across its volume. Moreover, pulsars behind supernova remnants can potentially also distort a smooth RM variation that is expected from a coherent large-scale field.

The latest developments in mapping the orientation of the large-scale magnetic field in the disc and halo come from a combination of a large number of extragalactic RMs and pulsar RMs or diffuse synchrotron emission. Using > 1,000 extragalactic RMs from the CGPS, SGPS and VLA Galactic plane surveys and combining them with 557 pulsar RMs from the literature, Van Eck et al. (2011) tested a number of popular large-scale, axisymmetric, bisymmetric and concentric-ring configurations (e.g. Brown et al. 2003, 2007; Weisberg et al. 2004; Han et al. 2006; Noutsos et al. 2008). This work concluded that the disc field is predominantly clockwise (as seen from the North), with the data being consistent with an axisymmetric spiral for the inner Galaxy, with only one large-scale reversal that appears to "spiral out" from the Galactic centre (see Fig. 2). Towards the outer Galaxy, the authors concluded that the pitch angle of the spiral field diminishes to resemble predominantly concentric rings.

The symmetry and structure of the halo magnetic fields is a long-standing question. Amongst the earliest work suggesting a coherent, large-scale structure in the halo RM sky was that by Han et al. (1997), who combined the few available pulsar RMs above 8° Galactic latitude with RMs of extragalactic sources to investigate the large-scale symmetry of the GMF in the halo. That work hinted at an anti-symmetric high-latitude sky, having azimuthal fields (of ~ 1 μG) above and below the Galactic plane with an opposite sense to each other. More recently, a similar conclusion was drawn from the work of Sun et al. (2008) and Sun and Reich (2009), who constructed 3D models of the Galactic total and polarised synchrotron emission from fits to the observed synchrotron maps and extragalactic RMs. The authors concluded that the best fit to the observations yielded a strong toroidal halo field ($\lesssim 10$ μG) that reverses its sense on opposite sides of the Galactic plane.

Despite the convincing evidence for an antisymmetric, large-scale Faraday sky at high latitudes, it has recently emerged from the work of Wolleben et al. (2010b) that the observed antisymmetry may largely be caused by a northern-sky, local H_I bubble, at a distance of ~ 100 pc. The strong, compressed magnetic fields in the shell of that local structure were found to be $\sim$20–34 μG and are plausibly strong to cause the apparent antisymmetry in the measured RMs. The influence of such foreground structures can be revealed, as was done in the above work, with high-resolution spectro-polarimetry and the application of RM Synthesis on the diffuse component of the Galactic emission.

In conclusion, it is becoming clear that only a small number, perhaps only one large-scale reversal is indeed present in the configuration of the disc field. In the view of the large number of reversals having been reported in precursor work, it is important to highlight the significance of the distortion to the picture of the large-scale GMF which the small-scale structure can induce. Equally important is the fact that the careful analysis, e.g. by eliminating unreliable data (as was done in recent work), and the combined use of large RM samples and other polarisation information can lead to the true form of the large-scale field.

5.4 Magnetic Fields in the Galactic Centre

The Galactic centre (GC) is particularly challenging observationally, as scattering and depolarisation are deleteriously diminishing the sensitivity of our searches for pulsars and, generally, polarised emission. The steep synchrotron spectra of pulsars are constraining in terms of how high in frequency we can observe before the flux becomes undetectable (Maron et al. 2000); on the other hand, scattering and depolarisation are quadratic and even quartic functions of λ, meaning that high-frequency observations (< 1 cm) are necessary in order to avoid smearing out the signal completely. So, understandably, there are only a few pulsar detections near the GC and, in fact, no Faraday rotation has been measured from them, yet (Deneva et al. 2009). In addition, only a few tens of polarised, extragalactic sources can be seen within a few degrees of Sgr A* (e.g. Roy et al. 2003).

One way of battling the effect of depolarisation in the direction to the GC is to conduct radio observations at high frequencies. This approach was followed in early observations of the GC region, with the Effelsberg telescope at 32 GHz (Reich 1990), in which a ≈ 400-pc long, highly-polarised filamentary structure (the "Arc") was found very near

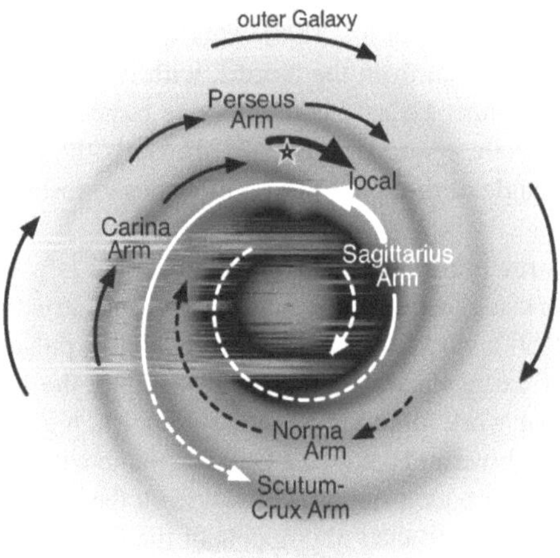

Fig. 2 (From Van Eck et al. 2011) Sketch of the large-scale field orientation of the Galactic magnetic field in the Milky Way disc, according to the contemporary view, overlaid on the NE2001 free-electron density model of Cordes and Lazio (2002). The *star* symbol corresponds to the Sun's position. The *thick, white arrow* indicates the only generally accepted large-scale reversal in Q1. The *thin, white arrow* indicates the large-scale reversal believed to spiral out from the Galactic centre. The *dashed arrows* in this figure correspond to poorly sampled regions, therefore corresponding to field directions that are uncertain

Sgr A* (≈ 30 pc away). The Arc, being nearly perpendicularly to the Galactic plane, was found to contain magnetic fields in the mG range, running almost parallel to its filamentary structure (Yusef-Zadeh and Morris 1987a, 1987b). It is worth noting that the Arc, given its extent and magnetic properties, is indeed a remarkable structure but not a unique one: further such filamentary structures near the GC have been discovered since (e.g. Reich 2003).

At γ-ray energies, the emission is practically unaffected by the aforementioned interstellar propagation effects that impact on radio waves; the recent work of Crocker et al. (2010) took advantage of this fact and placed a lower limit of 50 μG on the magnetic field strength (over 400-pc scales) near the GC, based on the observed EGRET γ-ray flux from that region. However, it should be stressed that stronger fields near the GC may be attributed to the increased synchrotron flux in the central regions of the Galaxy and therefore to stronger equipartition-field requirements—but this is not necessarily connected with the large-scale configuration of the GMF.

A very recent study of the central kpc of the inner Galaxy, using the Faraday rotation of the diffuse polarised emission, has revealed a poloidal magnetic field organised on 150-pc scales, directed from south to north (Law et al. 2011). It is also worth mentioning that the RM structure of the central 2° of the GC, as was measured in the above work, indicates an axial symmetry of the magnetic-field structure that is shifted by 50 pc to the west with respect to the dynamical centre of the Milky Way; the location of this axis of symmetry is consistent with, and possibly due to, the presence of a starburst outflow in the central 2° of the GC (Bland-Hawthorn and Cohen 2003). Meanwhile, optical polarimetry of dust grains has been used to bring to light the magnetic structure of the very central part of the Milky Way ($|b| < 0\overset{\circ}{.}4$; Nishiyama et al. 2010): a smooth transition from a toroidal to a poloidal magnetic field has been observed as one moves to higher latitudes, i.e. $|b| > 0\overset{\circ}{.}4$.

In conclusion, it appears that the efforts to uncover the large-scale magnetic-field structure of the GC, utilising different observables, have been dominated by magnetised filamentary structures, organised on a few 100-pc scales. However, there has been little evidence towards linking the magnetic field of those localised structures to a global form of the Galactic magnetic field, as would be e.g. a poloidal field predicted by some dynamo models.

A more focused review on magnetic fields near the GC can be found in Ferrière (2011).

6 Future Instrumentation and Prospects

The field of research dealing with the magnetic field of the Galaxy has gone through an initial phase of a few, promising—mainly qualitative—results (1970s, 1980s); in the 90s, after the pioneering measurements and before the large-scale pulsar surveys that followed, there was a relative paucity of data accumulation, but it meant that the theoretical investigations could flourish. A large part of the last decade saw the relative abundance of pulsar and extragalactic-source RMs (mainly due to a number of pulsar surveys, e.g. Parkes, GBT), but this led more often than not to confusing—or at least inconclusive—results from the respective analyses: as was well pointed out by Nota and Katgert (2010), "The various analyses of the large-scale Galactic magnetic field are often based on identical or very similar datasets. Still, they have led to different or even contradictory conclusions." However, during the last few years the dust has begun to settle and we are making confident conclusions about the strength and the direction of the large-scale GMF in a number of regions in the Galaxy. Moreover, we are beginning to understand the role of the turbulent ISM in our measurements and have even quantified—to some extent—the relative contribution of small-scale magnetic fields. This is not to say, of course, that we are close to a full description of the GMF. A number of exciting problems await resolution: the extent of the large-scale GMF in the halo has not been determined; how many large-scale reversals are there in total in the Galaxy? None of the external galaxies observed so far shows radial large-scale reversals: is our Galaxy unique in that respect? Also, a number of external galaxies (e.g. IC 342, NGC 6946) show the presence of magnetic arms between the optical arms (Beck 2011), where the field is stronger: is this also true for the Milky Way? What is the large-scale geometry of the GMF and how was it formed? At the moment, there are almost as many proposed models for the spiral field of the Galaxy as supporting publications for each of them. And last but not least, how do the small-scale turbulent fields affect our measurements? We need to characterise the observed fluctuations of the field: are these fluctuations positively or negatively correlated with the electron density?

6.1 Long-wavelength Phased Arrays

In the recent years, the revival of long-wavelength astronomy with the introduction of new, large-area phased arrays promises to be the next big leap in the field of Galactic—and not only—magnetic field studies. These arrays are composed of omnidirectional antennas, typically covering frequencies from a few tens to a few hundreds of MHz. Multiple of those antennas are arranged into stations that are spread across a large geographic area. The individual antennas observe the entire sky and directional beams can be formed by combining the signal from several antennas, and several stations, using powerful computer clusters.

Apart from the clear scientific value of these instruments, a driving force for their fast development has been the Square Kilometre Array (SKA), to which the long-wavelength arrays are considered pathfinder projects. In Europe, the Low Frequency Array (LOFAR; Stappers et al. 2011) is composed of 40 stations in the Netherlands and 12 international stations spread throughout Europe (i.e. Germany, France, UK, Sweden, Poland; more are

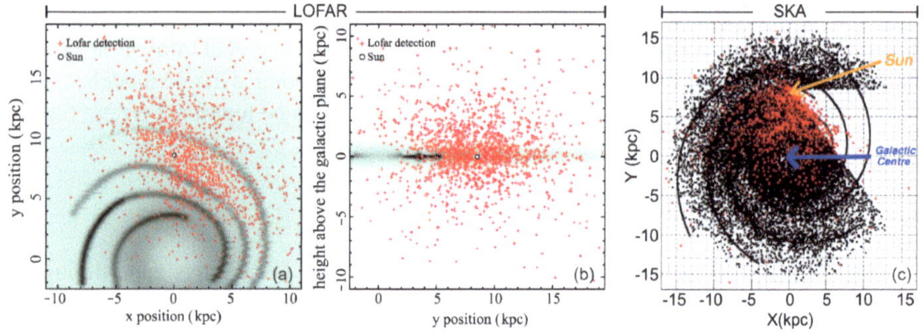

Fig. 3 (**a, b**) (From van Leeuwen and Stappers 2010) 1,100 simulated pulsar detections from a Galactic survey with LOFAR (1 h pointings). (**c**) (From M. Kramer, private communication) Simulated detections of pulsars from an SKA survey (*black points*); the *red points* correspond to the positions of all known pulsars

being funded). Each station is split between the Low Band Antennas (LBAs), covering frequencies of 10–80 MHz, and the High Band Antennas (HBAs), covering frequencies from 120–240 MHz. In Australia and the United States, the Murchinson Widefield Array (MWA; Lonsdale et al. 2009) and the Long Wavelength Array (LWA; Kassim et al. 2005), respectively, are similar projects that will consist of thousands of dipoles sensitive to frequencies from 80 to 300 MHz, with several thousands of m^2 of collecting area.

Many pulsars' flux spectra show a turnover at $\sim$ 100 MHz, which means that observing in the high band with LOFAR we expect to detect a new population of pulsars that are difficult to detect at higher frequencies, where they appear much weaker. Simulations by van Leeuwen and Stappers (2010) showed that a 2-month, all-sky survey with LOFAR, at 150 MHz, can produce an excess of 1,000 new, nearby and high-latitude pulsars (see Fig. 3a, b). An important conclusion from those simulations was that LOFAR will detect all the pulsars within 2 kpc from the Sun. Many of these pulsars may not have high polarisation, which will prevent us from using them to map the GMF. However, based on current statistics, 40% of the detected sample of pulsars have enough polarisation for RM determination (Men et al. 2008). This means that at least 400–500 new directions through the Galaxy will be sampled, which will greatly increase the detail of the GMF maps. But LOFAR has an additional advantage in that half of the pulsars found in the northern hemisphere are seen to increase their degree of linear polarisation towards low frequencies (Gould and Lyne 1998). Hence, we expect that a seemingly unpolarised or weakly polarised population of pulsars may appear very strongly polarised at 100 MHz.

Naturally, the advent of the SKA will give us the opportunity, with its unprecedented sensitivity, to discover all Galactic pulsars (Fig. 3c). It will then be truly possible to map the GMF across its entirety: from kiloparsec scales down to a few parsecs; from the central kiloparsec out to the halo.

6.2 Single Dishes

At higher frequencies, single dishes complement the low-frequency arrays in filling the RM sky. Ongoing surveys at 21 cm, with the Effelsberg 100-m telescope in the Northern hemisphere and the Parkes 64-m telescope in the southern hemisphere, promise to return hundreds of new pulsars; and because these observations are at a much higher frequency, scattering is less of a problem and those searches will uncover many undiscovered pulsars buried deep in the Galactic plane. Less severe scattering also means that millisecond pulsars are not be prohibitively scattered, which is the case at low frequencies; and exotic systems, like relativistic-binary pulsars, are more likely to be found in the Galactic plane, towards the denser Galactic centre. In addition, polarisation surveys of the Galactic diffuse emission are being carried out with Parkes at 2 GHz (S-PASS; Carretti et al. 2009b) and the 26-m Dominion Radio Astrophysical Observatory (DRAO; Wolleben et al. 2010a). In the future, a global project is planned at 300–1800 MHz, with DRAO, Effelsberg, Parkes and other single-dish telescopes, to map the diffuse synchrotron emission over the entire sky and hence provide invaluable information on the magneto-ionic medium of the Milky Way.

In conclusion, surveys with single dishes probe the lateral distribution of bright pulsars, limited off the plane by their collecting area, whereas phased arrays probe the vertical distribution of faint pulsars, limited in the plane by ISM scattering.

6.3 High Energies

The advent of γ-ray satellites with wide fields-of-view, like *Fermi* and *AGILE*, as well as the high sensitivities of ground-based Cherenkov arrays, like HESS, MAGIC and—in the future—the CTA, have opened new possibilities of either directly discovering new pulsars and pulsar wind nebulae (PWNs) or pointing to the right direction for follow-up work with radio telescopes. The *Fermi* γ-ray Space Telescope has already discovered tens of pulsars in blind searches and a much higher number of unidentified γ-ray sources. The latter are being searched with radio telescopes like Effelsberg, GBT and Parkes for radio emission. Searches such as these have led to the discovery of radio pulsars, which will be followed up with polarisation measurements and provide further RMs. In addition, Cherenkov arrays have conducted surveys of the γ-ray sky; their products, which may or may not be associated with radio pulsars, may still prove to be invaluable polarised radio sources themselves, which can be used with RM Synthesis to fill in the RM space between pulsars.

References

E. Barr, in *Radio Pulsars: an astrophysical key to unlock the secrets of the Universe*, ed. by M. Burgay, N. D'Amico, P. Esposito, A. Pellizzoni, A. Possenti. American Institute of Physics Conference Series, The Northern-sky High Time Resolution Universe Pulsar Survey, vol. 1357 (AIP, New York, 2011), p. 52

N. Bartel, M.F. Bietenholz, Adv. Space Res. **35**, 1057 (2005)

E. Battaner, E. Florido, Astron. Nachr. **328**, 92 (2007)

R. Beck, Astron. Astrophys. **470**, 539 (2007)

R. Beck, Space Sci. Rev. (2011). doi:10.1007/s11214-011-9782-z

R. Beck, A. Brandenburg, D. Moss, A. Shukurov, D. Sokoloff, Annu. Rev. Astron. Astrophys. **34**, 155 (1996)

R. Beck, A. Shukurov, D. Sokoloff, R. Wielebinski, Astron. Astrophys. **411**, 99 (2003)

E.M. Berkhuijsen, P. Müller, Astron. Astrophys. **490**, 179 (2008)

E.M. Berkhuijsen, D. Mitra, P. Mueller, Astron. Nachr. **327**, 82 (2006)

K. Beuermann, G. Kanbach, E.M. Berkhuijsen, Astron. Astrophys. **153**, 17 (1985)

J. Bland-Hawthorn, M. Cohen, Astrophys. J. **582**, 246 (2003)

A. Boulares, D.P. Cox, Astrophys. J. **365**, 544 (1990)

K. Bowers, H. Li, Phys. Rev. Lett. **98**, 035002 (2007)

M.A. Brentjens, A.G. de Bruyn, Astron. Astrophys. **441**, 1217 (2005)

J.C. Brown, A.R. Taylor, B.J. Jackel, Astrophys. J. Suppl. Ser. **145**, 213 (2003)

J.C. Brown, M. Haverkorn, B.M. Gaensler, A.R. Taylor, N.S. Bizunok, N.M. McClure-Griffiths, J.M. Dickey, A.J. Green, Astrophys. J. **663**, 258 (2007)

E. Carretti, B. Gaensler, L. Staveley-Smith, M. Haverkorn, M. Kesteven, S. Cortiglioni, G. Bernardi, S. Poppi, in *ATNF Proposal P560, Semester: April, 2009 S-band Polarization All Sky Survey (S-PASS)* (2009a), p. 2265

E. Carretti, M. Haverkorn, D. McConnell, G. Bernardi, S. Cortiglioni, N.M. McClure-Griffiths, S. Poppi, in *Magnetic Fields in the Universe II: From Laboratory and Stars to the Primordial Universe*, ed. by A. Esquivel, J. Franco, G. García-Segura, E.M. de Gouveia Dal Pino, A. Lazarian, S. Lizano, A. Raga. Serie de Conferencias, PGMS: to Study the Galactic Magnetism Out of the Galactic Plane, vol. 36 (Revista Mexicana de Astronomia y Astrofisica, Mexico, 2009b), p. 9
J.M. Cordes, T.J.W. Lazio, astro-ph/0207156 (2002)
D.P. Cox, Annu. Rev. Astron. Astrophys. **43**, 337 (2005)
R.M. Crocker, D.I. Jones, F. Melia, J. Ott, R.J. Protheroe, Nature **463**, 65 (2010)
A.G. de Bruyn, P. Katgert, M. Haverkorn, D.H.F.M. Schnitzeler, Astron. Nachr. **327**, 487 (2006)
A.T. Deller, S.J. Tingay, M. Bailes, J.E. Reynolds, Astrophys. J. **701**, 1243 (2009)
J.S. Deneva, J.M. Cordes, T.J.W. Lazio, Astrophys. J. **702**, L177 (2009)
M. Farge, Annu. Rev. Fluid Mech. **24**, 395 (1992)
K. Ferrière, in *Astrophysical Dynamics: from Stars to Galaxies*, ed. by N. Brummell, A.S. Brun, M.S. Miesch, Y. Ponty. Proceedings of the International Astronomical Union, The Interstellar Magnetic Field Near the Galactic Center, vol. 271 (Cambridge University Press, Cambridge, 2011), p. 170
V.L. Fish, M.J. Reid, Astrophys. J. Suppl. Ser. **164**, 99 (2006)
V.L. Fish, M.J. Reid, A.L. Argon, K.M. Menten, Astrophys. J. **596**, 328 (2003)
A. Fletcher, R. Beck, A. Shukurov, E.M. Berkhuijsen, C. Horellou, Mon. Not. R. Astron. Soc. **412**, 2396 (2011)
P. Fosalba, A. Lazarian, S. Prunet, J.A. Tauber, Astrophys. J. **564**, 762 (2002)
P. Frick, R. Stepanov, A. Shukurov, D. Sokoloff, Mon. Not. R. Astron. Soc. **325**, 649 (2001)
P. Frick, D. Sokoloff, R. Stepanov, R. Beck, Mon. Not. R. Astron. Soc. **401**, L24 (2010)
P. Frick, D. Sokoloff, R. Stepanov, R. Beck, Mon. Not. R. Astron. Soc. **414**, 2540 (2011)
B.M. Gaensler, J.M. Dickey, N.M. McClure-Griffiths, A.J. Green, M.H. Wieringa, R.F. Haynes, Astrophys. J. **549**, 959 (2001)
B.M. Gaensler, G.J. Madsen, S. Chatterjee, S.A. Mao, Publ. Astron. Soc. Aust. **25**, 184 (2008)
D.M. Gould, A.G. Lyne, Mon. Not. R. Astron. Soc. **301**, 235 (1998)
A. Grossman, J. Morlet, SIAM J. Math. Anal. **15**, 723 (1984)
J.L. Han, Astron. Astrophys. Suppl. Ser. **278**, 181 (2001)
J.L. Han, in *The Magnetized Interstellar Medium*, ed. by B. Uyaniker, W. Reich, R. Wielebinski. The Large-Scale Magnetic Field Structure of Our Galaxy: Efficiently Deduced from Pulsar Rotation Measures, vol. 1 (Copernicus GmbH, Katlenburg-Lindau, 2004), p. 3
J.L. Han, in *Astrophysical Masers and Their Environments*, ed. by J.M. Chapman, W.A. Baan. Proceedings of the International Astronomical Union, Magnetic Fields in Our Galaxy on Large and Small Scales, vol. 242 (Cambridge University Press, Cambridge, 2007), p. 55
J.L. Han, G.J. Qiao, Astron. Astrophys. **288**, 759 (1994)
J.L. Han, J.S. Zhang, Astron. Astrophys. **464**, 609 (2007)
J.L. Han, R.N. Manchester, E.M. Berkhuijsen, R. Beck, Astron. Astrophys. **322**, 98 (1997)
J.L. Han, R.N. Manchester, A.G. Lyne, G.J. Qiao, W. van Straten, Astrophys. J. **642**, 868 (2006)
J.L. Han, P.B. Demorest, W. van Straten, A.G. Lyne, Astrophys. J. Suppl. Ser. **181**, 557 (2009)
M. Hanasz, K. Otmianowska-Mazur, G. Kowal, H. Lesch, Astron. Astrophys. **498**, 335 (2009)
C.G.T. Haslam, U. Klein, C.J. Salter, H. Stoffel, W.E. Wilson, M.N. Cleary, D.J. Cooke, P. Thomasson, Astron. Astrophys. **100**, 209 (1981)
C.G.T. Haslam, C.J. Salter, H. Stoffel, W.E. Wilson, Astron. Astrophys. Suppl. Ser. **47**, 1 (1982)
M. Haverkorn, V. Heesen, Space Sci. Rev. (2011). doi:10.1007/s11214-011-9757-0
M. Haverkorn, P. Katgert, A.G. de Bruyn, Astron. Astrophys. **356**, L13 (2000)
M. Haverkorn, B.M. Gaensler, N.M. McClure-Griffiths, J.M. Dickey, A.J. Green, Astrophys. J. Suppl. Ser. **167**, 230 (2006)
M. Haverkorn, J.C. Brown, B.M. Gaensler, N.M. McClure-Griffiths, Astrophys. J. **680**, 362 (2008)
C. Heiles, Astrophys. J. **462**, 316 (1996)
C. Heiles, Astron. J. **119**, 923 (2000)
C. Heiles, R. Crutcher, in *Cosmic Magnetic Fields*, ed. by R. Wielebinski, R. Beck. Lecture Notes in Physics, Magnetic Fields in Diffuse HI and Molecular Clouds, vol. 664 (Springer, Berlin, 2005), p. 137
T. Hezareh, M. Houde, C. McCoey, H. Li, Astrophys. J. **720**, 603 (2010)
M. Holschneider, *Wavelets: An Analysis Tool* (Oxford University Press, Oxford, 1995)
T.R. Jaffe, J.P. Leahy, A.J. Banday, S.M. Leach, S.R. Lowe, A. Wilkinson, Mon. Not. R. Astron. Soc. **401**, 1013 (2010)
M. Kachelrieß, P.D. Serpico, M. Teshima, Astropart. Phys. **26**, 378 (2007)
N.E. Kassim, E.J. Polisensky, T.E. Clarke, B.C. Hicks, P.C. Crane, K.P. Stewart, P.S. Ray, K.W. Weiler, L.J. Rickard, T.J.W. Lazio, W.M. Lane, A.S. Cohen, M.E. Nord, W.C. Erickson, R.A. Perley, in *From Clark Lake to the Long Wavelength Array: Bill Erickson's Radio Science*, ed. by N. Kassim, M. Perez, W. Junor, P. Henning. ASP Conference Series, The Long Wavelength Array, vol. 345 (Astronomical Society of the Pacific, San Francisco, 2005), p. 392

M.J. Keith, A. Jameson, W. van Straten, M. Bailes, S. Johnston, M. Kramer, A. Possenti, S.D. Bates, N.D.R. Bhat, M. Burgay, S. Burke-Spolaor, N. D'Amico, L. Levin, P.L. McMahon, S. Milia, B.W. Stappers, Mon. Not. R. Astron. Soc. **409**, 619 (2010)
M. Krause, in *Magnetic Fields in the Universe II: From Laboratory and Stars to the Primordial Universe*, ed. by A. Esquivel, J. Franco, G. García-Segura, E.M. de Gouveia Dal Pino, A. Lazarian, S. Lizano, A. Raga. Serie de Conferencias, Magnetic Fields and Star Formation in Spiral Galaxies, vol. 36 (Revista Mexicana de Astronomia y Astrofisica, Mexico, 2009), p. 25
T.L. Landecker, Space Sci. Rev. (2011). doi:10.1007/s11214-011-9796-6
C.J. Law, M.A. Brentjens, G. Novak, Astrophys. J. **731**, 36 (2011)
H. Li, G.S. Griffin, M. Krejny, G. Novak, R.F. Loewenstein, M.G. Newcomb, P.G. Calisse, D.T. Chuss, Astrophys. J. **648**, 340 (2006)
C.J. Lonsdale, R.J. Cappallo, M.F. Morales et al., Proc. IEEE **97**, 1497 (2009)
A.G. Lyne, F.G. Smith, Mon. Not. R. Astron. Soc. **237**, 533 (1989)
R.N. Manchester, Astrophys. J. **172**, 43 (1972)
R.N. Manchester, Astrophys. J. **188**, 637 (1974)
R.N. Manchester, A.G. Lyne, F. Camilo, J.F. Bell, V.M. Kaspi, N. D'Amico, N.P.F. McKay, F. Crawford, I.H. Stairs, A. Possenti, M. Kramer, D.C. Sheppard, Mon. Not. R. Astron. Soc. **328**, 17 (2001)
R.N. Manchester, G.B. Hobbs, A. Teoh, M. Hobbs, Astron. J. **129**, 1993 (2005)
S.A. Mao, B.M. Gaensler, M. Haverkorn, E.G. Zweibel, G.J. Madsen, N.M. McClure-Griffiths, A. Shukurov, P.P. Kronberg, Astrophys. J. **714**, 1170 (2010)
O. Maron, J. Kijak, M. Kramer, R. Wielebinski, Astron. Astrophys. Suppl. Ser. **147**, 195 (2000)
N.M. McClure-Griffiths, J.M. Dickey, B.M. Gaensler, A.J. Green, M. Haverkorn, S. Strasser, Astrophys. J. Suppl. Ser. **158**, 178 (2005)
H. Men, K. Ferrière, J.L. Han, Astron. Astrophys. **486**, 819 (2008)
D. Mitra, R. Wielebinski, M. Kramer, A. Jessner, Astron. Astrophys. **398**, 993 (2003)
D.A. Moffett, T.H. Hankins, Astrophys. J. **522**, 1046 (1999)
S. Nishiyama, H. Hatano, M. Tamura, N. Matsunaga, T. Yoshikawa, T. Suenaga, J.H. Hough, K. Sugitani, T. Nagayama, D. Kato, T. Nagata, Astrophys. J. **722**, L23 (2010)
T. Nota, P. Katgert, Astron. Astrophys. **513**, A65 (2010)
A. Noutsos, S. Johnston, M. Kramer, A. Karastergiou, Mon. Not. R. Astron. Soc. **386**, 1881 (2008)
H. Ohno, S. Shibata, Mon. Not. R. Astron. Soc. **262**, 953 (1993)
L. Page, G. Hinshaw, E. Komatsu, M.R. Nolta, D.N. Spergel, C.L. Bennett, C. Barnes, R. Bean, O. Doré, J. Dunkley, M. Halpern, R.S. Hill, N. Jarosik, A. Kogut et al., Astrophys. J. Suppl. Ser. **170**, 335 (2007)
M. Prouza, R. Šmída, Astron. Astrophys. **410**, 1 (2003)
R.J. Rand, S.R. Kulkarni, Astrophys. J. **343**, 760 (1989)
R.J. Rand, A.G. Lyne, Mon. Not. R. Astron. Soc. **268**, 497 (1994)
J.M. Rankin, D.B. Campbell, R.B. Isaacman, R.R. Payne, Astron. Astrophys. **202**, 166 (1988)
M.J. Rees, Q. J. R. Astron. Soc. **28**, 197 (1987)
M.J. Rees, Astron. Nachr. **327**, 395 (2006)
W. Reich, in *Galactic and Intergalactic Magnetic Fields*, ed. by R. Beck, R. Wielebinski, P.P. Kronberg. Proceedings of the International Astronomical Union, The Large-Scale Magnetic Field Structure Near the Galactic Centre, vol. 140 (Kluwer Academic, Dordrecht, 1990), p. 369
W. Reich, Astron. Astrophys. **401**, 1023 (2003)
W. Reich, in *Cosmic Polarization*, ed. by R. Fabbri. Galactic Polarization Surveys, vol. 1 (Research Signpost, Kerala, 2006), p. 91
M.J. Reid, P.C. Myers, J.H. Bieging, Astrophys. J. **312**, 830 (1987)
S. Roy, A. Pramesh Rao, R. Subrahmanyan, Astron. Nachr. Suppl. **324**, 41 (2003)
A.A. Ruzmaikin, D.D. Sokolov, A.M. Shukurov, *Magnetic Fields of Galaxies* (Kluwer Academic, Dordrecht, 1988)
D.H.F.M. Schnitzeler, P. Katgert, A.G. de Bruyn, Astron. Astrophys. **494**, 611 (2009)
S. Sharpless, Astrophys. J. Suppl. Ser. **4**, 257 (1959)
A. Shukurov, in *Cosmic Magnetic Fields*, ed. by R. Wielebinski, R. Beck. Lecture Notes in Physics, Mesoscale Magnetic Structures in Spiral Galaxies, vol. 664 (Springer, Berlin, 2005), p. 113
B.W. Stappers, J.W.T. Hessels, A. Alexov, K. Anderson, T. Coenen, T. Hassall, A. Karastergiou, V.I. Kondratiev, M. Kramer, J. van Leeuwen, J.D. Mol, A. Noutsos, J.W. Romein, P. Weltevrede, R. Fender, R.A.M.J. Wijers et al., Astron. Astrophys. **530**, A80 (2011)
R. Stepanov, P. Frick, A. Shukurov, D. Sokoloff, Astron. Astrophys. **391**, 361 (2002)
R. Stepanov, T.G. Arshakian, R. Beck, P. Frick, M. Krause, Astron. Astrophys. **480**, 45 (2008)
X.H. Sun, W. Reich, Astron. Astrophys. **507**, 1087 (2009)
X.H. Sun, W. Reich, A. Waelkens, T.A. Enßlin, Astron. Astrophys. **477**, 573 (2008)
J.H. Taylor, J.M. Cordes, Astrophys. J. **411**, 674 (1993)

R.C. Thomson, A.H. Nelson, Mon. Not. R. Astron. Soc. **191**, 863 (1980)
P.G. Tinyakov, I.I. Tkachev, Astropart. Phys. **18**, 165 (2002)
C.L. Van Eck, J.C. Brown, J.M. Stil, K. Rae, S.A. Mao, B.M. Gaensler, A. Shukurov, A.R. Taylor, M. Haverkorn, P.P. Kronberg, N.M. McClure-Griffiths, Astrophys. J. **728**, 97 (2011)
J. van Leeuwen, B.W. Stappers, Astron. Astrophys. **509**, A7 (2010)
J.M. Weisberg, J.M. Cordes, B. Kuan, K.E. Devine, J.T. Green, D.C. Backer, Astrophys. J. Suppl. Ser. **150**, 317 (2004)
M. Wolleben, T.L. Landecker, W. Reich, R. Wielebinski, Astron. Astrophys. **448**, 411 (2006)
M. Wolleben, T.L. Landecker, G.J. Hovey, R. Messing, O.S. Davison, N.L. House, K.H.M.S. Somaratne, I. Tashev, Astron. J. **139**, 1681 (2010a)
M. Wolleben, A. Fletcher, T.L. Landecker, E. Carretti, J.M. Dickey, B.M. Gaensler, M. Haverkorn, N. McClure-Griffiths, W. Reich, A.R. Taylor, Astrophys. J. **724**, L48 (2010b)
F. Yusef-Zadeh, M. Morris, Astron. J. **94**, 1178 (1987a)
F. Yusef-Zadeh, M. Morris, Astrophys. J. **322**, 721 (1987b)

Space Sci Rev (2012) 169:27–72
DOI 10.1007/s11214-012-9895-z

Magnetic Fields in Astrophysical Jets: From Launch to Termination

Ralph E. Pudritz · Martin J. Hardcastle · Denise C. Gabuzda

Received: 24 December 2011 / Accepted: 8 May 2012 / Published online: 3 June 2012

Abstract Long-lived, stable jets are observed in a wide variety of systems, from protostars, through Galactic compact objects to active galactic nuclei (AGN). Magnetic fields play a central role in launching, accelerating, and collimating the jets through various media. The termination of jets in molecular clouds or the interstellar medium deposits enormous amounts of mechanical energy and momentum, and their interactions with the external medium, as well, in many cases, as the radiation processes by which they are observed, are intimately connected with the magnetic fields they carry. This review focuses on the properties and structures of magnetic fields in long-lived jets, from their launch from rotating magnetized young stars, black holes, and their accretion discs, to termination and beyond. We compare the results of theory, numerical simulations, and observations of these diverse systems and address similarities and differences between relativistic and non-relativistic jets in protostellar versus AGN systems. On the observational side, we focus primarily on jets driven by AGN because of the strong observational constraints on their magnetic field properties, and we discuss the links between the physics of these jets on all scales.

Keywords Magnetic fields · Jets · Polarization · Radio continuum: galaxies · X-rays: galaxies

R.E. Pudritz
Dept. of Physics and Astronomy, McMaster University, Hamilton, ON L8S 4M1, Canada
e-mail: pudritz@mcmaster.ca

M.J. Hardcastle (✉)
School of Physics, Astronomy & Mathematics, University of Hertfordshire, College Lane, Hatfield AL10 9AB, UK
e-mail: m.j.hardcastle@herts.ac.uk

D.C. Gabuzda
Physics Department, University College Cork, Cork, Republic of Ireland
e-mail: d.gabuzda@ucc.ie

1 Introduction

Astrophysical jets have been discovered in a very wide range of physical systems, from regions of star formation, to gamma-ray bursters and stellar black holes, on up in scale and energy to the nuclei of giant radio galaxies and quasars. They are often associated with accretion discs and their launch, internal dynamics, and impact upon their surroundings are of ongoing importance and significance over a whole range of astrophysical phenomena—from the parsec (pc) scales that characterize the impact of jets from young stellar objects within their natal molecular clouds, to the megaparsec (Mpc) scales of giant clusters of galaxies whose intercluster media are significantly stirred and heated by jets emanating from the cores of active galaxies (e.g., Dunn and Fabian 2006). On stellar scales, jets from young stellar objects (YSOs) span the breadth of the stellar mass spectrum: from massive stars such as W33A which shows a hot fast jet when observed in Br γ lines as well as an underlying accretion disc (Davies et al. 2010), to the well known solar-mass systems such as classical T-Tauri stars (TTS) (see Ray et al. 2007, for a review), and on down into the brown dwarfs (e.g., Whelan et al. 2009). There is good evidence that all of the protostellar systems are associated with discs. Stellar mass black holes, otherwise known as microquasars (Mirabel 2005) are also very active sources of relativistic jets. On extragalactic scales, jets are observed to be associated with accretion on to supermassive black holes (up to a few billion solar masses) in the centres of active galaxies (Blandford 2001). Models for gamma-ray bursts (GRBs), which are also located at cosmological distances, focus on either the collapse of massive stars or the merger of two compact stellar objects to form a black hole in a process that releases about a solar mass worth of gravitational potential energy and which generates a highly relativistic magnetized jet (Mészáros 2012).

Given the vast range in energies, scales, and details spanned by these diverse systems, it is intriguing that accreting systems seen in nature make such extensive and ubiquitous use of energetic outflows as part of their evolution. Is there a universal mechanism at work that almost guarantees that jets will be present? What is the basic physical mechanism that creates and drives jets across all of these scales? What role do they play in the evolution of such accreting compact systems? The fact that all these systems share common physical components—central gravitating sources, accretion discs, and magnetic fields—provides the unifying foundation on which theoretical models are constructed and observationally tested.

Magnetic fields play a critical role in creating, driving, and directing the evolution of jets. Indeed, the successful existing models of jets in these diverse systems invoke magnetic fields threading central objects such as gaseous accretion discs or rapidly rotating central objects themselves (see reviews by, e.g., Shu et al. 2000; Pudritz et al. 2007). The release of gravitational potential energy during accretion is a key power source for most of these jets. The great utility of magnetic torques is that they can readily tap into the shear energy available in the accretion discs that form as a consequence of gravitational collapse. Models of magnetized outflows from accretion discs in AGN, developed by Blandford and Payne (1982), were applied soon after to understanding jets in protostellar systems (Pudritz and Norman 1983, 1986). These theories envisage that material from a disc, in being forced to co-rotate with a threading magnetic field, is flung outwards along sufficiently inclined, open field lines. The gas continues to accelerate as long as the field is strong enough to maintain co-rotation with the central disc, or rotor. The inertia of this matter moving along the field line causes the field to wrap up in a toroidal-like structure—creating a helical field. It is this dominant, outer toroidal field component that exerts a pinching force upon the jet, collimating it towards the axis. A second major prediction of this basic theory is that jets should be extracting the angular momentum of their sources, and jet rotation is a key prediction of the models. Another abundant source of energy resides in the rotation available in rapidly rotating black holes or pulsars, and once again it is known that electromagnetic torques can tap rotational energy efficiently (Blandford and Znajek 1977). So for rapidly accreting or spinning objects, large scale magnetic fields appear to be nature's preferred tool in extracting energy and angular momentum and directing them in the formation of ordered, magnetically collimated, outflows.

Observations of crucial properties of jets, such as their densities, temperatures, velocity structure, and magnetic fields, vary quite considerably between protostellar and extragalactic systems. In the former, forbidden line emission (FLER) produced by shocks in the jet allows one to measure directly everything with the exception of the magnetic field. On the other hand, extragalactic jets emit synchrotron radiation which arises from the relativistic electrons that spiral around magnetic field lines. This radiation allows one directly to measure the direction of the magnetic field in the jet, and to estimate its strength. However, the absence of distinct emission lines from the outflowing material itself in AGN jets prevents one from measuring the other crucial physical variables. It is for this reason that protostellar systems have become such important laboratories for jet studies. Until 2010, these limitations prevented direct observational comparison of these classes of jets, and, as a result, meant that it was impossible to test whether or not jets have a universal magnetic field based mechanism. This all changed with the discovery of synchrotron radiation from the protostellar jet, HH 80-81 (Carrasco-González et al. 2010). Relativistic electrons in otherwise non-relativistic jets arise from acceleration in the jet's terminal shock. The results show that the magnetic structures of protostellar and AGN jets are similar—with a field parallel to the jet axis at the centre and a surrounding helical field towards the jet edges.

In this paper, we examine the structure and role of magnetic fields in the physics of jets—from their launch on discs or compact rotors, to their dynamics and evolution, and finally to a consideration of how such strongly magnetized flows affect their environments on these various scales. We begin our analysis with non-relativistic systems—the protostellar jets that are a ubiquitous and vital component of any star-forming region (Sect. 2). We also summarize the theory and simulations of relativistic jets in the context of AGNs. (We do not consider the impulsive, short-lived jets in gamma-ray bursts, although much of the same physics is expected to apply, as the observational challenges are very different; the reader is referred to reviews by Piran (2004) and Granot (2007).) In Sect. 3 we then turn to relativistic systems and discuss the measurement of the structure and dynamics of AGN jets on highly compact (pc) scales, probing the conditions near at the source of the jets using VLBI. The magnetic structure of jets on these scales is probably a reflection of their internal dynamics and not of the shocks which occur throughout the body of the jets. Indeed, the mounting evidence for helical magnetic fields in extragalactic jets, taken with the newly discovered evidence for helical fields in protostellar jets discussed above, strongly suggest that the physical mechanism of collimation may be the same. Finally, we go out to the scales of jet termination (Sect. 4), focussing on the evidence for magnetic field properties on these largest scales and exploring the connection between magnetic fields in jets in the hearts of AGNs with jets at the largest scales (1 kpc–1 Mpc). We conclude our study with a discussion of future prospects in this field (Sect. 5).

2 Magnetic Launching of Jets

2.1 Observed Properties of Protostellar Jets

Observations of the launch regions even of protostellar jets face significant observational challenges. For the nearest regions of star formation and jets in the Taurus molecular cloud (distance 140 pc), 1″ resolution corresponds to a physical scale of 140 AU. Theoretical models discussed below picture a jet acceleration region which is of the order of several tens of AU (0.1″), and the launch region in the very inner regions of the disc (several AU) that may be an order of magnitude smaller (0.01″). Jets around young stars have been studied on scales ranging from tens of AU to 5 pc (Bally et al. 2007).

Hubble Space Telescope (*HST*) observations of jets from YSOs have advanced our experimental understanding of jets enormously over the last decade, probing physical scales down to tens of AU. There are four basic types of observations that point towards the strong coupling of jets with underlying accretion discs—the correlation of jets with discs, the onion-like velocity structure of jets, jet rotation, and the strong link between the rates of mass transport in the jet to the accretion rate in the underlying disc. The advent of high resolution, spectro-imaging of jets by the *HST* and ground-based observatories equipped with adaptive optics has revealed their remarkable internal structure. Jets are strongly correlated with the presence of protostellar accretion discs around their host stars—at least during the first million years or so of their pre-main sequence evolution. Protostellar jet properties can be directly probed by observations of the forbidden line emission that they produce. Optical forbidden lines of oxygen, sulphur and nitrogen, for example, are produced in very specific ranges of density and temperature (see Ray et al. 2007, for a review). The spectro-imaging results indicate that jets have an "onion-like" velocity structure, with the highest-speed outflow component, up to 400 km s^{-1} in some sources, found at the core of the jet, while the lower-velocity components are found systematically farther from the outflow axis. The highest-resolution (*HST*) studies of the launch region of TTS jets (about 50 AU) show gas densities $> 2 \times 10^4$ cm^{-3}, modest electron temperatures ($T \simeq 2 \times 10^4$ K), and low ionization levels (0.03–0.3) (Coffey et al. 2008). As we shall see, the onion-like velocity structure seen close to the source is a reflection of the differential rotation of the underlying Keplerian disc. Higher-resolution observations, down to the 1-AU scales predicted for jet launch, would be highly desirable.

The discovery of jet rotation is perhaps one the most significant advances of the last decade (Bacciotti et al. 2002; Coffey et al. 2004). By orienting the slit of the spectrograph perpendicular to the jet axis at some distance along the jet, velocity asymmetry that is compatible with jet rotation can be observed. Shifts range from 5–25 km s^{-1} at positions of 25 AU along the jet. From the point of view of jet dynamics, these observations allow one to measure the amount of angular momentum that is being carried by the jet—which ranges from ~ 60 % to ~ 100 % of the angular momentum that would be carried through the disc by the observed disc accretion rates. The ratio of mass carried in the jet to that moving through the underlying accretion disc, $\dot{M}_{jet}/\dot{M}_a \simeq 0.01$–0.07 (Coffey et al. 2008).

The strength of the magnetic field in the HH80-81 jet is estimated from the synchrotron emission (on the basis of equipartition arguments: see Sect. 4.2), to be 0.2 milligauss (Carrasco-González et al. 2010). Even though the temperatures of jets are quite low, the mechanism of diffusive shock acceleration has been shown to be able to accelerate electrons moving at a few hundred km s^{-1} up to relativistic speeds. If this technique can be exploited, we may indeed be on the threshold of being able to measure all of the basic physical quantities that are needed to test the physics of protostellar jets.

2.2 Structure and Dynamics of Non-relativistic Jets

At the heart of the question of why jets are ubiquitous is the fact that rapidly rotating magnetized bodies undergo strong spin-down torques exerted by the threading fields. The basic theory of magnetized winds from rapidly rotating stars was first worked out in the seminal paper by Mestel (1968). This theory focuses on stationary solutions of axisymmetric flows from magnetized bodies—as applied in particular to rotating stars. It has turned out that this theoretical framework is very useful in analyzing the structure of jets and of the structure of their magnetic fields.

The overall dynamics of jets, and the physics of how they are launched, is complicated. However, significant progress has been made by the application of the theory of stationary, 2D (axisymmetric) MHD flows. The usefulness of this highly simplified treatment arises from the fact that we can derive conservation laws that turn out to be important even for understanding the behaviour of stationary jets in 3D. First we decompose vector quantities into poloidal and toroidal components (e.g. magnetic field $\mathbf{B} = \mathbf{B_p} + B_\phi \hat{\mathbf{e}}_\phi$). In axisymmetric conditions, the poloidal field $\mathbf{B_p}$ can be derived from a single scalar potential $a(r, z)$ whose individual values, $a = \text{const}$, define the surfaces of constant magnetic flux in the outflow and can be specified at the surface of the disc (Pelletier and Pudritz 1992).

The joint conservation of mass and magnetic flux along a field line can be combined into a single function k that is called the "mass load" of the wind, which is a constant along a magnetic field line:

$$\rho \mathbf{v_p} = k \mathbf{B_p}. \tag{1}$$

This function represents the mass load per unit time, per unit magnetic flux of the wind. For axisymmetric flows, its value is preserved on each ring of field lines emanating from the accretion disc. Its value on each field line is determined by physical conditions—including dissipative processes—near the surface of the rotor.

The mass load plays a central role in jet dynamics (see Sects. 3 and 4) and controls jet rotation, collimation, and angular momentum extraction. It may be more revealingly written as

$$k(a) = \frac{\rho v_p}{B_p} = \frac{d\dot{M}_\text{w}}{d\Psi}, \tag{2}$$

where $d\dot{M}_\text{w}$ is the mass flow rate through an annulus of cross-sectional area dA through the wind and $d\Psi$ is the amount of poloidal magnetic flux threading through this same annulus. The mass load profile (as a function of the footpoint radius r_0 of the wind on the disc) is determined by the physics of the underlying disc and imposes an important boundary condition for all aspects of jet physics.

The conservation of angular momentum along each field line leads to the conserved angular momentum per unit mass;

$$l(a) = r v_\phi - \frac{r B_\phi}{4\pi k} = \text{const}. \tag{3}$$

The form for l reveals that the total angular momentum is carried by both the rotating gas (first term) as well by the twisted field (second term), the relative proportion being determined by the mass load.

The value of $l(a)$ that is transported along each field line is fixed by the position of the Alfvén point in the flow, where the poloidal flow speed reaches the Alfvén speed for the first time ($m_\text{A} = 1$). Hence, the value of the specific angular momentum is found to be

$$l(a) = \Omega_0 r_\text{A}^2 = (r_\text{A}/r_0)^2 l_0, \tag{4}$$

where $l_0 = v_{K,0} r_0 = \Omega_0 r_0^2$ is the specific angular momentum of a Keplerian disc. (Note, however, that this treatment can easily be generalized to sub-Keplerian discs.) For a field line starting at a point r_0 on the rotor (disc in our case), the Alfvén radius is $r_A(r_0)$ and this constitutes a lever arm for the flow. The result shows that the angular momentum per unit mass that is being extracted from the disc by the outflow is a factor of $(r_A/r_0)^2$ greater than it is for gas in the disc. For typical lever arms, one particle in the outflow can carry the angular momentum of ten of its fellows left behind in the disc.

The conservation of energy along a field line is expressed as a generalized version of Bernoulli's theorem. Since the terminal speed $v_p = v_\infty$ of the wind is much greater than its rotational speed, and for cold flows, the pressure may also be ignored, one finds the result:

$$v_\infty \simeq 2^{1/2}\Omega_0 r_A = (r_A/r_0)v_{esc,0}. \tag{5}$$

There are several important consequences for jet dynamics that follow from this simple scaling. The first is that the terminal speed exceeds the *local* escape speed from its launch point on an accretion disc by the lever arm ratio. Furthermore, the scaling predicts that the terminal speed scales with the Kepler speed as a function of radius, so that the flow will have an onion-like layering of velocities, with the largest velocities inside and the smallest on larger scales. As we have already noted, such velocity structure is actually observed in the dynamical structure of well-resolved jets. Finally, the terminal speed depends on the depth of the local gravitational well at the footpoint of the flow (the dependence on the local Keplerian rotation speed in the case of a disc rotor)—implying that it is essentially scalable to flows from discs around YSOs of any mass and therefore universal.

2.3 Determining the Jet Launch Region

A very useful combination of the energy and angular momentum conservation that we derived above is given by the Jacobi constant along a field line (e.g., Pelletier and Pudritz 1992); $j(a) \equiv e(a) - \Omega_0 l(a)$. This expression has been used (Anderson et al. 2003) to constrain the launch region on the underlying Keplerian disc. Observers measure the jet rotation speed, $v_{\phi,\infty}$ at a radius r_∞, and the poloidal speed of the jet at this point, $v_{p,\infty}$. Evaluating j for a cold jet at infinity and noting that its value (calculated at the foot point) is $j(a_0) = -(3/2)v_{K,0}^2$, one can solve for the Kepler rotation at the point on the disc where this flow was launched:

$$\Omega_0 \simeq v_{p,\infty}^2/(2v_{\phi,\infty}r_\infty). \tag{6}$$

When this relation is applied to the observed rotation of the Large Velocity Component (LVC) of the jet DG Tau (Bacciotti et al. 2002), this yields a range of disc radii for the observed rotating material in the range of disc radii, 0.3–4 AU, and the magnetic lever arm is $r_A/r_0 \simeq 1.8$–2.6. These results show that an extended region of the disc can give rise to the jet, and that, moreover, the lever arm is in the predicted range for the efficient launching of jets.

2.4 The Disc/Jet Connection: Magnetic Torques on Discs

The magnetized wind that is accelerated off the rotor carries away the rotor's angular momentum. In the case of an accretion disc, the angular momentum equation for the accretion disc undergoing a purely external magnetic torque (i.e., viscous torque neglected) may be written:

$$\dot{M}_a \frac{d(r_0 v_0)}{dr_0} = -r_0^2 B_\phi B_z|_{r_0,H}, \tag{7}$$

where we have ignored transport by MRI turbulence or spiral waves. Note that the toroidal field plays a fundamental role in disc structure and accretion because it is so central to the action of a magnetic torque upon the disc. By using the relation between the poloidal field and outflow on the one hand, as well as the link between the toroidal field and rotation of the disc on the other, the angular momentum equation for the disc yields one of the most profound scaling relations in disc wind theory, namely, the link between disc accretion and mass outflow rate:

$$\dot{M}_{\mathrm{a}} \simeq (r_{\mathrm{A}}/r_0)^2 \dot{M}_{\mathrm{w}}. \tag{8}$$

The observationally well-known result that, in many systems, $\dot{M}_{\mathrm{w}}/\dot{M}_{\mathrm{a}} \simeq 0.1$ is a consequence of the fact that lever arms are often found in numerical and theoretical work to be $r_{\mathrm{A}}/r_0 \simeq 3$—the observations of DG Tau being a perfect example. Finally, we note that the angular momentum that is observed to be carried by these rotating flows (e.g., in DG Tau) is a consistent fraction of the excess disc angular momentum—from 60–100 % (Bacciotti et al. 2004), which is consistent with the high extraction efficiency discussed here.

In reality, the evolution of discs is dictated by both viscous as well as wind torques. The magnetic torque exerted on the underlying disc has important effects upon its structure (Ferreira 2002; Ferreira and Casse 2004). Compared to standard α-disk models, Combet and Ferreira (2008) show that outflow torques modify the underlying radial structure of discs, resulting in column density jumps of a couple of orders of magnitude in going from the outer SAD solution to the inner-jet dominated zone of the disc. The jet-dominated region is also cooler and thinner.

2.5 Jet Collimation and Toroidal Fields

In the standard picture of hydromagnetic winds, collimation of an outflow occurs because of the increasing toroidal magnetic field in the flow resulting from the inertia of the gas. Beyond the Alfvén surface, the ratio of the toroidal field to the poloidal field in the jet is of the order $B_\phi/B_p \simeq r/r_{\mathrm{A}} \gg 1$, so that the field becomes highly toroidal beyond the Alfvén surface. Collimation is achieved by the tension force associated with the toroidal field which leads to a radially inwards directed component of the Lorentz force (or "z-pinch"); $F_{\mathrm{Lorentz,r}} \simeq j_z B_\phi$.

The current carried by a jet can be easily written down:

$$I = 2\pi \int_0^r j_z(r', z') dr' = (c/2) r B_\phi. \tag{9}$$

The link between magnetic field structure and the collimation of jets was made by Heyvaerts and Norman (1989), where it was shown that two types of solution are possible, depending upon the asymptotic behaviour of the total current intensity in the jet. In the limit that $I \to 0$ as $r \to \infty$, the field lines are paraboloids which fill space. On the other hand, if the current is finite in this limit, then the flow is collimated to cylinders. The collimation of a jet therefore depends upon its current distribution—and hence on the radial distribution of its toroidal field.

It can be shown (Pudritz et al. 2006) that, for a power-law distribution of the magnetic field in the disc, $B_z(r_0, 0) \propto r_0^{\mu-1}$, with an injection speed at the base of a (polytropic) corona that scales as the Kepler speed, that the mass load takes the form $k \propto r_0^{-(1+\mu)}$. In this regime, the current takes the form $I(r, z) \propto r_0^{-(\mu+1/2)}$. Thus, the current goes to zero for models with $\mu < -1/2$, and these therefore must be wide angle flows. For models with $\mu > -1/2$, however, the current diverges, and the flow should collimate to cylinders.

These results predict that jets should show different degrees of collimation, depending on how they are mass loaded. As an example, neither the highly centrally concentrated, magnetic field lines associated with the initial split-monopole magnetic configuration used in simulations by Romanova et al. (1997), nor the similar field structure invoked in the X-wind model (Shu et al. 2000) should become collimated in this picture. On the other hand, less centrally (radially) concentrated magnetic configurations such as the potential configuration of Ouyed and Pudritz (1997) and Blandford and Payne (1982) should collimate to cylinders. By varying both the density profile of the disc as well as its magnetic structure, Fendt (2006) showed simulations in which highly collimated outflows from a "flat" radial disk structure produced unsteady knots in the jet.

The magnetic collimation picture, although it has considerable support from theory and simulations, has been questioned by Spruit (2010), who notes that the toroidal magnetic pressure in jets will force them to expand. Collimation in this picture results from magnetic fields and pressure in the external medium. Simulations do indeed provide evidence for a regime in which magnetic bubbles associated with simulations of protostellar outflows are observed to expand radially. These cases arise in simulations of disc winds that produce weak toroidal jet fields (Seifried et al. 2012). Strong toroidal fields in these jets give rise to well collimated flows.

2.6 Computer Simulations: Magnetic Field Structure of Jets and Their Stability

Numerical simulations have played an enormous role in exploring the origin and evolution of jets. The pioneering stage of simulations of jets from discs took place in the mid 1980s through the 1990s, with the advent of magnetohydrodynamics (MHD) codes such as ZEUS. These early simulations recognized that, since the mass load of the disc outflow is a key variable that controls jet dynamics and structure, then the disc can be taken as a boundary condition for the flow, with the initial magnetic field structure and mass loading defined across this boundary. Simulations demonstrated most aspects of the theory outlined above—the collimation of the jet, the concentration of jet density towards the axis of the flow, the acceleration to high speeds, the "onion-like" velocity structure of the jet, and the dominance of the toroidal magnetic field structure can be seen (e.g., Ouyed et al. 1997; Ouyed and Pudritz 1997; Romanova et al. 1997; Krasnopolsky et al. 1999; Fendt 2006). Highly episodic behaviour of jets arises as a consequence of sufficiently low mass loading of the jets (Ouyed et al. 1997).

Figure 1 shows ZEUS 3D simulations of a jet being launched from an underlying disc (Staff et al. 2010). The two panels show snapshots of the magnetic field line geometry and jet density that arises within two particular jet models—those of Ouyed and Pudritz (1997) and Blandford and Payne (1982). The field lines have a distinct toroidal component that dominates the structure of the overall field on larger scales. Poloidal magnetic field dominates towards the axis of the outflow. The bow shock created by the jet is clearly seen in both cases. The general structure of these models is rather similar. The densest part of the jet are surrounded by a diffuse cavity-like region that is dominated by the toroidal magnetic field. The more extended outer material moves more slowly than the jet core.

If jets are dominated by toroidal fields far from the jet axis, why are they stable? Theoretical studies of the linear stability of jets show that they should become unstable when their speeds exceed the local Alfvén velocity (Ray 1981). This was further demonstrated by 3D simulations of idealized radio jets by Hardee and Rosen (1999). It turns out, however, that toroidally dominated jets stabilize themselves through nonlinear processes. Simulations show that jet instability, particularly in the form of $m = 1$ helical modes, builds up beyond

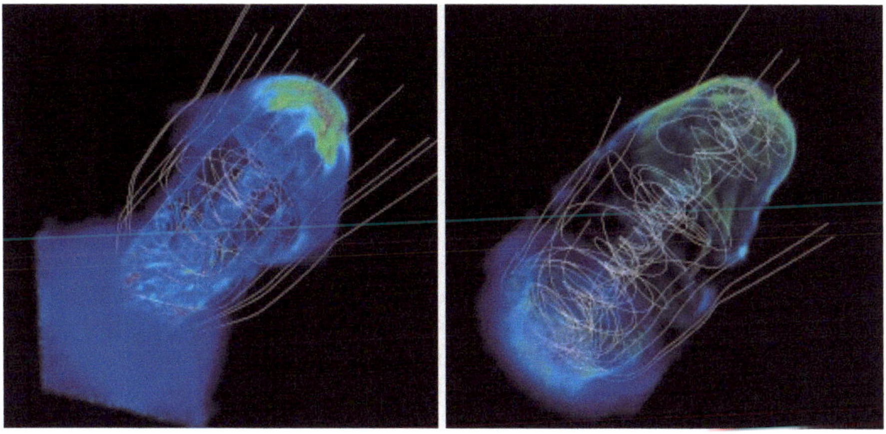

Fig. 1 Density and magnetic field lines for the Ouyed and Pudritz (1997) and Blandford and Payne (1982) models for the distribution of disc magnetic fields. The snapshots are shown as the jets have propagated out to 60 AU in the medium surrounding the magnetized disc. The protostar and the accretion disc are in the lower left corners of the panels, and hidden by the dense disc corona. Note that the core of the jet is dominated by poloidal field lines, while the outer regions are predominantly toroidal. Adapted from Staff et al. (2010)

the Alfvén surface (Ouyed et al. 2003). These are stabilized by the magnetic tension that stretches field in the linearly unstable region, which ultimately increases the local Alfvén speed again. The net result is a jet that is non-linearly stabilized by this mechanism, as well as by the "backbone" provided by the poloidal field that dominates near the outflow axis.

Since jets and accretion discs are highly coupled, jet torque and evolution will directly affect the disc. Hence, there has been a major effort over the last decade to simulate jets and their discs as a single dynamical system (e.g., Kudoh et al. 2002; Casse and Keppens 2002; von Rekowski and Brandenburg 2004). The important advantage of this approach is that one can avoid *ad hoc* descriptions of the mass loading and disc magnetic field line structure since these are self-consistently computed. These simulations show that jet structure and dynamics retain the basic features that the simpler simulations capture. Moreover, simulations by Zanni et al. (2007) which include the disk, and follow the evolution of jets using the Adaptive Mesh Refinement code FLASH, find that more than 90 % of the gravitational potential energy liberated in the accretion flow is released into the jet. Thus jets are highly efficient in extracting both energy and angular momentum from the underlying accreting systems.

2.7 Jets from Magnetized Star-Disc Systems

So far we have emphasized disks as the underlying magnetized engines for jets. However, collimated outflows may derive from both the central stars as well as their surrounding disks. T-Tauri stars (TTS) are well known to be significantly magnetized (Johns-Krull 2007). As evidence that this may be occurring, we focus on the fact that TTS are also known to rotate very slowly for the amount of angular momentum that they are known to be accreting. This has long been known to imply that there must be some external torque being exerted upon the rotating star, carrying off or intercepting the angular momentum. It has been proposed that this torque takes the form of an interaction of the star's magnetosphere and the surrounding disk through closed magnetic field lines (Königl 1991). Recent work suggests that most of the field in such interaction is readily opened up by shear in the disk, leading to very little spin-down torque and connection between star and disk (e.g., Matt and

Pudritz 2004). The interaction between inner disk and stellar magnetosphere has also been hypothesized to give rise to a so-called "X-wind", wherein the origin of jets and outflows is pictured to be an MHD outflow—based on the principles we discussed above—and originating from the very inner edge of the disk (Shu et al. 2000). In this picture, dipolar stellar magnetic field lines interacting with the inner edge of the disc are opened, leading to an outflow which intercepts angular momentum flowing inwards through the disc and driving it out of the system in the outflow. One difficulty with this models is in accounting for the huge amount of angular momentum that jets are observed to transport—60 % or more of the angular momentum transport through the disc at scales of 2 AU or so (Coffey et al. 2008); another issue is the difference between observed magnetic field strengths and the field strengths needed to drive the X-winds (Johns-Krull 2007). A third class of models has been proposed, wherein an accretion-powered wind from a magnetized, central star carries off the disk angular momentum (Matt and Pudritz 2005). The energy driving this wind is ultimately drawn from the gravitational energy that is released by the infalling material from the disc as it falls along magnetospheric field lines connecting the star and the disc. A strong flux of Alfvén waves stimulated by the impact of the accretion flow upon the star creates a wave-driven wind (Matt and Pudritz 2005; Matt and Pudritz 2008; Cranmer 2008).

Simulations of such combined magnetized star-disk systems have been carried out in which both the central rotating magnetized star and the surrounding disk are treated as boundary conditions for the flows (Fendt 2009). Compared to the case of pure disk winds, the overall outflow is initially less collimated. This changes with time, however, as a highly collimated flow emerges as a new dynamical state across the grid when a stellar wind is included. The magnetic interaction between the two wind components gives rise to reconnection and large-scale flares as well, and the time-scale for the flaring is long—of the order of hundreds of inner disc rotation periods. Finally, if the outflow is dominated by the stellar wind, the outflow is either too weak or too high for low and high mass loads respectively. The disk jet is required to have stable collimated flow out to large scales.

The interaction of spinning, magnetized stars with their surrounding disks have been extensively simulated for axisymmetric configurations (Romanova et al. 2002; Bessolaz et al. 2008). Magnetospheric accretion has been likened to "funnel flow" wherein matter falls onto the star along field lines that are connected to the star and co-rotate with it. The matter moves due to the gravitational force, which dominates over the centrifugal force. The lifting of material off of the disc and the onset of the funnel flow starts at the radius at which the disk is slowed due to interaction with the more slowly spinning magnetosphere. These simulations showed that the interaction is time-dependent, with bursts of accretion followed by quiescent periods in which material from the disc piles up against the magnetosphere. Models usually assume that the disks are characterized by a model "α-disc" viscosity. In self-consistent 3D simulations, accretion flow from a turbulent disk onto a tilted magnetized star, where turbulence is driven by the MRI instability, has been modelled (Romanova et al. 2012). The pattern of alternating compression due to accretion and reconnection is also seen in these 3D simulations. The magnetosphere truncates the disk at a few stellar radii. The magnetic structure of the disk outside of the stellar magnetosphere has a toroidal field B_ϕ that is a few times larger than the poloidal field component. As we have noted several times, the transport of angular momentum by twisted field lines is an important part of the outflow mechanism. In the star-disc interaction, the magnetic stresses dominate the matter stresses in the disk and the pattern is quite inhomogeneous. This goes together well with the predictions outlined in the basic theory above. Finally, recent simulations of star-disc systems report the launch of fast outflow at the interface in systems undergoing very high accretion rates—wherein the magnetic field in the star is strongly compressed (Lii et al. 2012). No X-winds are observed for the much lower accretion rates that dominate most of the star's accretion (Bessolaz et al. 2008).

2.8 Discs and Jet Formation During Gravitational Collapse

We now leave the domain of stationary outflows, and examine how they arise in the fully dynamical environments that characterize the birth of stars. Protostellar jets appear during the earliest stages of star formation and are the first obvious manifestation that gravitational collapse and star formation are underway. No 3D simulations of purely hydrodynamic collapse show that sustained jets can be produced in collapsing systems. This may be why they were missed in early hydrodynamical simulations of star formation. Add a threading magnetic field, however, and the launch of energetic jets and outflows is readily observed. The formation and evolution of jets takes place in a few related phases; (i) the formation and early evolution of a rotating, magnetized dense core, which is usually found embedded within larger scale filaments within molecular clouds; (ii) the extraction of angular momentum from the rotating core by torsional Alfvén waves during the early formation and collapse stages, which significantly de-spins the core; (iii) the gravitational collapse of a magnetized core with the ensuing formation of a protostellar accretion disc; (iv) the creation and launch of an outflow from the disc as it forms; and (v) the accretion of the disc onto the central star and with it, the gradual disappearance of the jet.

An important measure of the magnetization of the magnetized core is given by the so-called mass-to-flux ratio, which is the ratio of the gravitational to the magnetic energy of the system. Systems with supercritical ratios (values greater than unity) cannot be supported against collapse by their threading fields. The ratio can be written as (Mouschovias 1976)

$$\mu = \frac{M_{\mathrm{core}}}{\Phi_{\mathrm{core}}} \bigg/ \left(\frac{M}{\Phi}\right)_{\mathrm{crit}} = \frac{M_{\mathrm{core}}}{\int B_z dA} \bigg/ \frac{0.13}{\sqrt{G}}. \tag{10}$$

A second parameter of importance is the ratio of the rotational to the thermal energy of the core, known as β_{rot}. The phase space spanned by these two parameters turns out to divide up the physical properties of discs.

During the gravitational collapse of magnetized cores, the magnetic field threading the region is wrapped up—much like a twisting rubber band. As this happens, an outward flow of torsional Alfvén waves is created which transports angular momentum. Analytic solutions of the braking torque that arises for idealized, disc-like rotors have been written down. These have been found to fit the early stages of simulations of the collapse of a magnetized, Bonner-Ebert sphere (Banerjee and Pudritz 2006). As the collapse picks up, the winding of the field becomes particularly strong at the accretion shock. It is from this region that the first strong aspects of outflow begins. Soon after this happens, an inner region of outflow is launched which is centrifugally driven.

For fields that are sufficiently supercritical, the magnetic braking can be so significant that the disc material is left rotating at significantly sub-Keplerian levels. This has raised the question of whether or not discs can even form in the reasonably magnetized environments that are found in regions of star formation (Hennebelle and Fromang 2008; Mellon and Li 2008). The results depend quite sensitively on the initial conditions chosen—i.e., the mass-to-flux ratio. At values of $\mu \simeq 2$–10, the resulting systems are strongly affected. This intriguing result points to the importance of having appropriate astrophysical initial conditions. Recent simulations that feature long-term evolution of discs using "sink" particles find that in fact a reasonably Keplerian initial disc can form during the long-term evolution of the collapse (Duffin et al. 2012). A grid of models for the formation of massive stars shows that reasonably quickly rotating cores and lower magnetized cores produce Keplerian discs (Seifried et al. 2011).

Figure 2 shows a zoomed-in picture of the magnetic structure that arises in the collapse of a solar mass system of a low mass star formation (Duffin et al. 2012). The jet produced in this simulation reaches out to the extent of the initial Bonner-Ebert sphere, of about 10^4 AU in diameter. The gravitational collapse has pulled in the field lines which thread the disc. A transient, large disc of 2000 AU in extent has formed, which ultimately fragments into a ring and an inner 100 AU scale disc within an outer disc rotation period. The innermost region of the disc is seen to warp, which is a consequence of the torques induced by the magnetized wind. The outflow, as a consequence, precesses with the warp in the disc. Precessing jets are common and the simulation represented in this image shows, for the first time, that magnetic fields can cause this behaviour.

The launch mechanism of the outflows during collapse is under some current debate. For low-mass star formation ($M_{\rm core} \sim 1 M_\odot$), the generation, evolution and properties of protostellar outflows have been studied in great detail over the last years (e.g., Banerjee and Pudritz 2006; Mellon and Li 2008; Hennebelle and Fromang 2008; Duffin and Pudritz 2009). The two basic mechanisms that have been discussed in the literature focus on whether the outflow is driven by centrifugal acceleration (Blandford and Payne 1982; Pudritz and Norman 1986; Pelletier and Pudritz 1992; Ferreira 1997) or by the pressure of the toroidal magnetic field that was wrapped up in the initial collapse (Lynden-Bell 1996, 2003). The second mechanism—known as a tower flow—is highly transient and does not lead to sustained outflow moving out from the source. It is based on the study of equilibria of highly wound magnetic structures. General energy theorems demonstrate that they form tall magnetic towers, the height of which grows with every turn at a velocity related to the circular velocity in the accretion disc. The pinch effect amplifies the magnetic pressures toward the axis of the towers.

A more generalized criterion for outflow in disc wind theory, assuming that the gas is not strictly co-rotating with the field, has recently been proposed (Seifried et al. 2012). This model shows that disc-wind theories can have acceleration mechanisms that are driven by centrifugal forces as well as toroidal fields. Both magnetic tower and disc-wind theories can be commonly understood as a consequence of the same disc-wind models. It is of particular interest that the MHD flow equations discussed above actually have two aspects to them. Close to the outflow axis, it may be shown that the toroidal field generation is not significant, and there the outflow is driven like a centrifuge. Farther away, however, a toroidal field component is inescapably produced. It can be shown that the pressure gradient associated with this contributes to pushing a continuous outflow from the outer regions of the disc.

The two regions of the jet may be seen in Fig. 3 where we compare the size of the region controlled by purely centrifugal effects with those arising from toroidal pressure gradients. This illustrates that magnetic fields in jets play different roles depending upon where they are found. The difference between these two snapshots is the region in the jet where gas is driven primarily by toroidal field gradients. Note that these dominate in the outer parts of the jet, whereas the centrifugal driving predominates in the region close to the jet axis.

2.9 Relativistic Considerations and Relation to Protostellar Jets

There is nothing in the physics of jets that suggests that relativistic sources, such as microquasars and AGN, and should have radically different driving mechanisms from those

Fig. 2 Magnetic field structure in a collapsing magnetized core. A flattened structure is produced by the gravitational collapse. The *purple contours* correspond to 3.3×10^{-17} g cm^{-3} ($n = 8.5 \times 10^6$ cm^{-3}) and *black contours* to 1.3×10^{-15} g cm^{-3} ($n = 3.4 \times 10^8$ cm^{-3}). *Yellow contours* correspond to outflowing velocities $v_z > 1.5$ km s^{-1}, and extend to a height of 10^4 AU. *Black lines* are magnetic field lines. Adapted from Duffin et al. (2012)

of protostellar systems—indeed, the dominant magnetic models make use of similar MHD equations. The fact that both a sufficiently rapidly rotating black hole and/or the surrounding magnetized disc are sources for relativistic jets has an obvious parallel with the protostellar systems. The theory and simulations of non-relativistic jets, as applied to protostellar systems, show some remarkable connections with the observations of AGN jets, which are discussed in Sects. 3 and 4. There are some important differences, however, which we discuss here.

The theory of relativistic jets shows that the essential parameter is the *magnetization* parameter (Michel 1969; Camenzind 1986),

$$\sigma = \frac{\Psi^2 \Omega_{\mathrm{F}}^2}{4\dot{M}c^3}, \tag{11}$$

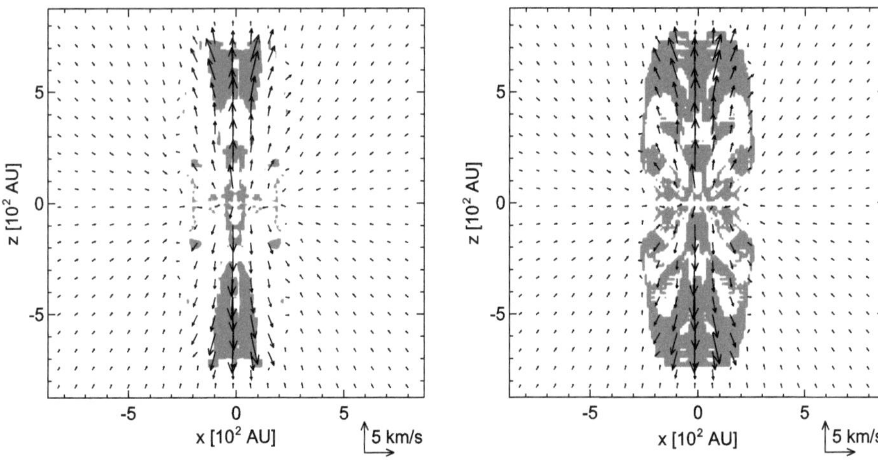

Fig. 3 *Left*: the region of jet that is driven by centrifugal acceleration—denoted in *grey*—in a simulation of a collapsing, solar mass, magnetized Bonner-Ebert sphere. *Right*: as *left*, but *grey regions* show the regions controlled by the full acceleration mechanism, including toroidal pressure gradients. Adapted from Duffin et al. (2012)

where the iso-rotation parameter $\Omega_{\rm F}$ is the angular velocity of the magnetic field lines. The function $\Psi = B_p r^2$ is a measure of the magnetic field distribution and $\dot{M} \equiv \pi \rho v_p R^2$ is the mass flow rate within the flux surface. Equation (11) demonstrates that the launch of a highly relativistic jet requires at least one of three conditions—rapid rotation, a strong magnetic field, and/or a comparatively low mass load. A strong magnetic field lowers the mass load in the jet, with the consequence that more Poynting flux can be converted into kinetic energy per unit mass flux.

In the case of a spherical outflow ($\Psi =$ constant) with negligible gas pressure one may derive the Michel scaling between the asymptotic Lorentz factor and the flow magnetization (Michel 1969),

$$\Gamma_\infty = \sigma^{1/3}. \tag{12}$$

Depending on the exact magnetic field distribution $\Psi(r,z)$, in a *collimating jet* the matter can be substantially accelerated beyond the fast magnetosonic point (Begelman and Li 1994; Fendt and Camenzind 1996). As a result, the power law index in Eq. (12) can be different from the Michel-scaling (Fendt and Camenzind 1996; Vlahakis and Königl 2003).

An essential difference between relativistic and non-relativistic flows is the role of electric fields. They are negligible in non-relativistic flows but are comparable to magnetic fields in the relativistic case. The issue is that these fields can lead to the decollimation of the jet. Current-carrying relativistic jets have, however, been demonstrated to collimate (Chiueh et al. 1991). Relativistic jets have another feature with no counterpart in non-relativistic systems, namely the *light cylinder*, located at the cylindrical radius $r_{\rm l} = c/\Omega_{\rm F}$. At the light cylinder the velocity of the magnetic field lines "rotating" with angular velocity $\Omega_{\rm F}$ coincides with the speed of light. Outside the light cylinder, the magnetic field lines "rotate" faster than the speed of light. As the field line is not a physical object, the laws of physics are not violated. The light cylinder is interpreted as the Alfvén surface in the limit of vanishing matter density (force-free limit). Crucially, *the location of the light cylinder determines the relativistic character of the magnetosphere. If the light cylinder is comparable to the dimensions of the object investigated, a relativistic treatment of MHD is required.*

Simulations of relativistic axisymmetric MHD disc winds have demonstrated that jets do indeed collimate, with half opening angles of 3–7° (Porth and Fendt 2010). These authors set up a relativistic analogue of the (Ouyed and Pudritz 1997) simulations, and emphasized flows far enough away from the central black hole that general relativistic effects could be ignored. The outflow has a mixed character—a combination of the Blandford-Payne, centrifugally driven wind with a toroidal pressure-dominated flow of the type discussed by Lynden-Bell (1996). Lorentz factors of the order $\Gamma = 6$ are produced. The outflow has an interesting substructure, with a narrow relativistic jet towards the axis that is surrounded by a sub-relativistic flow launched from further out in the disc. This structure agrees with the observations of AGNs on the pc scales, as we discuss in Sect. 3. Finally we note that the toroidal fields that are generated in 3D simulations of jets seem to act as a sheath that protects the spine of the jet core from losing momentum to the surrounding medium (Mignone et al. 2010).

The advent of 3D general relativistic, MHD (GRMHD) codes (e.g. McKinney 2006; McKinney et al. 2012) has opened up the study of the role of black holes in the origin and evolution of AGN jets. As for the purely relativistic case, one of the primary questions is to determine how relativistic the jets become, and to understand what maintains their stability and collimation. It is generally thought that there are two regimes for AGN jets: a magnetized disk surrounding a black hole, as described by the Blandford and Payne (1982) picture, and/or the spinning black hole whose ergosphere is threaded by magnetic field, as originally envisaged by Blandford and Znajek (1977). Most theories and the simulations suggest that it is the spinning black hole that produces the highly relativistic jets, while lower-speed outflows arise from the surrounding disk. Recent GRMHD simulations (McKinney and Blandford 2009) model the initial threading magnetic field geometries for the spinning black hole as either a dipole or quadrupole configuration, while a thick torus is used to model an AGN disk. In addition to the need for significant spin of the central hole (the spin parameter of the hole is $a/M > 0.4$), their results demonstrate that the initial magnetic configuration also plays a significant role. Jets produced by an initial dipole structure can achieve high Lorentz factors, ($\Gamma > 10$) and remain highly collimated with an opening half-angle of $\theta_j \simeq 5°$ at 10^3 gravitational radii. This is a stable jet with the substructure dominated by a stable $m = 1$ mode, probably excited in the underlying turbulent torus.

However, this does not occur for a more complex, quadrupolar, initial magnetic geometry, where the jet does not reach steady realistic speeds ($\Gamma < 3$). The quadrupolar field leads to mass loading which considerably slows the jet, and makes it prone to disruption. This outflow has a weak and disorganized poloidal field and a much stronger (by factor of 10–40) toroidal field. The authors speculate that these differences with the dipole jet may have something to do with the observed morphology of extragalactic jets; in particular, their classification into weakly relativistic "FR I-type" jets associated with AGNs in more clustered regions, and highly relativistic "FR II-type" jets associated with isolated objects. The latter would be associated with the spinning black holes and the former with the underlying accretion disks.

There are many similarities between these relativistic results and the non-relativistic work and theory described earlier. The stability of jets over a large variety of scales seems assured for both regimes—they are able to run with stable substructure, dominated by an $m = 1$ mode, without falling apart. Both types of systems can remain very highly collimated, and this is related to their mass loading which is connected with their magnetic geometry. Moreover, toroidal magnetic field dominates the non-relativistic jet structure, which again is a likely consequence of the theory we have outlined in this Section.

In AGN systems, there is a concerted effort to understand jets on the smallest scales, which correspond to a few pc, in order to test theoretical models. The analysis described in this Section has shown that magnetic fields in jets are poloidal near the jet axis, and dominated by toroidal field farther out. This is a natural consequence of the acceleration and collimation mechanisms that should also apply to non-relativistic (e.g. protostellar) and relativistic (AGN) accretion systems. We are now just beginning to be able to simulate the evolution of compact jets to larger scales (several pc) which allows direct comparison with observation. Simulations indicate that the field in protostellar jets becomes highly wrapped near the bow-shock of the jet (Staff et al. 2010), and this may have consequences for understanding AGN jets as well. In the next Section, we turn to the parsec-scale structure of AGN jets, to connect observations with the theory and simulations presented above.

3 Structure and Dynamics of Parsec-Scale AGN Jets

3.1 Backdrop for Current Studies

Radio studies of jets from Active Galactic Nuclei (AGNs) have primarily focused on parsec scales (with centimetre-wavelength Very Long Baseline Interferometry, VLBI, giving milliarcsecond resolution) and kiloparsec (kpc) scales (with arcsecond-resolution interferometers such as the Very Large Array and MERLIN), with relatively few observations probing intermediate scales. This means that, although the pc- and kpc-scale observations in some ways tell a coherent and self-consistent story, in others they seem to show discrepancies, with the connection between the behaviours observed on the two scales not always being clear. One reason for this may be that, although the parsec- and kpc-scale jets form different parts of a single structure, the observations may be determined by different contributions from "global" properties intrinsically related to the jet and "local" properties due to localized perturbations, interactions with the surrounding medium, variations in the magnetic field and density of the ambient medium, etc.

Throughout most of the period in which AGN jets have been studied with VLBI, since the early 1980s, interpretations of observed features and behaviour have focused on the possibility that the observed jet properties are associated primarily with local agents, in particular, relativistic shocks propagating in the jets. From a certain point of view, this was very natural, and was motivated by factors such as the visibly inhomogeneous appearance of the jets themselves (they are often dominated by distinct components) and the common occurrence of variability. A theoretical framework arose in which the jet components were primarily shocks propagating along the jet, with various processes related to these shocks giving rise to the observed very rapid variability. This picture seemed to receive support from the earliest VLBI polarization observations, which showed that an appreciable number of AGN jet components have magnetic fields that are predominantly transverse to their jets—this was interpreted as reflecting the compression of an initially tangled magnetic field by a transverse shock, causing the field to become aligned in the plane of compression (Gabuzda et al. 1992).

More recently, various theoretical and observational studies have begun to explore the possibility that at least some of the polarization structures detected on parsec scales are associated with the intrinsic magnetic fields of the jets themselves. One area of such research involves theoretical simulations and observational searches for evidence of helical or toroidal magnetic fields carried by the jets, which are expected to form as a result of the joint action of the rotation of the central black hole and its accretion disc and the jet outflow. The presence of toroidal or helical magnetic fields should give rise to characteristic transverse intensity and linear polarization structures across the jets, as well as transverse gradients in

Fig. 4 Example of a derived dependence of B-field strength on distance from the jet base, approximately $B \propto r^{-1}$ (O'Sullivan and Gabuzda 2009a)

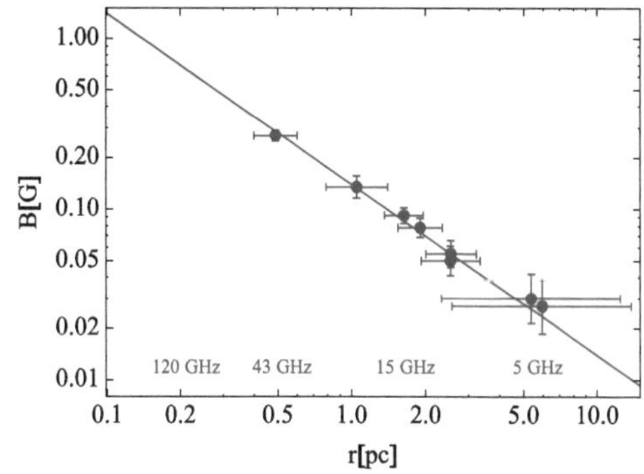

the observed Faraday rotation, due to the systematic change in the line-of-sight magnetic field across the jet.

Thus, one of the challenges currently faced by researchers studying AGN jets is to try to discern which features are primarily due to the action of local agents, such as shocks or interactions with the ambient medium, and which are primarily associated with the intrinsic magnetic field and other properties of the jets themselves. This Section will review recent observational results for high-resolution (VLBI) observations of AGN jets, together with some related theoretical simulations, in the framework of this challenge.

3.2 Estimates of VLBI-Core B-field Strengths

The direction of the magnetic field **B** giving rise to synchrotron radiation can be deduced from the direction of the associated observed electric vector position angle (EVPA), if the optical depth regime is known: in the absence of Faraday rotation, **B** is perpendicular to the EVPA in optically thin regions and parallel to the EVPA in sufficiently optically thick regions (see, e.g., Gabuzda and Gómez 2001). Thus, it is relatively straightforward to get an idea of the overall magnetic-field configuration even from VLBI observations at a single frequency, as long as the effects of Faraday rotation are not too severe. However, it is not possible to estimate the strength of the synchrotron B field directly using observations at a single frequency: multi-frequency observations are required, and even then, somewhat indirect techniques must be applied.

One of these is based on the frequency dependence of the position of the observed VLBI core in a Blandford–Königl jet (Blandford and Konigl 1979). In this picture, the VLBI core is essentially like a photosphere, and corresponds to the surface where the optical depth τ is equal to unity. The position of this $\tau = 1$ surface is located further down the jet at lower frequencies: $r \propto \nu^{1/k_r}$, where r is distance from the jet base, ν is frequency and k_r is a parameter that, in general, depends on the spectral index and the laws for the decline of the electron density and magnetic field with distance from the central engine (Konigl 1981). In practice, the observed VLBI core includes emission from both the vicinity of this $\tau = 1$ surface and the innermost jet, but the essence of this basic theoretical picture holds. In contrast to the behaviour of the core region, optically thin regions—in other words, regions in the jet—are expected to coincide at different frequencies. By aligning optically thin regions observed simultaneously at different frequencies and deriving the relative positions of the observed VLBI cores, the resulting "core shifts" can be used to derive estimates of the core magnetic fields, if reasonable estimates for the Doppler factor, viewing angle and jet opening angle are available (Lobanov 1998). This technique has only been applied to a handful of objects, but results obtained so far seem to give similar results for the cores of different AGNs, indicating that $k_r \approx 1$ in most cases (which corresponds to conditions not far from equipartition) and typical B fields at a distance of 1 pc from the jet base of about 0.15 Gauss (O'Sullivan and Gabuzda 2009a). Although these estimates are somewhat uncertain, due to uncertainties in the estimates of the Doppler factor and viewing angle, which cannot be derived directly from observations, the fact that similar core-region B fields are derived for various different AGNs suggests that these estimates are reasonably correct, at least to order of magnitude. These same studies have yielded evidence that the magnetic field falls off with distance from the jet base roughly as r^{-1} (Fig. 4).

3.3 Basic Observational Properties of the Polarization of AGN Jets

Although a wealth of polarization structure is observed among the variety of AGNs that have been studied with VLBI, it is possible to identify certain basic characteristics of the jet polarization. The most important of these is that, in a substantial majority of cases, the jet polarization is oriented close to parallel or perpendicular to the local jet direction (e.g., Lister and Homan 2005). The effect of Faraday rotation is usually modest outside the core regions, leading to polarization rotations of no more than 10–15° at centimetre wavelengths (see also Sect. 3.5). For this reason, this tendency for the jet polarization to be either parallel or perpendicular to the local jet direction was evident even in the earliest VLBI polarization observations carried out at a single wavelength (e.g., Gabuzda et al. 1992; Cawthorne et al. 1993). On average, there is a tendency for AGNs with relatively luminous optical line emission ("quasars") to display higher apparent superluminal speeds and to most often have primarily longitudinal jet magnetic fields, while those having relatively weak optical line emission ("BL Lac objects") generally display lower apparent jet speeds and most often have predominantly transverse jet magnetic fields (e.g., Gabuzda et al. 1994; Britzen et al. 2008).

Jet regions with oblique, or "misaligned", polarization (i.e., neither close to parallel or perpendicular to the local jet direction) that cannot be ascribed to Faraday rotation are sometimes observed, but are comparatively rare. Jets with extensive regions of longitudinal or transverse B field are observed. Little is known about the nature of "inter-knot" polarized emission located between bright jet features; VSOP space-VLBI polarization observations have revealed some cases in which longitudinal inter-knot polarization, implying transverse B fields associated with this underlying jet emission, are observed (Gabuzda 1999; Pushkarev et al. 2005).

Appreciable transverse polarization structure is observed in many jets, as becomes clear through a visual inspection of 15-GHz maps produced by the MOJAVE VLBA monitoring project (e.g., Lister and Homan 2005). In some cases, "spine+sheath" polarization structures are observed, with longitudinal polarization (orthogonal B field) near the central axis of the jet and orthogonal polarization (longitudinal B field) near the jet edges (e.g. Attridge et al. 1999; Pushkarev et al. 2005; Fig. 5). In other cases, the predominant polarization is orthogonal to the jet, and is offset appreciably from the central axis of the jet.

3.4 Interpretation of Observations

Jet knots whose dominant polarization is longitudinal, implying a transverse B field, have frequently been interpreted as transverse shocks that have compressed an initially tangled magnetic field so that it has become ordered in the plane of compression (Laing 1980; Hughes et al. 1989), while the presence of orthogonal polarization (longitudinal B field)

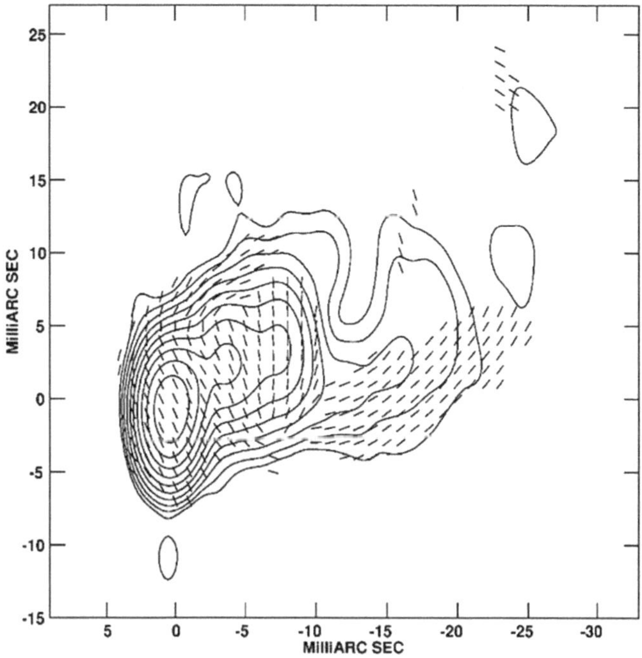

Fig. 5 Example of "spine+sheath" polarization structure across an the jet of the blazar 1055+018 (Attridge et al. 1999). The object is at a redshift $z = 0.888$, so 1 mas equals 7.78 pc. The vectors show the direction of the B-field

has been ascribed to the action of shear that has enhanced this component of the field in a layer surrounding the jet. Transverse polarization structures have similarly sometimes been interpreted as indicating the combined action of shocks and shear (e.g., Attridge et al. 1999). In the framework of the empirical connection between higher/lower apparent jet speeds and predominantly longitudinal/transverse jet magnetic fields, it has been suggested that transverse shocks may form more easily in jet outflows that have lower intrinsic speeds (Gabuzda et al. 1994; Duncan and Hughes 1994).

Alternately, all of these polarization structures can be understood as possible manifestations of helical magnetic fields carried by the jets (e.g., Lyutikov et al. 2005): for example, jets with predominantly orthogonal or longitudinal B fields could carry helical fields with relatively large or small pitch angles. The tendency for AGNs displaying higher/lower apparent superluminal speeds to be more likely to have primarily longitudinal/transverse jet magnetic fields is interesting in this connection. One possibility is that this could reflect a relationship between the pitch angle of a helical jet B field and the physical speed of the jet outflow: the ratio of the speed of the jet outflow to the rotational speed of the central black hole and accretion disc could be higher for quasars, leading to smaller pitch angles for the helical B fields threading their jets and a dominance of the longitudinal component of the helical field, whereas the low outflow speeds of BL Lac objects lead to higher pitch angles for their helical fields and a dominance of the toroidal component of the helical field.

The appearance of a "spine+sheath" transverse polarization structure or polarization, implying a longitudinal B field offset toward from the central axis of the jet, could likewise come about if the jet carries a helical magnetic field, with the azimuthal component dominating near the central axis of the jet and the longitudinal component becoming dominant near the edges. The different viewpoints advocated by these two different types of interpretation essentially reflect the challenge referred to above, of distinguishing between characteristics associated with the action of local agents and those associated with intrinsic properties of the jets themselves.

Considerable progress has also been made recently in using the results of theoretical computations and simulations of AGN jets to generate simulated VLBI intensity, spectral-index, polarization and Faraday-rotation distributions that can, in principle, be compared with observations. Examples include the studies carried out by Zakamska et al. (2008), Gracia et al. (2009), Mimica et al. (2009), Broderick and McKinney (2010), Porth and Fendt (2010), Porth et al. (2011) and Clausen-Brown et al. (2011). Mimica et al. (2009) and Mimica and Aloy (2010, 2012) have considered various observational signatures of the presence of shocks and shock-heated gas in relativistic jets, including jets carrying helical fields. Broderick and McKinney (2010) and Porth et al. (2011) have generated theoretical Faraday-rotation images based on computations and simulations of AGN jets with a helical magnetic-field component. Porth et al. (2011) also considered the behaviour of the polarization angle as a function of the wavelength λ squared in the (partially) optically thick core region, demonstrating deviations from a λ^2 relation away from a limited range of relatively short wavelengths; this effect appears to have been observed by O'Sullivan and Gabuzda (2009b). One limitation here is the need to use some scheme to extrapolate results for small scales to the scales accessible to observations with VLBI. However, such studies are beginning to enable much closer comparisons with observations than was possible previously, and are being used to identify a range of characteristic observational signatures of various types of jets and jet magnetic fields. It is clear that the ability to reliably measure transverse structure in intensity, spectral index, polarization and Faraday rotation across AGN jets will be key to probing the nature of the jet magnetic fields.

3.5 Observational Evidence for Helical or Toroidal Magnetic Fields

As is discussed in Sect. 2 above, helical jet magnetic fields are predicted by a wide range of theoretical models and numerical simulations (e.g. Blandford and Payne 1982; Ouyed and Pudritz 1997; Krasnopolsky et al. 1999; Fendt 2006), making observational searches for evidence for helical magnetic fields associated with AGN jets of considerable interest. The most promising approach in this area is searching for transverse gradients in the Faraday rotation measure distribution across AGN jets, as was first pointed out by Blandford (1993). The observed Faraday rotation is proportional to the integral along the line of sight of $n_e\mathbf{B}\cdot d\mathbf{l}$, where n_e and $\mathbf{B}$ are the electron density and magnetic field in the region of Faraday rotation and the length element $d\mathbf{l}$ points along the line of sight toward the observer. Therefore, the magnitude of the Faraday rotation depends on the line-of-sight component of the ambient magnetic field, and the sign of the Faraday rotation is determined by the direction of this component of the field. Therefore, if a jet and its immediate vicinity are threaded by a helical magnetic field, this field will give rise to a systematic transverse gradient in the observed Faraday rotation, due to the systematic change in the line-of-sight component of the helical B field across the jet. The detection of such transverse Faraday-rotation gradients is observationally challenging, since it involves a joint analysis of multi-frequency radio images, and the observing conditions are not optimal on either parsec or kpc scales: on parsec scales, the jets are very compact, so that they are sometimes only marginally resolved in regions of strong polarization, even when observed with VLBI, while, on kpc scales, the jets are better resolved, but the effect is weaker due to the increased distance from the base of the jet. To obtain increased sensitivity to Faraday rotation, one can observe at longer wavelengths, however, this comes at the expense of a corresponding loss of resolution.

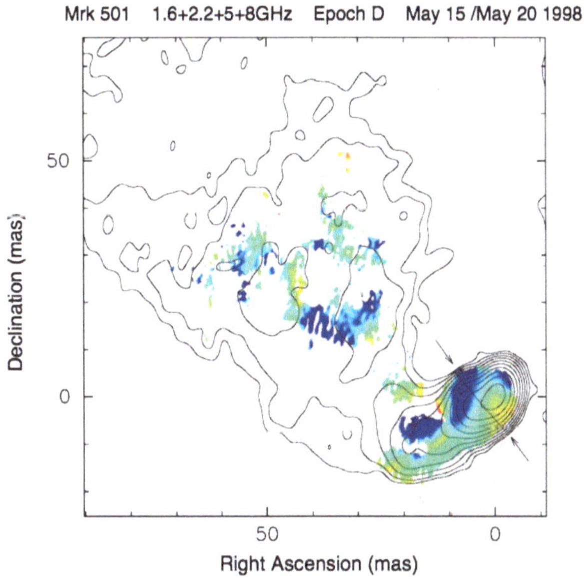

Fig. 6 Example of a transverse Faraday-rotation gradient detected across the jet of the blazar Mrk 501; the contours show the distribution of the radio intensity at 18 cm, and the colour scale the distribution of the observed Faraday rotation (Croke et al. 2010). The redshift of this object is 0.034, so that 1 mas is 0.66 pc

Nevertheless, firm detections of transverse Faraday rotation gradients have now been made for a number of AGNs. The first report of the detection of a transverse Faraday gradient across an AGN jet, interpreted as evidence for a helical jet B field, was made by Asada et al. (2002), based on Very Long Baseline Array (VLBA) observations of the nearby AGN 3C 273; this was later confirmed by Zavala and Taylor (2005), and the evolution of the gradient analyzed by Asada et al. (2008b). In the meantime, transverse Faraday-rotation gradients in several more objects were presented by Gabuzda et al. (2004). There have since then been at least ten further papers in refereed journals reporting the presence of transverse Faraday-rotation gradients across the jets of more than a dozen AGNs on parsec scales, interpreted as evidence that these jets carry helical magnetic fields (Asada et al., 2008a, 2008b, 2010; Gabuzda et al. 2008; Gómez et al. 2008; O'Sullivan and Gabuzda 2009b; Kharb et al. 2009; Mahmud et al. 2009; Croke et al. 2010; Kronberg et al. 2011); one example is shown in Fig. 6. A number of other tentative cases have been presented in conference proceedings or identified in previously published Faraday-rotation images in the literature (e.g., Contopoulos et al. 2009; Zavala and Taylor 2003, 2004).

This suggests that, when the observing conditions enable their detection, transverse Faraday-rotation gradients across AGN jets are not uncommon on parsec scales. This result is potentially of crucial importance for our understanding of AGN jets, since it provides direct evidence that these jets carry helical magnetic fields, which, in turn, has obvious implications for the jet-launching mechanism. In addition, the presence of an appreciable ordered toroidal field component would imply that the jets carry current. It is therefore important to try to ensure that the requirements for the reliable detection of transverse Faraday-rotation gradients are well understood and applied.

It has sometimes been suggested that apparently transverse Faraday-rotation gradients could come about due to gradients in the ambient electron density, and therefore not have anything to do with the presence of a helical jet B field. An interesting point here is that, if the viewing angle of the jet in the jet rest frame is not too far from 90° (the jet is viewed roughly from the side in the jet rest frame), the observed Faraday rotation associated with the jet's helical field should change sign across the jet; if the jet viewing angle in the jet rest frame is not close to 90°, the jet's helical field will still give rise to a transverse Faraday-rotation gradient, but there may not be a change in the sign of the Faraday rotation across the jet. Therefore, the clear detection of a change in the sign of the observed Faraday rotation across an AGN jet can easily be explained by the presence of a helical field, but not by electron-density gradients; such sign changes have been observed for a number of AGN jets (e.g., Asada et al. 2008a; Gabuzda et al. 2004, 2008; Mahmud et al. 2009).

Other observational problems that could potentially hinder the detection of transverse Faraday-rotation gradients associated with helical magnetic fields, or distort their interpretation, include the limited resolution available to centimeter-wavelength VLBI observations and optical depth effects in the region of the observed VLBI core (Taylor and Zavala 2010). The presence of non-monotonic and/or complex transverse Faraday-rotation gradients in the core region have been directly demonstrated by the simulations of Broderick and McKinney (2010) and Porth and Fendt (2010). These simulations indicate the difficulty of accurately deriving parameters of the jet and the helical field it is believed to carry based on observed transverse Faraday-rotation gradients. Note, however, that the direction of the simulated gradients is correct when they are monotonic; since most observational studies carried out thus far have focused on the detection of *monotonic* transverse gradients and their direction, the results of these studies are not greatly affected by the difficulties indicated by the simulation results.

Taylor and Zavala (2010) have proposed that a transverse Faraday-rotation gradient should span three "resolution elements" (taken to correspond to three beamwidths) in order for the gradient to be considered reliable; however, no theoretical basis was provided for this criterion. It is therefore of interest to develop more systematic and objective approaches to determining the resolution necessary to reliably distinguish transverse Faraday-rotation gradients, and results from a number of such studies have begun to appear. For example, Murphy and Gabuzda (2012) have calculated transverse intensity, polarization and Faraday-rotation distributions and subjected them to convolution with beams of various sizes. The initial results of these studies indicate that the detection of transverse polarization structure associated with helical jet B fields is much more robust to convolution with beams comparable to the jet width than is transverse intensity structure, although, of course, the convolution somewhat distorts the intrinsic polarization structure. Similarly, convolution of transverse Faraday-rotation gradients with beams that are comparable to or even larger than the transverse size of the jet reduces the magnitude of the Faraday-rotation gradient, but does not destroy the gradient completely. Hovatta et al. (2012) have recently presented the results of four-frequency VLBI polarization observations of roughly 200 AGNs, together with Monte Carlo simulations modeling the polarization uncertainties and how they are manifest in Faraday-rotation maps. These Monte Carlo simulations suggest that transverse Faraday-rotation gradients spanning as little as 1.5 beamwidths can be reliable, provided that the difference between the Faraday rotations on either side of the jet exceeds 3σ.

These results are also consistent with the general relativistic MHD simulations of Broderick and McKinney (2010), which directly demonstrate the presence of transverse Faraday-rotation gradients across regions that are only marginally resolved: transverse structures in their simulated Faraday-rotation distributions with intrinsic sizes less than 0.05 mas remain clearly visible even when convolved with a 0.9-mas beam. All this suggests that the minimum resolution required to reliably detect a transverse Faraday-rotation gradient across an AGN jet is considerably less than the three-beamwidth criterion suggested by Taylor and Zavala (2010), provided that the Faraday rotation measures measured on opposite sides of the jet can reliably be demonstrated to differ at more than the 3σ level. Of course, this concerns only the detection of the presence and direction of a transverse Faraday-rotation gradient; reliable measurement of the intrinsic magnitude of such a gradient would require much higher resolution.

Another issue that is currently being investigated is the question of how to accurately estimate the uncertainties at specific locations in VLBI images; this is non-trivial due to the complex computational procedures undertaken in the mapping process and the strong correlation between neighbouring pixels, due to convolution with the CLEAN beam. In the past, it has been usual to assign the uncertainty for a pixel in images of the Stokes parameters I, Q and U to be equal to the root-mean-square deviation about the mean calculated for a large region far from areas of source emission, but this is certainly an underestimate. Hovatta et al. (2012) and Mahmud et al. (2012) use Monte Carlo simulations to address this issue, motivated by the need to derive accurate estimates of uncertainties for individual locations in Faraday-rotation images.

Another interesting question in connection with the detection of transverse Faraday-rotation gradients across AGN jets is whether it is possible to observationally distinguish between a helical and a toroidal magnetic field. The point here is that the detection of the transverse gradient only demonstrates the presence of an ordered toroidal field component, without providing any information about the presence of a longitudinal field component. Both helical and toroidal jet B fields could give rise to "spine+sheath" transverse polarization structures, and a rise in the degree of polarization toward the edges of the jet, as is commonly observed. One means of distinguishing between these two possibilities is provided by the transverse intensity and polarization structure of the jet: a purely toroidal field can only produce symmetrical transverse intensity and polarization structures (e.g. Zakamska et al. 2008), whereas a helical field can produce either symmetrical or asymmetrical transverse structures, depending on the viewing angle of the jet and the pitch angle of the helical field (e.g., Murphy et al. 2010, 2012). Therefore, the joint detection of a transverse Faraday-rotation gradient and clearly asymmetrical transverse polarization structure would provide clear evidence for a helical, rather than a purely toroidal, jet B field.

Finally, note that Faraday rotation associated with detected transverse gradients observed can be external (not occurring throughout the radiating volume of the source), if it is associated with regions of helical magnetic field in the outer layers of the jet, or in the immediate vicinity of the jet.

3.6 Evidence for Reversals in the Direction of the Toroidal Field Component

One of the most unexpected results to come out of recent parsec-scale Faraday-rotation studies is the detection of reversals in the direction of transverse Faraday-rotation gradients across AGN jets, both with distance from the base of the jet Mahmud and Gabuzda (2008), Hallahan and Gabuzda (2008), Mahmud et al. (2012) and with time (Mahmud et al. 2009). This is at first glance difficult to understand, since the direction of the observed Faraday-rotation gradient is determined by the direction of the toroidal field component, which is, in turn, essentially determined by the direction of rotation of the central black hole and accretion disc, together with the direction of the net poloidal component of the initial "seed" field that is wound up. It is not physically plausible for the direction of the system's rotation to change on measurable time scales, and changing the direction of the jet poloidal component of the seed field would seem to require a change in the polarity of the field of the black hole, which likewise, seems implausible, at last on measurable time scales. Bisnovatyi-Kogan (2007) has suggested the possibility that, under certain circumstances, torsional oscillations can develop in a jet, which could give rise to changes in the direction of the toroidal B-field component; in this scenario, the direction of the observed transverse Faraday-rotation gradients would reverse from time to time when the direction of the torsional oscillation reverses, and this reversal pattern would presumably then propagate outward with the jet.

However, Mahmud et al. (2009, 2012) have suggested that a simpler and more likely explanation is a magnetic-tower-type model, with poloidal magnetic flux and poloidal current concentrated around the central axis (Lynden-Bell 1996; Nakamura et al. 2006). Fundamental physics dictates that the magnetic-field lines must close; in this picture, the magnetic field forms meridional loops that are anchored in the inner and outer parts of the accretion disc, which become twisted due to the differential rotation of the disc. This should essentially give rise to an "inner" helical B field near the jet axis and an "outer" helical field somewhat further from the jet axis. These two regions of helical field will give rise to oppositely directed Faraday-rotation gradients, and the total observed gradient will be determined by which region of helical field dominates the observed Faraday rotation. The presence of a change in the direction of the observed transverse Faraday-rotation gradient with distance from the jet base could represent a transition from dominance of the inner to dominance of the outer helical B fields in the total observed Faraday rotation. The reversals of the Faraday-rotation gradients with time observed for 1803+784 (Mahmud et al. 2009) could come about if the physical conditions in the jet changed such that there was a change in whether the inner or outer region of helical B field dominated the total observed Faraday rotation. If this interpretation is correct, these observations of reversals of the direction of transverse Faraday-rotation gradients across AGN jets represent the first observational evidence that the jet magnetic field closes in the outer accretion disc.

3.7 Evidence for the Action of a Cosmic Battery

Any transverse Faraday-rotation gradient can be described as being directed either clockwise (CW) or counter-clockwise (CCW) on the sky, relative to the base of the jet. Contopoulos et al. (2009) have reported a significant excess of CW transverse Faraday-rotation gradients for parsec-scale AGN jets, based on transverse Faraday-rotation gradients identified in maps from the literature. Considering the gradient closest to the VLBI core if two distinct regions with transverse gradients were present (as is described in the previous subsection) yielded 29 transverse Faraday-rotation gradients on parsec scales, of which 22 were CW and only 7 CCW, with the probability of this coming about by chance being less than 1 %. This is an extremely counterintuitive result, since the direction of the helical field threading an AGN jet should essentially be determined by the direction of rotation of the central black hole and accretion disc, together with the direction of the poloidal "seed" field that is wound up, and our instincts tell us that both of these should be random.

Contopoulos et al. (2009) suggest that this seemingly bizarre result can be explained in a straightforward way via the action of a mechanism they call the "Poynting–Robertson cosmic battery." The essence of this mechanism is the Poynting–Robertson drag experienced by charges in the accretion disc, which absorb energy emitted by the active nucleus and re-radiate this energy isotropically in their own rest frames. Because these charges are rotating with the accretion disc, this radiation will be beamed in the forward direction of their motion, i.e., in the direction of the disc rotation. Due to conservation of momentum, the charges then feel a reaction force opposite to the direction of their motion; since the magnitude of this force exhibits an inverse dependence on the mass of the charge, this leads to a difference in the deceleration experienced by the protons and electrons in the disc. Since the electrons are decelerated more strongly, this leads to a net current in the disc, in the direction of rotation. This current, in turn, gives rise to a net poloidal magnetic field whose direction is

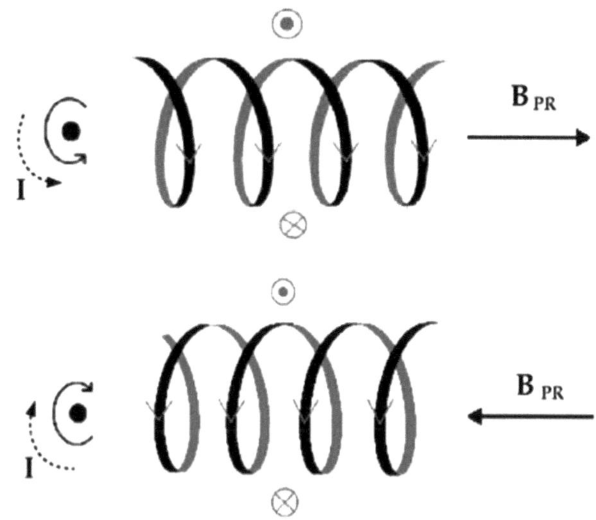

Fig. 7 Illustration of how the action of the Poynting–Robertson battery couples the direction of rotation of the accretion disc and the direction of the poloidal seed field that is wound up, leading to the creation of Faraday-rotation gradients associated with the resulting helical *B* field that are CW on the sky, independent of the direction of disc rotation (Mahmud 2010)

coupled to the direction of the current in the disc, i.e., to the direction of the disc rotation. This coupling of the disc rotation and the direction of the poloidal field that is "wound up" breaks the symmetry in the direction of the toroidal field component, and predicts that the observed Faraday-rotation gradients should be predominantly CW on the sky, independent of the direction of the disc rotation as seen by the observer (Fig. 7). In a "nested helical field" type picture such as that described in the previous subsection, the inner/outer regions of helical field should give rise to CW/CCW Faraday-rotation gradients; therefore, the observed excess of CW implies that the inner region of helical *B* field dominates on parsec scales.

If this excess of CW Faraday-rotation gradients is confirmed by further studies, this will have cardinal implications for our understanding of AGN jets. The development of approaches to carrying out objective searches for reliable transverse Faraday-rotation gradients for as many AGNs as possible is of interest in this connection. Independent of the physical origin of the predominance of a particular orientation, the results of Contopoulos et al. (2009) essentially rule out the possibility that most of the observed gradients are spurious, and not associated with the toroidal field components of the jets, since spurious (incorrectly identified) gradients should certainly be randomly oriented on the sky. If the Poynting–Robertson battery does not operate sufficiently efficiently to provide the observed excess of CW gradients, then some other mechanism that can couple the direction of rotation of the accretion disc and the direction of the poloidal field that is wound up by differential rotation must be identified.

3.8 Connection with Kiloparsec Scales

Recently, Christodoulou et al. (2012) (see also Gabuzda et al. 2012) searched the literature for transverse Faraday-rotation gradients across the kpc-scale jets and lobes of AGNs and radio galaxies. This yielded six AGNs displaying continuous, monotonic RM gradients across their jets, oriented roughly orthogonal to the local jet direction. The most fundamental implication of this result is that transverse Faraday-rotation gradients that can most naturally be interpreted as evidence for toroidal or helical magnetic fields threading the jets are also present on kpc scales. This is an important result, because, if the jets of AGNs carry helical magnetic fields, these should be present on essentially all scales where the jets propagate, provided that the intrinsic field structure of the jet is not disrupted by interactions with the surrounding environment.

Another intriguing aspect of the analysis of Christodoulou et al. (2012) is that all six of the firm transverse kpc-scale Faraday-rotation gradients identified are oriented CCW on the sky, relative to their jet bases. This is on the verge of being a firm result statistically: based on a simple unweighted binomial probability function, the probability for six out of six gradients to be CCW by chance is about 1.5 %. However, the significance of this result is tentatively supported by VLBA observations of 3C 120 (Coughlan et al. 2010), Mrk501 (Croke et al. 2010), and 1749+701 (Hallahan and Gabuzda 2008): in each case, transverse CCW Faraday-rotation gradients are found on scales of tens of pc. Further, all AGNs for which reversals of their transverse Faraday-rotation gradients have been observed display CCW Faraday-rotation gradients further from the jet base (Mahmud and Gabuzda 2008; Mahmud et al. 2012; Reichstein and Gabuzda 2010; Coughlan et al. 2010). This growing evidence for a predominance of CW/CCW Faraday-rotation gradients on parsec/kpc scales can be interpreted as a consequence of a nested-helical-field configuration, as is described above, with the inner region of helical field dominating on parsec scales and the outer region of helical field dominating on kpc scales. This picture is consistent with theoretical studies of the jet launching mechanism (Blandford and Payne 1982; Contopoulos and Lovelace 1994; Contopoulos 1995; Spruit 2010), which suggest that the field is effectively wound up only beyond the Alfvén distance, which is $\sim$10 times the radial extent of the outflow at its base. Thus, the theoretical models suggest that the magnetic field in the inner jet and the outer extended accretion disc will develop significant toroidal components on distances beyond $\sim$10 AU and $\sim$10 pc, respectively, from the AGN centres; this would predict a predominance of CW Faraday-rotation gradients on observed scales less than $\sim$10 pc and of CCW gradients on scales greater than $\sim$10 pc, consistent with the results of Contopoulos et al. (2009) and Christodoulou et al. (2012). Thus, although the results of Christodoulou et al. (2012) must clearly be confirmed based on a larger number of AGNs, they have the potential to considerably influence our understanding of the overall magnetic-field configurations of AGN jets.

The analysis of Christodoulou et al. (2012) suggests that the occurrence of reliably detectable transverse Faraday-rotation gradients is appreciably lower on kpc than on parsec scales: they were able to find reasonably clear transverse Faraday-rotation gradients in only six of about 85 objects. In fact, this is quite natural. There is an appreciable turbulent, inhomogeneous component to the thermal ambient media surrounding the jets on kpc scales, which superposes a more or less random pattern over the systematic pattern due to the helical fields (see the discussion in Sect. 4.3). This random component in the Faraday-rotation distribution may dominate in the majority of cases. Thus, it is natural that most of the observed Faraday-rotation distributions appear random and patchy, but the overall pattern due to the helical fields sometimes comes through. This suggests that, on average, it is easier to detect the systematic Faraday-rotation component due to helical jet magnetic fields on parsec scales, where the ordered inner field is more dominant.

The additional complications imposed by the large-scale interaction with the jet and the ambient medium, and the structures (lobes, hotspots etc.) generated by this interaction, are discussed in the following Section.

4 Interaction with Surrounding Medium and Termination

4.1 Introduction

The best-studied regions of jet termination and the best examples of interactions of jets with the surrounding medium are furnished by radio-loud AGN, The parsec-scale AGN jets discussed in Sect. 3 generally feed into kpc-scale structures—kpc-scale jets, plumes, lobes and hotspots. The magnetic fields in these structures must exhibit some continuity with the fields set by the jet launching process (Sect. 2) and exhibited on the parsec scale (Sect. 3), particularly in the inner parts of kpc-scale jets. However, on the hundred-kpc scales of the jet termination and the downstream structures such as lobes and plumes, it seems likely that much of the initial structure has been erased. As is the case on parsec scales, the intrinsic polarization of synchrotron emission gives us information about the direction and degree of ordering of the magnetic field, though we always need to bear in mind that what we can measure is a projected, polarized-emission-weighted line-of-sight average of the intrinsic degree of polarization and the E- or B-vector angle. In addition, Faraday rotation both external and internal to the emitting medium has an effect on what we observe. A good deal is known about the polarization properties of radio-loud AGN on the largest scales, and we summarize the observational situation in Sects. 4.3, 4.4 and 4.5. We then go on to discuss the relationship between observations and modelling in Sect. 4.6 and the situation in the comparable structures in stellar-scale outflows in Sect. 4.7. Before that, however, we discuss the methods by which magnetic field strength can be measured in AGN on these scales.

4.2 Measuring Magnetic Field Strengths on the kpc Scale

Unfortunately, synchrotron radiation on large scales can tell us little or nothing about the magnetic field *strength*, although this is clearly a key piece of information in understanding the dynamics and energetics of the jet and its environment, without significant (and possibly incorrect) additional assumptions. The methods discussed in Sect. 3.2 can only be applied in regions with a well-known structure and a non-negligible optical depth, and kpc-scale structures are almost always optically thin at accessible frequencies.

The standard equipartition/minimum energy assumptions (Burbidge 1956) are widely used, but even at best only provide us with a plausible order of magnitude estimate of the magnetic field strength: there is no a priori reason to suppose that the plasma reaches the minimum energy or equipartition conditions, and even if it does, these assumptions require us to guess the appropriate values for the filling factor and the fraction of energy in non-radiating particles, ϕ and κ in the standard notation for equipartition:

$$\frac{B^2}{2\mu_0} = \frac{1+\kappa}{\phi} \int EN(E)\mathrm{d}E \tag{13}$$

Finally, long-standing practice in the radio-galaxy community is to carry out the integral in Eq. (13) over a fixed range in *observed frequency*, not in electron energy, although the latter seems more physically reasonable (Myers and Spangler 1985; Brunetti et al. 1997; Hardcastle et al. 1998) and the differences between these two sets of assumptions can lead to significantly different results (Beck and Krause 2005).

If equipartition and minimum energy are unsatisfactory, what else is available? Another widely used approach that gives at least some constraints is to consider the properties of a medium external to the radio source, which may be easier to measure. For example, thermal bremsstrahlung in the X-ray from the environments of large-scale jets and lobes can give us measurements of the temperature and density, and therefore pressure, of the external environment. On the reasonable assumption that the pressure in a radio-emitting component cannot be much less than that in the medium in which it is embedded, we therefore have a constraint on the total energy density in the source. Combining this with the synchrotron measurements, we can derive values for the magnetic field strength. However, these are not unique (for a given pressure greater than the minimum pressure there are two possible values of B to choose from) and the method is still dependent on assumptions about κ and ϕ as well as field geometry.

By far the best method of measuring magnetic field strength in extended components of radio-loud AGN is the use of inverse-Compton emission. The volume emissivity from the inverse-Compton process depends essentially on the electron and photon number densities as a function of energy. If the properties of the photon field are known, a detection of inverse-Compton emission gives us a constraint on the normalization of the electron energy spectrum. We can use this in combination with a measurement of synchrotron emissivity to estimate the B-field in a very robust way: assuming we know the geometry from radio observations, the only model dependence comes from our assumptions about the shape of the electron energy spectrum, since typically we will be observing energetically different electron populations using the two processes. For very simple spatial and spectral properties of the electron populations, analytical estimates of B from inverse-Compton detections can be derived (e.g. Harris and Grindlay 1979); for more complex situations, numerical codes must be used to integrate the relevant equations (e.g. Hardcastle et al. 1998, 2002; Brunetti et al. 2001).

Inverse-Compton emission from large-scale components of radio-loud AGN was first discovered in the X-ray (Harris et al. 1994; Feigelson et al. 1995) and X-ray observations dominate the current applications of this technique of measuring B. A complication of this is that various other processes, including thermal bremsstrahlung from the hot phase of the IGM and synchrotron emission from high-energy electrons, can produce X-ray emission from the radio sources and their environments. In the following subsections we will discuss the available constraints on magnetic field strengths using this method, bearing in mind these observational limitations.

4.3 Magnetic Fields in kpc-Scale Jets

Kpc-scale jets are often strongly polarized, as discussed in Sect. 3.8. In the most powerful jets, the apparent magnetic field direction after correction for Faraday rotation is normally along the jet, often following bends in the jet very closely (e.g. Bridle et al. 1994). In the less powerful jets of FRI radio galaxies, there is often a transition between a parallel field configuration in the inner few kpc to a perpendicular field in the centre of the jet further out: parallel fields may be observed at the edge of the jet (e.g. Bridle and Perley 1984). These FRI jets are the jets that have been studied in the greatest detail to date, because their proximity and relative brightness allows sensitive measurements to be made in regions of the jet that are transversely resolved on scales from sub-arcsec to tens of arcsec (hundreds of pc to a few kpc, at typical distances). In addition, both a jet and counterjet are routinely detected in FRIs, and the assumption that the jet and counterjet are intrinsically symmetrical is a viable one, allowing the effects of relativistic beaming and aberration to be separated from the intrinsic (rest-frame) behaviour of the source. Detailed modelling of these FRI jets (e.g. Laing and Bridle 2002; Laing et al. 2006) shows that the observed polarization structures can be modelled in terms of a magnetic field that is locally random but anisotropic. The anisotropic component of the field is mainly a combination of toroidal and longitudinal components, with the toroidal component dominating both at the edges of the jets and at large distances from the nucleus. Globally ordered helical models for the magnetic field (cf. Sect. 3.5) have been ruled out in some of the best-studied cases (Canvin et al. 2005; Laing et al. 2006): if the field is initially helical, it must evolve away from such a configuration by the kpc scale.

Measurement of B-field strengths in jets is more difficult; although resolved X-ray emission from jets is common (see e.g. Harris and Krawczynski 2006 for a review), there are two fairly serious problems in using them to determine B. The first is that some of the X-ray emission seen from these structures, particularly in low-power objects, is unequivocally known to be synchrotron in origin (e.g. Hardcastle et al. 2001b). Where synchrotron emission is dominant, we only have a (usually uninteresting) lower limit on B in the jet. The second is that the jets of the most powerful objects are known to be at least mildly relativistic on kpc scales (Bridle et al. 1994; Wardle and Aaron 1997; Hardcastle et al. 1999; Mullin and Hardcastle 2009) and this introduces additional model dependence into any estimates of B, since the observed properties of the jet now depend on the bulk Lorentz factor Γ and the angle to the line of sight θ. A good deal of interest was triggered by the discovery that the X-ray emission from the powerful source PKS 0637-752 (Schwartz et al. 2000), which could not be explained by a one-zone synchrotron model, could be explained as inverse-Compton scattering of the CMB with a magnetic field close to equipartition, using a Lorentz factor ($\Gamma \sim 10$) and angle to the line of sight close to those implied by VLBI observations in the nucleus (Tavecchio et al. 2000; Celotti et al. 2001), which would imply no jet deceleration from pc to hundred-kpc scales. However, this certainly does not seem to be the case for all such powerful quasars (Hardcastle 2006) and later work on the broad-band SED of the knots in the best-studied example of this class, 3C 273, suggests that the data are inconsistent with an inverse-Compton model anyway (Jester et al. 2002, 2005). As yet there is therefore no likelihood of being able to determine large-scale magnetic fields from inverse-Compton emission in the jets of the most powerful objects, though in (bright, nearby) lower-power objects we may hope for inverse-Compton emission at higher energies (see Sect. 5.3).

4.4 Jet Termination

The bright compact regions at the terminations of powerful jets, known as hotspots, are believed to be shocks at which the jet material interacts with plasma already present in the lobes. They are well studied in the radio because of their high surface brightness, and polarization observations at high resolution (e.g. Carilli et al. 1999) show them to have complex polarization structures (e.g. Fig. 8), presumably because of the complex hydrodynamics in this part of the source. There is a general tendency for the inferred magnetic field angle to be perpendicular to the jet direction on entry to the hotspot (e.g. Leahy et al. 1997; Hardcastle et al. 1997), possibly because of field compression at the shock.

Because the hotspots are often the brightest compact features in sources in which they occur, they would be expected *a priori* to be good sources of so-called synchrotron self-Compton (SSC) emission, where synchrotron emission provides the seed photons for inverse-Compton scattering (Fig. 9). Observational constraints suggest that the bulk flow through hotspots is at best mildly relativistic (e.g. Mullin et al. 2008) and so beaming is less important. Early observations of X-ray emission from hotspots were consistent in many cases with the X-rays being SSC and the magnetic field strengths implied being close to equipartition (Harris et al. 1994, 2000; Hardcastle et al. 2001a, 2002). However, at the same time, a number of hotspots showed X-ray emission that was much brighter than the expectation from SSC at equipartition and (in some cases) was spectrally inconsistent with being

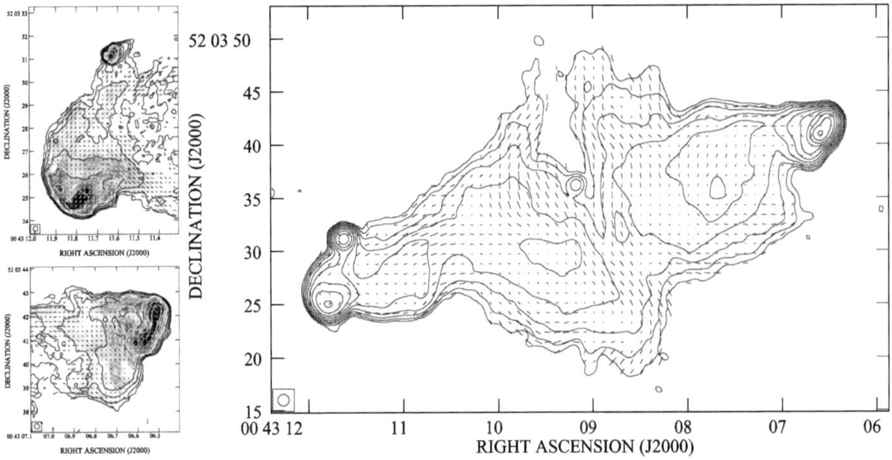

Fig. 8 Complex polarization structure in the lobes and hotspots of the FRII radio galaxy 3C 20 at 8 GHz. *Right*: the overall structure of the source. *Top left*: the E hotspot. *Bottom left*: the W hotspot. Polarization vectors have length representing fractional polarization and are plotted at 90° to the direction of the polarization E-vector, thus giving a representation of the magnetic field direction. Contours show total intensity, increasing logarithmically by a factor of 2 at each step. The greyscales shown for the hotspots indicate polarized intensity to give an idea of the complex filamentary structure that is visible in polarized emission. Data from Hardcastle et al. (1997)

SSC at all; this X-ray emission seemed likely to be partly or wholly synchrotron in origin (e.g. Harris et al. 1998; Wilson et al. 2001). Hardcastle et al. (2004), based on the large amount of hotspot data collected in the first few years of the *Chandra* mission, showed that it was plausible that the difference was controlled by the luminosity, and hence the magnetic field strength, of the hotspots (cf. Brunetti et al. 2003); the most luminous hotspots have magnetic field strengths high enough that strong synchrotron losses inhibit the acceleration of electrons to energies at which X-ray synchrotron emission would be detected. If this is the case, then it is plausible that all hotspots, not just the subset for which SSC has been detected, have magnetic fields close to the equipartition value (Hardcastle et al. 2004; Kataoka and Stawarz 2005). Remarkably, therefore, it seems that there is a rapidly-acting (since hotspots are short-lived), small-scale physical process that acts to amplify magnetic field strengths up to values within a factor of a few of their equipartition values. These field strengths may be up to 20 nT (200 μG) in the brightest sources.

4.5 Lobes

Beyond the end of the jet and the jet termination shock, if present, are the structures fed by the jet. The most common structure seen is large-scale lobes of synchrotron-emitting material extending back towards the centre of the host galaxy, although extended diffuse plumes are seen in a minority of low-power radio galaxies.

The lobes are often strongly polarized, with the typical apparent magnetic field direction being parallel to contours of constant total intensity surface brightness (e.g. Fig. 8). Because of the large spatial scales, strong polarization and large sizes of magnetic field structures, lobes are the best places for detailed studies of Faraday rotation intrinsic to the source or its local environment. The change in polarization position angle $\Delta\theta$ produced by a foreground

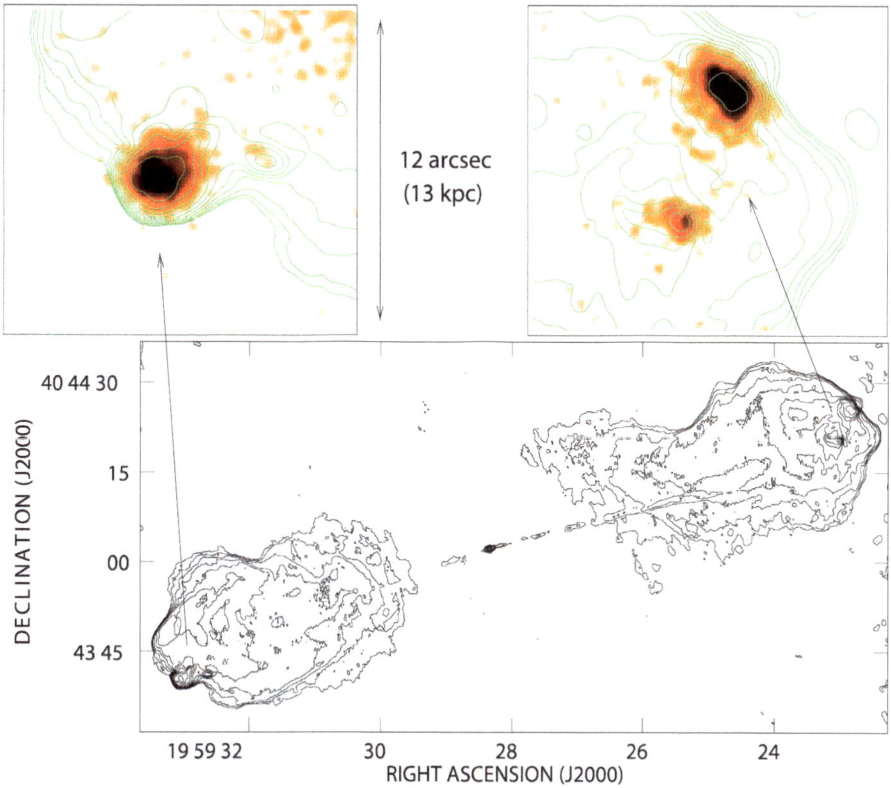

Fig. 9 Synchrotron self-Compton X-ray emission from the hotspots of Cygnus A (3C 405), as first reported by Harris et al. (1994). *Lower panel*: contours show the 4.5-GHz radio emission from the whole source, from the image of Carilli et al. (1991). *Upper panel* shows the same contours overlaid on smoothed *Chandra* data in the 0.5–5.0 keV range from the dataset of Wilson et al. (2006). Note the close morphological agreement between radio and X-ray emission. The level of the inverse-Compton emission implies field strengths close to the equipartition value in the hotspots

Faraday screen is given by $\Delta\theta = RM\lambda^2$, where RM is the rotation measure,

$$RM = C \int n_e \mathbf{B} \cdot \mathrm{d}\mathbf{l} \tag{14}$$

Thus in principle, with multi-wavelength observations to give the required λ^2 coverage, we can measure both RM and the intrinsic polarization angle and so constrain the electron density and field strength in the Faraday screen. In practice, there are a number of complications (e.g. Laing 1984). Faraday-active material (thermal electrons) *inside* the synchrotron-emitting region will induce depolarization and departures from a λ^2 law; in fact, the absence of this effect gives very strong, albeit model-dependent, limits on the possible densities of thermal electrons inside the lobes (e.g. Dreher et al. 1987). Even if the Faraday screen is external to the lobes, its characteristic size scale must be fully resolved by the radio observations, otherwise beam depolarization results and the problem of determining n_e and B becomes much harder (though in some circumstances it is possible, see e.g. Hardcastle et al. 2010). When Faraday rotation external to the radio galaxy is measured, the derived RM values are consistent with being due to the hot phase of the intracluster medium, and

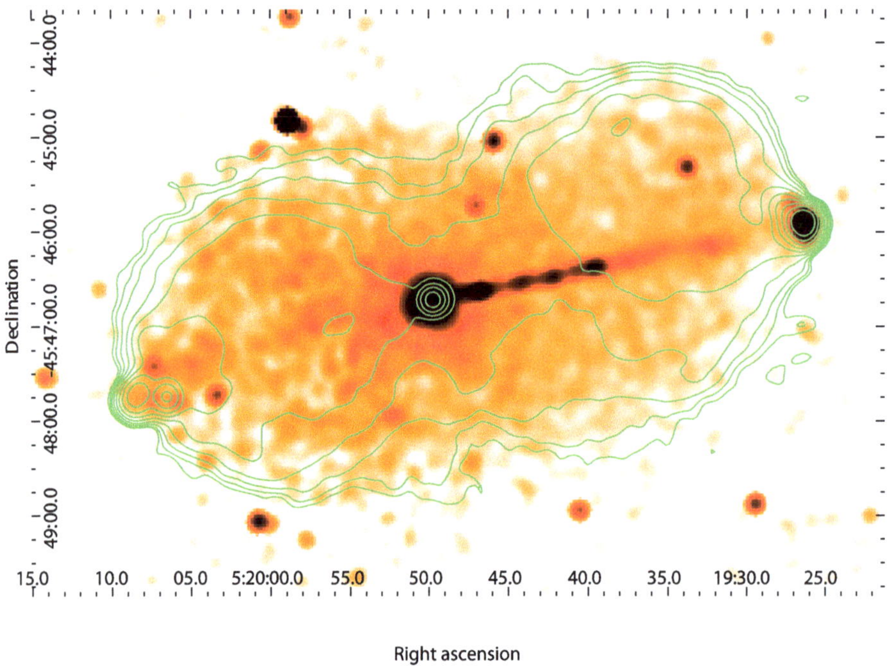

Fig. 10 X-ray emission from the nearby FRII radio galaxy Pictor A. The colour scale shows Gaussian-smoothed *Chandra* counts in the 0.5–5.0 keV range, using the data described by Marshall et al. (2010): contours are from the 1.4-GHz radio map of Perley et al. (1997). Note the X-ray emission (most likely synchrotron, see Hardcastle and Croston 2005) from the jet and hotspot, the excellent match between the boundaries of the X-ray and radio emission, and the relatively constant X-ray surface brightness of the lobes, in contrast to the strongly varying radio surface brightness (contours are logarithmic, increasing by a factor 2 at each step)

with constraints on the external density the field strengths may be estimated. Observed *RM* values range from $\pm 100\ \mathrm{rad\,m^{-2}}$ in typical radio galaxies to $\pm 4000\ \mathrm{rad\,m^{-2}}$ in Cygnus A, at the centre of a rich cluster (Dreher et al. 1987) and the estimated energy densities in magnetic fields are then comparable to the thermal energy densities of the plasma. The structures observed in Faraday rotation may also be used to constrain the power spectrum of magnetic field fluctuations in the external medium (Laing et al. 2008). However, in the absence of *internal* Faraday rotation, for which there is no conclusive evidence in any object on kpc scales, these estimates of B tell us nothing about the field strength or its variations in the lobes themselves.

The lobes are expected to be particularly good sources of inverse-Compton emission from scattering of diffuse photon populations such as the CMB and the extragalactic background light (Fig. 10). Because their geometry and synchrotron properties are also relatively well constrained, because it is certain that they do not have relativistic bulk motions, and because there is no evidence in general for significant ongoing particle acceleration, they represent the best cases for modelling and hence inverse-Compton observations give the cleanest constraints on magnetic field strengths. Early observations with *ROSAT* and *ASCA* allowed some field measurements to be made (e.g. Feigelson et al. 1995), but again it was *Chandra* and to some extent *XMM-Newton* that gave us comparatively large samples (Hardcastle et al. 2002; Croston et al. 2004; Croston et al. 2005; Kataoka and Stawarz 2005). These provide a very clear picture: both the detections and the non-detections of inverse-Compton emission are consistent with a picture in which the characteristic magnetic field in the lobes is close to, but in general slightly below, the equipartition value. The actual field strength value depends strongly on the lobe size and power, but might be of order 0.1–few nT (1 to a few tens of μG) in large, bright sources, with the energy density in electrons being perhaps an order of magnitude higher. Thus it seems that the mechanism that ensures (approximate) equipartition in the hotspot regions may also operate on the much larger, hundred-kpc scales of the lobes.

We can go slightly further in the best-studied systems, since the fact that the synchrotron and inverse-Compton emission can be *resolved* with *Chandra* and *XMM-Newton* means that we are not restricted to considering one-zone models, as first realised by Isobe et al. (2002). Hardcastle and Croston (2005) showed that in fact the magnetic field strength has to vary by a factor of up to 2 throughout the lobe of the best-studied lobe IC source, Pictor A (Fig. 10), in the sense that the field strength decreases with distance from the hotspot. More recently, Goodger et al. (in prep.) have used inverse-Compton data to argue that the strongly filamentary structure seen in synchrotron observations of many powerful radio sources must be a result of strong small-scale variations of the magnetic field strength (if it were the electron energy density that was varying, there would be a stronger correlation between the synchrotron and inverse-Compton emission than is observed). Although this detailed work is at the limits of what is presently achievable with X-ray observations, it may be the only way in which the detailed relationship between field and electrons in the lobes can be studied.

The limitation of the lobe inverse-Compton technique is that it can only really be used to study sources in which the diffuse X-ray emission is not dominated by thermal bremsstrahlung from the IGM; i.e. objects in relatively poor environments. Among other things, this means that with the exception of Cen A (see below, Sect. 5.3) there is no inverse-Compton estimate of the magnetic field strength in the lobes of any low-power (FRI) source, although inverse-Compton limits have been used to constrain the particle content in such objects (e.g. Jetha et al. 2008; Hardcastle and Croston 2010).

4.6 Modelling of Large-Scale Fields

Analytical modelling of the dynamics of the lobes of radio galaxies has since its beginning focussed on a pure hydrodynamical approach (e.g. Scheuer 1974; Kaiser and Alexander 1997) and numerical modelling has often followed suit (e.g. Basson and Alexander 2003; Krause 2005). To some extent, this approach may be (retrospectively) justifiable by the inverse-Compton observations discussed above (Sect. 4.5), which tell us that magnetic fields cannot dominate the large-scale dynamics. Nevertheless, numerical modelling involving magnetic fields is important for three reasons; firstly, even synthetic synchrotron mapping requires some idea of the magnetic field strength distribution, while reproducing observations of polarization and RM/depolarization structures (see above) is impossible without a model of the magnetic field both in the radio source and the environment; secondly, while the magnetic field may not dominate the dynamics of the large-scale structures, it may still have a dynamical role to play, for example in suppressing Kelvin-Helmholtz instabilities; and thirdly, such modelling in principle allows us to test assumptions about the transport of the magnetic field structures imposed at jet generation and observed in the parsec-scale jet (Sects. 2 and 3).

The earliest simulations of lobe dynamics involving magnetic field structure (Clarke et al. 1989; Matthews and Scheuer 1990) involved passive transport of the field, but more recently it has been possible to numerically solve the full MHD equations for semi-realistic density, jet speed and magnetization regimes. Some compromises have generally been necessary to allow such simulations. For example, the work of Jones et al. (1999) and Tregillis et al. (2001, 2004) considered electron transport and radiative losses, but operated in a uniform environment. MHD simulations involving realistic large-scale atmospheres, matched to the information derived from X-ray observations (e.g. O'Neill et al. 2005; Gaibler et al. 2009; O'Neill and Jones 2010) tend to neglect electron radiative losses and also, in earlier work, tend to consider rather low jet density contrasts. Increased computational power has started to allow the merger of these two approaches (e.g. Mendygral et al. 2012). Key results from this modelling to date include the broad reproduction of the lobe dynamics expected from hydrodynamical analytic or numerical modelling; the suppression of large-scale Kelvin-Helmholtz instabilities and of the entrainment of external gas seen in pure hydrodynamic modelling (Gaibler et al. 2009), and the emergence of complex small-scale structure in the magnetic field and hence the synchrotron emissivity (e.g. Tregillis et al. 2001). 3D MHD simulations which can simultaneously capture realistic large-scale dynamics and have resolution good enough to be matched to the detailed observations of synchrotron and inverse-Compton emission in radio galaxy lobes, so as to help with questions such as the origin of filamentary structures seen in synchrotron emission (Sect. 4.5) can perhaps be expected in the next few years. Numerical modelling is also perhaps starting to shed some light on the *origins* of the field structures on large scales; for example, Gaibler et al. (2009) put in as an input condition a helical field as observed in parsec-scale AGN jets (Sect. 3), and show that this initial field is amplified by shear processes to reach much larger values than expected from flux conservation. On the other hand, Huarte-Espinosa et al. (2011) are able to reproduce many features of real radio galaxy lobes with a field which is initially *random* (in both the jet and the environment into which it propagates). It will be of great interest to see whether high-resolution simulations of large-scale jet-environment interactions can give predictions which would allow us to distinguish between different models of the small-scale jet magnetic field structure.

4.7 Large-Scale Magnetic Fields in Stellar-Scale Outflows

The study of the terminations and the equivalents of lobes in stellar-scale outflows such as protostellar jets and microquasars is much less well developed. These structures require persistent jets, and so impulsive ejection events such as those seen in some microquasars are not expected to produce them. Where jets are persistent or at least recurrent on short timescales,, there is evidence in a few well-studied microquasars for lobe-like synchrotron emission (e.g. Mirabel et al. 1992; Fabrika 2004; Hardcastle 2005) and even both lobes and hotspots (Soria et al. 2010), as well as indirect evidence for the existence of lobes whose synchrotron emission is not seen in some other cases (Gallo et al. 2005). The lack of lobes in general in these systems has been attributed to the low-density environments in which they reside (Heinz 2002). Even when found, though, these lobes are generally faint, so that even polarization measurements are generally not possible, and techniques such as magnetic field strength measurements by inverse-Compton emission are completely out of the question, even setting aside the complex technical challenges of doing so in the presence of an often strong central X-ray source and of a complex and poorly constrained photon field. Authors wishing to estimate magnetic field strengths use the equipartition assumption, despite its limitations (Sect. 4.2).

Turning to protostellar outflows, models for these systems fall into two general categories: a jet-driven bow shock picture analoguous to the dynamics of FRII radio galaxies, and a wind-driven shell picture in which the molecular gas is driven by an underlying wide-angle wind component, such as is given by the X-wind (Cabrit et al. 1997). A survey of molecular outflows by Lee et al. (2000) found that both mechanisms are needed in order to explain the full set of systems observed. The advent of Spitzer observations of protostellar jets has led to spectacular advances in our understanding of jet morphology, shock structure, and jet termination. For example, the jet in the classic T-Tauri star system HH46/47, imaged in the infrared with high resolution techniques (down to sub-arcsec levels), is bright and highly collimated and has both a wide-angle outflow cavity as well as a collimated jet that extends through all scales from the protostellar scale up to the termination point on the bow shock (Velusamy et al. 2007), showing a morphology which is remarkably similar to those of powerful radio galaxies. The jet is well collimated with a width of 1–1.5″ and length of 10″ (4500 AU). One of the most prominent features are bright "hotspots" that are coincident with the head of the jet and appear to be analogous to the hotspots of radio galaxies. Cool molecular gas is detected in the wide-angle flow that surrounds this jet. Similarly, observations of the jet Cep E associated with a more massive forming star (with up to 4 $M_\odot$), using the same techniques, show shocked molecular hydrogen H_2 emission—which requires the existence of a magnetic precursor (C-shocks) in order to be excited (Velusamy et al. 2011). Here again, the hottest emission comes from atomic/ionic gas produced at the hot spot at the bow shock of the jet. However, despite this important evidence for magnetization of the shocked region, and unlike the AGN case, there is as yet little direct evidence for the role of magnetic fields in the production of "hotspots" and lobes, since there are few observational diagnostics of field strength in this regime and synchrotron radiation is seldom observed. In the case of protostellar jets, the state of the art, as discussed above (Carrasco-González et al. 2010) is the detection of radio polarization, confirming a synchrotron origin for some of the detected radio continuum, but this is a detection of the jet, rather than of a lobe-like structure, and here again equipartition assumptions must be used to estimate a magnetic field strength. Next-generation radio and X-ray facilities will be required for the study of these systems to catch up with that of extragalactic large-scale lobes.

5 New Directions

5.1 Magnetic Field Strengths and Directions in Protostellar Jets

The Expanded VLA (EVLA) and the extended Multi-Element Radio Linked Interferometer (e-MERLIN) will have an order of magnitude more sensitivity than their predecessors. Much larger parts of the radio spectrum will be measured as a consequence, and this will have a huge impact on our ability to measure synchrotron radiation from protostellar systems. The one detected source HH80-81 is one of the most powerful TTS jets known. If the weaker jets turn out to have measurable synchrotron emission, this will herald a new era in the study of jets in general, and of their magnetic field structure in particular.

A second major advance in understanding the launch of jets will come with the ALMA observatory. ALMA observations will allow one to resolve the 1-AU scale of nearby protostellar discs for the first time. The important point is that ALMA will also be able to measure linearly polarized emission. It is expected that this will allow one to map the structure of magnetic fields on the disc, and, in this way, to study the structure of the disc field and its connection to the launch mechanism.

The bottom line is that the era of real magnetic field measurements in protostellar systems is very near—and this will help revolutionize our knowledge of how jets are launched and collimated.

5.2 Parsec-Scale Magnetic Fields in AGN: Physics of the Launching, Propagation and Confinement of the Jets

Observations with the Very Long Baseline Array over the past decade have made the importance of high-resolution, multi-frequency, polarization-sensitive radio studies with VLBI abundantly clear. The availability of spectral information can help identify locations of particle re-acceleration and low-frequency absorption in the jets, and the Faraday rotation distribution can provide information about both the thermal gas in the vicinity of the AGN and the intrinsic magnetic fields of the jets themselves. The question of whether there is an appreciable helical or toroidal component is of crucial importance for our understanding of the launching and confinement of the jets, and will be a key aspect of future observational studies. Multi-frequency polarization VLBI observations of more AGNs are very much needed for these reasons.

Faraday-rotation studies on parsec scales with high resolution are of particular interest; however, work of this kind has been hindered by the fact that increasing resolution requires observing at higher frequencies, whereas increasing the sensitivity to Faraday rotation requires observing at lower frequencies. One promising way to overcome this limitation is through space VLBI observations at centimetre wavelengths. The currently flying RadioAstron mission is likely to be of only limited use for this purpose, since it has limited imaging capability due to the elongated orbit of the space antenna. The Japanese-led VSOP2 mission would have provided much better imaging capability, but, unfortunately, this mission has recently been cancelled. It will be important to ensure that any other future space-VLBI missions include frequencies low enough (8, 5 and possibly even 1.6 GHz) to make these power tools for high-resolution Faraday-rotation studies. On the other hand, polarization observations at 3 mm–1.3 cm sampling subparsec scales could also prove valuable for future Faraday-rotation studies, since these will probe regions with high electron densities and high magnetic fields in the innermost jets, where very high Faraday rotations may be reached.

Another area in which future VLBI observations can play an important role is in relatively low-frequency multi-frequency observations with global arrays such as the VLBA. Such observations can help link up the parsec scales studied this far with VLBI and the kpc scales that will be studied with high sensitivity using the EVLA and e-MERLIN. One such project is a recently completed set of four-wavelength 18–22 cm polarization VLBA observations of the 135 AGNs in the main MOJAVE sample, which typically provide information about the jet structures out to scales of tens of parsec. Observations of additional AGNs at these and longer wavelengths would be of considerable interest. It is thus extremely important to develop techniques for accurate intensity, polarization and polarization position angle calibration of such long-wavelength VLBI data.

5.3 Kiloparsec-Scale AGN: New Approaches to Field Direction and Strength

The EVLA, and other next-generation radio telescopes with observing bandwidths of order GHz rather than tens of MHz, will have a great impact on the study of Faraday rotation in the large-scale polarized components of radio galaxies. Rather than requiring many monochromatic multi-frequency observations, it will in many cases be possible to make a good measurement of *RM* and depolarization with a single broad-band observation. This should allow the results obtained from detailed studies of bright objects in rich environments (Dreher et al. 1987; Perley and Carilli 1996) to be generalized to more typical systems. The sensitivity and resolution provided by the EVLA and e-MERLIN will allow the detailed, quantitative study of polarization structures in jets (Sect. 4.3) to be extended to more objects, to smaller scales (connecting 100-pc and pc-scale jets) and possibly to the large-scale jets in FRII sources. At lower frequencies, instruments such as LOFAR with the capability to measure polarization sensitively at low frequencies may allow us to probe the *RM* due to very tenuous, large-scale gas seen in front of the polarized outer regions of giant radio galaxies, giving constraints on n_e and B on these scales that are currently unobtainable in any other way.

Progress in inverse-Compton measurements of B will also have to come from new observational developments. Currently one problem with the method is the extrapolation between the properties of the $\gamma \approx 1000$ electrons responsible for inverse-Compton scattering CMB photons into the X-ray band and the $\gamma \approx 10^4$ electrons responsible for radio synchrotron emission at GHz frequencies. This situation will be improved significantly with the advent of next-generation low-frequency radio telescopes such as LOFAR. We can also hope, with less certainty, for imaging missions working at harder X-ray energies, such as NuSTAR, which will not only probe somewhat higher-energy electrons but, by operating above the cutoff energies for thermal bremsstrahlung, will make it much easier to study inverse-Compton emission from sources in rich environments. Another ambiguity in models is the correct modelling of the very low-energy electron population: as shown by Colafrancesco (2008), ALMA may in principle be able to detect the Sunyaev-Zel'dovich effect from radio galaxy lobes, which is a very strong probe of the total number of relativistic electrons and hence of the low-energy cutoff (although see Hardcastle and Looney 2008 for some caveats). Finally, inverse-Compton studies at very high energies have some promise. *Fermi* recently made the first detection of what appears to be inverse-Compton scattering of the CMB to $\sim$ GeV energies, in the nearby, and therefore very large in angular size, radio galaxy Cen A (Abdo et al. 2010): taken at face value the detection implies a magnetic field strength in the lobes that is again close to the equipartition value. *Fermi* can resolve only a very few nearby and (necessarily) atypical radio galaxies, but even limits from this process may be useful. At higher energies still, existing detections with TeV imaging instruments such as HESS and MAGIC place limits on inverse-Compton scattering (predominantly of starlight: Stawarz et al. 2003) by the TeV electrons in the jets of low-power radio galaxies: for example, existing HESS detections of Cen A, whose synchrotron properties are very well known (e.g. Hardcastle et al. 2007 and refs therein) already require that the magnetic field cannot be too much lower than the equipartition value, although detailed modelling is complex (Hardcastle and Croston 2011). The resolution and sensitivity that will be achieved with instruments such as the planned Cerenkov Telescope Array (CTA) will actually allow imaging of the TeV inverse-Compton jets in objects such as Cen A, if $B \approx B_{\text{eq}}$, and so will permit mapping of the B-field in an X-ray synchrotron jet for the first time. In the very long term, X-ray telescopes with very high sensitivity (e.g., the currently planned *Athena* mission) may allow us to measure inverse-Compton emission from the lobes of Galactic microquasars and synchrotron-emitting protostellar jets (Sect. 4.7).

6 Conclusion

In conclusion, this review has covered a vast amount of observational, theoretical and computational phase space with the purpose of drawing out and analyzing the importance of magnetic fields for the origin and structure of astrophysical jets. The underlying engines for all of these systems involves rotating magnetized objects such as stars, black holes, and accretion discs. Relativistic and even general relativistic MHD leads to similar conclusions about jet dynamics to those deduced (and in some cases observed) in protostellar systems. While measurements of densities, temperatures, and velocities is straightforward in the latter type of system, magnetic measurements are very difficult. Exactly the opposite is the case for AGN jets. Encompassing these two complementary worlds lies the complete picture of astrophysical jets which is still in a very exciting stage of exploration. It will be important therefore to keep probing and extending our ability to measure magnetic field strengths and geometries in these diverse systems, since these provide insight into the origin of one nature's most marvellous creations.

Acknowledgements We thank André Balogh and the organizers of an ISSI workshop on "Large Scale Magnetic Fields in the Universe" for their invitation to review this topic. This paper has benefited from helpful comments from an anonymous referee. The research of REP is supported by a Discovery Grant from the National Science and Engineering Research Council of Canada. MJH acknowledges support from the Royal Society and the UK Science and Technology Funding Council.

References

A.A. Abdo, M. Ackermann, M. Ajello, W.B. Atwood, L. Baldini, J. Ballet, G. Barbiellini, D. Bastieri, B.M. Baughman, K. Bechtol, R. Bellazzini, R. Berenji, R.D. Blandford, E.D. Bloom, E. Bonamente, A.W. Borgland, J. Bregeon, A. Brez, M. Brigida, P. Bruel, T.H. Burnett, S. Buson, G.A. Caliandro, R.A. Cameron, P.A. Caraveo, J.M. Casandjian, E. Cavazzuti, C. Cecchi, Ö. Çelik, A. Chekhtman, C.C. Cheung, J. Chiang, S. Ciprini, R. Claus, J. Cohen-Tanugi, S. Colafrancesco, L.R. Cominsky, J. Conrad, L. Costamante, S. Cutini, D.S. Davis, C.D. Dermer, A. de Angelis, F. de Palma, S.W. Digel, E. do Couto e Silva, P.S. Drell, R. Dubois, D. Dumora, C. Farnier, C. Favuzzi, S.J. Fegan, J. Finke, W.B. Focke, P. Fortin, Y. Fukazawa, S. Funk, P. Fusco, F. Gargano, D. Gasparrini, N. Gehrels, M. Georganopoulos, S. Germani, B. Giebels, N. Giglietto, F. Giordano, M. Giroletti, T. Glanzman, G. Godfrey, I.A. Grenier, J.E. Grove, L. Guillemot, S. Guiriec, Y. Hanabata, A.K. Harding, M. Hayashida, E. Hays, R.E. Hughes, M.S. Jackson, G. Jóhannesson, S. Johnson, T.J. Johnson, W.N. Johnson, T. Kamae, H. Katagiri, J. Kataoka, N. Kawai, M. Kerr, J. Knödlseder, M.L. Kocian, M. Kuss, J. Lande, L. Latronico, M. Lemoine-Goumard, F. Longo, F. Loparco, B. Lott, M.N. Lovellette, P. Lubrano, G.M. Madejski, A. Makeev, M.N. Mazziotta, W. McConville, J.E. McEnery, C. Meurer, P.F. Michelson, W. Mitthumsiri, T. Mizuno, A.A. Moiseev, C. Monte, M.E. Monzani, A. Morselli, I.V. Moskalenko, S. Murgia, P.L. Nolan, J.P. Norris, E. Nuss, T. Ohsugi, N. Omodei, E. Orlando, J.F. Ormes, D. Paneque, D. Parent, V. Pelassa, M. Pepe, M. Pesce-Rollins, F. Piron, T.A. Porter, S. Rainò, R. Rando, M. Razzano, S. Razzaque, A. Reimer, O. Reimer, T. Reposeur, S. Ritz, L.S. Rochester, A.Y. Rodriguez, R.W. Romani, M. Roth, F. Ryde, H.F.-W. Sadrozinski, R. Sambruna, D. Sanchez, A. Sander, P.M. Saz Parkinson, J. D. Scargle, C. Sgrò, J. Siskind, D.A. Smith, P.D. Smith, G. Spandre, P. Spinelli, J.-L. Starck, L. Stawarz, M.S. Strickman, D.J. Suson, H. Tajima, H. Takahashi, T. Takahashi, T. Tanaka, J.B. Thayer, J.G. Thayer, D.J. Thompson, L. Tibaldo, D.F. Torres, G. Tosti, A. Tramacere, Y. Uchiyama, T.L. Usher, V. Vasileiou, N. Vilchez, V. Vitale, A.P. Waite, E. Wallace, P. Wang, B.L. Winer, K.S. Wood, T. Ylinen, M. Ziegler, M.J. Hardcastle, D. Kazanas, Fermi LAT Collaboration, Fermi gamma-ray imaging of a radio galaxy. Science **328**, 725 (2010). doi:10.1126/science.1184656

J.M. Anderson, Z.Y. Li, R. Krasnopolsky, R.D. Blandford, Locating the launching region of T Tauri winds: The case of DG Tauri. Astrophys. J. Lett. **590**, 107–110 (2003). doi:10.1086/376824

K. Asada, M. Inoue, Y. Uchida, S. Kameno, K. Fujisawa, S. Iguchi, M. Mutoh, A helical magnetic field in the jet of 3C 273. Proc. Astron. Soc. Jpn. **54**, 39–43 (2002)

K. Asada, M. Inoue, M. Nakamura, S. Kameno, H. Nagai, Multifrequency polarimetry of the NRAO 140 jet: Possible detection of a helical magnetic field and constraints on its pitch angle. Astrophys. J. **682**, 798–802 (2008a). doi:10.1086/588573

K. Asada, M. Inoue, S. Kameno, H. Nagai, Time variation of the rotation measure gradient in the 3C 273 jet. Astrophys. J. **675**, 79–82 (2008b). doi:10.1086/524000

K. Asada, M. Nakamura, M. Inoue, S. Kameno, H. Nagai, Multi-frequency polarimetry toward S5 0836+710: A possible spine-sheath structure for the jet. Astrophys. J. **720**, 41–45 (2010). doi:10.1088/0004-637X/720/1/41

J.M. Attridge, D.H. Roberts, J.F.C. Wardle, Radio jet-ambient medium interactions on parsec scales in the blazar 1055+018. Astrophys. J. Lett. **518**, 87–90 (1999). doi:10.1086/312078

F. Bacciotti, T.P. Ray, R. Mundt, J. Eislöffel, J. Solf, Hubble space telescope/STIS spectroscopy of the optical outflow from DG Tauri: Indications for rotation in the initial jet channel. Astrophys. J. **576**, 222–231 (2002). doi:10.1086/341725

F. Bacciotti, T.P. Ray, D. Coffey, J. Eislöffel, J. Woitas, Testing the models for jet generation with Hubble space telescope observations. Astrophys. Space Sci. **292**, 651–658 (2004). doi:10.1023/B:ASTR.0000045071.52869.a4

J. Bally, B. Reipurth, C.J. Davis, Observations of jets and outflows from young stars, in *Protostars and Planets V*, ed. by B. Reipurth, D. Jewitt, K. Keil (2007), pp. 215–230

R. Banerjee, R.E. Pudritz, Outflows and jets from collapsing magnetized cloud cores. Astrophys. J. **641**, 949–960 (2006). doi:10.1086/500496

J.F. Basson, P. Alexander, The long-term effect of radio sources on the intracluster medium. Mon. Not. R. Astron. Soc. **339**, 353–359 (2003). doi:10.1046/j.1365-8711.2003.06069.x

R. Beck, M. Krause, Revised equipartition and minimum energy formula for magnetic field strength estimates from radio synchrotron observations. Astron. Nachr. **326**, 414–427 (2005). doi:10.1002/asna.200510366

M.C. Begelman, Z.Y. Li, Asymptotic domination of cold relativistic MHD winds by kinetic energy flux. Astrophys. J. **426**, 269–278 (1994). doi:10.1086/174061

N. Bessolaz, C. Zanni, J. Ferreira, R. Keppens, J. Bouvier, Accretion funnels onto weakly magnetized young stars. Astron. Astrophys. **478**, 155–162 (2008). doi:10.1051/0004-6361:20078328

G.S. Bisnovatyi-Kogan, Dynamic confinement of jets by magnetotorsional oscillations. Mon. Not. R. Astron. Soc. **376**, 457–464 (2007). doi:10.1111/j.1365-2966.2007.11452.x

R. Blandford, Acceleration and collimation mechanisms in jets, in *Astrophysics and Space Science Library*, vol. 103 (1993), pp. 15–33

R.D. Blandford, Black holes and relativistic jets. Prog. Theor. Phys. Suppl. **143**, 182–201 (2001). doi:10.1143/PTPS.143.182

R.D. Blandford, A. Konigl, Relativistic jets as compact radio sources. Astrophys. J. **232**, 34–48 (1979). doi:10.1086/157262

R.D. Blandford, D.G. Payne, Hydromagnetic flows from accretion discs and the production of radio jets. Mon. Not. R. Astron. Soc. **199**, 883–903 (1982)

R.D. Blandford, R.L. Znajek, Electromagnetic extraction of energy from Kerr black holes. Mon. Not. R. Astron. Soc. **179**, 433–456 (1977)

A.H. Bridle, R.A. Perley, Extragalactic radio jets. Annu. Rev. Astron. Astrophys. **22**, 319–358 (1984). doi:10.1146/annurev.aa.22.090184.001535

A.H. Bridle, D.H. Hough, C.J. Lonsdale, J.O. Burns, R.A. Laing, Deep VLA imaging of twelve extended 3CR quasars. Astron. J. **108**, 766–820 (1994). doi:10.1086/117112

S. Britzen, R.C. Vermeulen, R.M. Campbell, G.B. Taylor, T.J. Pearson, A.C.S. Readhead, W. Xu, I.W. Browne, D.R. Henstock, P. Wilkinson, A multi-epoch VLBI survey of the kinematics of CFJ sources. II. Analysis of the kinematics. Astron. Astrophys. **484**, 119–142 (2008). doi:10.1051/0004-6361:20077717

A.E. Broderick, J.C. McKinney, Parsec-scale Faraday rotation measures from general relativistic magnetohydrodynamic simulations of active galactic nucleus jets. Astrophys. J. **725**, 750–773 (2010). doi:10.1088/0004-637X/725/1/750

G. Brunetti, G. Setti, A. Comastri, Inverse Compton X-rays from strong FRII radio-galaxies. Astron. Astrophys. **325**, 898–910 (1997)

G. Brunetti, M. Cappi, G. Setti, L. Feretti, D.E. Harris, Anisotropic inverse Compton scattering in powerful radio galaxies: The case of 3C 295. Astron. Astrophys. **372**, 755–767 (2001). doi:10.1051/0004-6361:20010484

G. Brunetti, K.H. Mack, M.A. Prieto, S. Varano, In-situ particle acceleration in extragalactic radio hot spots: observations meet expectations. Mon. Not. R. Astron. Soc. **345**, 40–44 (2003). doi:10.1046/j.1365-8711.2003.07185.x

G.R. Burbidge, On synchrotron radiation from Messier 87. Astrophys. J. **124**, 416 (1956). doi:10.1086/146237

S. Cabrit, A. Raga, F. Gueth, Models of bipolar molecular outflows, in *Herbig-Haro Flows and the Birth of Stars*, ed. by B. Reipurth, C. Bertout. IAU Symposium, vol. 182 (1997), pp. 163–180

M. Camenzind, Hydromagnetic flows from rapidly rotating compact objects. I—Cold relativistic flows from rapid rotators. Astron. Astrophys. **162**, 32–44 (1986)

J.R. Canvin, R.A. Laing, A.H. Bridle, W.D. Cotton, A relativistic model of the radio jets in NGC 315. Mon. Not. R. Astron. Soc. **363**, 1223–1240 (2005). doi:10.1111/j.1365-2966.2005.09537.x

C.L. Carilli, R.A. Perley, J.W. Dreher, J.P. Leahy, Multifrequency radio observations of Cygnus A—Spectral aging in powerful radio galaxies. Astrophys. J. **383**, 554–573 (1991). doi:10.1086/170813

C.L. Carilli, J.D. Kurk, P.P. van der Werf, R.A. Perley, G.K. Miley, High-resolution millimeter and infrared observations of the hot spots of Cygnus A. Astron. J. **118**, 2581–2591 (1999). doi:10.1086/301137

C. Carrasco-González, L.F. Rodríguez, G. Anglada, J. Martí, J.M. Torrelles, M. Osorio, A magnetized jet from a massive protostar. Science **330**, 1209 (2010). doi:10.1126/science.1195589

F. Casse, R. Keppens, Magnetized accretion-ejection structures: 2.5-dimensional magnetohydrodynamic simulations of continuous ideal jet launching from resistive accretion disks. Astrophys. J. **581**, 988–1001 (2002). doi:10.1086/344340
T.V. Cawthorne, J.F.C. Wardle, D.H. Roberts, D.C. Gabuzda, L.F. Brown, Milliarcsecond polarization structure of 24 objects from the Pearson-Readhead sample of bright extragalactic radio sources. I. The images. Astrophys. J. **416**, 496 (1993). doi:10.1086/173253
A. Celotti, G. Ghisellini, M. Chiaberge, Large-scale jets in active galactic nuclei: multiwavelength mapping. Mon. Not. R. Astron. Soc. **321**, 1–5 (2001). doi:10.1046/j.1365-8711.2001.04160.x
T. Chiueh, Z.Y. Li, M.C. Begelman, Asymptotic structure of hydromagnetically driven relativistic winds. Astrophys. J. **377**, 462–466 (1991). doi:10.1086/170375
D. Christodoulou, D. Gabuzda, I. Contopoulos, D. Kazanas, Transverse Faraday rotation measure gradients across the kiloparsec-scale radio structures of active galactic nuclei. Astron. Astrophys. (2012, submitted)
D.A. Clarke, J.O. Burns, M.L. Norman, Numerical observations of a simulated jet with a passive helical magnetic field. Astrophys. J. **342**, 700–717 (1989). doi:10.1086/167631
E. Clausen-Brown, M. Lyutikov, P. Kharb, Signatures of large-scale magnetic fields in active galactic nuclei jets: transverse asymmetries. Mon. Not. R. Astron. Soc. **415**, 2081–2092 (2011). doi:10.1111/j.1365-2966.2011.18757.x
D. Coffey, F. Bacciotti, J. Woitas, T.P. Ray, J. Eislöffel, Rotation of jets from young stars: New clues from the Hubble space telescope imaging spectrograph. Astrophys. J. **604**, 758–765 (2004). doi:10.1086/382019
D. Coffey, F. Bacciotti, L. Podio, T Tauri jet physics resolved near the launching region with the Hubble space telescope. Astrophys. J. **689**, 1112–1126 (2008). doi:10.1086/592343
S. Colafrancesco, SZ effect from radio-galaxy lobes: astrophysical and cosmological relevance. Mon. Not. R. Astron. Soc. **385**, 2041–2048 (2008). doi:10.1111/j.1365-2966.2008.12961.x
C. Combet, J. Ferreira, The radial structure of protostellar accretion disks: influence of jets. Astron. Astrophys. **479**, 481–491 (2008). doi:10.1051/0004-6361:20078734
I. Contopoulos, D.M. Christodoulou, D. Kazanas, D.C. Gabuzda, The invariant twist of magnetic fields in the relativistic jets of active galactic nuclei. Astrophys. J. Lett. **702**, 148–152 (2009). doi:10.1088/0004-637X/702/2/L148
J. Contopoulos, A simple type of magnetically driven jets: An astrophysical plasma gun. Astrophys. J. **450**, 616 (1995). doi:10.1086/176170
J. Contopoulos, R.V.E. Lovelace, Magnetically driven jets and winds: Exact solutions. Astrophys. J. **429**, 139–152 (1994). doi:10.1086/174307
C. Coughlan, R. Murphy, K. McEnery, H. Patrick, R. Hallahan, D. Gabuzda, First results from 18–22 cm VLBA polarisation observations of the MOJAVE-I AGNs, in *Proceedings of the 10th European VLBI Network Symposium and EVN Users Meeting: VLBI and the New Generation of Radio Arrays*, 20–24 September 2010, Manchester, UK (2010). Published online at http://pos.sissa.it/cgi-bin/reader/conf.cgi?confid=125,id.46
S.R. Cranmer, Turbulence-driven polar winds from T Tauri stars energized by magnetospheric accretion. Astrophys. J. **689**, 316–334 (2008). doi:10.1086/592566
S.M. Croke, S.P. O'Sullivan, D.C. Gabuzda, The parsec-scale distributions of intensity, linear polarization and Faraday rotation in the core and jet of Mrk501 at 8.4–1.6 GHz. Mon. Not. R. Astron. Soc. **402**, 259–270 (2010). doi:10.1111/j.1365-2966.2009.15923.x
J.H. Croston, M. Birkinshaw, M.J. Hardcastle, D.M. Worrall, X-ray emission from the nuclei, lobes and hot-gas environments of two FR II radio galaxies. Mon. Not. R. Astron. Soc. **353**, 879–889 (2004). doi:10.1111/j.1365-2966.2004.08118.x
J.H. Croston, M.J. Hardcastle, D.E. Harris, E. Belsole, M. Birkinshaw, D.M. Worrall, An X-ray study of magnetic field strengths and particle content in the lobes of FR II radio sources. Astrophys. J. **626**, 733–747 (2005). doi:10.1086/430170
B. Davies, S.L. Lumsden, M.G. Hoare, R.D. Oudmaijer, W.J. de Wit, The circumstellar disc, envelope and bipolar outflow of the massive young stellar object W33A. Mon. Not. R. Astron. Soc. **402**, 1504–1515 (2010). doi:10.1111/j.1365-2966.2009.16077.x
J.W. Dreher, C.L. Carilli, R.A. Perley, The Faraday rotation of Cygnus A—Magnetic fields in cluster gas. Astrophys. J. **316**, 611–625 (1987). doi:10.1086/165229
D.F. Duffin, R.E. Pudritz, The early history of protostellar disks, outflows, and binary stars. Astrophys. J. Lett. **706**, 46–51 (2009). doi:10.1088/0004-637X/706/1/L46
D.F. Duffin et al. (2012, in preparation)
G.C. Duncan, P.A. Hughes, Simulations of relativistic extragalactic jets. Astrophys. J. Lett. **436**, 119 (1994). doi:10.1086/187647
R.J.H. Dunn, A.C. Fabian, Investigating AGN heating in a sample of nearby clusters. Mon. Not. R. Astron. Soc. **373**, 959–971 (2006). doi:10.1111/j.1365-2966.2006.11080.x

S. Fabrika, The jets and supercritical accretion disk in SS433. Astrophys. Space Phys. Rev. **12**, 1–152 (2004)
E.D. Feigelson, S.A. Laurent-Muehleisen, R.I. Kollgaard, E.B. Fomalont, Discovery of inverse-Compton X-rays in radio lobes. Astrophys. J. Lett. **449**, 149 (1995). doi:10.1086/309642
C. Fendt, Collimation of astrophysical jets: The role of the accretion disk magnetic field distribution. Astrophys. J. **651**, 272–287 (2006). doi:10.1086/507976
C. Fendt, Formation of protostellar jets as two-component outflows from star-disk magnetospheres. Astrophys. J. **692**, 346–363 (2009). doi:10.1088/0004-637X/692/1/346
C. Fendt, M. Camenzind, On collimated stellar jet magnetospheres. II. Dynamical structure of collimating wind flows. Astron. Astrophys. **313**, 591–604 (1996)
J. Ferreira, Magnetically-driven jets from Keplerian accretion discs. Astron. Astrophys. **319**, 340–359 (1997)
J. Ferreira, Theory of magnetized accretion discs driving jets, in *EAS Publications Series*, vol. 3, ed. by J. Bouvier, J.-P. Zahn (2002), pp. 229–277
J. Ferreira, F. Casse, Stationary accretion disks launching super-fast-magnetosonic magnetohydrodynamic jets. Astrophys. J. Lett. **601**, 139–142 (2004). doi:10.1086/381804
D.C. Gabuzda, VSOP observations of the compact BL Lacertae object 1803+784. New Astron. Rev. **43**, 691–694 (1999). doi:10.1016/S1387-6473(99)00079-2
D.C. Gabuzda, J.L. Gómez, VSOP polarization observations of the BL Lacertae object OJ 287. Mon. Not. R. Astron. Soc. **320**, 49–54 (2001). doi:10.1046/j.1365-8711.2001.04147.x
D.C. Gabuzda, T.V. Cawthorne, D.H. Roberts, J.F.C. Wardle, A survey of the milliarcsecond polarization properties of BL Lacertae objects at 5 GHz. Astrophys. J. **388**, 40–54 (1992). doi:10.1086/171128
D.C. Gabuzda, C.M. Mullan, T.V. Cawthorne, J.F.C. Wardle, D.H. Roberts, Evolution of the milliarcsecond total intensity and polarization structures of BL Lacertae objects. Astrophys. J. **435**, 140–161 (1994). doi:10.1086/174801
D.C. Gabuzda, É. Murray, P. Cronin, Helical magnetic fields associated with the relativistic jets of four BL Lac objects. Mon. Not. R. Astron. Soc. **351**, 89–93 (2004). doi:10.1111/j.1365-2966.2004.08037.x
D.C. Gabuzda, V.M. Vitrishchak, M. Mahmud, S.P. O'Sullivan, Radio circular polarization produced in helical magnetic fields in eight active galactic nuclei. Mon. Not. R. Astron. Soc. **384**, 1003–1014 (2008). doi:10.1111/j.1365-2966.2007.12773.x
D. Gabuzda, D. Christodoulou, I. Contopoulos, D. Kazanas, Evidence for helical magnetic fields in kiloparsec-scale AGN jets and the action of a cosmic battery, in *Proceedings of the Conference Beamed and Unbeamed Gamma-Rays from Radio Galaxies*, Muonio, Finland, 11–15 April 2011 (2012, to be published)
V. Gaibler, M. Krause, M. Camenzind, Very light magnetized jets on large scales—I. Evolution and magnetic fields. Mon. Not. R. Astron. Soc. **400**, 1785–1802 (2009). doi:10.1111/j.1365-2966.2009.15625.x
E. Gallo, R. Fender, C. Kaiser, D. Russell, R. Morganti, T. Oosterloo, S. Heinz, A dark jet dominates the power output of the stellar black hole Cygnus X-1. Nature **436**, 819–821 (2005). doi:10.1038/nature03879
J.L. Gómez, A.P. Marscher, S.G. Jorstad, I. Agudo, M. Roca-Sogorb, Faraday rotation and polarization gradients in the jet of 3C 120: Interaction with the external medium and a helical magnetic field? Astrophys. J. Lett. **681**, 69–72 (2008). doi:10.1086/590388
J. Gracia, N. Vlahakis, I. Agudo, K. Tsinganos, S.V. Bogovalov, Synthetic synchrotron emission maps from MHD models for the jet of M87. Astrophys. J. **695**, 503–510 (2009). doi:10.1088/0004-637X/695/1/503
J. Granot, The structure and dynamics of GRB jets, in *Revista Mexicana de Astronomia y Astrofisica*, vol. 27 (2007), pp. 140–165
R. Hallahan, D. Gabuzda, The deca-parsec scale radio structures of BL Lac objects, in *The role of VLBI in the Golden Age for Radio Astronomy* (2008)
M.J. Hardcastle, Low-frequency constraints on the spectra of the lobes of the microquasar GRS 1758 258. Astron. Astrophys. **434**, 35–39 (2005). doi:10.1051/0004-6361:20041735
M.J. Hardcastle, Testing the beamed inverse-Compton model for jet X-ray emission: velocity structure and deceleration. Mon. Not. R. Astron. Soc. **366**, 1465–1474 (2006). doi:10.1111/j.1365-2966.2005.09923.x
M.J. Hardcastle, J.H. Croston, The Chandra view of extended X-ray emission from Pictor A. Mon. Not. R. Astron. Soc. **363**, 649–660 (2005). doi:10.1111/j.1365-2966.2005.09469.x
M.J. Hardcastle, J.H. Croston, Searching for the inverse-Compton emission from bright cluster-centre radio galaxies. Mon. Not. R. Astron. Soc. **404**, 2018–2027 (2010). doi:10.1111/j.1365-2966.2010.16420.x
M.J. Hardcastle, J.H. Croston, Modelling TeV γ-ray emission from the kiloparsec-scale jets of Centaurus A and M87. Mon. Not. R. Astron. Soc. **415**, 133–142 (2011). doi:10.1111/j.1365-2966.2011.18678.x
M.J. Hardcastle, L.W. Looney, The properties of powerful radio sources at 90 GHz. Mon. Not. R. Astron. Soc. **388**, 176–186 (2008). doi:10.1111/j.1365-2966.2008.13370.x
M.J. Hardcastle, P. Alexander, G.G. Pooley, J.M. Riley, High-resolution observations at 3.6 cm of seventeen FRII radio galaxies with $0.15 < z < 0.30$. Mon. Not. R. Astron. Soc. **288**, 859–890 (1997)

M.J. Hardcastle, M. Birkinshaw, D.M. Worrall, Magnetic field strengths in the hotspots of 3C 33 and 111. Mon. Not. R. Astron. Soc. **294**, 615 (1998). doi:10.1046/j.1365-8711.1998.01159.x
M.J. Hardcastle, P. Alexander, G.G. Pooley, J.M. Riley, FR II radio galaxies with $z < 0.3$—II. Beaming and unification. Mon. Not. R. Astron. Soc. **304**, 135–144 (1999). doi:10.1046/j.1365-8711.1999.02298.x
M.J. Hardcastle, M. Birkinshaw, D.M. Worrall, A Chandra detection of the radio hotspot of 3C123. Mon. Not. R. Astron. Soc. **323**, 17–22 (2001a). doi:10.1046/j.1365-8711.2001.04341.x
M.J. Hardcastle, M. Birkinshaw, D.M. Worrall, Chandra observations of the X-ray jet in 3C 66B. Mon. Not. R. Astron. Soc. **326**, 1499–1507 (2001b). doi:10.1046/j.1365-8711.2001.04699.x
M.J. Hardcastle, M. Birkinshaw, R.A. Cameron, D.E. Harris, L.W. Looney, D.M. Worrall, Magnetic field strengths in the hot spots and lobes of three powerful Fanaroff-Riley type II radio sources. Astrophys. J. **581**, 948–973 (2002). doi:10.1086/344409
M.J. Hardcastle, D.E. Harris, D.M. Worrall, M. Birkinshaw, The origins of X-ray emission from the hot spots of FR II radio sources. Astrophys. J. **612**, 729–748 (2004). doi:10.1086/422808
M.J. Hardcastle, R.P. Kraft, G.R. Sivakoff, J.L. Goodger, J.H. Croston, A. Jordán, D.A. Evans, D.M. Worrall, M. Birkinshaw, S. Raychaudhury, N.J. Brassington, W.R. Forman, W.E. Harris, C. Jones, A.M. Juett, S.S. Murray, P.E.J. Nulsen, C.L. Sarazin, K.A. Woodley, New results on particle acceleration in the Centaurus A jet and counterjet from a deep Chandra observation. Astrophys. J. Lett. **670**, 81–84 (2007). doi:10.1086/524197
M.J. Hardcastle, F. Massaro, D.E. Harris, X-ray emission from the extended emission-line region of the powerful radio galaxy 3C171. Mon. Not. R. Astron. Soc. **401**, 2697–2705 (2010). doi:10.1111/j.1365-2966.2009.15855.x
P.E. Hardee, A. Rosen, On the dynamics and structure of three-dimensional trans-Alfvénic jets. Astrophys. J. **524**, 650–666 (1999). doi:10.1086/307833
D.E. Harris, J.E. Grindlay, The prospects for X-ray detection of inverse-Compton emission from radio source electrons and photons of the microwave background. Mon. Not. R. Astron. Soc. **188**, 25–37 (1979)
D.E. Harris, H. Krawczynski, X-ray emission from extragalactic jets. Annu. Rev. Astron. Astrophys. **44**, 463–506 (2006). doi:10.1146/annurev.astro.44.051905.092446
D.E. Harris, C.L. Carilli, R.A. Perley, X-ray emission from the radio hotspots of Cygnus A. Nature **367**, 713–716 (1994). doi:10.1038/367713a0
D.E. Harris, K.M. Leighly, J.P. Leahy, X-ray emission from a radio hot spot in 3C 390.3: Evidence for the deflection of a radio jet by a neighboring galaxy. Astrophys. J. Lett. **499**, 149 (1998). doi:10.1086/311365
D.E. Harris, P.E.J. Nulsen, T.J. Ponman, M. Bautz, R.A. Cameron, L.P. David, R.H. Donnelly, W.R. Forman, L. Grego, M.J. Hardcastle, J.P. Henry, C. Jones, J.P. Leahy, M. Markevitch, A.R. Martel, B.R. McNamara, P. Mazzotta, W. Tucker, S.N. Virani, J. Vrtilek, Chandra X-ray detection of the radio hot spots of 3C 295. Astrophys. J. Lett. **530**, 81–84 (2000). doi:10.1086/312503
S. Heinz, Radio lobe dynamics and the environment of microquasars. Astron. Astrophys. **388**, 40–43 (2002). doi:10.1051/0004-6361:20020402
P. Hennebelle, S. Fromang, Magnetic processes in a collapsing dense core. I. Accretion and ejection. Astron. Astrophys. **477**, 9–24 (2008). doi:10.1051/0004-6361:20078309
J. Heyvaerts, C. Norman, The collimation of magnetized winds. Astrophys. J. **347**, 1055–1081 (1989). doi:10.1086/168195
T. Hovatta, M. Lister, M. Aller, H. Aller, D. Homan, Y. Kovalev, A. Pushkarev, MOJAVE: monitoring of jets in active galactic nuclei with VLBA experiments. VIII. Faraday rotation in parsec-scale AGN jets. Astrophys. J. (2012, submitted)
M. Huarte-Espinosa, M. Krause, P. Alexander, 3D magnetohydrodynamic simulations of the evolution of magnetic fields in Fanaroff-Riley class II radio sources. Mon. Not. R. Astron. Soc. **417**, 382–399 (2011). doi:10.1111/j.1365-2966.2011.19271.x
P.A. Hughes, H.D. Aller, M.F. Aller, Synchrotron emission from shocked relativistic jets. II. A model for the centimeter wave band quiescent and burst emission from BL Lacertae. Astrophys. J. **341**, 68 (1989). doi:10.1086/167472
N. Isobe, M. Tashiro, K. Makishima, N. Iyomoto, M. Suzuki, M.M. Murakami, M. Mori, K. Abe, A Chandra detection of diffuse hard X-ray emission associated with the lobes of the radio galaxy 3C 452. Astrophys. J. Lett. **580**, 111–115 (2002). doi:10.1086/345658
S. Jester, H.J. Röser, K. Meisenheimer, R. Perley, X-rays from the jet in 3C 273: Clues from the radio-optical spectra. Astron. Astrophys. **385**, 27–30 (2002). doi:10.1051/0004-6361:20020278
S. Jester, H.J. Röser, K. Meisenheimer, R. Perley, The radio-ultraviolet spectral energy distribution of the jet in 3C 273. Astron. Astrophys. **431**, 477–502 (2005). doi:10.1051/0004-6361:20047021
N.N. Jetha, M.J. Hardcastle, A. Babul, E. O'Sullivan, T.J. Ponman, S. Raychaudhury, J. Vrtilek, The nature of the ghost cavity in the NGC 741 group. Mon. Not. R. Astron. Soc. **384**, 1344–1354 (2008). doi:10.1111/j.1365-2966.2007.12829.x

C.M. Johns-Krull, The magnetic fields of classical T Tauri stars. Astrophys. J. **664**, 975–985 (2007). doi:10.1086/519017

T.W. Jones, D. Ryu, A. Engel, Simulating electron transport and synchrotron emission in radio galaxies: Shock acceleration and synchrotron aging in axisymmetric flows. Astrophys. J. **512**, 105–124 (1999). doi:10.1086/306772

C.R. Kaiser, P. Alexander, A self-similar model for extragalactic radio sources. Mon. Not. R. Astron. Soc. **286**, 215–222 (1997)

J. Kataoka, Ł. Stawarz, X-ray emission properties of large-scale jets, hot spots, and lobes in active galactic nuclei. Astrophys. J. **622**, 797–810 (2005). doi:10.1086/428083

P. Kharb, D.C. Gabuzda, C.P. O'Dea, P. Shastri, S.A. Baum, Rotation measures across parsec-scale jets of Fanaroff-Riley type I radio galaxies. Astrophys. J. **694**, 1485–1497 (2009). doi:10.1088/0004-637X/694/2/1485

A. Konigl, Relativistic jets as X-ray and gamma-ray sources. Astrophys. J. **243**, 700–709 (1981). doi:10.1086/158638

A. Königl, Disk accretion onto magnetic T Tauri stars. Astrophys. J. Lett. **370**, 39–43 (1991). doi:10.1086/185972

R. Krasnopolsky, Z.Y. Li, R. Blandford, Magnetocentrifugal launching of jets from accretion disks. I. Cold axisymmetric flows. Astrophys. J. **526**, 631–642 (1999). doi:10.1086/308023

M. Krause, Very light jets II: Bipolar large scale simulations in King atmospheres. Astron. Astrophys. **431**, 45–64 (2005). doi:10.1051/0004-6361:20041191

P.P. Kronberg, R.V.E. Lovelace, G. Lapenta, S.A. Colgate, Measurement of the electric current in a kpc-scale jet. Astrophys. J. Lett. **741**, 15 (2011). doi:10.1088/2041-8205/741/1/L15

T. Kudoh, R. Matsumoto, K. Shibata, Are jets ejected from locally magnetized accretion disks? Proc. Astron. Soc. Jpn. **54**, 267–274 (2002)

R.A. Laing, A model for the magnetic-field structure in extended radio sources. Mon. Not. R. Astron. Soc. **193**, 439–449 (1980)

R.A. Laing, Interpretation of radio polarization data, in *Physics of Energy Transport in Extragalactic Radio Sources*, ed. by A.H. Bridle (1984), p. 90

R.A. Laing, A.H. Bridle, Relativistic models and the jet velocity field in the radio galaxy 3C 31. Mon. Not. R. Astron. Soc. **336**, 328–352 (2002). doi:10.1046/j.1365-8711.2002.05756.x

R.A. Laing, J.R. Canvin, A.H. Bridle, M.J. Hardcastle, A relativistic model of the radio jets in 3C296. Mon. Not. R. Astron. Soc. **372**, 510–536 (2006). doi:10.1111/j.1365-2966.2006.10903.x

R.A. Laing, A.H. Bridle, P. Parma, M. Murgia, Structures of the magnetoionic media around the Fanaroff-Riley Class I radio galaxies 3C31 and Hydra A. Mon. Not. R. Astron. Soc. **391**, 521–549 (2008). doi:10.1111/j.1365-2966.2008.13895.x

J.P. Leahy, A.R.S. Black, J. Dennett-Thorpe, M.J. Hardcastle, S. Komissarov, R.A. Perley, J.M. Riley, P.A.G. Scheuer, A study of FRII radio galaxies with $z < 0.15$—II. High-resolution maps of 11 sources at 3.6 CM. Mon. Not. R. Astron. Soc. **291**, 20–53 (1997)

C.F. Lee, L.G. Mundy, B. Reipurth, E.C. Ostriker, J.M. Stone, CO outflows from young stars: Confronting the jet and wind models. Astrophys. J. **542**, 925–945 (2000). doi:10.1086/317056

P. Lii, M. Romanova, R. Lovelace, Magnetic launching and collimation of jets from the disc-magnetosphere boundary: 2.5 D MHD simulations. Mon. Not. R. Astron. Soc. **420**, 2020–2033 (2012). doi:10.1111/j.1365-2966.2011.20133.x

M.L. Lister, D.C. Homan, MOJAVE: Monitoring of jets in active galactic nuclei with VLBA experiments. I. First-epoch 15 GHz linear polarization images. Astron. J. **130**, 1389–1417 (2005). doi:10.1086/432969

A.P. Lobanov, Ultracompact jets in active galactic nuclei. Astron. Astrophys. **330**, 79–89 (1998)

D. Lynden-Bell, Magnetic collimation by accretion discs of quasars and stars. Mon. Not. R. Astron. Soc. **279**, 389–401 (1996)

D. Lynden-Bell, On why discs generate magnetic towers and collimate jets. Mon. Not. R. Astron. Soc. **341**, 1360–1372 (2003). doi:10.1046/j.1365-8711.2003.06506.x

M. Lyutikov, V.I. Pariev, D.C. Gabuzda, Polarization and structure of relativistic parsec-scale AGN jets. Mon. Not. R. Astron. Soc. **360**, 869–891 (2005). doi:10.1111/j.1365-2966.2005.08954.x

M. Mahmud, Searching for helical magnetic fields in active galactic nuclei. PhD thesis, University College Cork (2010)

M. Mahmud, D. Gabuzda, Using Faraday rotation gradients to probe magnetic tower models, in *The Role of VLBI in the Golden Age for Radio Astronomy* (2008)

M. Mahmud, D.C. Gabuzda, V. Bezrukovs, Surprising evolution of the parsec-scale Faraday rotation gradients in the jet of the BL Lac object B1803+784. Mon. Not. R. Astron. Soc. **400**, 2–12 (2009). doi:10.1111/j.1365-2966.2009.15013.x

M. Mahmud, D. Gabuzda, C. Coughlan, D. Hallahan, Connecting magnetic towers with Faraday rotation gradients in active galactic nuclei jets. Mon. Not. R. Astron. Soc. (2012, submitted)

H.L. Marshall, M.J. Hardcastle, M. Birkinshaw, J. Croston, D. Evans, H. Landt, E. Lenc, F. Massaro, E.S. Perlman, D.A. Schwartz, A. Siemiginowska, Ł. Stawarz, C.M. Urry, D.M. Worrall, A flare in the jet of Pictor A. Astrophys. J. Lett. **714**, 213–216 (2010). doi:10.1088/2041-8205/714/2/L213
S. Matt, R.E. Pudritz, Does disk locking solve the stellar angular momentum problem? Astrophys. J. Lett. **607**, 43–46 (2004). doi:10.1086/421351
S. Matt, R.E. Pudritz, Accretion-powered stellar winds as a solution to the stellar angular momentum problem. Astrophys. J. Lett. **632**, 135–138 (2005). doi:10.1086/498066
S. Matt, R.E. Pudritz, Accretion-powered stellar winds. II. Numerical solutions for stellar wind torques. Astrophys. J. **678**, 1109–1118 (2008). doi:10.1086/533428
A.P. Matthews, P.A.G. Scheuer, Models of radio galaxies with tangled magnetic fields—Part two–Numerical simulations and their interpretation. Mon. Not. R. Astron. Soc. **242**, 623 (1990)
J.C. McKinney, General relativistic magnetohydrodynamic simulations of the jet formation and large-scale propagation from black hole accretion systems. Mon. Not. R. Astron. Soc. **368**, 1561–1582 (2006). doi:10.1111/j.1365-2966.2006.10256.x
J.C. McKinney, R.D. Blandford, Stability of relativistic jets from rotating, accreting black holes via fully three-dimensional magnetohydrodynamic simulations. Mon. Not. R. Astron. Soc. **394**, 126–130 (2009). doi:10.1111/j.1745-3933.2009.00625.x
J.C. McKinney, A. Tchekhovskoy, R.D. Blandford, General relativistic magnetohydrodynamic simulations of magnetically choked accretion flows around black holes. Mon. Not. R. Astron. Soc. (2012, in press)
R.R. Mellon, Z.Y. Li, Magnetic braking and protostellar disk formation: The ideal MHD limit. Astrophys. J. **681**, 1356–1376 (2008). doi:10.1086/587542
P. Mendygral, T. Jones, K. Dolag, MHD simulations of AGN jets in a dynamic galaxy cluster medium. Astrophys. J. (2012). doi:10.1088/0004-637X/750/2/166
L. Mestel, Magnetic braking by a stellar wind-I. Mon. Not. R. Astron. Soc. **138**, 359 (1968)
P. Mészáros, Gamma ray bursts. Astropart. Phys. (2012, in press)
F.C. Michel, Relativistic stellar-wind torques. Astrophys. J. **158**, 727 (1969). doi:10.1086/150233
A. Mignone, P. Rossi, G. Bodo, A. Ferrari, S. Massaglia, High-resolution 3D relativistic MHD simulations of jets. Mon. Not. R. Astron. Soc. **402**, 7–12 (2010). doi:10.1111/j.1365-2966.2009.15642.x
P. Mimica, M.A. Aloy, On the dynamic efficiency of internal shocks in magnetized relativistic outflows. Mon. Not. R. Astron. Soc. **401**, 525–532 (2010). doi:10.1111/j.1365-2966.2009.15669.x
P. Mimica, M.A. Aloy, Radiative signature of magnetic fields in internal shocks. Mon. Not. R. Astron. Soc. **421**, 2635–2647 (2012). doi:10.1111/j.1365-2966.2012.20495.x
P. Mimica, M.A. Aloy, I. Agudo, J.M. Martí, J.L. Gómez, J.A. Miralles, Spectral evolution of superluminal components in parsec-scale jets. Astrophys. J. **696**, 1142–1163 (2009). doi:10.1088/0004-637X/696/2/1142
I.F. Mirabel, Black hole jet sources, in *The Tenth Marcel Grossmann Meeting. On Recent Developments in Theoretical and Experimental General Relativity, Gravitation and Relativistic Field Theories*, ed. by M. Novello, S. Perez Bergliaffa, R. Ruffini (2005), p. 606. doi:10.1142/9789812704030_0037
I.F. Mirabel, L.F. Rodriguez, B. Cordier, J. Paul, F. Lebrun, A double-sided radio jet from the compact Galactic Centre annihilator 1E140.7-2942. Nature **358**, 215–217 (1992). doi:10.1038/358215a0
T.C. Mouschovias, Nonhomologous contraction and equilibria of self-gravitating, magnetic interstellar clouds embedded in an intercloud medium: Star formation. I Formulation of the problem and method of solution. Astrophys. J. **206**, 753–767 (1976). doi:10.1086/154436
L.M. Mullin, M.J. Hardcastle, Bayesian inference of jet bulk-flow speeds in Fanaroff-Riley type II radio sources. Mon. Not. R. Astron. Soc. **398**, 1989–2004 (2009). doi:10.1111/j.1365-2966.2009.15232.x
L.M. Mullin, J.M. Riley, M.J. Hardcastle, Observed properties of FRII quasars and radio galaxies at $z < 1.0$. Mon. Not. R. Astron. Soc. **390**, 595–621 (2008). doi:10.1111/j.1365-2966.2008.13534.x
E. Murphy, D.C. Gabuzda, Investigating the effects of finite resolution on observed transverse jet profiles. J. Phys. Conf. Ser. **355**(1), 012009 (2012). doi:10.1088/1742-6596/355/1/012009
E. Murphy, D. Gabuzda, T. Cawthorne, Gleaning secrets from the transverse poles of AGN jets, in *Proceedings of the 10th European VLBI Network Symposium and EVN Users Meeting: VLBI and the New Generation of Radio Arrays*, 20–24 September 2010, Manchester, UK (2010). Published online at http://pos.sissa.it/cgi-bin/reader/conf.cgi?confid=125id.41
E. Murphy, D. Gabuzda, T. Cawthorne, Analysing the transverse structure of the relativistic jets of AGN. Mon. Not. R. Astron. Soc. (2012, submitted)
S.T. Myers, S.R. Spangler, Synchrotron aging in the lobes of luminous radio galaxies. Astrophys. J. **291**, 52–62 (1985). doi:10.1086/163040
M. Nakamura, H. Li, S. Li, Structure of magnetic tower jets in stratified atmospheres. Astrophys. J. **652**, 1059–1067 (2006). doi:10.1086/508338
S.M. O'Neill, T.W. Jones, Three-dimensional simulations of bi-directed magnetohydrodynamic jets interacting with cluster environments. Astrophys. J. **710**, 180–196 (2010). doi:10.1088/0004-637X/710/1/180

S.M. O'Neill, I.L. Tregillis, T.W. Jones, D. Ryu, Three-dimensional simulations of MHD jet propagation through uniform and stratified external environments. Astrophys. J. **633**, 717–732 (2005). doi:10.1086/491618

S.P. O'Sullivan, D.C. Gabuzda, Magnetic field strength and spectral distribution of six parsec-scale active galactic nuclei jets. Mon. Not. R. Astron. Soc. **400**, 26–42 (2009a). doi:10.1111/j.1365-2966.2009.15428.x

S.P. O'Sullivan, D.C. Gabuzda, Three-dimensional magnetic field structure of six parsec-scale active galactic nuclei jets. Mon. Not. R. Astron. Soc. **393**, 429–456 (2009b). doi:10.1111/j.1365-2966.2008.14213.x

R. Ouyed, R.E. Pudritz, Numerical simulations of astrophysical jets from Keplerian disks. I. Stationary models. Astrophys. J. **482**, 712 (1997). doi:10.1086/304170

R. Ouyed, D.A. Clarke, R.E. Pudritz, Three-dimensional simulations of jets from Keplerian disks: Self-regulatory stability. Astrophys. J. **582**, 292–319 (2003). doi:10.1086/344507

R. Ouyed, R.E. Pudritz, J.M. Stone, Episodic jets from black holes and protostars. Nature **385**, 409–414 (1997). doi:10.1038/385409a0

G. Pelletier, R.E. Pudritz, Hydromagnetic disk winds in young stellar objects and active galactic nuclei. Astrophys. J. **394**, 117–138 (1992). doi:10.1086/171565

R.A. Perley, C.L. Carilli, The structure and polarization of Cygnus A at λ 3.6 cm, ed. by C.L. Carilli, D.E. Harris (1996), p. 168

R.A. Perley, H.J. Roser, K. Meisenheimer, The radio galaxy PictorA—a study with the VLA. Astron. Astrophys. **328**, 12–32 (1997)

T. Piran, The physics of gamma-ray bursts. Rev. Mod. Phys. **76**, 1143–1210 (2004). doi:10.1103/RevModPhys.76.1143

O. Porth, C. Fendt, Acceleration and collimation of relativistic magnetohydrodynamic disk winds. Astrophys. J. **709**, 1100–1118 (2010). doi:10.1088/0004-637X/709/2/1100

O. Porth, C. Fendt, Z. Meliani, B. Vaidya, Synchrotron radiation of self-collimating relativistic magnetohydrodynamic jets. Astrophys. J. **737**, 42 (2011). doi:10.1088/0004-637X/737/1/42

R.E. Pudritz, C.A. Norman, Centrifugally driven winds from contracting molecular disks. Astrophys. J. **274**, 677–697 (1983). doi:10.1086/161481

R.E. Pudritz, C.A. Norman, Bipolar hydromagnetic winds from disks around protostellar objects. Astrophys. J. **301**, 571–586 (1986). doi:10.1086/163924

R.E. Pudritz, C.S. Rogers, R. Ouyed, Controlling the collimation and rotation of hydromagnetic disc winds. Mon. Not. R. Astron. Soc. **365**, 1131–1148 (2006). doi:10.1111/j.1365-2966.2005.09766.x

R.E. Pudritz, R. Ouyed, C. Fendt, A. Brandenburg, Disk winds, jets, and outflows: theoretical and computational foundations, in *Protostars and Planets V*, ed. by B. Reipurth, D. Jewitt, K. Keil (2007), pp. 277–294

A.B. Pushkarev, D.C. Gabuzda, Y.N. Vetukhnovskaya, V.E. Yakimov, Spine-sheath polarization structures in four active galactic nuclei jets. Mon. Not. R. Astron. Soc. **356**, 859–871 (2005). doi:10.1111/j.1365-2966.2004.08535.x

T.P. Ray, Kelvin-Helmholtz instabilities in radio jets. Mon. Not. R. Astron. Soc. **196**, 195–207 (1981)

T. Ray, C. Dougados, F. Bacciotti, J. Eislöffel, A. Chrysostomou, Toward resolving the outflow engine: An observational perspective, in *Protostars and Planets V*, ed. by B. Reipurth, D. Jewitt, K. Keil (2007), pp. 231–244

A. Reichstein, D. Gabuzda, Studying magnetic fields in several parsec-scale AGN jets using Faraday rotation, in *Proceedings of the 10th European VLBI Network Symposium and EVN Users Meeting: VLBI and the New Generation of Radio Arrays*, 20–24 September 2010, Manchester, UK (2010). Published online at http://pos.sissa.it/cgi-bin/reader/conf.cgi?confid=125id.44

M.M. Romanova, G.V. Ustyugova, A.V. Koldoba, V.M. Chechetkin, R.V.E. Lovelace, Formation of stationary magnetohydrodynamic outflows from a disk by time-dependent simulations. Astrophys. J. **482**, 708 (1997). doi:10.1086/304199

M.M. Romanova, G.V. Ustyugova, A.V. Koldoba, R.V.E. Lovelace, Magnetohydrodynamic simulations of disk-magnetized star interactions in the quiescent regime: Funnel flows and angular momentum transport. Astrophys. J. **578**, 420–438 (2002). doi:10.1086/342464

M.M. Romanova, G.V. Ustyugova, A.V. Koldoba, R.V.E. Lovelace, MRI-driven accretion on to magnetized stars: Global 3D MHD simulations of magnetospheric and boundary layer regimes. Mon. Not. R. Astron. Soc. **421**, 63–77 (2012). doi:10.1111/j.1365-2966.2011.20055.x

P.A.G. Scheuer, Models of extragalactic radio sources with a continuous energy supply from a central object. Mon. Not. R. Astron. Soc. **166**, 513–528 (1974)

D.A. Schwartz, H.L. Marshall, J.E.J. Lovell, B.G. Piner, S.J. Tingay, M. Birkinshaw, G. Chartas, M. Elvis, E.D. Feigelson, K.K. Ghosh, D.E. Harris, H. Hirabayashi, E.J. Hooper, D.L. Jauncey, K.M. Lanzetta, S. Mathur, R.A. Preston, W.H. Tucker, S. Virani, B. Wilkes, D.M. Worrall, Chandra discovery of a 100 kiloparsec X-ray jet in PKS 0637-752. Astrophys. J. Lett. **540**, 69 (2000). doi:10.1086/312875

D. Seifried, R. Banerjee, R.S. Klessen, D. Duffin, R.E. Pudritz, Magnetic fields during the early stages of massive star formation—I. Accretion and disc evolution. Mon. Not. R. Astron. Soc. **417**, 1054–1073 (2011). doi:10.1111/j.1365-2966.2011.19320.x

D. Seifried, R.E. Pudritz, R. Banerjee, D. Duffin, R.S. Klessen, Magnetic fields during the early stages of massive star formation—II. A generalised outflow criterion. Mon. Not. R. Astron. Soc. (2012). doi:10.1111/j.1365-2966.2012.20610.x

F.H. Shu, J.R. Najita, H. Shang, Z.Y. Li, X-winds theory and observations, in *Protostars and Planets IV*, ed. by V. Mannings, A. Boss, S. Russell (2000), p. 789

R. Soria, M.W. Pakull, J.W. Broderick, S. Corbel, C. Motch, Radio lobes and X-ray hotspots in the microquasar S26. Mon. Not. R. Astron. Soc. **409**, 541–551 (2010). doi:10.1111/j.1365-2966.2010.17360.x

H.C. Spruit, Theory of magnetically powered jets, in *Lecture Notes in Physics*, vol. 794, ed. by T. Belloni (Springer, Berlin, 2010), p. 233. doi:10.1007/978-3-540-76937-8_9

J.E. Staff, B.P. Niebergal, R. Ouyed, R.E. Pudritz, K. Cai, Confronting three-dimensional time-dependent jet simulations with Hubble space telescope observations. Astrophys. J. **722**, 1325–1332 (2010). doi:10.1088/0004-637X/722/2/1325

Ł. Stawarz, M. Sikora, M. Ostrowski, High-energy gamma rays from FR I jets. Astrophys. J. **597**, 186–201 (2003). doi:10.1086/378290

F. Tavecchio, L. Maraschi, R.M. Sambruna, C.M. Urry, The X-ray jet of PKS 0637-752: Inverse Compton radiation from the cosmic microwave background? Astrophys. J. Lett. **544**, 23–26 (2000). doi:10.1086/317292

G.B. Taylor, R. Zavala, Are there rotation measure gradients across active galactic nuclei jets? Astrophys. J. Lett. **722**, 183–187 (2010). doi:10.1088/2041-8205/722/2/L183

I.L. Tregillis, T.W. Jones, D. Ryu, Simulating electron transport and synchrotron emission in radio galaxies: Shock acceleration and synchrotron aging in three-dimensional flows. Astrophys. J. **557**, 475–491 (2001). doi:10.1086/321657

I.L. Tregillis, T.W. Jones, D. Ryu, Synthetic observations of simulated radio galaxies. I. Radio and X-ray analysis. Astrophys. J. **601**, 778–797 (2004). doi:10.1086/380756

T. Velusamy, W.D. Langer, K.A. Marsh, Highly collimated jets and wide-angle outflows in HH 46/47: New evidence from Spitzer infrared images. Astrophys. J. Lett. **668**, 159–162 (2007). doi:10.1086/522929

T. Velusamy, W.D. Langer, M.S.N. Kumar, J.M.C. Grave, Jets and wide-angle outflows in Cepheus E: New evidence from Spitzer. Astrophys. J. **741**, 60 (2011). doi:10.1088/0004-637X/741/1/60

N. Vlahakis, A. Königl, Relativistic magnetohydrodynamics with application to gamma-ray burst outflows. I. Theory and semianalytic trans-Alfvénic solutions. Astrophys. J. **596**, 1080–1103 (2003). doi:10.1086/378226

B. von Rekowski, A. Brandenburg, Structured, dynamo driven stellar and disc winds, in *Asymmetrical Planetary Nebulae III: Winds, Structure and the Thunderbird*, ed. by M. Meixner, J.H. Kastner, B. Balick, N. Soker. Astronomical Society of the Pacific Conference Series, vol. 313 (2004), p. 476

J.F.C. Wardle, S.E. Aaron, How fast are the large-scale jets in quasars? Constraints on both Doppler beaming and intrinsic asymmetries. Mon. Not. R. Astron. Soc. **286**, 425–435 (1997)

E.T. Whelan, T.P. Ray, L. Podio, F. Bacciotti, S. Randich, Classical T Tauri-like outflow activity in the brown dwarf mass regime. Astrophys. J. **706**, 1054–1068 (2009). doi:10.1088/0004-637X/706/2/1054

A.S. Wilson, A.J. Young, P.L. Shopbell, Chandra X-ray observations of Pictor A: High-energy cosmic rays in a radio galaxy. Astrophys. J. **547**, 740–753 (2001). doi:10.1086/318412

A.S. Wilson, D.A. Smith, A.J. Young, The cavity of Cygnus A. Astrophys. J. Lett. **644**, 9–12 (2006). doi:10.1086/504108

N.L. Zakamska, M.C. Begelman, R.D. Blandford, Hot self-similar relativistic magnetohydrodynamic flows. Astrophys. J. **679**, 990–999 (2008). doi:10.1086/587870

C. Zanni, A. Ferrari, R. Rosner, G. Bodo, S. Massaglia, MHD simulations of jet acceleration from Keplerian accretion disks. The effects of disk resistivity. Astron. Astrophys. **469**, 811–828 (2007). doi:10.1051/0004-6361:20066400

R.T. Zavala, G.B. Taylor, A view through Faraday's fog: Parsec-scale rotation measures in active galactic nuclei. Astrophys. J. **589**, 126–146 (2003). doi:10.1086/374619

R.T. Zavala, G.B. Taylor, A view through Faraday's fog. II. Parsec-scale rotation measures in 40 active galactic nuclei. Astrophys. J. **612**, 749–779 (2004). doi:10.1086/422741

R.T. Zavala, G.B. Taylor, Faraday rotation measure gradients from a helical magnetic field in 3C 273. Astrophys. J. Lett. **626**, 73–76 (2005). doi:10.1086/431901

Space Sci Rev (2012) 169:123–157
DOI 10.1007/s11214-012-9909-x

Current Status of Turbulent Dynamo Theory

From Large-Scale to Small-Scale Dynamos

Axel Brandenburg · Dmitry Sokoloff · Kandaswamy Subramanian

Received: 28 March 2012 / Accepted: 6 June 2012 / Published online: 5 July 2012

Abstract Several recent advances in turbulent dynamo theory are reviewed. High resolution simulations of small-scale and large-scale dynamo action in periodic domains are compared with each other and contrasted with similar results at low magnetic Prandtl numbers. It is argued that all the different cases show similarities at intermediate length scales. On the other hand, in the presence of helicity of the turbulence, power develops on large scales, which is not present in non-helical small-scale turbulent dynamos. At small length scales, differences occur in connection with the dissipation cutoff scales associated with the respective value of the magnetic Prandtl number. These differences are found to be independent of whether or not there is large-scale dynamo action. However, large-scale dynamos in homogeneous systems are shown to suffer from resistive slow-down even at intermediate length scales. The results from simulations are connected to mean field theory and its applications. Recent work on magnetic helicity fluxes to alleviate large-scale dynamo quenching, shear dynamos, nonlocal effects and magnetic structures from strong density stratification are highlighted. Several insights which arise from analytic considerations of small-scale dynamos are discussed.

Keywords Magnetic fields · Turbulence · Sun: Magnetic fields

A. Brandenburg (✉)
Nordita, Royal Institute of Technology and Stockholm University, Roslagstullsbacken 23, 10691 Stockholm, Sweden
e-mail: brandenb@nordita.org

A. Brandenburg
Department of Astronomy, Stockholm University, 10691 Stockholm, Sweden

D. Sokoloff
Department of Physics, Moscow University, 119992 Moscow, Russia
e-mail: d_sokoloff@hotmail.com

K. Subramanian
Inter-University Centre for Astronomy and Astrophysics, Post Bag 4, Ganeshkhind, Pune 411 007, India
e-mail: kandu@iucaa.ernet.in

1 Introduction

Dynamos are now generally believed to be the main source of magnetic fields in many astrophysical environments. There used to be a time when dynamo critics tried to explain the magnetic fields in galaxies and galaxy clusters through winding up a primordial magnetic field. However, maintaining tangled magnetic fields seen in these systems against turbulent decay requires dynamo action; otherwise the Lorentz force due to the field would rapidly transfer magnetic energy to kinetic energy, which in turn can be dissipated by viscosity or by driving turbulent motions. Thus it becomes hard to make a convincing case for purely primordial magnetic fields, without any dynamo action, to explain observed cosmic magnetism. An exception is perhaps the space between clusters of galaxies, in particular the voids which are empty of galaxies. The weak fields recently claimed to be detected in such void regions from Gamma Ray astronomy (Neronov and Vovk 2010), would, if substantiated, be difficult (but not impossible) to explain by purely astrophysical mechanisms, and could point to primordial fields.

Obvious examples where only dynamos can explain the observed fields are planetary dynamos such as the geodynamo, for which the resistive decay time would be around 1000 years. In stars the global field is of large enough scale so that the resistive time would be comparable with their life time, but the Sun's convection zone is turbulent, and the turbulent decay time is much shorter, just enough to explain the 22 year activity cycle of the Sun's magnetic field. Similar arguments also apply to galaxies, which are also turbulent and have a thickness small enough for turbulent diffusion to destroy the field within a few billion years. Clusters of galaxies come in great varieties and their magnetic fields vary correspondingly. The suspicion is that their magnetic fields are reinforced by mergers with other clusters that tend to occur on time scales of about 2 Gyr.

Dynamos convert kinetic energy into magnetic energy. In the astrophysical context, one always means by a dynamo a self-excited device, except that the conductivity is uniformly distributed and not limited to conducting wires in an insulating container, as in laboratory dynamos. Since the beginning of the space age, it is well understood that such a device can work, at least in principle. The discovery of the first rigorously proven example by Herzenberg (1958) was significant, even though the particular configuration studied by him was not of immediate astrophysical interest. The Herzenberg configuration consists of a uniformly conducting solid medium (e.g., copper) with two or more conducting rotors in electrical contact with the rest. Dynamo action is possible when the rotors spin faster than a certain critical value that depends on the angle between the rotors (Gailitis 1973; Gibson and Roberts 1967; Moffatt 1978). For angles between 90 and 180 degrees the solutions are non-oscillatory, while for angles between 0 and 90 degrees there are oscillatory solutions that were discovered only more recently (Brandenburg et al. 1998).

The Herzenberg dynamo belongs to the class of "slow" dynamos, for which the dynamo growth rate is maximum for intermediate values of the conductivity, but it returns to zero in the limit of perfect conductivity. Another example of a slow dynamo is the Roberts flow, which consists of a steady two-dimensional flow pattern, $\boldsymbol{u} = \boldsymbol{u}(x, y)$, but all three flow components are non-vanishing. This dynamo was first studied by Roberts (1970, 1972). It provides an important benchmark of a large-scale dynamo, where magnetic field on scales larger than the scale of motions are produced by a mechanism called the α effect. Here, α refers to the name of the coefficient in the relation between mean magnetic field $\overline{\boldsymbol{B}}$ and mean electromotive force $\overline{\boldsymbol{\mathcal{E}}}$ in a turbulent flow. The basic idea goes back to Parker (1955) who proposed a relation of the form $\overline{\boldsymbol{\mathcal{E}}} = \alpha\overline{\boldsymbol{B}}$. Later, Steenbeck et al. (1966) computed a tensorial relation of the form $\overline{\mathcal{E}}_i = \alpha_{ij}\overline{B}_j$ for rotating stratified turbulence. It was clear that higher derivatives of the magnetic field would also enter, so a more general expression is

$$\overline{\mathcal{E}}_i = \alpha_{ij} \circ \overline{B}_j + \eta_{ijk} \circ \overline{B}_{j,k}, \tag{1}$$

where the isotropic part of the tensor η_{ijk}, $\eta_{ijk} = \eta_t \epsilon_{ijk}$, corresponds to turbulent magnetic, diffusion where η_t is the *turbulent* magnetic diffusivity. The circles indicate convolution over time and space, which is however commonly replaced by a multiplication in the limit in which the integral kernels are well approximated by δ functions in space and time. The work of Steenbeck et al. (1966) marked the beginning of mean-field electrodynamics, which is still the leading theory to explain the amplification of magnetic fields on length scales that are larger than the scale of the energy-carrying turbulent eddies. Such systems are therefore also referred to as large-scale dynamos. Essential here is that the trace of the α tensor is non-vanishing. This typically requires helicity of streamlines (at low conductivity) or of vortex lines (at high conductivity).

In this review we discuss our current knowledge of dynamos covering both large-scale and small-scale turbulent dynamos. In practice, all dynamos of astrophysical relevance tend to be "fast" and thus have a finite growth rate also in the limit of perfect conductivity. There are examples of predetermined flows for which fast dynamo action is indicated by numerical simulations, for example the ABC flow (Galloway and Frisch 1986) and the Galloway and Proctor (1992) flow, but the convergence of the α effect with increasing magnetic Reynolds number can be quite slow or non-existent (see, e.g., Courvoisier et al. 2006). By contrast, when relaxing the assumption of a predetermined kinematic flow pattern and adopting a turbulent flow field, the α effect appears to be converged for magnetic Reynolds numbers exceeding a critical value of the order of unity up to values of about 200 probed in the simulations (Sur et al. 2008).

We begin by discussing examples of numerical realizations of turbulent dynamos and then turn to some astrophysical examples. Large-scale dynamos produce fields that are well characterized by suitable averages. A theory for describing the evolution of such averaged fields is mean-field theory that is obtained by averaging the governing equations, most notably the induction equation. Large-scale dynamos can then also be referred to as mean-field dynamos (MFD). Under certain conditions, making suitable assumptions about the spatial variation of the α effect, solutions of the mean-field induction equation have been used to characterize the large-scale magnetic fields seen in astrophysical bodies like the Sun and some spiral galaxies. Mean-field theory is also applied to the momentum equation. This leads to a number of effects including turbulent viscosity, the Λ effect (responsible for driving differential solar rotation; see Rüdiger 1980, 1989), the anisotropic kinetic alpha effect (Frisch et al. 1987), as well as the negative effective magnetic pressure effect (Kleeorin et al. 1989). In this paper, we also discuss important effects resulting from the mean-field momentum equation, namely the spontaneous formation of magnetic flux concentrations in a strongly stratified layer. Such results may be relevant to explaining the formation of active regions in the Sun.

There is now increased interest in what is known as small-scale or fluctuation dynamos. These are dynamos that can work already under fully isotropic conditions and were anticipated by Batchelor (1950) and others (Biermann and Schlüter 1951; Elsasser 1956), but the now accepted theory was provided by Kazantsev (1967). Fluctuation (or small-scale) dynamos are important in cosmic objects because they are generic to any random flow of a sufficiently conducting plasma. Furthermore, the growth rate is fast and random fields can grow in the beginning exponentially on the eddy turnover timescale of the smallest eddies (Subramanian 1998), and later linearly on the turnover timescale of the larger eddies

(Beresnyak 2012), which are both typically much shorter than the age of the system. This is particularly true of galaxy clusters, where the typical scale and velocity of the turbulent eddies are respectively around 100 kpc and 100 $\mathrm{km\,s^{-1}}$ giving a growth time $\sim 10^8$ yr. Thus, small-scale dynamos should operate and are crucial to explaining the observed magnetic fields on the scale of tens of kiloparsecs. Indeed, it would be difficult to explain magnetic fields on larger scales, because the conditions for large-scale dynamo action are probably absent (Subramanian et al. 2006).

Small-scale dynamos are nowadays also invoked to describe the small-scale magnetic field at the solar surface. However, in many contexts both small-scale and large-scale dynamos go together. And then it is not clear whether one can (or should) distinguish between a small-scale field from a small-scale dynamo from that associated with the fluctuations that are inherent to any large-scale dynamo and that can be caused by tangling and amplification of the large-scale field.

2 Turbulent Dynamos in Various Astrophysical Settings

Small-scale dynamos are excited whenever the magnetic Reynolds number exceeds a certain critical value. Throughout this paper, we adopt the following definition for this number:

$$\mathrm{Re}_M = u_{\mathrm{rms}}/\eta k_{\mathrm{f}}, \tag{2}$$

where $u_{\mathrm{rms}} = \langle \boldsymbol{u}^2 \rangle^{1/2}$ is the root mean square (rms) value of the turbulent velocity, η is the magnetic diffusivity, and k_{f} is the wavenumber of energy injection. A similar definition is adopted for the fluid Reynolds number, $\mathrm{Re} = u_{\mathrm{rms}}/\nu k_{\mathrm{f}}$, where ν is the kinematic viscosity. We also define the magnetic Prandtl number, $\mathrm{Pr}_M = \nu/\eta = \mathrm{Re}_M/\mathrm{Re}$.

The values of Re_M, Re, and Pr_M have been compiled by Brandenburg and Subramanian (2005a) for several astrophysical bodies using Spitzer values for both magnetic diffusivity and kinematic viscosity. Both Re_M and Re are generally quite large—much larger than what can be dealt with in numerical simulations. However, the ratio of both numbers, Pr_M, is usually less than unity in denser environments such as stars and larger than unity in tenuous gases such as galaxies. The Spitzer resistivity decreases with temperature proportional to the 3/2 power, while the kinematic viscosity increases with temperature to the 5/2 power, so the magnetic Prandtl number increases with temperature to the fourth power. It is however also inversely proportional to the density, but because the temperature is generally around 10^4 K both in the warm interstellar medium and in the upper layers of the Sun, Pr_M would be unity or larger when

$$\rho \le 1.1 \times 10^{-13}\,\mathrm{g\,cm^{-3}} \left(\frac{T}{10^4\,\mathrm{K}}\right)^4 \left(\frac{\ln \Lambda}{20}\right)^{-2} \quad \text{for } \mathrm{Pr}_M \ge 1, \tag{3}$$

where $\ln \Lambda$ is in the range between 5 and 20. Since a density of $10^{-13}\,\mathrm{g\,cm^{-3}}$ is rather atypical of any astrophysical settings, we do not expect Pr_M to be of the order of unity except in numerical simulations where this value is the easiest one to deal with. This will be discussed in more detail in Sect. 3.

3 Numerical Realizations of Turbulent Dynamos

Next, we discuss some numerical realizations of turbulent dynamos. By "turbulence" we mean here flows that are solutions to the Navier-Stokes equations that are irregular in space and time, subject to energy injection at large length scales and energy dissipation at short length scales. The ratio between forcing scale to dissipation scale is quantified by the Reynolds number.

At small Reynolds numbers, $\mathrm{Re} \ll 1$, the flow is determined by viscosity and has qualitatively different properties from flows with $\mathrm{Re} \gg 1$. The break point is close to unity, so, for example, a flow with $\mathrm{Re} = 5$ would be called turbulent (or at least mildly turbulent), because it begins to show certain asymptotic scaling properties that are also found for fully turbulent flows. The most famous aspect of fluid turbulence is the Kolmogorov $E_{\mathrm{K}}(k) \sim k^{-5/3}$ energy spectrum, which is normalized such that $\int E_{\mathrm{K}}\,\mathrm{d}k = u_{\mathrm{rms}}^2/2$. Density is omitted in this definition, which requires that density fluctuations are unimportant and the flow is nearly incompressible. In this review we also discuss stratified flows in which density varies strongly due to gravity. In those cases the incompressibility condition $\nabla \cdot \boldsymbol{u} = 0$ has to be replaced by $\nabla \cdot \rho \boldsymbol{u} = 0$, which is valid as long as the flows are slow compared with the sound speed and the typical scales of variation smaller than a scale height. However, in the following we do not make such assumptions and consider fully compressible flows.

3.1 Small-Scale and Large-Scale Dynamos

We begin by discussing first the difference between small-scale and large-scale dynamos. A small-scale dynamo is one that generates magnetic field at scales much smaller than that of the energy-carrying eddies, while a large-scale dynamo generates field at scales larger than that of the energy-carrying eddies. In many practical circumstances, the difference is of somewhat academic interest, because the small-scale dynamo is always excited when the magnetic Reynolds number is large, which is the case in many astrophysical applications. Furthermore, the conditions for the excitation of large-scale dynamo action are met in many situations of interest. However, in the case of isotropic turbulence such a distinction can be made by considering helical and non-helical turbulence. In both cases the system can be homogeneous and it makes sense to compute spectra of magnetic and kinetic energy, $E_{\mathrm{M}}(k)$ and $E_{\mathrm{K}}(k)$, respectively; see Fig. 1 for $\mathrm{Pr}_M = 1$. The early evolution of such a dynamo is quite similar: both dynamos have a $k^{3/2}$ power spectrum at small scales. Such a scaling was predicted by Kazantsev (1967) (see also Kulsrud and Anderson (1992)) for a single scale non-helical flow which was δ-correlated in time, but seems to be obtained in the simulations more generally. However, helical turbulence allows the development of an inverse cascade (Frisch et al. 1975) on a longer resistive timescale (see below).

We see that, at least in the saturated state, large-scale dynamos produce and sustain magnetic fields at scales larger than the energy injection scale, while small-scale dynamos produce and sustain magnetic fields at scales smaller than the energy injection scale. The lack of similar behavior in the linear regime could be interpreted as evidence that the underlying mechanism for producing large-scale fields must be nonlinear in nature (Herault et al. 2011). Alternatively, one could interpret the resulting dynamo as a combination (or even a superposition) of large-scale and small-scale dynamo action, where the latter has a larger growth rate such that in the kinematic regime the field is dominated by the small-scale field, although the large-scale dynamo does still operate in the background. As a consequence of such a scenario, it gets the chance to dominate only when the small-scale dynamo has already saturated. Even in this case, and if Re_M is large, we will argue that one needs to additionally dissipate small-scale magnetic helicity before the largest scale field can appear.

Both large-scale and small-scale dynamos show that the spectral magnetic energy exceeds the kinetic spectral energy in the beginning of the inertial range, i.e., for wavenumbers larger than that of the magnetic energy maximum. This difference is slightly weaker

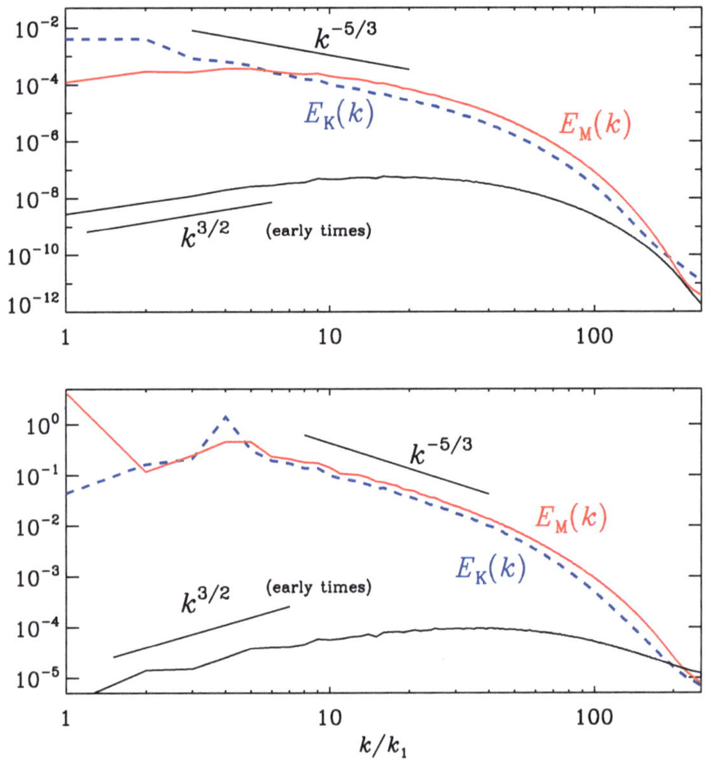

Fig. 1 Kinetic and magnetic energy spectra in a turbulence simulation without net helicity (*up*) and with net helicity (*bottom*) for a magnetic Prandtl number of unity and a mesh size is 512^3 meshpoints. Notice the pronounced peak of $E_M(k)$ at $k = k_1$ in the case with helicity. The energy input wavenumbers are $k_f = 1.5k_1$ in the non-helical case (*upper panel*, $\mathrm{Re}_M = 600$, $\mathrm{Pr}_M = 1$) and $k_f/k_1 = 4$ in the helical case (*lower plot*, $\mathrm{Re}_M = 450$, $\mathrm{Pr}_M = 1$). Adapted from Brandenburg and Subramanian (2005a) and Brandenburg et al. (2008)

for large-scale dynamos, but this might be an artifact of the Reynolds number still not being large enough. Such a difference used to be completely absent at Reynolds numbers previously reported; see, for example Brandenburg (2001). The slight super-equipartition was quite evident when simulations at a resolution of 1024^3 meshpoints became first available (Haugen et al. 2003), although this feature can already be seen in earlier simulations (Meneguzzi et al. 1981; Kida et al. 1991). This spectral excess of magnetic fields is expected to diminish as one proceeds further down the inertial range. Such behavior was indeed seen in simulations of Haugen and Brandenburg (2006) when using a combination of Smagorinsky viscosity for the velocity field and hyperresistivity for the magnetic field; see Fig. 2. This implies that the two spectra cannot be parallel to each other at intermediate length scales, and that the slope of $E_M(k)$ must be slightly steeper than that of $E_K(k)$. This difference in the two slopes at intermediate wavenumbers is now associated with the observed differences in the spectral exponents in the solar wind; see Boldyrev et al. (2011), who find steeper spectra for the magnetic field than the velocity field both from simulations and solar wind observations.

In the spectra of Fig. 1 we see a remarkable difference between small-scale and large-scale dynamos. In particular, large-scale dynamos are capable of producing a peak of magnetic energy at the smallest possible wavenumber, while small-scale dynamos do not. On

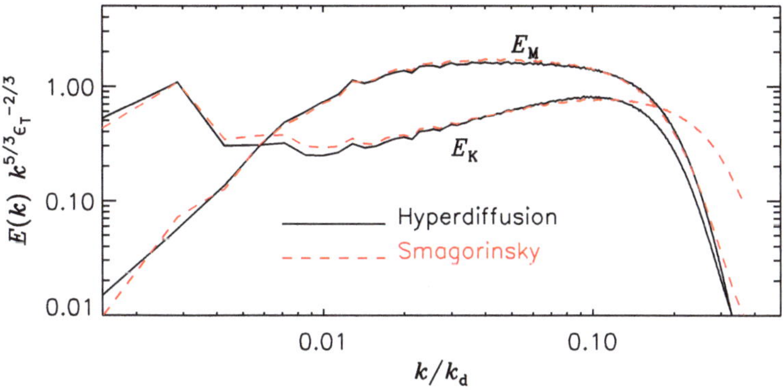

Fig. 2 Magnetic and kinetic energy spectra for runs with 512^3 meshpoints and hyperviscosity with hyperresistivity (*solid line*) and Smagorinsky viscosity with hyperresistivity (*red, dashed line*). Note the mutual approach of kinetic and magnetic energy spectra before entering the dissipative subrange. Adapted from Haugen and Brandenburg (2006)

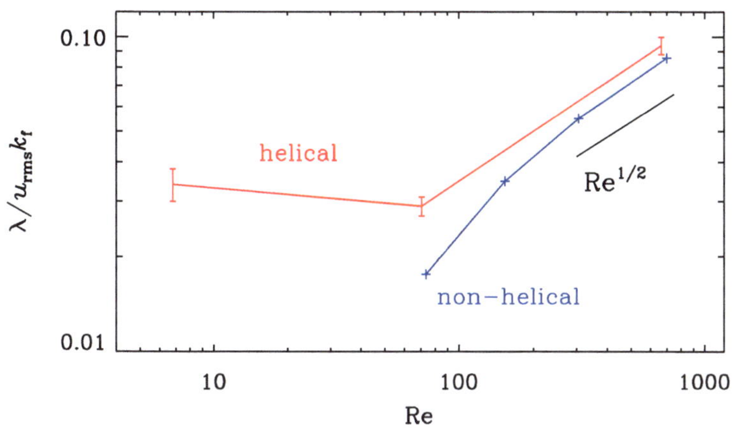

Fig. 3 Dependence of dynamo growth rates of the rms magnetic field on Re_M for helical and non-helical turbulence. Adapted from Brandenburg (2009)

the other hand, we have stated earlier that small-scale and large-scale dynamos are difficult to distinguish in the early stage. In Fig. 3 we show the critical values of Re_M ($=$ Re) for dynamo action both for small-scale (non-helical) and large-scale (helical) dynamos. Note that for $Re_M > 35$ the growth rate attains a $Re^{1/2}$ scaling both for small-scale and large-scale dynamos. The same growth rates are also obtained for dynamos driven by convection; see Fig. 15 of Käpylä et al. (2008). This $Re^{1/2}$ scaling of the growth rate of the rms magnetic field, implies that the growth rate is not controlled by the turnover time of the energy-carrying eddies, but of eddies at the dissipation scale (Schekochihin et al. 2002, 2004a). At low values of Re_M, only large-scale dynamo action remains possible. Its excitation condition is determined by the requirement that a certain dynamo number exceeds a critical value. This usually translates to the condition that the degree of scale separation is large enough; see (5) of Brandenburg (2009).

An important issue for astrophysical applications is how coherent are the fields generated by small-scale dynamos (Subramanian et al. 2006; Cho and Ryu 2009). On the basis of simulations done with large Pr_M, but small Re, Schekochihin et al. (2004b) argued that the field saturates with a folded structure, where the field reverses at the folds such that power concentrates on resistive scales. The simulations of Haugen et al. (2003, 2004a) with $\mathrm{Re}_M = \mathrm{Re} \gg 1$, found the magnetic correlation lengths $\sim 1/6$th the velocity correlation length, but much larger than the resistive scale. This seems consistent with the simple Subramanian (1999) model for nonlinear saturation of small-scale dynamos. What happens at large Re and large Pr_M, which is representative of galactic and cluster plasmas, is not easy to capture in simulations. The $\mathrm{Pr}_M = 50$, $\mathrm{Re} = 80$ simulation described in Brandenburg and Subramanian (2005a), showed evidence for strong field regions with folds, but equally, regions with strong fields and no folds, illustrating that such structures need not be volume filling. Moreover, the 'spontaneous stochasticity', that applies to highly turbulent flows (Eyink 2011), suggests that the dynamics of small-scale dynamos could be quite different in turbulent compared to laminar high-Pr_M systems. Furthermore, in galaxy clusters the viscosity may be set by plasma effects (Schekochihin et al. 2005a). It could also be highly anisotropic owing to the presence of magnetic fields (Parrish et al. 2012). In addition, heat conduction is also very anisotropic, giving rise to magnetothermal and heat flux-driven buoyancy instabilities (Parrish and Stone 2005; Parrish and Quataert 2008). Further work on these aspects is desirable, using both semi-analytical ideas and high resolution simulations.

3.2 Low Magnetic Prandtl Number

In many astrophysical settings, like solar, stellar, or accretion disk plasmas, the magnetic Prandtl number is rather low ($\sim 10^{-4}$ or less). This becomes numerically hard to handle, especially if the magnetic Reynolds number should still be large enough to support dynamo action. In Fig. 4 we compare non-helical and helical runs with magnetic Prandtl numbers of 0.02 and 0.01, respectively. In the former case the magnetic Reynolds number Re_M is 230, which is weakly supercritical; the critical value of Re_M for small values of Pr_M is $R_{\mathrm{m,crit}} \approx 150$ (Brandenburg 2011a), compared with $R_{\mathrm{m,crit}} \approx 35$ for $\mathrm{Pr}_M = 1$ (Haugen et al. 2004a). These values agree with those of earlier work (Schekochihin et al. 2005b, 2007; Iskakov et al. 2007). An increase of the critical value of Re_M follows also from considerations of the Kazantsev model (Rogachevskii and Kleeorin 1997; Boldyrev and Cattaneo 2004; Schober et al. 2012). So, for $\mathrm{Re}_M = 230$ and $\mathrm{Pr}_M = 0.02$, we have $\mathrm{Re} = \mathrm{Re}_M/\mathrm{Pr}_M = 11,500$. Normally, this would require a numerical resolution of about $10,000^3$ meshpoints, but it turns out that at low values of Pr_M, most of the energy is dissipated resistively, leaving thus very little kinetic energy to be cascaded, terminating therefore the kinetic energy cascade earlier than at $\mathrm{Pr}_M = 1$.

At intermediate length scales, kinetic and magnetic energy spectra are close to each other. The magnetic energy spectrum no longer exceeds the kinetic energy spectrum, as was found for $\mathrm{Pr}_M = 1$; see Fig. 1. Again, this might be a consequence of still insufficiently large Reynolds numbers and limited resolution. In fact, it is plausible that, in the limit of large fluid and magnetic Reynolds numbers, kinetic and magnetic energy spectra coincide, even if Pr_M is small. And only at much smaller scales, the magnetic energy spectrum turns into a dissipative subrange, and goes below the kinetic energy spectrum, due to stronger Ohmic dissipation. The slopes of the $k^{-11/3}$ spectrum of Golitsyn (1960) and Moffatt (1961) and the scale-invariant k^{-1} spectrum (Ruzmaikin and Shukurov 1982; Kleeorin and Rogachevskii 1994; Kleeorin et al. 1996) are shown for comparison. However, the kinetic energy spectrum tends to be steeper than $k^{-5/3}$ and is closer to $k^{-1.9}$ and $k^{-7/3}$ in the non-helical and helical cases.

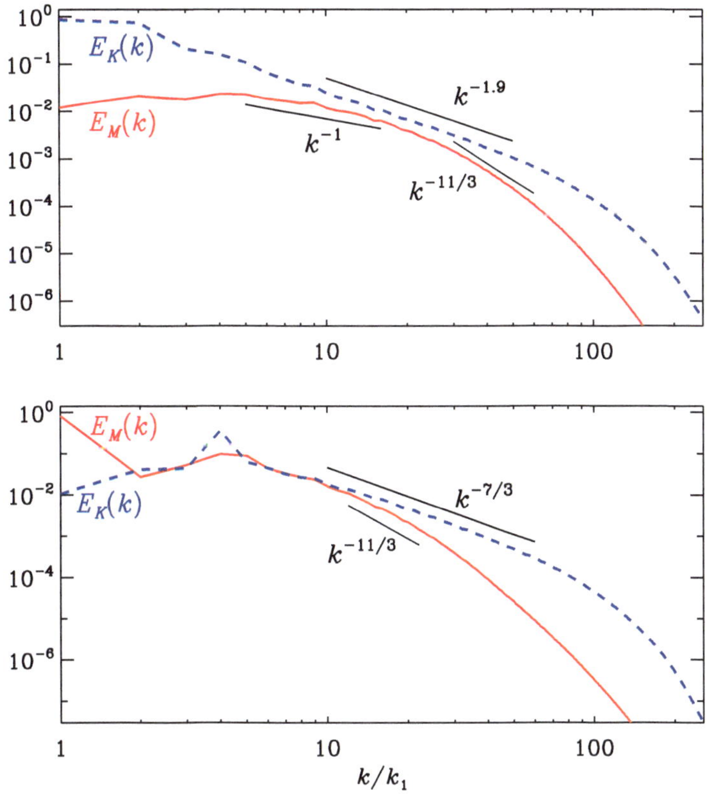

Fig. 4 Kinetic and magnetic energy spectra for non-helical (*top*) and helical (*bottom*) turbulence at low magnetic Prandtl numbers of $\mathrm{Pr}_M = 0.02$ and 0.01, respectively. Here, $\mathrm{Re}_M = 230$ with $\mathrm{Re} = 11,500$ in the non-helical case, adapted from Brandenburg (2011a), and $\mathrm{Re}_M = 23$ with $\mathrm{Re} = 2300$ in the helical case adapted from Brandenburg (2009). In both cases the resolution is 512^3 mesh points

3.3 Inverse Transfer and α Effect

The dynamics of a large-scale dynamo is most dramatic when the scale separation ratio is large, i.e., $k_f/k_1 \gg 1$. In Fig. 1 it was only 4, but now we consider a case where $k_f/k_1 = 15$. In Fig. 5 we show visualizations of one component of the field for different times. Evidently, a large-scale field develops that varies in the y direction. This particular large-scale field is best described by xz averages. In Fig. 6 we show the evolution of the mean energy density of this large-scale magnetic field, $\langle \overline{\boldsymbol{B}}^2 \rangle$, and compare it with that of the small-scale field, $\langle \boldsymbol{b}^2 \rangle = \langle \boldsymbol{B}^2 \rangle - \langle \overline{\boldsymbol{B}}^2 \rangle$. Note that the small-scale field reaches its final saturation value during the time span considered, while the large-scale field has not yet saturated and is expected to do so on the diffusive time scale of the box such that $t\eta k_1^2 = O(1)$. It is also interesting to note that the large-scale field starts becoming noticeable only when $t\eta k_1^2 \sim 0.02$, or when $(t/t_d) \sim 4.5$, where $t_d = (\eta k_f^2)^{-1}$ is the resistive timescale at the forcing scale. In other words one needs several resistive (diffusive) times at the forcing scale, before the large-scale field can grow. We will interpret this result below in terms of the resistive dissipation of small-scale magnetic helicity which alleviates α quenching.

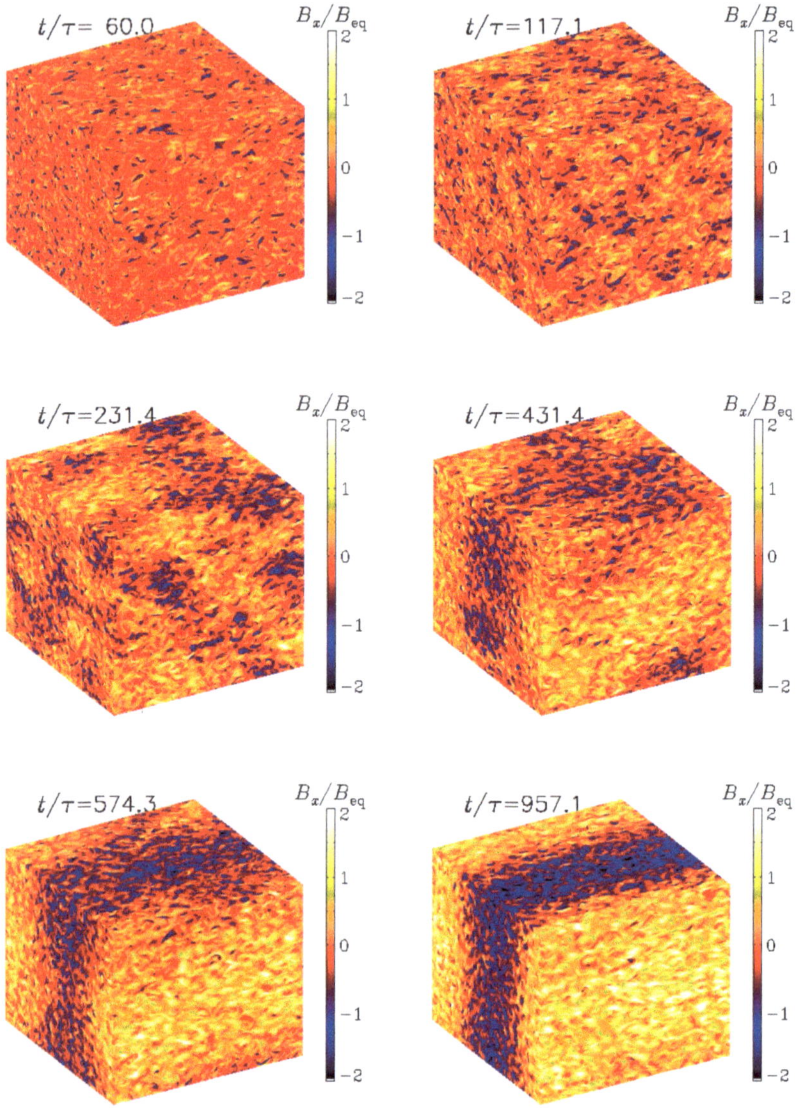

Fig. 5 Visualizations of $B_x/B_{\rm eq}$ on the periphery of the domain at six times during the late saturation stage of the dynamo when a large-scale field is gradually building up. The small-scale field has reached its final value after $t/\tau \approx 100$ turnover times. Here, $B_{\rm eq} = \sqrt{\mu_0\rho_0}\,u_{\rm rms}$ is the equipartition field strength where kinetic and magnetic energy densities are comparable, and ρ_0 is the mean density. Note that the maximum field strength is about twice $B_{\rm eq}$

The evolution of the magnetic energy spectrum for this case is shown in Fig. 7, where we see several stages during the early phase of the dynamo, and in a separate panel the later saturation behavior. Clearly, a large-scale field is already present for $t/\tau > 100$, but the field is then still fairly isotropic and therefore not very pronounced in visualizations shown in Fig. 5. Spectra of magnetic energy and rescaled magnetic helicity are shown in Fig. 8 for the saturated state. Here, magnetic helicity spectra $H_{\rm M}(k)$ are normalized such

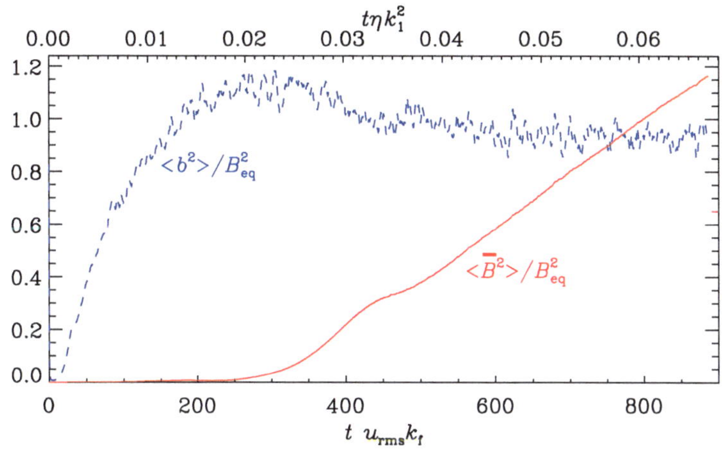

Fig. 6 Saturation of the small-scale magnetic energy density and continued increase of the large-scale magnetic energy density. Here, $k_f/k_1 = 15$ and $\mathrm{Re}_M \approx 60$

that $\int H_M\,\mathrm{d}k = \langle \boldsymbol{A} \cdot \boldsymbol{B}\rangle$, where $\boldsymbol{B} = \nabla \times \boldsymbol{A}$ is the magnetic field expressed in terms of its vector potential. Note that at early times, the magnetic field shows the Kazantsev $k^{3/2}$ slope ($t/\tau = 10$) in the range $7 \le k/k_1 \le 25$. However, already at times $t/\tau = 20$ and 30 one sees a small hump at $k_f/2$, which is significant in view of an interpretation of these results in terms of a so-called α^2 dynamo, which will be discussed in Sect. 3.4.

The temporal increase of the typical scale of the large-scale field can be determined by monitoring the quantity

$$\ell_m(t) \equiv k_m^{-1}(t) = \int k^{-1} E_M(k)\,\mathrm{d}k \Big/ \int E_M(k)\,\mathrm{d}k. \tag{4}$$

In Fig. 9 we plot $k_m^{-1}(t)$ for different values Re_M. Here, we normalize with respect to the minimal wavenumber $k_1 = 2\pi/L$ in the domain of size L, so we plot k_1/k_m. There is a limited scaling range that allows us to determine the increase of $\ell_m(t)$ for different values of Re_M. We measure the speed at which the bump travels toward larger scales by the quantity $\alpha_{\mathrm{trav}} = \mathrm{d}\ell_m/\mathrm{d}t$. In the right-hand panel of Fig. 9 we plot the resulting values of α_{trav} as a function of Re_M. The results are compatible with a resistively limited increase whose speed diminishes like $\mathrm{Re}_M^{-1/2}$. This behavior was first seen in simulations of Brandenburg et al. (2002) and then more convincingly at larger resolution in simulations of Brandenburg (2011b). Such a behavior further reinforces our earlier remark that the large-scale field can only grow on the slow resistive timescale in periodic boxes.

3.4 Connection with Mean-Field Theory

There exists a close analogy between the inverse transfer described above and mean-field dynamo theory in that both are able to predict the occurrence of large-scale fields with similar properties and excitation conditions. In mean-field theory one splits the velocity field $\boldsymbol{U}$ and magnetic field $\boldsymbol{B}$ into the sum of mean, large-scale components ($\overline{\boldsymbol{U}}$ and $\overline{\boldsymbol{B}}$) and turbulent, stochastic components ($\boldsymbol{u}$ and $\boldsymbol{b}$); that is $\boldsymbol{U} = \overline{\boldsymbol{U}} + \boldsymbol{u}$ and $\boldsymbol{B} = \overline{\boldsymbol{B}} + \boldsymbol{b}$. One then solves the averaged induction equation,

$$\frac{\partial \overline{\boldsymbol{B}}}{\partial t} = \nabla \times (\overline{\boldsymbol{U}} \times \overline{\boldsymbol{B}} + \overline{\boldsymbol{\mathcal{E}}} - \eta\mu_0 \overline{\boldsymbol{J}}), \tag{5}$$

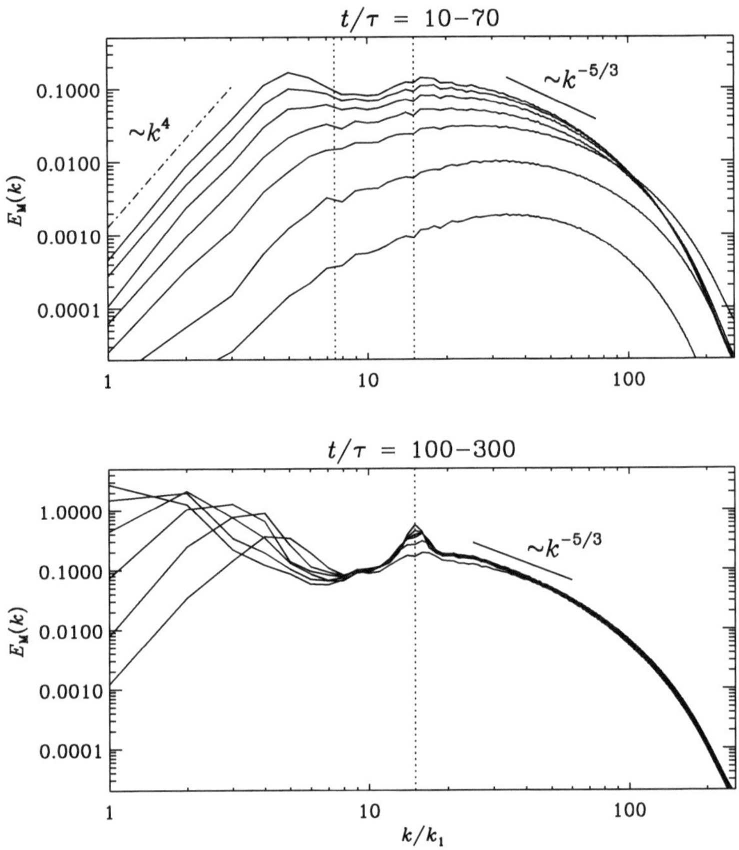

Fig. 7 Magnetic energy spectra $E_{\mathrm{M}}(k)$, at earlier (*top*) and later (*bottom*) times. The scale separation ratio is $k_{\mathrm{f}}/k_1 = 15$. The range of time t is given in units of the turnover time, $\tau = 1/u_{\mathrm{rms}}k_{\mathrm{f}}$. At small wavenumbers, the $E_{\mathrm{M}}(k)$ spectrum is proportional to k^4, while to the right of $k_{\mathrm{f}}/k_1 = 15$ there is a short range with a $k^{-5/3}$ spectrum

where $\overline{\boldsymbol{\mathcal{E}}} = \overline{\boldsymbol{u} \times \boldsymbol{b}}$ is the mean electromotive force that we discussed already in connection with (1). Under the assumption of isotropy and sufficient scale separation in space and time, we have just $\overline{\boldsymbol{\mathcal{E}}} = \alpha \overline{\boldsymbol{B}} - \eta_{\mathrm{t}}\mu_0 \overline{\boldsymbol{J}}$, where α and η_{t} are a pseudo-scalar and a scalar respectively. For the case when there is no mean flow, a stability analysis gives the dispersion relation for the growth rate λ as

$$\lambda = \alpha k - (\eta_{\mathrm{t}} + \eta)k^2, \tag{6}$$

and the eigenfunctions are force-free solutions with $\eta_{\mathrm{t}}\mu_0 \overline{\boldsymbol{J}} = \alpha \overline{\boldsymbol{B}}$, which are plane polarized waves, just like in Fig. 5, where the large-scale field can be approximated by $\overline{\boldsymbol{B}} \propto (\sin ky, 0, \cos ky)$, ignoring here an arbitrary phase shift in the y direction. The dynamo is excited when $C_\alpha \equiv \alpha/(\eta_{\mathrm{t}} + \eta)k > 1$, where C_α is the relevant dynamo number in this context. The fastest growing mode occurs at wavenumber $k = k_{\max} = \alpha/[2(\eta_{\mathrm{t}} + \eta)]$. Furthermore, using estimates for the high-conductivity limit,

$$\alpha = \alpha_{\mathrm{K}} = -\frac{1}{3}\tau \langle \boldsymbol{\omega} \cdot \boldsymbol{u} \rangle \quad \text{and} \quad \eta_{\mathrm{t}} = \frac{1}{3}\tau \langle \boldsymbol{u}^2 \rangle, \tag{7}$$

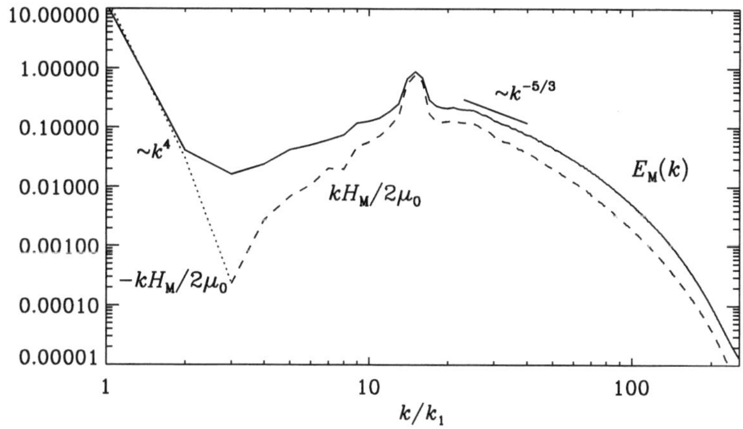

Fig. 8 Spectra of magnetic energy, $E_{\rm M}(k)$, and rescaled magnetic helicity, $\pm k H_{\rm M}(k)/2\mu_0$

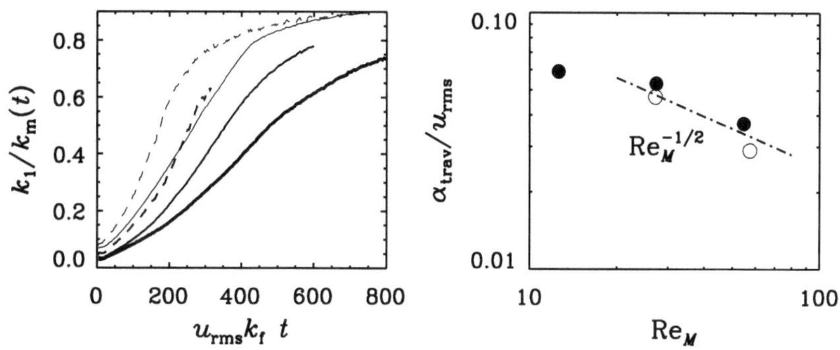

Fig. 9 *Left panel*: time dependence of the peak wavenumber for scale separation ratios of 15 (*dashed*) and 30 (*solid lines*) at Re_M of 12, 27, and around 57 (increasing line thickness). *Right panel*: Re_M dependence of the cascade speed for scale separation ratios of 15 (*open symbols*) and 30 filled symbols. The *straight lines* give the $\mathrm{Re}_M^{-1/2}$ (*dotted*) and Re_M^{-1} (*dashed*) dependences. Adapted from Brandenburg (2011b)

we find that $C_\alpha = \epsilon_{\rm f} k_{\rm f}/k_1$ (Brandenburg et al. 2002), where $\epsilon_{\rm f} \le 1$ is the fractional helicity, and $\eta \ll \eta_{\rm t}$ has been assumed. We can now return to our discussion in connection with Fig. 7, where we notice that at early times the field growth occurs at wavenumber $k_{\rm f}/2$. This is indeed the value expected from our simple estimate, since for fully helical turbulence, we expect $\langle \boldsymbol{\omega} \cdot \boldsymbol{u} \rangle = k_{\rm f} \langle \boldsymbol{u}^2 \rangle$, and thus $k_{\max} = \alpha/2\eta_{\rm t} = k_{\rm f}/2$.

It is of interest at this point to comment on the validity of mean-field dynamo concepts. In condensed matter physics for example mean-field theory is generally valid when applied to systems where fluctuations are assumed small. In high Re_M turbulent systems on the other hand, the fluctuations grow more rapidly than the mean-field, due to small scale dynamo action. Thus, even in the kinematic stage when Lorentz forces are small, one needs a closure theory to calculate the mean-field turbulent coefficients like α and $\eta_{\rm t}$. Traditionally these coefficients have been derived by still making a quasilinear approximation (strictly valid for small fluctuating fields), which is also known as the first order smoothing approximation (FOSA) (Moffatt 1978; Krause and Rädler 1980; Brandenburg and Subramanian 2005a). This is sometimes also referred to as the second order correlation approximation. Some improvements to this can be made by adopting closure approximations like the minimal tau approximation whereby triple correlations involving fluctuating fields are taken to provide a damping term proportional to the quadratic correlations (Blackman and Field 2002; Rädler et al. 2003; Brandenburg and Subramanian 2005a). There are also a few cases, like δ correlated flows (Kazantsev 1967; Zeldovich et al. 1983) or renovating flows (Dittrich et al. 1984; Gilbert and Bayly 1992) which provide analytically solvable models, where the form of α and η_t given by (7) is recovered.

In this context, direct simulations as discussed above, which can be interpreted in terms of mean-field concepts, lend some validity to the theory. This applies also to the interpretation of results in the nonlinear regime to be discussed below. Moreover, when the α and η_t have been measured directly in simulations of isotropic turbulence, one gets results remarkably close to the estimates of FOSA given in (7) (Sur et al. 2008). This suggests that the strong magnetic field fluctuations produced by small-scale dynamo action do not contribute a systematic large-scale component to the mean emf $\overline{\boldsymbol{\mathcal{E}}} = \overline{\boldsymbol{u} \times \boldsymbol{b}}$, correlated with the mean field. They do make the mean-field coefficients noisy. However, the fact that we can still use the mean-field concept in understanding the results of direct simulations implies that this noise does not have a crucial effect, perhaps after the small-scale dynamo has nearly saturated. The saturation of the dynamo will be discussed further in Sect. 3.6.

3.5 Shear Dynamos

Remarkably, not all large-scale dynamos require an α effect. In fact, large-scale dynamo action has been seen in simulations with just shear and no helicity; see Brandenburg (2005a) for simulations using a shear profile motivated by that of the Sun. An obvious candidate at the time was the so-called shear–current effect (Rogachevskii and Kleeorin 2003, 2004), which requires $\eta_{ij}\overline{U}_{i,j} > 0$, where η_{ij} is the part of the magnetic diffusivity tensor that multiplies $\overline{\boldsymbol{J}}$ such that $\overline{\mathcal{E}}_i = \cdots - \mu_0\eta_{ij}\overline{J}_j$, and $\overline{U}_{i,j}$ is the velocity shear. Already early calculations using the test-field method showed that the relevant component of η_{ij} has the wrong sign (Brandenburg 2005b). This confirmed the results of quasilinear calculations (Rädler and Stepanov 2006; Rüdiger and Kitchatinov 2006; Sridhar and Singh 2010; Singh and Sridhar 2011). Moreover in the large Re_M limit using FOSA, but for arbitrarily strong shear, the corresponding cross coupling implied by the shear current effect was shown to be absent (Sridhar and Subramanian 2009a, 2009b). The issue of how the mean-field grows in nonhelical turbulence in the presence of shear remained open in view of other possible contenders.

One possibility is the incoherent α–shear dynamo that lives from the combination of shear and fluctuations of the α effect and was originally invoked by Vishniac and Brandenburg (1997) to explain the magnetic field of unstratified shearing box simulations of accretion disc turbulence (Hawley et al. 1995). This mechanism has received considerable attention in the following years (Sokolov 1997; Silant'ev 2000; Fedotov et al. 2006; Proctor 2007; Kleeorin and Rogachevskii 2008; Sur and Subramanian 2009). Meanwhile, evidence for the existence of shear dynamos in simple shearing box simulations was mounting (Yousef et al. 2008a, 2008b; Brandenburg et al. 2008; Herault et al. 2011). Although the underlying mechanism may have appeared to be a new one, there is now quantitative evidence that this can be explained by an incoherent α-shear dynamo (Brandenburg et al. 2008; Heinemann et al. 2011; Mitra and Brandenburg 2012). This is remarkable given the unconventional nature of the approach whereby one uses mean-field theory over two spatial directions and considers the fluctuations that remain in time and the third coordinate direction as physically meaningful.

3.6 α-Quenching

A fully satisfactory theory for the magnetic feedback on the α effect does not exist. What we do know is that for strong mean fields $\overline{\boldsymbol{B}}$, this α becomes a tensor of the form $\alpha_{ij} = \alpha_1(\overline{B})\delta_{ij} - \alpha_2(\overline{B})\hat{B}_i\hat{B}_j$, where $\hat{\mathbf{B}} = \overline{\boldsymbol{B}}/\overline{B}$ are unit vectors, and $\overline{B} = |\overline{\boldsymbol{B}}|$ is the modulus. However, if this tensor is applied to the $\overline{\boldsymbol{B}}$ field, we have $\alpha_{ij}\overline{B}_j = (\alpha_1 - \alpha_2)\overline{B}_i$, which suggests that α is just like a scalar. We also know that at least part of the quenching acts in such a way that the total field (small-scale and large-scale) obeys the magnetic helicity evolution equation. This was derived some time ago in a certain approximation by Pouquet et al. (1976) and was then applied to derive an equation for the quenching of α (Kleeorin and Ruzmaikin 1982; Gruzinov and Diamond 1994; Kleeorin et al. 2000; Field and Blackman 2002; Blackman and Brandenburg 2002; Subramanian 2002; Brandenburg and Subramanian 2005a).

The crucial starting point is the realization of Pouquet et al. (1976) that under the influence of Lorentz forces, the α effect has an additional component, $\alpha_{\rm M} = \frac{1}{3}\tau\overline{\boldsymbol{j}\cdot\boldsymbol{b}}/\rho_0$, where $\overline{\boldsymbol{j}\cdot\boldsymbol{b}}$ is the current helicity associated with the small-scale field and $\alpha = \alpha_{\rm K} + \alpha_{\rm M}$ is the sum of kinetic and magnetic α effects. Interestingly, Pouquet et al. (1976) also showed that $\eta_{\rm t}$ does not get renormalized under the same approximation. Under locally isotropic conditions, in the Coulomb gauge, $\overline{\boldsymbol{j}\cdot\boldsymbol{b}}$ can be approximated by $k_{\rm f}^2\overline{\boldsymbol{a}\cdot\boldsymbol{b}}/\mu_0$, where $\overline{\boldsymbol{a}\cdot\boldsymbol{b}} \equiv \overline{h}_{\rm f}$ is the magnetic helicity of the small-scale fields. In order to write an evolution equation for the magnetic helicity density one can fix a gauge for the vector potential. One could also work in terms of the evolution equation for the current helicity (Subramanian and Brandenburg 2004). Perhaps more elegant is to write this evolution in terms of a gauge invariant magnetic helicity density, defined as the density of correlated links of $\boldsymbol{b}$, and which is most closely related to $\overline{h}_{\rm f}$ in the Coulomb gauge (Subramanian and Brandenburg 2006). The evolution equation for $\overline{h}_{\rm f}$ is

$$\frac{\partial\overline{h}_{\rm f}}{\partial t} = -2\overline{\boldsymbol{\mathcal{E}}}\cdot\overline{\boldsymbol{B}} - 2\eta k_{\rm f}^2\overline{h}_{\rm f} - \nabla\cdot\overline{\boldsymbol{F}}_{\rm f}, \tag{8}$$

where $\overline{\boldsymbol{F}}_{\rm f}$ is the magnetic helicity flux of the small-scale field. This equation shows that the α effect produces magnetic helicity at a rate $-2\overline{\boldsymbol{\mathcal{E}}}\cdot\overline{\boldsymbol{B}} = -2\alpha_{\rm red}\overline{\boldsymbol{B}}^2$, where $\alpha_{\rm red} = \alpha - \eta_{\rm t}k_{\rm m}$ is the reduced α effect and $k_{\rm m} = \mu_0\overline{\boldsymbol{J}}\cdot\overline{\boldsymbol{B}}/\overline{\boldsymbol{B}}^2$ is the effective wavenumber of the mean field. In a supercritical dynamo, the sign of $\alpha_{\rm red}$ agrees with that of α (the $\eta_{\rm t}$ term is subdominant). Then, starting with a specific sign for the kinetic $\alpha_{\rm K}$ and zero magnetic $\alpha_{\rm M}$, this produces $\alpha_{\rm M}$ of opposite sign, which quenches the total α and the dynamo progressively with increasing field strength. In the absence of a magnetic helicity flux, this process happens on a resistive time scale, which is what is seen in Fig. 6, where final saturation is not even remotely in sight. We recall that a rapid evolution of the energy of the mean field up to $k_1/k_{\rm f}$ times the equipartition value is expected on theoretical grounds (Blackman and Brandenburg 2002). In practice, this is hard to verify because at early times the mean field has not yet reached the scale of the system and modes of different orientation are still competing. Nevertheless, by splitting the magnetic field into its positively and negatively polarized contributions, $E_{\rm M}^{\pm}(k) = \frac{1}{2}[E_{\rm M}(k) \pm kH_{\rm M}(k)]$, it is possible to separate large-scale and small-scale fields (Brandenburg et al. 2002; Brandenburg and Subramanian 2005a; Brandenburg 2011b). In Fig. 10 we clearly see a faster build-up of the large-scale field through $E_{\rm M}^{-}(k)$ compared with the small-scale field through $E_{\rm M}^{+}(k)$. As we have argued before, this build-up of large-scale fields is still resistively slow, but it is important to realize that the demonstrated existence of large-scale fields in the kinematic stage provides support for the usefulness of the mean-field approach.

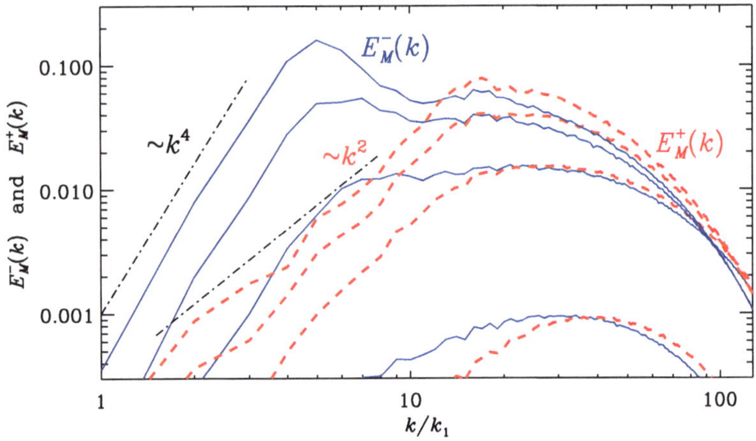

Fig. 10 Magnetic energy spectra $E_{\mathrm{M}}^{\pm}(k)$ of positively (*red*) and negatively (*blue*) polarized parts at earlier times. Note the preferred build-up of the large-scale field at $E_{\mathrm{M}}^{-}(k_{\mathrm{f}}/2)$ relative to the slower growth of $E_{\mathrm{M}}^{+}(k_{\mathrm{f}})$. Slopes proportional to k^2 for E_{M}^{+} and k^4 for E_{M}^{-} are shown

The final saturation value, in periodic box simulations, can be estimated simply by noting that in the absence of magnetic helicity fluxes, the total current helicity must vanish in the steady saturated case, i.e., $\overline{\boldsymbol{J}} \cdot \overline{\boldsymbol{B}} + \overline{\boldsymbol{j} \cdot \boldsymbol{b}} = 0$. This is because the total current helicity drives the change of magnetic helicity, which in steady state must be zero. Such a state can be obtained in a nontrivial manner with helical forcing when current helicity has opposite signs at large and small length scales. For example, if the kinetic helicity at small scales is positive (that is α_{K} is negative), then the generated α_{M} is positive. In that case the current helicity of small-scale fields is also positive with $\mu_0 \overline{\boldsymbol{j} \cdot \boldsymbol{b}} = k_{\mathrm{f}} \overline{\boldsymbol{b}^2}$ and hence $\mu_0 \overline{\boldsymbol{J}} \cdot \overline{\boldsymbol{B}} = -k_1 \overline{\boldsymbol{B}}^2$ is negative. (This implies that $k_{\mathrm{m}} = -k_1$.) Assuming furthermore equipartition between kinetic and magnetic energies at small scales, i.e., $\overline{\boldsymbol{b}^2} \approx \mu_0 \overline{\rho \boldsymbol{u}^2} \equiv B_{\mathrm{eq}}^2$, we obtain (Brandenburg 2001)

$$\overline{\boldsymbol{B}}^2 / B_{\mathrm{eq}}^2 \approx k_{\mathrm{f}} / k_1 \tag{9}$$

in the final state. We recall that in the run shown in Fig. 6, the scale separation ratio is $k_{\mathrm{f}}/k_1 = 15$, so it is understandable that there is not yet any sign of saturation of the large-scale field; see Brandenburg (2001) for early results on the resistively slow saturation.

We do not expect resistively slow saturation behavior to occur in stars or galaxies, because the $\nabla \cdot \overline{\boldsymbol{F}}_{\mathrm{f}}$ term can usually not be neglected (Blackman and Field 2000a, 2000b; Kleeorin et al. 2000; Brandenburg and Subramanian 2005b; Shukurov et al. 2006). These results were obtained by solving the mean-field equations. Subsequent simulations have shown that turbulent diffusive fluxes exist that would constitute a sufficient contribution to $\nabla \cdot \overline{\boldsymbol{F}}_{\mathrm{f}}$ (Mitra et al. 2010; Hubbard and Brandenburg 2010), especially when Re_M is larger than a critical value around 10^4. This flux term can then dominate over the $2\eta \overline{\boldsymbol{j} \cdot \boldsymbol{b}}$ term (Candelaresi et al. 2011). Under certain considerations it is possible that a flux of the form $\overline{\boldsymbol{\mathcal{E}}} \times \overline{\boldsymbol{A}}$ from the mean electromotive force contributes to the flux, but by solving an evolution equation for the magnetic helicity density of the total (small- and large-scale) field, consideration of this term can be avoided (Hubbard and Brandenburg 2012). This approach is also suited to deal with fluxes associated with gauges that can introduce artificial fluxes in shearing environments; see Hubbard and Brandenburg (2011). They find no evidence for a flux resulting from shear that were previously argued to be important (Vishniac and Cho 2001). Using simulations with anisotropic non-helical forcing in the presence of shear, Shapovalov and Vishniac (2011) argue that large-scale dynamos might even live entirely due to helicity fluxes, although the exact origin of this flux remains to be clarified. Another natural contribution to the flux term is just advection of both small-scale and large-scale fields, along with the associated magnetic helicity (Shukurov et al. 2006; Subramanian and Brandenburg 2006). This will naturally arise from coronal mass ejections in the solar context, or supernovae driven outflows in galaxies (Blackman and Brandenburg 2003; Shukurov et al. 2006; Sur et al. 2007; Warnecke et al. 2011).

Large-scale dynamos therefore seem to need helicity fluxes to work efficiently. This conclusion can be understood more physically as follows. As the large-scale mean field grows, the turbulent emf $\overline{\mathcal{E}}$ is transferring helicity between the small- and large-scale fields. The large-scale helicity is in the links of the mean poloidal and toroidal fields of the astrophysical system like the Sun or the Galaxy, while the small-scale helicity is in what can be described as 'twist helicity' (or simply twist) of the small-scale field, produced by helical motions. Lorentz forces associated with the 'twisted' small-scale field would like to untwist the field. This would lead to an effective magnetic $\alpha_{\rm M}$ effect which opposes the kinetic $\alpha_{\rm K}$ produced by the helical turbulence. The cancellation of the total α effect can lead to catastrophic quenching of the dynamo. This quenching can be avoided if there is some way of dissipating the twist (which is slow in high-Re_M system) or transferring the twists in the small-scale field out of the region of dynamo action, or canceling it between two hemispheres. That is, if there are helicity fluxes out of the system or between different parts of the system, the large-scale field can grow to observable strengths.

3.7 Application of Mean-Field Theory to Galaxies

The mean-field theory described above has been applied extensively to understand magnetic fields of disk galaxies. The mean-field dynamo equations allow substantial simplification provided a suitable parameterization of turbulent transport coefficients is chosen. Of course, this parameterization presumes a suitable closure for nonlinear effects to arrive at a closed set of nonlinear mean-field dynamo equations. Such an approach does not necessarily imply a deep understanding of the physical processes involved in the magnetic field evolution. However, it appears to be sufficient for pragmatic modeling of magnetic field configurations in particular galaxies to be compared with the observational data of polarized radio emission.

A first example of such parameterization and further drastic simplification of mean-field dynamo equations was presented by Parker (1955). Early simplified models for the galactic dynamo, which allow analytic or quasi-analytic investigations, can be found in Ruzmaikin et al. (1988), and more recent reviews by Beck et al. (1996), Shukurov (2004), and Kulsrud and Zweibel (2008). Next, one can suggest, as the most pragmatic contender, the so-called no-z model (Subramanian and Mestel 1993; Moss 1995) as a simple mean-field model for galactic dynamo. The idea of this model is to present azimuthal and radial components of the mean galactic magnetic field by their quantities at the galactic equator and average the mean-field equations with respect to the coordinate z perpendicular to the galactic plane. The third component of the magnetic field can be reconstructed from the condition $\mathrm{div}\,\overline{\boldsymbol{B}} = 0$. This approach is an obvious oversimplification, but it appears still adequate for modeling magnetic field evolution including the helicity fluxes discussed above (Sur et al. 2007; Chamandy et al. 2012). It also allows one to model magnetic configurations for design studies of new generations of radio telescopes such as the Square Kilometer Array; cf. Moss et al. (2012).

3.8 Application to Mean-Field Dynamos of Stars and the Sun

In comparison with galactic dynamos, the status of the theory of solar and stellar dynamos is less satisfactory. Early models by Steenbeck and Krause (1969) provided numerical solutions to the full two-dimensional axisymmetric mean-field equations under the assumption of an assumed profile of the internal angular velocity $\Omega(r,\theta)$, whose radial gradient, $\partial\Omega/\partial r$, was negative, and variations in colatitude θ were ignored in most of their models. This yielded cyclic magnetic fields with equatorward migration under the assumption that the α effect is predominantly positive in the northern hemisphere. Subsequently, helioseismology delivered detailed contours of $\Omega(r,\theta)$, which excluded the previously assumed Ω profiles and suggested that $\partial\Omega/\partial r > 0$ at low latitudes where strong magnetic flux belts are observed to propagate equatorward during the course of the 11 year sunspot cycle.

Various solutions have been offered to this dilemma (Parker 1987). The most prominent one is the flux transport dynamo (Choudhuri et al. 1995; Dikpati and Charbonneau 1999), whereby meridional circulation causes the dynamo wave to turn around in the opposite direction. This model operates under the assumption that the turbulent magnetic diffusivity operating on the toroidal field is much lower than the value suggested by mixing length theory, and the α effect is assumed to work only at the surface. The other proposal is that the dynamo wave obeys equatorward migration owing to a narrow near-surface shear layer (Brandenburg 2005a), where $\partial\Omega/\partial r$ is indeed strongly negative; see Fig. 4 of Benevolenskaya et al. (1999). This proposal still lacks detailed modeling. In view of the shortcomings in the treatment of mean-field dynamo theory (e.g., our ignorance concerning nonlinearity discussed in Sect. 3.6 or the neglect of finite scale separation discussed below in Sect. 3.10), the ground for speculation remains fertile. Magnetic helicity fluxes in interface and flux transport dynamos have already been looked at Chatterjee et al. (2011), but finite scale separation effects are neglected. Indeed, the Sun is strongly stratified with its scale height changing rapidly with depth, making it hard to imagine that simple-minded approaches that ignore this can be meaningful at all. A cornerstone for the proposal that solar activity is a shallow phenomenon may lie in the success of explaining the formation of active regions and perhaps even sunspots as a result of spontaneous formation of flux concentrations by convective flux collapse (Kitchatinov and Mazur 2000) or the negative effective magnetic pressure instability; see Sect. 3.11 below.

3.9 Magnetic Helicity from the Solar Dynamo Exterior

There is now explicit evidence for the presence of magnetic helicity in the exterior of dynamos. In particular, it has been possible to detect magnetic helicity of opposite signs at small and large length scales. This has been possible through measurements of magnetic helicity spectra in the solar wind away from the equatorial plane (Brandenburg et al. 2011). This data pointed for the first time to a *reversal* of magnetic helicity density between interior and exterior of the dynamo. Such reversals have now been confirmed in numerical simulations of dynamos coupled to an exterior. There are first of all the simulations of Warnecke et al. (2011) showing coronal mass ejections from a turbulent dynamo in a wedge of a spherical shell, where the reversal occurred some distance away from the dynamo. Next, there are related simulations in Cartesian geometry where a reversal can be found immediately above the surface; see Fig. 12 of Warnecke and Brandenburg (2010). Finally, there are earlier mean-field simulations showing such a reversal as well; see the lower panel of Fig. 7 of Brandenburg et al. (2009).

The occurrence of such a reversal is now well understood in terms of the magnetic helicity equation for the small-scale field shown in (8). In the dynamo interior, the α effect dominates over turbulent magnetic diffusion and produces magnetic helicity of a sign opposite to that of α. In the northern hemisphere, α is positive, so this produces negative $\overline{h}_\mathrm{f}$. Turbulent magnetic diffusivity opposes this effect, but it is still subdominant and can therefore not change the sign of $\overline{h}_\mathrm{f}$. This is different in the solar wind where α becomes subdominant compared with turbulent diffusion, which is the reason the sign of $\overline{h}_\mathrm{f}$ is now different (Brandenburg et al. 2011).

3.10 Scale Separation in Space and Time

In connection with (1) we noted that the relation between $\overline{\boldsymbol{\mathcal{E}}}$ and $\overline{\boldsymbol{B}}$ does, in general, involve a convolution in space and time. This becomes important when the variations of $\overline{\boldsymbol{B}}$ occur on time and length scales comparable with those of the turbulence, whose turnover time is ℓ/u_rms and its typical scale is $\ell = k_\mathrm{f}^{-1}$. The properties of the integral kernel are often determined in Fourier space, in which case useful approximation of it is $\hat{\alpha}_{ij}(k,\omega) = \alpha_{ij}^{(0)}\hat{K}(k,\omega)$ and $\hat{\eta}_{ijk}(k,\omega) = \eta_{ijk}^{(0)}\hat{K}(k,\omega)$, where

$$\hat{K}(k,\omega) = \frac{1}{1 - \mathrm{i}\omega\tau + k^2\ell^2}. \tag{10}$$

Such an integral kernel has recently been obtained with the test-field method applied to passive scalar diffusion (Rheinhardt and Brandenburg 2012), and in limiting cases (either with $\omega = 0$ or with $k = 0$) for α and η_t; see Brandenburg et al. (2008) and Hubbard and Brandenburg (2009) for details and applications to spatial and temporal nonlocalities, respectively. The test-field method allows one to determine the turbulent transport coefficients by solving an extra set of equations that describe the evolution of the fluctuating magnetic field for each test field, which is a predetermined mean field. Under some conditions it is also necessary to solve corresponding evolution equations for velocity perturbations (Rheinhardt and Brandenburg 2010). The combined presence of spatio-temporal nonlocality was first considered by Rheinhardt and Brandenburg (2012), who proposed (10) and used it to reformulate (1) as a simple differential equation of the form

$$\left(1 + \tau\frac{\partial}{\partial t} - \ell^2\frac{\partial^2}{\partial z^2}\right)\overline{\mathcal{E}}_i = \alpha_{ij}^{(0)}\overline{B}_j + \eta_{ijk}^{(0)}\overline{B}_{j,k}. \tag{11}$$

Such an equation is quite easy to implement and represents an improvement in terms of physical realism. This representation avoids not only the problem of causality associated with the infinite speed of signal propagation in the absence of the time derivative (Brandenburg et al. 2004), but it also prevents the development of artificially small scales in the mean field.

The application of this new technique is still in its infancy, and it needs to be seen to what extent spatio-temporal nonlocality can substantially alter the nature of the solutions. As an example we note that a finite τ has been found to lower the critical dynamo number for oscillatory solutions by a factor of about 2 (Rheinhardt and Brandenburg 2012). In the context of disk galaxies, such non-locality in time can also lead to phase shifts between the spiral forcing of the dynamo by matter arms, and the resulting magnetic spirals, as seen in the galaxy NGC6946 (Chamandy et al. 2012).

3.11 Magnetic Structures Resulting from Strong Density Stratification

Finally, let us discuss a mean-field effect that occurs under the condition of strong density stratification. It has been theoretically anticipated long ago (Kleeorin et al. 1989, 1990, 1993,

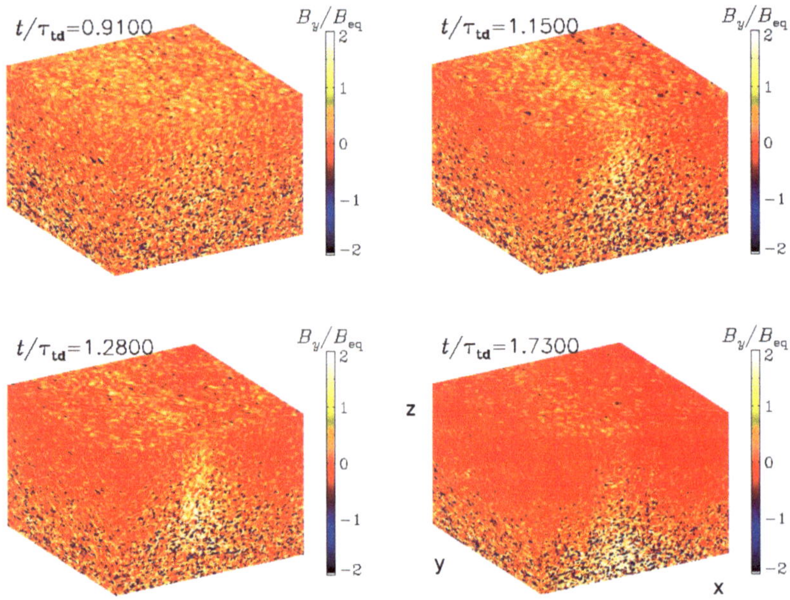

Fig. 11 Visualizations of B_y for a simulation with strong density stratification and a weak imposed magnetic field $(0, B_0, 0)$ with $B_0/B_{eq0} = 0.01$, $\mathrm{Re}_M = 18$ and $\mathrm{Pr}_M = 0.5$. Time is given in units of turbulent–diffusive times. Note the gradual emergence of a large-scale magnetic flux concentration which then sinks as a result of negative effective magnetic buoyancy. Adapted from Kemel et al. (2012)

1996; Kleeorin and Rogachevskii 1994; Rogachevskii and Kleeorin 2007), but only more recently was it also seen in numerical simulations of the mean-field equations (Brandenburg et al. 2010) and then in direct numerical simulations (Brandenburg et al. 2011; Kemel et al. 2012).

The essence of this effect is the suppression of turbulent intensity by a mean magnetic field. This means that the effective pressure caused by $\overline{B}$ is not just $\overline{\boldsymbol{B}}^2/2\mu_0$, but there must also be an additional contribution from the suppression of the turbulence, which leads to a negative contribution (Kleeorin et al. 1989, 1990, 1996; Kleeorin and Rogachevskii 1994; Rogachevskii and Kleeorin 2007). This modifies the nature of the magnetic buoyancy instability in such a way that magnetic structures become heavier than their surroundings and sink. This has been demonstrated using both mean-field simulations (Brandenburg et al. 2010) as well as direct numerical simulations (Brandenburg et al. 2011). As an example we show in Fig. 11 a snapshot from a direct simulation where a weak ($B_0/B_{eq0} = 0.01$) magnetic field is imposed in the y direction. The density stratification is isothermal, so the density scale height is constant in the direction of gravity (the negative z direction). The total density contrast from bottom to top is about 540. The instability grows at a rate which scales with $\eta_{t0}k^2$, where $\eta_{t0} = u_{rms}/3k_f$ is an estimate for the turbulent magnetic diffusivity, which is well reproduced by simulations using the test-field method (Sur et al. 2008).

The study of this negative effective magnetic pressure is still very much in progress. In particular, it has not yet been studied how this negative effective magnetic pressure instability interacts with the mean-field dynamo. It is envisaged that this instability might produce local magnetic field enhancements in the surface layers that resemble active regions. Such regions are long lived ($\sim 1/2$ year). Traditionally such long timescales have not been associated with the surface layers. However, the time scale of the negative effective magnetic pressure instability is the turbulent–diffusive time, $\tau_{\rm td} = (\eta_{\rm t0} k_1^2)^{-1}$, where $\eta_{\rm t0} = u_{\rm rms}/3k_{\rm f}$ is the estimated turbulent magnetic diffusivity. This time can be much longer than the local turnover time, $\tau_{\rm to} = (u_{\rm rms} k_{\rm f})^{-1}$, which is about 1 day at the bottom of the near-surface shear layer at 40 Mm depth. The ratio of these time scales is $\tau_{\rm td}/\tau_{\rm to} = 3(k_{\rm f}/k_1)^2$, which can be around 300 for a scale separation ratio of just 10.

In the rest of this paper we focus on small-scale magnetic fields that occur over a range of different astrophysical settings. They are believed to be important in understanding small-scale magnetic fields in the surface layers of the Sun, although that part of the field might also be a consequence of tangling the large-scale magnetic field. Another possible astrophysical application of small-scale dynamos might be the clusters of galaxies, because in the absence of rotation it is difficult to motivate any form of large-scale dynamo action.

4 Analytical Approaches to Small-Scale Turbulent Dynamos

Direct numerical simulations are now the most straightforward way to understand turbulent dynamos. However, more traditional analytical method provide useful support for them. Analytic methods have provided particular insights both into large-scale and small-scale dynamos. We discuss some specific analytic considerations of small-scale dynamos further below.

4.1 Correlation Tensor and Small-Scale Dynamo

A natural approach here is to introduce the second-order correlation tensor of the magnetic field $B_i(\boldsymbol{x}, t)$ as

$$\mathcal{B}_{ij}(\boldsymbol{x}, \boldsymbol{y}, t_1, t_2) = \langle B_i(\boldsymbol{x}, t_1) B_j(\boldsymbol{y}, t_2)\rangle, \tag{12}$$

taken at two spatial points $\boldsymbol{x}$ and $\boldsymbol{y}$ and at two instants t_1 and t_2. Here $\langle \ldots \rangle$ denotes averaging over an ensemble of turbulent velocity field fluctuations which can be described by the velocity field correlation tensor $\mathcal{V}_{ij}(\boldsymbol{x}, \boldsymbol{y}, t_1, t_2)$ constructed in the same way as $\mathcal{B}_{ij}$.

For a particular model of turbulence (short-correlated random flow), Kazantsev (1967), and simultaneously Kraichnan and Nagarajan (1967) for a slightly different model, obtained a governing equation known now as the Kazantsev equation. In particular Kazantsev (1967) assumed $\mathcal{V}_{ij}(\boldsymbol{x}, \boldsymbol{y}, t_1, t_2) = \bar{\mathcal{V}}_{ij}(\boldsymbol{x}, \boldsymbol{y})\delta(t_1 - t_2)$, and derived an evolution equation for the magnetic field correlation tensor calculated at two simultaneous instants $\mathcal{B}_{ij}(\boldsymbol{x}, \boldsymbol{y}, t) = \mathcal{B}_{ij}(\boldsymbol{x}, \boldsymbol{y}, t, t)$, with $t_1 = t_2 = t$. This reads

$$\frac{\partial \mathcal{B}_{ij}}{\partial t} = \hat{L}_{ijkm} \mathcal{B}_{km} \tag{13}$$

where $\hat{L}_{ijkm}$ is a second-order differential operator with coefficients depending on $\bar{\mathcal{V}}_{ij}$, its spatial derivatives and coefficient of magnetic diffusion η. In some sense, the Kazantsev equation is similar to the famous Steenbeck-Krause-Rädler α effect equation in mean-field electrodynamics. In practice however the latter equation provided much more astronomically fruitful results than the first one. The reason is presumably two-fold. First of all, the Kazantsev equation requires more algebra for its solution than the mean-field equations. We only take here some new points isolated recently in this bulky algebra and refer to a detailed review given by Zeldovich et al. (1990). The other reason is that the physical interpretation of solutions of Eq. (13) is more delicate than that for the mean-field equation. This is the main issue presented in following sections.

Original insight which led Kazantsev to Eq. (13) required some mathematics from quantum field theory that can hardly be considered 'user friendly' for a person with an ordinary MHD background. A more familiar approach, which is however rather bulky, can be found elsewhere (Zeldovich et al. 1990; Subramanian 1997; Brandenburg and Subramanian 2005a). The particular form of the Kazantsev equation (13) is associated with a specific model of turbulence assumed, in particular the assumption of δ-correlated (in time) velocity fluctuations. This model is quite restrictive and does not allow one to represent adequately some basic properties of the Kolmogorov cascade. Various attempts to derive these equations for more realistic models were undertaken (see, e.g., Kleeorin et al. 2002). For example, incorporating a finite correlation time results in quite complicated albeit beautiful mathematics and gives rise to integral equations. As far as it is known however the results are more or less the same if just applied to Eq. (13) with $\bar{\mathcal{V}}_{ij}$ taken from a suitable model of turbulence ignoring the fact that this very model is incompatible with the derivation of Eq. (13).

A reasonable way to simplify Eq. (13) to a useful level is to consider statistically homogeneous, isotropic and mirror-symmetric turbulence and look for solutions having the same symmetry. This means that we exploit a velocity correlation tensor of the form

$$\bar{\mathcal{V}}_{ij} = A(r) r_i r_j + B(r)\delta_{ij} \tag{14}$$

($\boldsymbol{r} = \boldsymbol{y} - \boldsymbol{x}$) and look for the magnetic field correlation tensor in a similar form. The incompressibility condition means that the functions A and B depend on a single function, say, $F(r)$ while the solenoidality condition of the magnetic field means that $\mathcal{B}_{ij}$ depends on a single function, say, $W(r,t)$. Then Eq. (13) can be reduced to a single second-order ordinary differential equation for a single function which depends on W with coefficients depending on F. In fact this is the equation which was obtained by Kazantsev (1967). The algebra here remains quite bulky and we avoid to present it here in detail; see, for example, the detailed discussions in Zeldovich et al. (1990) and Brandenburg and Subramanian (2005a).

There is no problem to solve the Kazantsev equation in the homogeneous and isotropic case for a particular choice of F numerically or by analytical approximations. In fact the Kazantsev equation can be reformulated as a Schrödinger type equation for a particle with variable mass $m(r)$ in a potential $U(r)$ (both of which depend on $F(r)$), whose bound states correspond to exponentially growing $W(r,t)$. It is possible to develop a WKB-like method for an approximate solution in the limit of large magnetic Reynolds numbers Re_M ($\sim \eta^{-1}$).

A general result following from these solutions can be summarized as follows. For a sufficiently large Re_M, the magnetic field correlation tensor (i.e. W) grows exponentially with a growth rate γ_2 which is determined by l/v where l is the turbulence correlation time and v is its rms velocity. The critical magnetic Reynolds number is of order $R_{\mathrm{m,crit}} \approx 10^2$, with $R_{\mathrm{m,crit}} = 26$ in the most simple example with $F = \exp -(r/l)^2$. For this illustrative example, the function W which determines the magnetic field correlation properties has quite a complicated form which contains spatial scales from l up to $l\mathrm{Re}_M^{-1/2}$.

A plausible scenario for the nonlinear saturation of the growth of the magnetic field correlation tensor governing the Kazantsev equation was suggested by Subramanian (1999). The main idea here is that nonlinearity results in an effective suppression of Re_M up to $R_{\mathrm{m,crit}}$. It is quite straightforward to consider Kazantsev equation for homogeneous, isotropic and mirror-asymmetric turbulence and to combine concepts of the second-order correlation tensor with the mean-field approach (Subramanian 1999; Gabov and Sokoloff 2004). Interestingly, the Kazantsev equation in the presence of kinetic helicity, can be reformulated into a tunneling type quantum mechanical problem, whereby the bound states of the small scale dynamo can 'tunnel' to develop long range correlations (Subramanian 1999; Brandenburg and Subramanian 2000; Boldyrev et al. 2005). It is also possible to solve the

Kazantsev equation for locally homogeneous and isotropic turbulence in a finite body of size $L \gg l$ (Maslova et al. 1987; Belyanin et al. 1993).

It has been argued that the small-scale dynamo at large Reynolds numbers exhibits a universal behavior in the sense that the magnetic energy grows at a constant rate (of about 5 % as determined from simulations) of the total turbulent dissipation rate (Beresnyak 2012). Numerical simulations have also shown that when the turbulence becomes transonic, the critical value of Re_M for small-scale dynamo action increases by about a factor of two (Haugen et al. 2004b). Similar results have also been obtained by Federrath et al. (2011) who find a decrease in the growth rate when the Mach number Ma crosses unity, but a subsequent increase proportional to $\mathrm{Ma}^{1/3}$.

4.2 Random Magnetic Fields in Cosmology

The description of a random magnetic field by a second order correlation function is widely exploited in cosmology. To be specific, Chernin (1966) considered at first a cosmological model with a random magnetic field and discussed cosmological evolution of its magnetic energy. We have to stress however that the concepts of statistical hydrodynamics, which are a mathematical basis for the Kazantsev approach, need some modification to be applicable for curved spaces of General Relativity. The point is that Eq. (12) considers the product of two vectors applied at different spatial points while such option is absent in Riemannian geometry. We have to consider a geodesic line connecting points $\boldsymbol{x}$ and $\boldsymbol{y}$, transport a field from the second point to the first one and consider the product of the two fields applied at one spatial point. This recipe presumes that a geodesic line which leaves the point $\boldsymbol{x}$ does no longer cross other geodesic lines which leave this point. In other words this means that geodesic lines do not contain conjugated points and the space-time has no gravitational lenses. If they do exist we have to elaborate somehow the concept of the correlation tensor and no general recommendations are suggested until now.

Under statistical homogeneity, isotropy and mirror symmetry in a curved spatial section of a Friedmann cosmological model, the magnetic correlation function reads

$$\mathcal{B}^{ij} = C(r)n^i n^j + D(r)g^{ij}, \tag{15}$$

i.e. we have to distinguish upper and lower indices and use tangent vector n^i of the geodesic line connecting $\boldsymbol{x}$ and $\boldsymbol{y}$ instead of the vector $\boldsymbol{r}$ which does not exist in curved space. The other point is that one deals with a curved space formulation of the solenoidality condition to reduce C and D to one function. Inspired by earlier work of Garcia de Andrade (2010), Rubashny and Sokoloff (2010) performed a corresponding analysis to show that for the Lobachevsky space with negative curvature,

$$C = -R\frac{\mathrm{th}\,\frac{r}{R}}{2}F', \qquad D = F + R\frac{\mathrm{th}\,\frac{r}{R}}{2}F', \tag{16}$$

where R is the curvature radius of the spatial section and r is the distance between $\boldsymbol{x}$ and $\boldsymbol{y}$. It is instructive to compare this representation with that for Euclidean space

$$C = -\frac{r}{2}F', \qquad D = F + \frac{r}{2}F'. \tag{17}$$

Because $\mathrm{th}\,\frac{r}{R}$ has a finite limit at $r/R \to \infty$, correlations decay slightly slower for the Lobachevsky space than in Euclidean space (of course F is the same in both cases). More significantly, however, the volume of a sphere with radius r grows exponentially in the Lobachevsky space and not as a power law, like in Euclidean space. It means that F has to decay much faster in Lobachevsky space than in Euclidean space to get convergence of various spatial means which are based on the correlation tensor.

For spherical space (closed cosmological model) one obtains

$$C = -R\frac{\operatorname{tg}\frac{r}{R}}{2}F', \qquad D = F + R\frac{\operatorname{tg}\frac{r}{R}}{2}F'. \tag{18}$$

This representation gives $C(\pi R) = 0$, i.e., C vanishes if $\boldsymbol{y}$ is just the opposite point to $\boldsymbol{x}$. This looks reasonable because n^i is not determined uniquely for the opposite points. $\operatorname{tg}\frac{r}{R}$ diverges for $r = \pi R/2$ so the finiteness of correlations means that $F'(\frac{\pi R}{2}) = 0$. This condition is specific for spherical geometry and has no direct analogues for the Euclidean one. Remarkably, a closed universe admits another homogeneous and isotropic topological structure of spatial section, namely an elliptical space which is twice smaller than the spherical one. Instead of the above condition we find here $F''(\pi R/2) = 0$. Moreover, an elliptical space does not admit orientation, so one cannot distinguish between left- and right-handed coordinate system there. It means that quantities such as helicities, α effect and other pseudo-scalar quantities cannot be introduced there. Such topological constrains on the magnetic field correlation properties look rather strange. Fortunately they do not affect substantially physically interesting conclusions because the correlation length l is usually much smaller than the curvature radius R. The problem however is that both quantities as well as the horizon radius vary strongly during the course of cosmological evolution, that is, a given length scale l being negligible in the present day cosmological scale might be very large in scales of the Early Universe. A more severe problem arises if we are going to consider a homogeneous and isotropic ensemble of random gravitational waves (Ivanova and Sokoloff 2008), which are often discussed in cosmological context.

4.3 Higher Statistical Moments and Intermittency

To get more detailed information concerning a dynamo-generated small-scale magnetic field, it is useful to consider higher statistical moments which are introduced as *ensemble* averages of a product of p magnetic field vectors (p is the order of the statistical moment). Following the Kazantsev approach one can obtain the governing equations for these quantities and demonstrate that the moments grow provided Re_M is high enough (see, e.g., Kleeorin et al. 2002). Of course, the algebra becomes more bulky as m increases. The problem is that the higher moments grow faster than the lower ones in the sense that

$$\gamma_2/2 < \gamma_4/4 < \gamma_6/6 \ldots . \tag{19}$$

Of course, this fact can be supported by a direct calculation. However, it is much more instructive to demonstrate the phenomenon at a qualitative level (e.g. Molchanov et al. 1988). Let us consider a flow with a memory time τ so the magnetic field $\boldsymbol{B}(n\tau)$ at instant $n\tau$ can be considered to be developed from the initial field $\boldsymbol{B}(0)$ which is affected by n independent random transport operators $\hat{T}_i$

$$\boldsymbol{B}(n\tau) = \hat{T}_n\hat{T}_{n-1}\ldots T_1\boldsymbol{B}(0). \tag{20}$$

As a matter of fact, progressive growth of higher statistical moments in Eq. (19) does not depend critically on the fine structure of operators $\hat{T}_i$ so we can illustrate the phenomenon by considering the simplest operators $\hat{T}_i$, i.e. just independent random numbers T_i.

For the sake of definiteness, let $\ln T_i$ have Gaussian distribution with zero mean and standard deviation σ. Then

$$T = T_n T_{n-1}\ldots T_1 \tag{21}$$

is a log-normal random quantity and $\ln T$ has zero mean and standard deviation $\sqrt{n}\sigma = \sigma\sqrt{t/\tau}$. A straightforward calculation shows that

$$\langle T^p \rangle = \exp(\sigma^2 p^2 t/2\tau), \tag{22}$$

so the normalized growth rates of the moments $\gamma_p/p = \sigma^2 p/2\tau$ grow linearly with the degree p of statistical moment. The other point is that the growth is determined not by a typical value of T_i which is of order σ, but by strong deviations which are of order $\sigma\sqrt{tp/\tau}$. The probability density of a Gaussian quantity to achieve the level $\sigma\sqrt{tp/\tau}$ is of order $\exp(-tp/2\tau)$ so the size N of a statistical ensemble should be as large as $N^*(t) = \exp(tp/2\tau)$ to include such rare events.

Note that the above analysis presumes that the statistical ensemble is infinitely large. If the ensemble is only large but finite, its size N should exceed a critical value $N^*(t)$ which grows in time exponentially. For large t we obtain $N < N^*(t)$ and the above estimates becomes inapplicable.

If we consider a medium of independent cells of size l renovating after a memory time τ, then the critical size of the system which allows one to recognize the growth governed by Eq. (22) is given by

$$L^*(t) = lN^{1/3} = l\exp(tp/6\tau). \tag{23}$$

This means that the behavior of the statistical moment of the order p is determined by very rare cells and Eq. (23) gives an estimate for the distance to the nearest cell which determines the moment at a given point. The phenomenon of a random field whose properties are determined by rare and remote events is known as intermittency. The wording comes from medicine and means a state when a person is near death, but his/her heart still works from time to time. These rare events of the heart activity determine the fact that the person is still alive.

Note that if we calculated a PDF of T based on a limited sample with $N < N^*(t)$ it is practically impossible to recognize the existence of the above mentioned rare events which do not contribute to the PDF calculated. Of course, the importance of the result depends on how large t should be to make the intermittency recognizable and how large the corresponding value N^* is in comparison with N typical of celestial bodies.

It is natural to address this point based on a simple physical example rather than just a product of random operators. A simple example of this kind accessible for simple numerics has been suggested by Zeldovich (1964) in a cosmological context. Let a remote object have a (small) angular size θ, let x be the distance to the object and $y = \theta x$ its linear size. $y(x)$ is known in the Riemannian geometry as Jacobi field and is governed by the so-called Jacobi equation

$$y'' + Ky = 0, \tag{24}$$

where a prime means the derivative taken with respect to x and K is the spatial (sectional) curvature. Zeldovich (1964) recognized the importance of density and then curvature fluctuations on the evolution of y along the line of sight. In other words, we consider K as a random, say, Gaussian quantity.

It is quite easy to simulate many independent solutions of Eq. (24) and determine experimentally how large $N^*(t)$ is. It appears (Artyushkova and Sokoloff 2005) that one needs $N \approx 5 \times 10^5 \ldots 10^6$ to recognize the difference between $\gamma_2/2$ and $\gamma_4/4$ for $t \approx 10^2$. Of course, a simulation of 10^6 independent 3D cells for a hundred turnover times becomes prohibitive given the purely computational problems, in addition to the problems associated with the further data processing of the results. On the other hand, the number of independent turbulent cells in a typical galaxy is of the order of 10^6, so the effects of intermittency can contribute in mean quantities of interest for galactic dynamos. In practice however the difference between $\gamma_2/2$ and $\gamma_4/4$ is for the intermittent fields, as far as it is known, not very large.

The other point here is that the growing magnetic field becomes sooner or later dynamically important. Of course, it is important to know which happens first, whether the field becomes dynamically important first or whether the size of the domain becomes too small to contain the intermittent structure? Unfortunately, the abilities of simple models like Eq. (24) to reproduce the stage of nonlinear dynamo saturation are limited. A numerical experiment with a simple model shows that the number N required to reproduce the behavior of higher statistical moments declines strongly when the solution becomes dynamically important such that exponential growth of statistical moments saturates.

Yet another point which can be recognized from the experience with simple models is that ensemble averaging is not the only option to describe the behavior of the growing solution. Lamburt et al. (2003) demonstrated that the quantity

$$\frac{\ln|y|}{t} \to \gamma > 0 \quad (\text{for } t \to \infty) \tag{25}$$

for almost all realizations (with probability 1, or "almost sure" in the wording of probability theory) where no averaging is taken at all. Quantities such as γ are known as self-averaging quantities and γ in particular is known as Lyapunov exponent. The phenomenon becomes clearer if one introduces a 2D row-vector (y, y') and rewrites Eq. (24) as a vectorial equation

$$z' = z\hat{A}, \tag{26}$$

where $\hat{A}$ is a random matrix process with vanishing trace (i.e. $\mathrm{Tr}\,\hat{A} = 0$). Then the evolution of z from an initial condition z_0 can be represented as a product of independent random unimodular matrices $B_i = \exp(\hat{A}_i\tau)$, $\det B_i = 1$, where $\hat{A}_i$ is a realization at a given interval of renovation of the random matrix process $\hat{A}$.

The product of independent random matrices is quite well investigated in probability theory (so-called Furstenberg theory). Zeldovich et al. (1984) stressed the importance of this theory for small-scale dynamos (here the elements of $\hat{A}$ are $\partial v_i/\partial x_j$). Molchanov et al. (1984) argue that magnetic field generated by a small-scale dynamo grows such that

$$\frac{\ln|\boldsymbol{B}|(\boldsymbol{x},t)}{t} \to \gamma > 0 \quad \text{for } t \to \infty, \tag{27}$$

where γ is a positive constant. A numerical experiment for the Jacobi equation supports this interpretation and shows that t should be of the order of a hundred memory times to get this behavior (Artyushkova and Sokoloff 2005). The PDF of the dynamo-excited magnetic field is investigated by Chertkov et al. (1999).

Unfortunately, this approach of simple models cannot mimic the α effect. The point is that one cannot produce a pseudoscalar quantity α based on correlations of A_{ij} and A_{mn} and a Levi-Civita tensor e_{pqr}.

Generally speaking, the analytical results discussed above show that detailed direct numerical simulations of small-scale dynamos at the kinematic stage can be nontrivial to interpret due to the rapidly growing intermittency. Fortunately, the magnetic field becomes dynamically important quite rapidly (at least at small scales), so the dynamo becomes nonlinear and mathematical distinctions between properties of various statistical moments become less and less important. Artyushkova and Sokoloff (2005) investigated several simple models of how catastrophic intermittency typical of the kinematic stage gradually evaporate when nonlinear effects become important. Presumably, something like that happens for the much more complicated full small-scale dynamo equations.

4.4 Small-Scale Magnetic Field and Shell Models

The Kazantsev model is developed for the kinematic or weakly nonlinear stage of the dynamo and its ability to describe the strongly nonlinear stage of a dynamo is obviously quite limited. However, it remains useful to have a simple model of strongly nonlinear dynamos in terms of ordinary (instead of partial) differential equations. An option of this kind is given by so-called shell models of MHD-turbulence.

The starting point of the shell model approach is to note that for random fields and flows, one can hardly reproduce in a numerical simulation the actual realization of the random field which are obtained in a given celestial body. In practice we are interested in some spectral properties of the field of interest. If we can get such properties without solving the full set of equations, we would be happy with such a result.

Shell models are designed to describe the cascade process over a large range of scales (wavenumbers) by a chain of variables $u_n(t)$, $b_n(t)$, each of them characterizing velocity and magnetic field oscillations with wavenumbers k in the range from $k_n = k_0\lambda^n$ to k_{n+1} i.e., a shell of wavenumbers. The parameter λ characterizes the ratio of two adjacent scales (the width of the shell) and usually $\lambda \leq 2$. The model includes a corresponding set of ordinary differential equations, which should reproduce the basic properties of the equation of motion. In particular, the model has to reproduce the type of nonlinearity of the primitive equations and to retain the same integrals of motion in the dissipationless limit. Let us note that shell models can possess positively defined integrals of motion (energy, enstrophy in two-dimensional turbulence, and the square of magnetic potential in 2D MHD-turbulence), as well as quadratic integrals with an arbitrary sign (the integrals of this kind are usually called 'helicities'). The signs of the helicities are defined by the balance between the contributions of odd and even shells to corresponding quantity.

The shell models were suggested by Kolmogorov's school to describe the spectral energy transfer (Gledzer 1973; Desnianskii and Novikov 1974). After numerous refinements they became an effective tool for description of the spectral properties of the small-scale turbulence (see for review Bohr et al. 1998). The shell models for MHD turbulence were introduced by Frick (1984), Gloaguen et al. (1985), Brandenburg et al. (1996), and Frick and Sokoloff (1998). This approach reveals many intrinsic features of small-scale dynamo action in fully developed turbulence of conducting fluids (for review see Biferale 2003). In particular, the shell model suggested by Frick and Sokoloff (1998) gives a fast growth of small-scale magnetic fields (on the timescale l/v) and its saturation at the equipartition level as well as non-Gaussian (similar to lognormal) PDF for small-scale magnetic field in the saturated state. In some cases shell models give a hint concerning dynamo action in the parametric domain inaccessible to direct numerical simulations. In particular, Stepanov and Plunian (2006) and Frick et al. (2006) argue, based on simulations of MHD-shell models, that the critical magnetic Reynolds number for small-scale dynamo action remains moderate $R_{\mathrm{m,crit}} \approx 80$ in the case of low Prandtl numbers. This is also confirmed by the simulations discussed in Sect. 3.2. General speaking, shell models seem to provide an effective way to investigate small-scale dynamos.

It may be possible to combine shell models as a tool to describe small-scale variables in a dynamo with grid or spectral methods for large-scale variables in mean-field equations. However only the first steps in this direction have been made until now (Frick et al. 2002, 2006; Nigro and Veltri 2011).

4.5 Dynamical Chaos and Small-Scale Dynamo

One more point of comparison between numerical and analytical approaches to the small-scale dynamos is as follows. The Kazantsev model and its related investigations consider turbulence as a truly random field and apply concepts of probability theory in full extent. In contrast, direct numerical simulations and shell models consider turbulence as a chaotic behavior of solutions for deterministic equations of motion. It is quite risky to insist *a priori* that dynamical chaos reproduces all properties of random flows required for analytical approaches. It is even less obvious that interstellar turbulence driven by supernova explosions (e.g. Korpi et al. 1999; Gressel et al. 2008) provides a truly random velocity field. On the other hand, dynamo models based on steady flows with stochastic flow lines such as ABC flows excite magnetic fields which look rather different from the field discussed for turbulent dynamos. Zeldovich (see Zeldovich et al. 1983) suggested that dynamo action in nonstationary flows (say, when parameters A, B and C in the ABC flow fluctuate in time) becomes much more similar to the dynamo action in random flows than dynamos in stationary flows. Recent work of Kleeorin et al. (2009) supports this idea and demonstrates that in a fluctuating ABC flow, a large-scale magnetic field can indeed grow in a way similar to what is supposed to grow in a random mirror-asymmetric (helical) turbulent flow.

5 Conclusions

In this review we have discussed our current understanding of both large-scale and small-scale dynamos that are relevant in astrophysics. In particular, we have illustrated differences and similarities between them and have compared them in terms of the energy spectra with the corresponding cases at low magnetic Prandtl numbers. We have also briefly highlighted the resistively slow saturation phenomenon as well as catastrophic quenching of helicity-driven large-scale dynamos. Finally, we have discussed new issues in connection with small-scale dynamos and their intermittency.

Relative to earlier reviews (e.g. Brandenburg and Subramanian 2005a) there have been some unexpected advances regarding the nature of magnetic helicity fluxes and direct observational evidence for magnetic helicity in the solar wind. Another completely unexpected development concerns the numerical detection of the negative effective magnetic pressure instability in simulations of strongly stratified turbulence. It is expected that these developments will contribute to an improved understanding of the magnetic field generation in astrophysical bodies. On the technical side, Brandenburg and Subramanian (2005a) discussed just the basics of the test-field method, but now this technique has provided significant insights into issues such as non-locality in space and time, as well as the nonlinear quenching of dynamo coefficients.

In this review we have barely touched upon applications to actual astrophysical bodies. In fact, a lot of progress has been made by trying to model the Sun. Direct numerical simulations of convection in spherical shells has shown signs of cyclic large-scale fields (Brown et al. 2010, 2011; Käpylä et al. 2010; Ghizaru et al. 2010; Racine et al. 2011), but only for systems that are rotating at least 3 times faster than the Sun; see Brown et al. (2011) for simulations with otherwise realistic solar parameters. Similar results have also been obtained for local simulations of the galactic dynamo, which only appears to be excited when the rotation speed is artificially enhanced (Gressel et al. 2008). This might well indicate that one is on the right track, but that the turbulence present in the system is exerting too much effective diffusion owing to it being dominated by rather large eddies. It is conceivable that in reality the turbulent eddies would be smaller, lowering thereby the effective turbulent diffusion, which can at the moment (with the larger eddies) only be emulated by adopting faster rotation. Similar results have recently also been seen in models of the negative effective magnetic pressure instability, where direct numerical simulations showed the development of the instability only when there were enough turbulent eddies in the domain and thereby the turbulent diffusivity sufficiently small on that scale; see Brandenburg et al. (2011) and, in particular, Fig. 17 of Brandenburg et al. (2012). Thus, much has been learned about turbulent dynamos and their relevance for astrophysical systems but as usual, much remains to be elucidated.

Acknowledgements We thank Andre Balogh for providing an inspiring atmosphere at the International Space Science Institute in Bern in 2010, which has led to new collaborations and scientific progress. Computing resources were provided by the Swedish National Allocations Committee at the Center for Parallel Computers at the Royal Institute of Technology in Stockholm and the High Performance Computing Center North in Umeå. This work was supported in part by the European Research Council under the AstroDyn Research Project No. 227952 and the Swedish Research Council under the project grant 621-2011-5076.

References

M.E. Artyushkova, D.D. Sokoloff, Numerical modelling of the solutions of the Jacobi equation on a geodesic with random curvature. Astron. Rep. **49**, 520–525 (2005)
G.K. Batchelor, On the spontaneous magnetic field in a conducting liquid in turbulent motion. Proc. R. Soc. Lond. Ser. A, Math. Phys. Sci. A **201**, 405–416 (1950)
R. Beck, A. Brandenburg, D. Moss, A. Shukurov, D. Sokoloff, Galactic magnetism: recent developments and perspectives. Annu. Rev. Astron. Astrophys. **34**, 155–206 (1996)
M.P. Belyanin, D. Sokoloff, A. Shukurov, Simple models of nonlinear fluctuation dynamo. Geophys. Astrophys. Fluid Dyn. **68**, 237–261 (1993)
E.E. Benevolenskaya, J.T. Hoeksema, A.G. Kosovichev, P.H. Scherrer, The interaction of new and old magnetic fluxes at the beginning of solar cycle 23. Astrophys. J. **517**, L163–L166 (1999)
A. Beresnyak, Universal nonlinear small-scale dynamo. Phys. Rev. Lett. **108**, 035002 (2012)
L. Biermann, A. Schlüter, Cosmic radiation and cosmic magnetic fields. II. Origin of cosmic magnetic fields. Phys. Rev. **82**, 863–868 (1951)
L. Biferale, Shell models of energy cascade in turbulence. Ann. Rev. Fluid Dyn. **35**, 441–468 (2003)
E.G. Blackman, G.B. Field, Constraints on the magnitude of α in dynamo theory. Astrophys. J. **534**, 984–988 (2000a)
E.G. Blackman, G.B. Field, Coronal activity from dynamos in astrophysical rotators. Mon. Not. R. Astron. Soc. **318**, 724–732 (2000b)
E.G. Blackman, A. Brandenburg, Dynamic nonlinearity in large scale dynamos with shear. Astrophys. J. **579**, 359–373 (2002)
E.G. Blackman, A. Brandenburg, Doubly helical coronal ejections from dynamos and their role in sustaining the solar cycle. Astrophys. J. Lett. **584**, L99–L102 (2003)
E.G. Blackman, G.B. Field, New dynamical mean-field dynamo theory and closure approach. Phys. Rev. Lett. **89**, 265007 (2002)
T. Bohr, M.H. Jensen, G. Paladin, A. Vulpiani, *Dynamical Systems Approach to Turbulence* (Cambridge University Press, Cambridge, 1998)
S.A. Boldyrev, F. Cattaneo, Magnetic-field generation in Kolmogorov turbulence. Phys. Rev. Lett. **92**, 144501 (2004)
S. Boldyrev, F. Cattaneo, R. Rosner, Magnetic-field generation in helical turbulence. Phys. Rev. Lett. **95**, 255001 (2005)
S. Boldyrev, J.C. Perez, J.E. Borovsky, J.J. Podesta, Spectral scaling laws in magnetohydrodynamic turbulence simulations and in the solar wind. Astrophys. J. **741**, L19 (2011)
A. Brandenburg, The inverse cascade and nonlinear alpha-effect in simulations of isotropic helical hydromagnetic turbulence. Astrophys. J. **550**, 824–840 (2001)
A. Brandenburg, The case for a distributed solar dynamo shaped by near-surface shear. Astrophys. J. **625**, 539–547 (2005a)
A. Brandenburg, Turbulence and its parameterization in accretion discs. Astron. Nachr. **326**, 787–797 (2005b)

A. Brandenburg, Large-scale dynamos at low magnetic Prandtl numbers. Astrophys. J. **697**, 1206–1213 (2009)
A. Brandenburg, Nonlinear small-scale dynamos at low magnetic Prandtl numbers. Astrophys. J. **741**, 92 (2011a)
A. Brandenburg, Chandrasekhar-Kendall functions in astrophysical dynamos. Pramana J. Phys. **77**, 67–76 (2011b)
A. Brandenburg, K. Subramanian, Large scale dynamos with ambipolar diffusion nonlinearity. Astron. Astrophys. **361**, L33–L36 (2000)
A. Brandenburg, K. Subramanian, Astrophysical magnetic fields and nonlinear dynamo theory. Phys. Rep. **417**, 1–209 (2005a)
A. Brandenburg, K. Subramanian, Strong mean field dynamos require supercritical helicity fluxes. Astron. Nachr. **326**, 400–408 (2005b)
A. Brandenburg, S. Candelaresi, P. Chatterjee, Small-scale magnetic helicity losses from a mean-field dynamo. Mon. Not. R. Astron. Soc. **398**, 1414–1422 (2009)
A. Brandenburg, W. Dobler, K. Subramanian, Magnetic helicity in stellar dynamos: new numerical experiments. Astron. Nachr. **323**, 99–122 (2002)
A. Brandenburg, K. Enqvist, P. Olesen, Large-scale magnetic fields from hydromagnetic turbulence in the very early universe. Phys. Rev. D **54**, 1291–1300 (1996)
A. Brandenburg, P. Käpylä, A. Mohammed, Non-Fickian diffusion and tau-approximation from numerical turbulence. Phys. Fluids **16**, 1020–1027 (2004)
A. Brandenburg, K. Kemel, N. Kleeorin, D. Mitra, I. Rogachevskii, Detection of negative effective magnetic pressure instability in turbulence simulations. Astrophys. J. **740**, L50 (2011)
A. Brandenburg, K. Kemel, N. Kleeorin, I. Rogachevskii, The negative effective magnetic pressure in stratified forced turbulence. Astrophys. J. **749**, 179 (2012)
A. Brandenburg, N. Kleeorin, I. Rogachevskii, Large-scale magnetic flux concentrations from turbulent stresses. Astron. Nachr. **331**, 5–13 (2010)
A. Brandenburg, D. Moss, A.M. Soward, New results for the Herzenberg dynamo: steady and oscillatory solutions. Proc. R. Soc. A, Math. Phys. Eng. Sci. **454**, 1283–1300 (1998)
A. Brandenburg, K.-H. Rädler, M. Rheinhardt, P.J. Käpylä, Magnetic diffusivity tensor and dynamo effects in rotating and shearing turbulence. Astrophys. J. **676**, 740–751 (2008)
A. Brandenburg, K.-H. Rädler, M. Rheinhardt, K. Subramanian, Magnetic quenching of alpha and diffusivity tensors in helical turbulence. Astrophys. J. **676**, 740–L52 (2008)
A. Brandenburg, K.-H. Rädler, M. Schrinner, Scale dependence of alpha effect and turbulent diffusivity. Astron. Astrophys. **482**, 739–746 (2008)
A. Brandenburg, K. Subramanian, A. Balogh, M.L. Goldstein, Scale-dependence of magnetic helicity in the solar wind. Astrophys. J. **734**, 9 (2011)
B.P. Brown, M.K. Browning, A.S. Brun, M.S. Miesch, J. Toomre, Persistent magnetic wreaths in a rapidly rotating Sun. Astrophys. J. **711**, 424–438 (2010)
B.P. Brown, M.S. Miesch, M.K. Browning, A.S. Brun, J. Toomre, Magnetic cycles in a convective dynamo simulation of a young solar-type star. Astrophys. J. **731**, 69 (2011)
S. Candelaresi, A. Hubbard, A. Brandenburg, D. Mitra, Magnetic helicity transport in the advective gauge family. Phys. Plasmas **18**, 012903 (2011)
L. Chamandy, K. Subramanian, A. Shukurov, Galactic spirals and dynamo action: slaved non-axisymmetric modes (2012, in preparation)
P. Chatterjee, G. Guerrero, A. Brandenburg, Magnetic helicity fluxes in interface and flux transport dynamos. Astron. Astrophys. **525**, A5 (2011)
A.D. Chernin, A cosmological model with a disordered magnetic field. Astron. Zh. **43**, 797–801 (1966)
M. Chertkov, G. Falkovich, I. Kolokolov, M. Vergassola, Small-scale turbulent dynamo. Phys. Rev. Lett. **83**, 4065–4068 (1999)
J. Cho, D. Ryu, Characteristic lengths of magnetic field in magnetohydrodynamic turbulence. Astrophys. J. **705**, L90–L94 (2009)
A.R. Choudhuri, M. Schüssler, M. Dikpati, The solar dynamo with meridional circulation. Astron. Astrophys. **303**, L29–L32 (1995)
A. Courvoisier, D.W. Hughes, S.M. Tobias, α-Effect in a family of chaotic flows. Phys. Rev. Lett. **96**, 034503 (2006)
V.N. Desnianskii, E.A. Novikov, Simulation of cascade processes in turbulent flows. Prikl. Mat. Meh. **38**, 507–512 (1974)
M. Dikpati, P. Charbonneau, A Babcock-Leighton flux transport dynamo with solar-like differential rotation. Astrophys. J. **518**, 508–520 (1999)
P. Dittrich, S.A. Molchanov, D.D. Sokolov, A.A. Ruzmaikin, Mean magnetic field in renovating random flow. Astron. Nachr. **305**, 119–125 (1984)

W.M. Elsasser, Hydromagnetic dynamo theory. Rev. Mod. Phys. **28**, 135–163 (1956)
G.L. Eyink, Stochastic flux freezing and magnetic dynamo. Phys. Rev. **83**, 056405 (2011)
C. Federrath, G. Chabrier, J. Schober, R. Banerjee, R.S. Klessen, D.R.G. Schleicher, Mach number dependence of turbulent magnetic field amplification: solenoidal versus compressive flows. Phys. Rev. Lett. **107**, 114504 (2011)
S. Fedotov, I. Bashkirtseva, L. Ryashko, Memory effects in a turbulent dynamo: generation and propagation of a large-scale magnetic field. Phys. Rev. **73**, 066307 (2006)
G.B. Field, E.G. Blackman, Dynamical quenching of the α^2 dynamo. Astrophys. J. **572**, 685–692 (2002)
P.G. Frick, Two-dimensional MHD turbulence. Hierarchical model. Magnetohydrodynamics **20**, 262–267 (1984)
P. Frick, D. Sokoloff, Cascade and dynamo action in a shell model of magnetohydrodynamic turbulence. Phys. Rev. E **57**, 4155–4164 (1998)
P. Frick, M. Reshetnyak, D. Sokoloff, Combined grid-shell approach for convection in a rotating spherical layer. Europhys. Lett. **59**, 212–217 (2002)
P. Frick, R. Stepanov, D. Sokoloff, Large- and small-scale interactions and quenching in an α^2-dynamo. Phys. Rev. **74**, 066310 (2006)
U. Frisch, A. Pouquet, J. Léorat, A. Mazure, Possibility of an inverse cascade of magnetic helicity in hydrodynamic turbulence. J. Fluid Mech. **68**, 769–778 (1975)
U. Frisch, Z.-S. She, P.L. Sulem, Large-scale flow driven by the anisotropic kinetic alpha effect. Physica D **28**, 382–392 (1987)
S.A. Gabov, D.D. Sokoloff, Current helicity and the small-scale dynamo. Astron. Rep. **48**, 949–953 (2004)
D. Galloway, U. Frisch, Dynamo action in a family of flows with chaotic streamlines. Geophys. Astrophys. Fluid Dyn. **36**, 53–84 (1986)
D.J. Galloway, M.R.E. Proctor, Numerical calculations of fast dynamos in smooth velocity fields with realistic diffusion. Nature **356**, 691–693 (1992)
A.K. Gailitis, On the theory of Herzenberg's dynamo. Magn. Gidrodin. **4**, 12–16 (1973)
L.C. Garcia de Andrade, Primordial magnetic fields constrained from CMB anisotropies on dynamo cosmology. Astrophys. Space Sci. **330**, 347–351 (2010)
M. Ghizaru, P. Charbonneau, P.K. Smolarkiewicz, Magnetic cycles in global large-eddy simulations of solar convection. Astrophys. J. **715**, L133–L137 (2010)
R.D. Gibson, P.H. Roberts, Some comments on the theory of homogeneous dynamos, in *Magnetism in the cosmos*, ed. by W.R. Hindmarsh, F.J. Lowes, P.H. Roberts, S.K. Runcorn (Oliver & Boyd, Edinburgh, 1967), pp. 108–120
A.D. Gilbert, B.J. Bayly, Magnetic field intermittency and fast dynamo action in random helical flows. J. Fluid Mech. **241**, 199–214 (1992)
E.B. Gledzer, System of hydrodynamic type admitting two quadratic integrals of motion. Sov. Phys. Dokl. **18**, 216–221 (1973)
C. Gloaguen, J. Léorat, A. Pouquet, R. Grappin, A scalar model for MHD turbulence. Physica D **17**, 154–182 (1985)
G.S. Golitsyn, Fluctuations of the magnetic field and current density in a turbulent flow of a weakly conducting fluid. Sov. Phys. Dokl. **5**, 536–539 (1960)
O. Gressel, D. Elstner, U. Ziegler, G. Rüdiger, Direct simulations of a supernova-driven galactic dynamo. Astron. Astrophys. **486**, L35–L38 (2008)
A.V. Gruzinov, P.H. Diamond, Self-consistent theory of mean-field electrodynamics. Phys. Rev. Lett. **72**, 1651–1653 (1994)
N.E.L. Haugen, A. Brandenburg, Hydrodynamic and hydromagnetic energy spectra from large eddy simulations. Phys. Fluids **18**, 075106 (2006)
N.E.L. Haugen, A. Brandenburg, W. Dobler, Is nonhelical hydromagnetic turbulence peaked at small scales? Astrophys. J. **597**, L141–L144 (2003)
N.E.L. Haugen, A. Brandenburg, W. Dobler, Simulations of nonhelical hydromagnetic turbulence. Phys. Rev. **70**, 016308 (2004a)
N.E.L. Haugen, A. Brandenburg, A.J. Mee, Mach number dependence of the onset of dynamo action. Mon. Not. R. Astron. Soc. **353**, 947–952 (2004b)
J.F. Hawley, C.F. Gammie, S.A. Balbus, Local three dimensional simulations of an accretion disk hydromagnetic dynamo. Astrophys. J. **440**, 742–763 (1995)
T. Heinemann, J.C. McWilliams, A.A. Schekochihin, Magnetic-field generation by randomly forced shearing waves. Phys. Rev. Lett. **107**, 255004 (2011)
J. Herault, F. Rincon, C. Cossu, G. Lesur, G.I. Ogilvie, P.-Y. Longaretti, Periodic magnetorotational dynamo action as a prototype of nonlinear magnetic-field generation in shear flows. Phys. Rev. **84**, 036321 (2011)
A. Herzenberg, Geomagnetic dynamos. Proc. R. Soc. Lond. Ser. A, Math. Phys. Sci. A **250**, 543–583 (1958)

A. Hubbard, A. Brandenburg, Memory effects in turbulent transport. Astrophys. J. **706**, 712–726 (2009)
A. Hubbard, A. Brandenburg, Magnetic helicity fluxes in an α^2 dynamo embedded in a halo. Geophys. Astrophys. Fluid Dyn. **104**, 577–590 (2010)
A. Hubbard, A. Brandenburg, Magnetic helicity flux in the presence of shear. Astrophys. J. **727**, 11 (2011)
A. Hubbard, A. Brandenburg, Catastrophic quenching in $\alpha\Omega$ dynamos revisited. Astrophys. J. **748**, 51 (2012)
A.B. Iskakov, A.A. Schekochihin, S.C. Cowley, J.C. McWilliams, M.R.E. Proctor, Numerical demonstration of fluctuation dynamo at low magnetic Prandtl numbers. Phys. Rev. Lett. **98**, 208501 (2007)
E.V. Ivanova, D.D. Sokoloff, Statistically homogeneous and isotropic curvature fluctuations in general relativity. Moscow Univ. Phys. Bull. **63**, 109–111 (2008)
P.J. Käpylä, M.J. Korpi, A. Brandenburg, Large-scale dynamos in turbulent convection with shear. Astron. Astrophys. **491**, 353–362 (2008)
P.J. Käpylä, M.J. Korpi, A. Brandenburg, D. Mitra, R. Tavakol, Convective dynamos in spherical wedge geometry. Astron. Nachr. **331**, 73–81 (2010)
A.P. Kazantsev, Enhancement of a magnetic field by a conducting fluid. Sov. Phys. JETP **53**, 1806–1809 (1967) [English translation, JETP 26, 1031 (1968)]
K. Kemel, A. Brandenburg, N. Kleeorin, D. Mitra, I. Rogachevskii, Spontaneous formation of magnetic flux concentrations in stratified turbulence. Sol. Phys. (2012). doi:10.1007/s11207-012-9949-0
S. Kida, S. Yanase, J. Mizushima, Statistical properties of MHD turbulence and turbulent dynamo. Phys. Fluids A **3**, 457–465 (1991)
L.L. Kitchatinov, M.V. Mazur, Stability and equilibrium of emerged magnetic flux. Sol. Phys. **191**, 325–340 (2000)
N. Kleeorin, I. Rogachevskii, Effective Ampère force in developed magnetohydrodynamic turbulence. Phys. Rev. E **50**, 2716–2730 (1994)
N. Kleeorin, I. Rogachevskii, Mean-field dynamo in a turbulence with shear and kinetic helicity fluctuations. Phys. Rev. **77**, 036307 (2008)
N.I. Kleeorin, A.A. Ruzmaikin, Dynamics of the average turbulent helicity in a magnetic field. Magnetohydrodynamics **18**, 116–122 (1982) [Translation from Magnitnaya Gidrodinamika, 2, pp. 17–24 (1982)]
N.I. Kleeorin, I.V. Rogachevskii, A.A. Ruzmaikin, Negative magnetic pressure as a trigger of large-scale magnetic instability in the solar convective zone. Sov. Astron. Lett. **15**, 274–277 (1989)
N.I. Kleeorin, I.V. Rogachevskii, A.A. Ruzmaikin, Magnetic force reversal and instability in a plasma with advanced magnetohydrodynamic turbulence. Sov. Phys. JETP **70**, 878–883 (1990)
N. Kleeorin, M. Mond, I. Rogachevskii, Magnetohydrodynamic instabilities in developed small-scale turbulence. Phys. Fluids B **5**, 4128–4134 (1993)
N. Kleeorin, M. Mond, I. Rogachevskii, Magnetohydrodynamic turbulence in the solar convective zone as a source of oscillations and sunspots formation. Astron. Astrophys. **307**, 293–309 (1996)
N. Kleeorin, D. Moss, I. Rogachevskii, D. Sokoloff, Helicity balance and steady-state strength of the dynamo generated galactic magnetic field. Astron. Astrophys. **361**, L5–L8 (2000)
N. Kleeorin, I. Rogachevskii, D. Sokoloff, Magnetic fluctuations with a zero mean field in a random fluid flow with a finite correlation time and a small magnetic diffusion. Phys. Rev. **65**, 036303 (2002)
N. Kleeorin, I. Rogachevskii, D. Sokoloff, D. Tomin, Mean-field dynamos in random Arnold-Beltrami-Childress and Roberts flows. Phys. Rev. **79**, 046302 (2009)
M.J. Korpi, A. Brandenburg, A. Shukurov, I. Tuominen, Å. Nordlund, A supernova regulated interstellar medium: simulations of the turbulent multiphase medium. Astrophys. J. **514**, L99–L102 (1999)
R.H. Kraichnan, S. Nagarajan, Growth of turbulent magnetic fields. Phys. Fluids **10**, 859–870 (1967)
F. Krause, K.-H. Rädler, *Mean-field Magnetohydrodynamics and Dynamo Theory* (Pergamon, Oxford, 1980)
R.M. Kulsrud, S.W. Anderson, The spectrum of random magnetic fields in the mean field dynamo theory of the galactic magnetic field. Astrophys. J. **396**, 606–630 (1992)
R.M. Kulsrud, E.G. Zweibel, On the origin of cosmic magnetic fields. Rep. Prog. Phys. **71**, 046901 (2008)
V.G. Lamburt, V.N. Tutubalin, D.D. Sokoloff, Jacobi fields along a geodesic with random curvature. Math. Notae **74**, 393–400 (2003)
T.B. Maslova, T.S. Shumkina, A.A. Ruzmaikin, D.D. Sokoloff, Self-excitation of fluctuation magnetic fields in a space contained by a random stream. Sov. Phys. Dokl. **32**, 520–521 (1987)
M. Meneguzzi, U. Frisch, A. Pouquet, Helical and nonhelical turbulent dynamos. Phys. Rev. Lett. **47**, 1060–1064 (1981)
D. Mitra, A. Brandenburg, Scaling and intermittency in incoherent α-shear dynamo. Mon. Not. R. Astron. Soc. **420**, 2170–2177 (2012)
D. Mitra, S. Candelaresi, P. Chatterjee, R. Tavakol, A. Brandenburg, Equatorial magnetic helicity flux in simulations with different gauges. Astron. Nachr. **331**, 130–135 (2010)
H.K. Moffatt, The amplification of a weak applied magnetic field by turbulence in fluids of moderate conductivity. J. Fluid Mech. **11**, 625–635 (1961)

H.K. Moffatt, *Magnetic Field Generation in Electrically Conducting Fluids* (Cambridge University Press, Cambridge, 1978)
S.A. Molchanov, A.A. Ruzmaikin, D.D. Sokoloff, A dynamo theorem. Geophys. Astrophys. Fluid Dyn. **30**, 241–259 (1984)
S.A. Molchanov, A.A. Ruzmaikin, D.D. Sokoloff, Short-correlated random flow as a fast dynamo. Sov. Phys. Dokl. **32**, 569–570 (1988)
D. Moss, On the generation of bisymmetric magnetic field structures in spiral galaxies by tidal interactions. Mon. Not. R. Astron. Soc. **275**, 191–194 (1995)
D. Moss, R. Stepanov, T.G. Arshakian, R. Beck, M. Krause, D. Sokoloff, Multiscale magnetic fields in spiral galaxies: evolution and reversals. Astron. Astrophys. **537**, A68 (2012)
A. Neronov, I. Vovk, Evidence for strong extragalactic magnetic fields from Fermi observations of TeV blazars. Science **328**, 73–75 (2010)
G. Nigro, P. Veltri, A study of the dynamo transition in a self-consistent nonlinear dynamo model. Astrophys. J. **740**, L37 (2011)
E.N. Parker, Hydromagnetic dynamo models. Astrophys. J. **122**, 293–314 (1955)
E.N. Parker, The dynamo dilemma. Sol. Phys. **110**, 11–21 (1987)
I.J. Parrish, E. Quataert, Nonlinear simulations of the heat-flux-driven buoyancy instability and its implications for galaxy clusters. Astrophys. J. **677**, L9–L12 (2008)
I.J. Parrish, J.M. Stone, Nonlinear evolution of the magnetothermal instability in two dimensions. Astrophys. J. **633**, 334–348 (2005)
I.J. Parrish, M. McCourt, E. Quataert, P. Sharma, The effects of anisotropic viscosity on turbulence and heat transport in the intracluster medium. Mon. Not. R. Astron. Soc. (2012). doi:10.1111/j.1365-2966.2012.20650.x
A. Pouquet, U. Frisch, J. Léorat, Strong MHD helical turbulence and the nonlinear dynamo effect. J. Fluid Mech. **77**, 321–354 (1976)
M.R.E. Proctor, Effects of fluctuation on alpha-omega dynamo models. Mon. Not. R. Astron. Soc. **382**, L39–L42 (2007)
É. Racine, P. Charbonneau, M. Ghizaru, A. Bouchat, P.K. Smolarkiewicz, On the mode of dynamo action in a global large-eddy simulation of solar convection. Astrophys. J. **735**, 46 (2011)
K.-H. Rädler, R. Stepanov, Mean electromotive force due to turbulence of a conducting fluid in the presence of mean flow. Phys. Rev. **73**, 056311 (2006)
K.-H. Rädler, N. Kleeorin, I. Rogachevskii, The mean electromotive force for MHD turbulence: the case of a weak mean magnetic field and slow rotation. Geophys. Astrophys. Fluid Dyn. **97**, 249–274 (2003)
M. Rheinhardt, A. Brandenburg, Test-field method for mean-field coefficients with MHD background. Astron. Astrophys. **520**, A28 (2010)
M. Rheinhardt, A. Brandenburg, Modeling spatio-temporal nonlocality in mean-field dynamos. Astron. Nachr. **333**, 71–77 (2012)
G.O. Roberts, Spatially periodic dynamos. Philos. Trans. R. Soc. A **266**, 535–558 (1970)
G.O. Roberts, Dynamo action of fluid motions with two-dimensional periodicity. Philos. Trans. R. Soc. A **271**, 411–454 (1972)
I. Rogachevskii, N. Kleeorin, Intermittency and anomalous scaling for magnetic fluctuations. Phys. Rev. E **56**, 417–426 (1997)
I. Rogachevskii, N. Kleeorin, Electromotive force and large-scale magnetic dynamo in a turbulent flow with a mean shear. Phys. Rev. **68**, 036301 (2003)
I. Rogachevskii, N. Kleeorin, Nonlinear theory of a 'shear–current' effect and mean-field magnetic dynamos. Phys. Rev. **70**, 046310 (2004)
I. Rogachevskii, N. Kleeorin, Magnetic fluctuations and formation of large-scale inhomogeneous magnetic structures in a turbulent convection. Phys. Rev. **76**, 056307 (2007)
A.S. Rubashny, D.D. Sokoloff, Fully developed turbulent dynamo at low magnetic Prandtl numbers. Moscow Univ. Phys. Bull. **58**(2), 5–87 (2010)
G. Rüdiger, Reynolds stresses and differential rotation I. On recent calculations of zonal fluxes in slowly rotating stars. Geophys. Astrophys. Fluid Dyn. **16**, 239–261 (1980)
G. Rüdiger, *Differential Rotation and Stellar Convection: Sun and Solar-Type Stars* (Gordon & Breach, New York, 1989)
G. Rüdiger, L.L. Kitchatinov, Do mean-field dynamos in nonrotating turbulent shear-flows exist? Astron. Nachr. **327**, 298–303 (2006)
A.A. Ruzmaikin, A.M. Shukurov, Spectrum of the galactic magnetic field. Astrophys. Space Sci. **82**, 397–407 (1982)
A.A. Ruzmaikin, D.D. Sokoloff, A.M. Shukurov, *Magnetic Fields of Galaxies* (Kluwer, Dordrecht, 1988)
A.A. Schekochihin, J.L. Maron, S.C. Cowley, J.C. McWilliams, The small-scale structure of magnetohydrodynamic turbulence with large magnetic Prandtl numbers. Astrophys. J. **576**, 806–813 (2002)

A.A. Schekochihin, S.C. Cowley, J.L. Maron, J.C. McWilliams, Critical magnetic Prandtl number for small-scale dynamo. Phys. Rev. Lett. **92**, 054502 (2004a)
A.A. Schekochihin, S.C. Cowley, S.F. Taylor, J.L. Maron, J.C. McWilliams, Simulations of the small scale turbulent dynamo. Astrophys. J. **612**, 276–307 (2004b)
A.A. Schekochihin, S.C. Cowley, R.M. Kulsrud, G.W. Hammett, P. Sharma, Plasma instabilities and magnetic field growth in clusters of galaxies. Astrophys. J. **629**, 139–142 (2005a)
A.A. Schekochihin, N.E.L. Haugen, A. Brandenburg, S.C. Cowley, J.L. Maron, J.C. McWilliams, Onset of small scale dynamo at small magnetic Prandtl numbers. Astrophys. J. **625**, L115–L118 (2005b)
A.A. Schekochihin, A.B. Iskakov, S.C. Cowley, J.C. McWilliams, M.R.E. Proctor, T.A. Yousef, Fluctuation dynamo and turbulent induction at low magnetic Prandtl numbers. New J. Phys. **9**, 300 (2007)
J. Schober, D. Schleicher, C. Federrath, R. Klessen, R. Banerjee, Magnetic field amplification by small-scale dynamo action: Dependence on turbulence models and Reynolds and Prandtl numbers. Phys. Rev. **85**, 026303 (2012)
D.S. Shapovalov, E.T. Vishniac, Simulations of turbulent dynamos driven by the magnetic helicity flux. Astrophys. J. **738**, 66 (2011)
A. Shukurov, Introduction to galactic dynamos, in *Mathematical Aspects of Natural Dynamos*. EDP Press, Paris. arXiv:astro-ph/0411739 (2004)
A. Shukurov, D. Sokoloff, K. Subramanian, A. Brandenburg, Galactic dynamo and helicity losses through fountain flow. Astron. Astrophys. **448**, L33–L36 (2006)
N.A. Silant'ev, Magnetic dynamo due to turbulent helicity fluctuations. Astron. Astrophys. **364**, 339–347 (2000)
N.K. Singh, S. Sridhar, Transport coefficients for the shear dynamo problem at small Reynolds numbers. Phys. Rev. **83**, 056309 (2011)
D.D. Sokolov, The disk dynamo with fluctuating spirality. Astron. Rep. **41**, 68–72 (1997)
S. Sridhar, N.K. Singh, The shear dynamo problem for small magnetic Reynolds numbers. J. Fluid Mech. **664**, 265–285 (2010)
S. Sridhar, K. Subramanian, Shear dynamo problem: quasilinear kinematic theory. Phys. Rev. **79**, 045305R (2009a)
S. Sridhar, K. Subramanian, Nonperturbative quasilinear approach to the shear dynamo problem. Phys. Rev. **80**, 066315 (2009b)
K. Subramanian, Dynamics of fluctuating magnetic fields in turbulent dynamos incorporating ambipolar drifts, arXiv:astro-ph/9708216 (1997)
K. Subramanian, Can the turbulent galactic dynamo generate large-scale magnetic fields? Mon. Not. R. Astron. Soc. **294**, 718–728 (1998)
K. Subramanian, Unified treatment of small- and large-scale dynamos in helical turbulence. Phys. Rev. Lett. **83**, 2957–2960 (1999)
K. Subramanian, Magnetic helicity in galactic dynamos. Bull. Astron. Soc. India **30**, 715–721 (2002)
K. Subramanian, A. Brandenburg, Nonlinear current helicity fluxes in turbulent dynamos and alpha quenching. Phys. Rev. Lett. **93**, 205001 (2004)
K. Subramanian, A. Brandenburg, Magnetic helicity density and its flux in weakly inhomogeneous turbulence. Astrophys. J. **648**, L71–L74 (2006)
K. Subramanian, L. Mestel, Galactic dynamos and density wave theory. II. An alternative treatment for strong non-axisymmetry. Mon. Not. R. Astron. Soc. **265**, 649–654 (1993)
K. Subramanian, A. Shukurov, N.E.L. Haugen, Evolving turbulence and magnetic fields in galaxy clusters. Mon. Not. R. Astron. Soc. **366**, 1437–1454 (2006)
M. Steenbeck, F. Krause, Zur Dynamotheorie stellarer und planetarer Magnetfelder. I. Berechnung sonnenähnlicher Wechselfeldgeneratoren. Astron. Nachr. **291**, 49–84 (1969). See also the translation in Roberts & Stix, The turbulent dynamo, Tech. note 60, NCAR, Boulder, CO (1971)
M. Steenbeck, F. Krause, K.-H. Rädler, Berechnung der mittleren Lorentz-Feldstärke $\overline{\boldsymbol{v} \times \boldsymbol{B}}$ für ein elektrisch leitendes Medium in turbulenter, durch Coriolis-Kräfte beeinflußter Bewegung. Z. Naturforsch. **21a**, 369–376 (1966). See also the translation in Roberts & Stix, The turbulent dynamo, Tech. Note 60, NCAR, Boulder, CO (1971)
R. Stepanov, F. Plunian, Fully developed turbulent dynamo at low magnetic Prandtl numbers. J. Turbul. **7**, N39 (2006)
S. Sur, K. Subramanian, Galactic dynamo action in presence of stochastic α and shear. Mon. Not. R. Astron. Soc. **392**, L6–L10 (2009)
S. Sur, A. Shukurov, K. Subramanian, Galactic dynamos supported by magnetic helicity fluxes. Mon. Not. R. Astron. Soc. **377**, 874–882 (2007)
S. Sur, A. Brandenburg, K. Subramanian, Kinematic alpha effect in isotropic turbulence simulations. Mon. Not. R. Astron. Soc. **385**, L15–L19 (2008)

E.T. Vishniac, A. Brandenburg, An incoherent α–Ω dynamo in accretion disks. Astrophys. J. **475**, 263–274 (1997)
E.T. Vishniac, J. Cho, Magnetic helicity conservation and astrophysical dynamos. Astrophys. J. **550**, 752–760 (2001)
J. Warnecke, A. Brandenburg, Surface appearance of dynamo-generated large-scale fields. Astron. Astrophys. **523**, A19 (2010)
J. Warnecke, A. Brandenburg, D. Mitra, Dynamo-driven plasmoid ejections above a spherical surface. Astron. Astrophys. **534**, A11 (2011)
T.A. Yousef, T. Heinemann, A.A. Schekochihin, N. Kleeorin et al., Generation of magnetic field by combined action of turbulence and shear. Phys. Rev. Lett. **100**, 184501 (2008a)
T.A. Yousef, T. Heinemann, F. Rincon, A.A. Schekochihin et al., Numerical experiments on dynamo action in sheared and rotating turbulence. Astron. Nachr. **329**, 737–749 (2008b)
Ya.B. Zeldovich, Observations in a Universe homogeneous in the mean. Astron. Zh. **41**, 19–23 (1964) [Sov. Astron. 8, 13–17 (1964)]
Ya.B. Zeldovich, A.A. Ruzmaikin, D.D. Sokoloff, *Magnetic Fields in Astrophysics* (Gordon & Breach, New York, 1983)
Ya.B. Zeldovich, A.A. Ruzmaikin, S.A. Molchanov, D.D. Sokoloff, Kinematic dynamo problem in a linear velocity field. J. Fluid Mech. **144**, 1–11 (1984)
Ya.B. Zeldovich, A.A. Ruzmaikin, D.D. Sokoloff, *The Almighty Chance* (World Scientific, Singapore, 1990)

FSC
www.fsc.org
MIX
Papier aus verantwortungsvollen Quellen
Paper from responsible sources
FSC® C105338

If you have any concerns about our products,
you can contact us on
ProductSafety@springernature.com

In case Publisher is established outside the EU,
the EU authorized representative is:
Springer Nature Customer Service Center GmbH
Europaplatz 3, 69115 Heidelberg, Germany

Printed by Libri Plureos GmbH
in Hamburg, Germany